Base SAS® 9.4 Procedures Guide
Statistical Procedures
Fifth Edition

SAS® Documentation

The correct bibliographic citation for this manual is as follows: SAS Institute Inc. 2016. *Base SAS® 9.4 Procedures Guide: Statistical Procedures, Fifth Edition*. Cary, NC: SAS Institute Inc.

Base SAS® 9.4 Procedures Guide: Statistical Procedures, Fifth Edition

Copyright © 2016, SAS Institute Inc., Cary, NC, USA

All Rights Reserved. Produced in the United States of America.

For a hard-copy book: No part of this publication may be reproduced, stored in a retrieval system, or transmitted, in any form or by any means, electronic, mechanical, photocopying, or otherwise, without the prior written permission of the publisher, SAS Institute Inc.

For a web download or e-book: Your use of this publication shall be governed by the terms established by the vendor at the time you acquire this publication.

The scanning, uploading, and distribution of this book via the Internet or any other means without the permission of the publisher is illegal and punishable by law. Please purchase only authorized electronic editions and do not participate in or encourage electronic piracy of copyrighted materials. Your support of others' rights is appreciated.

U.S. Government License Rights; Restricted Rights: The Software and its documentation is commercial computer software developed at private expense and is provided with RESTRICTED RIGHTS to the United States Government. Use, duplication, or disclosure of the Software by the United States Government is subject to the license terms of this Agreement pursuant to, as applicable, FAR 12.212, DFAR 227.7202-1(a), DFAR 227.7202-3(a), and DFAR 227.7202-4, and, to the extent required under U.S. federal law, the minimum restricted rights as set out in FAR 52.227-19 (DEC 2007). If FAR 52.227-19 is applicable, this provision serves as notice under clause (c) thereof and no other notice is required to be affixed to the Software or documentation. The Government's rights in Software and documentation shall be only those set forth in this Agreement.

SAS Institute Inc., SAS Campus Drive, Cary, NC 27513-2414

September 2017

SAS® and all other SAS Institute Inc. product or service names are registered trademarks or trademarks of SAS Institute Inc. in the USA and other countries. ® indicates USA registration.

Other brand and product names are trademarks of their respective companies.

SAS software may be provided with certain third-party software, including but not limited to open-source software, which is licensed under its applicable third-party software license agreement. For license information about third-party software distributed with SAS software, refer to `http://support.sas.com/thirdpartylicenses`.

Contents

Chapter 1.	What's New in the Base SAS 9.4 Statistical Procedures	1
Chapter 2.	The CORR Procedure	3
Chapter 3.	The FREQ Procedure	63
Chapter 4.	The UNIVARIATE Procedure	277

Subject Index   **541**

Syntax Index   **551**

Chapter 1
What's New in the Base SAS 9.4 Statistical Procedures

Overview

Base SAS 9.4M5 includes enhancements to the FREQ statistical procedure. In addition, some defaults have changed.

FREQ Procedure Enhancements

The COMMONRISKDIFF option in the TABLES statement provides estimates, confidence limits, and tests for the overall risk (proportion) difference for multiway 2×2 tables. New statistics for the common risk difference include minimum risk confidence limits, the minimum risk test (Mehrotra and Railkar 2000), and stratified Newcombe confidence limits that are computed by using minimum risk weights.

The new METHOD=SCORE2 option is available for exact confidence limits for the risk difference and the relative risk (which you can request by specifying the RISKDIFF or RELRISK option, respectively, in the EXACT statement). METHOD=SCORE2 computes exact confidence limits by inverting a single two-sided exact test that is based on a score statistic (Agresti and Min 2001).

The EXACT SYMMETRY option provides an exact version of Bowker's symmetry test.

The "Kappa Details" table now includes the B_n measure (Bangdiwala 1988; Bangdiwala et al. 2008).

The POLYCHORIC(ADJUST) option adjusts zero-frequency table cells in the polychoric correlation computation.

FREQ Procedure Changes

The default method for computing exact confidence limits for the risk difference and the relative risk has changed. Beginning in SAS 9.4M5, the default is METHOD=SCORE, which inverts two separate one-sided exact tests that are based on the score statistic (Chan and Zhang 1999). In releases before SAS 9.4M5, the default method is based on the unstandardized statistic (risk difference or relative risk), which you can now request by specifying RISKDIFF(METHOD=NOSCORE) or RELRISK(METHOD=NOSCORE), respectively, in the EXACT statement.

Beginning in SAS 9.4M5, PROC FREQ displays all agreement tables (which are produced by the AGREE option in the TABLES statement) in tabular format. You can specify the AGREE(TABLES=RESTORE) option to display these tables in factoid (label-value) format, which is their format in releases before SAS 9.4M4.

References

Agresti, A., and Min, Y. (2001). "On Small-Sample Confidence Intervals for Parameters in Discrete Distributions." *Biometrics* 57:963–971.

Bangdiwala, S. I. (1988). *The Agreement Chart*. Technical report, Department of Biostatistics, University of North Carolina at Chapel Hill.

Bangdiwala, S. I., Haedo, A. S., Natal, M. L., and Villaveces, A. (2008). "The Agreement Chart as an Alternative to the Receiver-Operating Characteristic Curve for Diagnostic Tests." *Journal of Clinical Epidemiology* 61:866–874.

Chan, I. S. F., and Zhang, Z. (1999). "Test-Based Exact Confidence Intervals for the Difference of Two Binomial Proportions." *Biometrics* 55:1202–1209.

Mehrotra, D. V., and Railkar, R. (2000). "Minimum Risk Weights for Comparing Treatments in Stratified Binomial Trials." *Statistics in Medicine* 19:811–825.

Chapter 2
The CORR Procedure

Contents

Overview: CORR Procedure	**4**
Getting Started: CORR Procedure	**5**
Syntax: CORR Procedure	**8**
PROC CORR Statement	8
BY Statement	16
FREQ Statement	16
ID Statement	17
PARTIAL Statement	17
VAR Statement	17
WEIGHT Statement	17
WITH Statement	18
Details: CORR Procedure	**18**
Pearson Product-Moment Correlation	18
Spearman Rank-Order Correlation	20
Kendall's Tau-b Correlation Coefficient	21
Hoeffding Dependence Coefficient	22
Partial Correlation	22
Fisher's z Transformation	24
Polychoric Correlation	27
Polyserial Correlation	27
Cronbach's Coefficient Alpha	29
Confidence and Prediction Ellipses	31
Missing Values	32
In-Database Computation	32
Output Tables	33
Output Data Sets	34
ODS Table Names	35
ODS Graphics	36
Examples: CORR Procedure	**37**
Example 2.1: Computing Four Measures of Association	37
Example 2.2: Computing Correlations between Two Sets of Variables	41
Example 2.3: Analysis Using Fisher's z Transformation	45
Example 2.4: Applications of Fisher's z Transformation	46
Example 2.5: Computing Polyserial Correlations	49
Example 2.6: Computing Cronbach's Coefficient Alpha	51
Example 2.7: Saving Correlations in an Output Data Set	53

> Example 2.8: Creating Scatter Plots 55
> Example 2.9: Computing Partial Correlations 59
> References **62**

Overview: CORR Procedure

The CORR procedure computes Pearson correlation coefficients, three nonparametric measures of association, polyserial correlation coefficients, and the probabilities associated with these statistics. The correlation statistics include the following:

- Pearson product-moment correlation
- Spearman rank-order correlation
- Kendall's tau-b coefficient
- Hoeffding's measure of dependence, D
- Pearson, Spearman, and Kendall partial correlation
- polychoric correlation
- polyserial correlation

Pearson product-moment correlation is a parametric measure of a linear relationship between two variables. For nonparametric measures of association, Spearman rank-order correlation uses the ranks of the data values and Kendall's tau-b uses the number of concordances and discordances in paired observations. Hoeffding's measure of dependence is another nonparametric measure of association that detects more general departures from independence. A partial correlation provides a measure of the correlation between two variables after controlling the effects of other variables.

Polyserial correlation measures the correlation between two continuous variables with a bivariate normal distribution, where only one variable is observed directly. Information about the unobserved variable is obtained through an observed ordinal variable that is derived from the unobserved variable by classifying its values into a finite set of discrete, ordered values.

A related type of correlation, polychoric correlation, measures the correlation between two unobserved variables with a bivariate normal distribution. Information about these variables is obtained through two corresponding observed ordinal variables that are derived from the unobserved variables by classifying their values into finite sets of discrete, ordered values.

When only one set of analysis variables is specified, the default correlation analysis includes descriptive statistics for each analysis variable and pairwise Pearson correlation statistics for these variables. You can also compute Cronbach's coefficient alpha for estimating reliability.

When two sets of analysis variables are specified, the default correlation analysis includes descriptive statistics for each analysis variable and pairwise Pearson correlation statistics between the two sets of variables.

For a Pearson or Spearman correlation, the Fisher's z transformation can be used to derive its confidence limits and a p-value under a specified null hypothesis $H_0: \rho = \rho_0$. Either a one-sided or a two-sided alternative is used for these statistics.

When the relationship between two variables is nonlinear or when outliers are present, the correlation coefficient might incorrectly estimate the strength of the relationship. Plotting the data enables you to verify the linear relationship and to identify the potential outliers. If ODS Graphics is enabled, scatter plots and a scatter plot matrix can be created via the Output Delivery System (ODS). Confidence and prediction ellipses can also be added to the scatter plot. See the section "Confidence and Prediction Ellipses" on page 31 for a detailed description of the ellipses.

You can save the correlation statistics in a SAS data set for use with other statistical and reporting procedures.

Getting Started: CORR Procedure

The following statements create the data set Fitness, which has been altered to contain some missing values:

```
*----------------- Data on Physical Fitness -----------------*
| These measurements were made on men involved in a physical |
| fitness course at N.C. State University.                   |
| The variables are Age (years), Weight (kg),                |
| Runtime (time to run 1.5 miles in minutes), and            |
| Oxygen (oxygen intake, ml per kg body weight per minute)   |
| Certain values were changed to missing for the analysis.   |
*------------------------------------------------------------*;
data Fitness;
   input Age Weight Oxygen RunTime @@;
   datalines;
44 89.47 44.609 11.37    40 75.07 45.313 10.07
44 85.84 54.297  8.65    42 68.15 59.571  8.17
38 89.02 49.874    .     47 77.45 44.811 11.63
40 75.98 45.681 11.95    43 81.19 49.091 10.85
44 81.42 39.442 13.08    38 81.87 60.055  8.63
44 73.03 50.541 10.13    45 87.66 37.388 14.03
45 66.45 44.754 11.12    47 79.15 47.273 10.60
54 83.12 51.855 10.33    49 81.42 49.156  8.95
51 69.63 40.836 10.95    51 77.91 46.672 10.00
48 91.63 46.774 10.25    49 73.37    .   10.08
57 73.37 39.407 12.63    54 79.38 46.080 11.17
52 76.32 45.441  9.63    50 70.87 54.625  8.92
51 67.25 45.118 11.08    54 91.63 39.203 12.88
51 73.71 45.790 10.47    57 59.08 50.545  9.93
49 76.32    .      .     48 61.24 47.920 11.50
52 82.78 47.467 10.50
;
```

The following statements invoke the CORR procedure and request a correlation analysis:

```
ods graphics on;
proc corr data=Fitness plots=matrix(histogram);
run;
```

The "Simple Statistics" table in Figure 2.1 displays univariate statistics for the analysis variables.

Figure 2.1 Univariate Statistics

The CORR Procedure

4 **Variables:** Age   Weight Oxygen RunTime

Variable	N	Mean	Std Dev	Sum	Minimum	Maximum
Age	31	47.67742	5.21144	1478	38.00000	57.00000
Weight	31	77.44452	8.32857	2401	59.08000	91.63000
Oxygen	29	47.22721	5.47718	1370	37.38800	60.05500
RunTime	29	10.67414	1.39194	309.55000	8.17000	14.03000

By default, all numeric variables not listed in other statements are used in the analysis. Observations with nonmissing values for each variable are used to derive the univariate statistics for that variable.

The "Pearson Correlation Coefficients" table in Figure 2.2 displays the Pearson correlation, the *p*-value under the null hypothesis of zero correlation, and the number of nonmissing observations for each pair of variables.

Figure 2.2 Pearson Correlation Coefficients

Pearson Correlation Coefficients
Prob > |r| under H0: Rho=0
Number of Observations

	Age	Weight	Oxygen	RunTime
Age	1.00000	-0.23354	-0.31474	0.14478
		0.2061	0.0963	0.4536
	31	31	29	29
Weight	-0.23354	1.00000	-0.15358	0.20072
	0.2061		0.4264	0.2965
	31	31	29	29
Oxygen	-0.31474	-0.15358	1.00000	-0.86843
	0.0963	0.4264		<.0001
	29	29	29	28
RunTime	0.14478	0.20072	-0.86843	1.00000
	0.4536	0.2965	<.0001	
	29	29	28	29

By default, Pearson correlation statistics are computed from observations with nonmissing values for each pair of analysis variables. Figure 2.2 displays a correlation of –0.86843 between Runtime and Oxygen, which is significant with a *p*-value less than 0.0001. That is, there exists an inverse linear relationship between these two variables. As Runtime (time to run 1.5 miles in minutes) increases, Oxygen (oxygen intake, ml per kg body weight per minute) decreases.

When you use the PLOTS=MATRIX(HISTOGRAM) option, the CORR procedure displays a symmetric matrix plot for the analysis variables in Figure 2.3. The histograms for these analysis variables are also displayed on the diagonal of the matrix plot. This inverse linear relationship between the two variables, Oxygen and Runtime, is also shown in the plot.

Note that ODS Graphics must be enabled and you must specify the PLOTS= option to produce graphs. For more information about ODS Graphics, see Chapter 21, "Statistical Graphics Using ODS" (*SAS/STAT User's Guide*).

Figure 2.3 Symmetric Matrix Plot

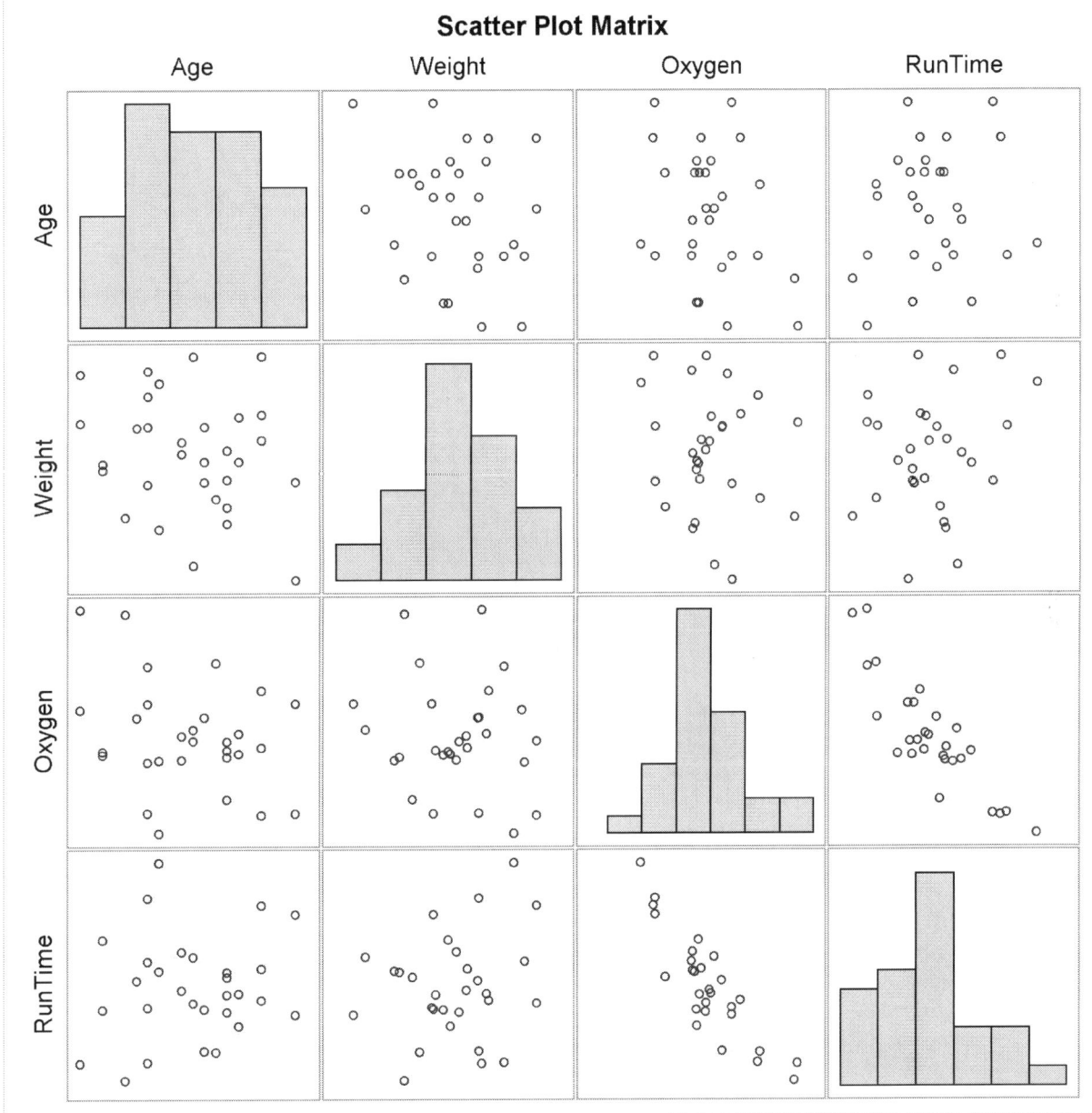

Syntax: CORR Procedure

The following statements are available in PROC CORR:

PROC CORR < *options* > ;
   BY *variables* ;
   FREQ *variable* ;
   ID *variables* ;
   PARTIAL *variables* ;
   VAR *variables* ;
   WEIGHT *variable* ;
   WITH *variables* ;

The BY statement specifies groups in which separate correlation analyses are performed.

The FREQ statement specifies the variable that represents the frequency of occurrence for other values in the observation.

The ID statement specifies one or more additional tip variables to identify observations in scatter plots and scatter plot matrices.

The PARTIAL statement identifies controlling variables to compute Pearson, Spearman, or Kendall partial-correlation coefficients.

The VAR statement lists the numeric variables to be analyzed and their order in the correlation matrix. If you omit the VAR statement, all numeric variables not listed in other statements are used.

The WEIGHT statement identifies the variable whose values weight each observation to compute Pearson product-moment correlation.

The WITH statement lists the numeric variables with which correlations are to be computed.

The PROC CORR statement is the only required statement for the CORR procedure. The rest of this section provides detailed syntax information for each of these statements, beginning with the PROC CORR statement. The remaining statements are presented in alphabetical order.

PROC CORR Statement

PROC CORR < *options* > ;

Table 2.1 summarizes the options available in the PROC CORR statement.

Table 2.1 Summary of PROC CORR Options

Option	Description
Data Sets	
DATA=	Specifies the input data set
OUTH=	Specifies the output data set with Hoeffding's *D* statistics
OUTK=	Specifies the output data set with Kendall correlation statistics
OUTP=	Specifies the output data set with Pearson correlation statistics

Table 2.1 continued

Option	Description
OUTPLC=	Specifies the output data set with polychoric correlation statistics
OUTPLS=	Specifies the output data set with polyserial correlation statistics
OUTS=	Specifies the output data set with Spearman correlation statistics
Statistical Analysis	
EXCLNPWGT	Excludes observations with nonpositive weight values from the analysis
FISHER	Requests correlation statistics using Fisher's z transformation
HOEFFDING	Requests Hoeffding's measure of dependence, D
KENDALL	Requests Kendall's tau-b
NOMISS	Excludes observations with missing analysis values from the analysis
PEARSON	Requests Pearson product-moment correlation
POLYCHORIC	Requests polychoric correlation
POLYSERIAL	Requests polyserial correlation
SPEARMAN	Requests Spearman rank-order correlation
Pearson Correlation Statistics	
ALPHA	Computes Cronbach's coefficient alpha
COV	Computes covariances
CSSCP	Computes corrected sums of squares and crossproducts
FISHER	Computes correlation statistics based on Fisher's z transformation
SINGULAR=	Specifies the singularity criterion
SSCP	Computes sums of squares and crossproducts
VARDEF=	Specifies the divisor for variance calculations
ODS Output Graphics	
PLOTS=MATRIX	Displays the scatter plot matrix
PLOTS=SCATTER	Displays scatter plots for pairs of variables
Printed Output	
BEST=	Displays the specified number of ordered correlation coefficients
NOCORR	Suppresses Pearson correlations
NOPRINT	Suppresses all printed output
NOPROB	Suppresses p-values
NOSIMPLE	Suppresses descriptive statistics
RANK	Displays ordered correlation coefficients

The following options can be used in the PROC CORR statement. They are listed in alphabetical order.

ALPHA

calculates and prints Cronbach's coefficient alpha. PROC CORR computes separate coefficients using raw and standardized values (scaling the variables to a unit variance of 1). For each VAR statement variable, PROC CORR computes the correlation between the variable and the total of the remaining variables. It also computes Cronbach's coefficient alpha by using only the remaining variables.

If a WITH statement is specified, the ALPHA option is invalid. When you specify the ALPHA option, the Pearson correlations will also be displayed. If you specify the OUTP= option, the output data set also contains observations with Cronbach's coefficient alpha. If you use the PARTIAL statement, PROC CORR calculates Cronbach's coefficient alpha for partialled variables. See the section "Partial Correlation" on page 22 for details.

BEST=n

prints the n highest correlation coefficients for each variable, $n \geq 1$. Correlations are ordered from highest to lowest in absolute value. Otherwise, PROC CORR prints correlations in a rectangular table, using the variable names as row and column labels.

If you specify the HOEFFDING option, PROC CORR displays the D statistics in order from highest to lowest.

COV

displays the variance and covariance matrix. When you specify the COV option, the Pearson correlations will also be displayed. If you specify the OUTP= option, the output data set also contains the covariance matrix with the corresponding _TYPE_ variable value 'COV.' If you use the PARTIAL statement, PROC CORR computes a partial covariance matrix.

CSSCP

displays a table of the corrected sums of squares and crossproducts. When you specify the CSSCP option, the Pearson correlations will also be displayed. If you specify the OUTP= option, the output data set also contains a CSSCP matrix with the corresponding _TYPE_ variable value 'CSSCP.' If you use a PARTIAL statement, PROC CORR prints both an unpartial and a partial CSSCP matrix, and the output data set contains a partial CSSCP matrix.

DATA=SAS-data-set

names the SAS data set to be analyzed by PROC CORR. By default, the procedure uses the most recently created SAS data set.

EXCLNPWGT

EXCLNPWGTS

excludes observations with nonpositive weight values from the analysis. By default, PROC CORR treats observations with negative weights like those with zero weights and counts them in the total number of observations.

FISHER < (fisher-options) >

requests confidence limits and p-values under a specified null hypothesis, $H_0: \rho = \rho_0$, for correlation coefficients by using Fisher's z transformation. These correlations include the Pearson correlations and Spearman correlations.

The following fisher-options are available:

ALPHA=α

specifies the level of the confidence limits for the correlation, $100(1 - \alpha)\%$. The value of the ALPHA= option must be between 0 and 1, and the default is ALPHA=0.05.

BIASADJ=YES | NO

specifies whether or not the bias adjustment is used in constructing confidence limits. The BIASADJ=YES option also produces a new correlation estimate that uses the bias adjustment. By default, BIASADJ=YES.

RHO0=ρ_0

specifies the value ρ_0 in the null hypothesis $H_0: \rho = \rho_0$, where $-1 < \rho_0 < 1$. By default, RHO0=0.

TYPE=LOWER | UPPER | TWOSIDED

specifies the type of confidence limits. The TYPE=LOWER option requests a lower confidence limit for the test of the one-sided hypothesis $H_0: \rho \leq \rho_0$ against the alternative hypothesis $H_1: \rho > \rho_0$, the TYPE=UPPER option requests an upper confidence limit for the test of the one-sided hypothesis $H_0: \rho \geq \rho_0$ against the alternative hypothesis $H_1: \rho < \rho_0$, and the default TYPE=TWOSIDED option requests two-sided confidence limits for the test of the hypothesis $H_0: \rho = \rho_0$.

HOEFFDING

requests a table of Hoeffding's D statistics. This D statistic is 30 times larger than the usual definition and scales the range between –0.5 and 1 so that large positive values indicate dependence. The HOEFFDING option is invalid if a WEIGHT or PARTIAL statement is used.

KENDALL

requests a table of Kendall's tau-b coefficients based on the number of concordant and discordant pairs of observations. Kendall's tau-b ranges from –1 to 1.

The KENDALL option is invalid if a WEIGHT statement is used. If you use a PARTIAL statement, probability values for Kendall's partial tau-b are not available.

NOCORR

suppresses displaying of Pearson correlations. If you specify the OUTP= option, the data set type remains CORR. To change the data set type to COV, CSSCP, or SSCP, use the TYPE_ data set option.

NOMISS

excludes observations with missing values from the analysis. Otherwise, PROC CORR computes correlation statistics by using all of the nonmissing pairs of variables. Using the NOMISS option is computationally more efficient.

NOPRINT

suppresses all displayed output, which also includes output produced with ODS Graphics. Use the NOPRINT option if you want to create an output data set only.

NOPROB

suppresses displaying the probabilities associated with each correlation coefficient.

NOSIMPLE

suppresses printing simple descriptive statistics for each variable. However, if you request an output data set, the output data set still contains simple descriptive statistics for the variables.

OUTH=*output-data-set*

creates an output data set that contains Hoeffding's D statistics. The contents of the output data set are similar to those of the OUTP= data set. When you specify the OUTH= option, the Hoeffding's D statistics will be displayed.

OUTK=*output-data-set*
>   creates an output data set that contains Kendall correlation statistics. The contents of the output data set are similar to those of the OUTP= data set. When you specify the OUTK= option, the Kendall correlation statistics will be displayed.

OUTP=*output-data-set*

OUT=*output-data-set*
>   creates an output data set that contains Pearson correlation statistics. This data set also includes means, standard deviations, and the number of observations. The value of the _TYPE_ variable is 'CORR.' When you specify the OUTP= option, the Pearson correlations will also be displayed. If you specify the ALPHA option, the output data set also contains six observations with Cronbach's coefficient alpha.

OUTPLC=*output-data-set*
>   creates an output data set that contains polychoric correlation statistics. (Polychoric correlation between two observed binary variables is also known as tetrachoric correlation.) This data set also includes means, standard deviations, and the number of observations. The value of the _TYPE_ variable is 'CORR.'

OUTPLS=*output-data-set*
>   creates an output data set that contains polyserial correlation statistics. The contents of the output data set are similar to those of the OUTPLC= data set.

OUTS=*SAS-data-set*
>   creates an output data set that contains Spearman correlation coefficients. The contents of the output data set are similar to those of the OUTP= data set. When you specify the OUTS= option, the Spearman correlation coefficients will be displayed.

PEARSON
>   requests a table of Pearson product-moment correlations. The correlations range from –1 to 1. If you do not specify the HOEFFDING, KENDALL, SPEARMAN, POLYCHORIC, POLYSERIAL, OUTH=, OUTK=, or OUTS= option, the CORR procedure produces Pearson product-moment correlations by default. Otherwise, you must specify the PEARSON, ALPHA, COV, CSSCP, SSCP, or OUT= option for Pearson correlations. Also, if a scatter plot or a scatter plot matrix is requested, the Pearson correlations will be displayed.

PLOTS <(MAXPOINTS=NONE | *n* **) > =** *plot-request*

PLOTS <(MAXPOINTS=NONE | *n* **) > = (** *plot-request* **<...** *plot-request* **>)**
>   requests statistical graphics via the Output Delivery System (ODS).
>
>   ODS Graphics must be enabled before plots can be requested. For example:
>
>   ```
>   ods graphics on;
>   proc corr data=Fitness plots=matrix(histogram);
>   run;
>   ```
>
>   For more information about enabling and disabling ODS Graphics, see the section "Enabling and Disabling ODS Graphics" in Chapter 21, "Statistical Graphics Using ODS" (*SAS/STAT User's Guide*).
>
>   The global plot option MAXPOINTS= specifies that plots with elements that require processing more than *n* points be suppressed. The default is MAXPOINTS=5000. This limit is ignored if you specify MAXPOINTS=NONE. The plot request options include the following:

ALL
    produces all appropriate plots.

MATRIX < (*matrix-options*) >
    requests a scatter plot matrix for variables. That is, the procedure displays a symmetric matrix plot with variables in the VAR list if a WITH statement is not specified. Otherwise, the procedure displays a rectangular matrix plot with the WITH variables appearing down the side and the VAR variables appearing across the top.

NONE
    suppresses all plots.

SCATTER < (*scatter-options*) >
    requests scatter plots for pairs of variables. That is, the procedure displays a scatter plot for each applicable pair of distinct variables from the VAR list if a WITH statement is not specified. Otherwise, the procedure displays a scatter plot for each applicable pair of variables, one from the WITH list and the other from the VAR list.

When a scatter plot or a scatter plot matrix is requested, the Pearson correlations will also be displayed.

The available *matrix-options* are as follows:

HIST | HISTOGRAM
    displays histograms of variables in the VAR list (specified in the VAR statement) in the symmetric matrix plot.

NVAR=ALL | *n*
    specifies the maximum number of variables in the VAR list to be displayed in the matrix plot, where $n > 0$. The NVAR=ALL option uses all variables in the VAR list. By default, NVAR=5.

NWITH=ALL | *n*
    specifies the maximum number of variables in the WITH list (specified in the WITH statement) to be displayed in the matrix plot, where $n > 0$. The NWITH=ALL option uses all variables in the WITH list. By default, NWITH=5.

If the resulting maximum number of variables in the VAR or WITH list is greater than 10, only the first 10 variables in the list are displayed in the scatter plot matrix.

The available *scatter-options* are as follows:

ALPHA=α
    specifies the α values for the confidence or prediction ellipses to be displayed in the scatter plots, where $0 < \alpha < 1$. For each α value specified, a $(1 - \alpha)$ confidence or prediction ellipse is created. By default, $\alpha = 0.05$.

ELLIPSE=PREDICTION | CONFIDENCE | NONE
    requests prediction ellipses for new observations (ELLIPSE=PREDICTION), confidence ellipses for the mean (ELLIPSE=CONFIDENCE), or no ellipses (ELLIPSE=NONE) to be created in the scatter plots. By default, ELLIPSE=PREDICTION.

NOINSET
> suppresses the default inset of summary information for the scatter plot. The inset table contains the number of observations (Observations) and correlation.

NVAR=ALL | n
> specifies the maximum number of variables in the VAR list (specified in the VAR statement) to be displayed in the plots, where n > 0. The NVAR=ALL option uses all variables in the VAR list. By default, NVAR=5.

NWITH=ALL | n
> specifies the maximum number of variables in the WITH list (specified in the WITH statement) to be displayed in the plots, where n > 0. The NWITH=ALL option uses all variables in the WITH list. By default, NWITH=5.

If the resulting maximum number of variables in the VAR or WITH list is greater than 10, only the first 10 variables in the list are displayed in the scatter plots.

POLYCHORIC <(*options*)>
> requests a table of polychoric correlation coefficients. (Polychoric correlation between two observed binary variables is also known as tetrachoric correlation.) A polychoric correlation measures the correlation between two unobserved, continuous variables that have a bivariate normal distribution. Information about each unobserved variable is obtained through an observed ordinal variable that is derived from the unobserved variable by classifying its values into a finite set of discrete, ordered values. If you specify a WEIGHT statement, the POLYCHORIC option is not applicable.

You can specify the following *options* for computing polychoric correlation:

CONVERGE=p
> specifies the convergence criterion. The value *p* must be between 0 and 1. The iterations are considered to have converged when the absolute change in the parameter estimates between iteration steps is less than *p* for each parameter—that is, for the correlation and the thresholds for the unobserved continuous variable that define the categories for the ordinal variable. By default, CONVERGE=0.0001.

MAXITER=number
> specifies the maximum number of iterations. The iterations stop when the number of iterations exceeds *number*. By default, MAXITER=200.

NGROUPS=ALL | n
> specifies the maximum number of groups allowed for each ordinal variable, where n > 1. NGROUPS=ALL allows an unlimited number of groups in each ordinal variable. Otherwise, if the number of groups exceeds the specified number *n*, polychoric correlations are not computed for the affected pairs of variables. By default, NGROUPS=20.

POLYSERIAL <(*options*)>
> requests a table of polyserial correlation coefficients. A polyserial correlation measures the correlation between two continuous variables with a bivariate normal distribution, where one variable is observed and the other is unobserved. Information about the unobserved variable is obtained through an observed ordinal variable that is derived from the unobserved variable by classifying its values into a finite set of discrete, ordered values. If you specify a WEIGHT statement, the POLYSERIAL option is not applicable.

You can specify the following *options* for computing polyserial correlation:

CONVERGE=p
> specifies the convergence criterion. The value p must be between 0 and 1. The iterations are considered to have converged when the absolute change in the parameter estimates between iteration steps is less than p for each parameter—that is, for the correlation and the thresholds for the unobserved continuous variable that define the categories for the ordinal variable. By default, CONVERGE=0.0001.

MAXITER=number
> specifies the maximum number of iterations. The iterations stop when the number of iterations exceeds *number*. By default, MAXITER=200.

**NGROUPS=ALL | **n
> specifies the maximum number of groups allowed for each ordinal variable, where $n > 1$. NGROUPS=ALL allows an unlimited number of groups in each ordinal variable. Otherwise, if the number of groups exceeds the specified number *n*, polyserial correlations are not computed for the affected pairs of variables. By default, NGROUPS=20.

ORDINAL=WITH | VAR
> specifies the ordinal variable list. The ORDINAL=WITH option specifies that the ordinal variables are provided in the WITH statement, and the continuous variables are provided in the VAR statement. The ORDINAL=VAR option specifies that the ordinal variables are provided in the VAR statement, and the continuous variables are provided in the WITH statement. By default, ORDINAL=WITH.

RANK
> displays the ordered correlation coefficients for each variable. Correlations are ordered from highest to lowest in absolute value. If you specify the HOEFFDING option, the *D* statistics are displayed in order from highest to lowest.

SINGULAR=p
> specifies the criterion for determining the singularity of a variable if you use a PARTIAL statement. A variable is considered singular if its corresponding diagonal element after Cholesky decomposition has a value less than p times the original unpartialled value of that variable. By default, SINGULAR=1E−8. The range of p is between 0 and 1.

SPEARMAN
> requests a table of Spearman correlation coefficients based on the ranks of the variables. The correlations range from −1 to 1. If you specify a WEIGHT statement, the SPEARMAN option is invalid.

SSCP
> displays a table of the sums of squares and crossproducts. When you specify the SSCP option, the Pearson correlations are also displayed. If you specify the OUTP= option, the output data set contains a SSCP matrix and the corresponding _TYPE_ variable value is 'SSCP.' If you use a PARTIAL statement, the unpartial SSCP matrix is displayed, and the output data set does not contain an SSCP matrix.

VARDEF=DF | N | WDF | WEIGHT | WGT
> specifies the variance divisor in the calculation of variances and covariances. The default is VARDEF=DF.

Table 2.2 displays available values and associated divisors for the VARDEF= option, where n is the number of nonmissing observations, k is the number of variables specified in the PARTIAL statement, and w_j is the weight associated with the jth nonmissing observation.

Table 2.2 Possible Values for the VARDEF= Option

Value	Description	Divisor
DF	Degrees of freedom	$n - k - 1$
N	Number of observations	n
WDF	Sum of weights minus one	$\sum_{j}^{n} w_j - k - 1$
WEIGHT | WGT	Sum of weights	$\sum_{j}^{n} w_j$

BY Statement

BY *variables* ;

You can specify a BY statement with PROC CORR to obtain separate analyses of observations in groups that are defined by the BY variables. When a BY statement appears, the procedure expects the input data set to be sorted in order of the BY variables. If you specify more than one BY statement, only the last one specified is used.

If your input data set is not sorted in ascending order, use one of the following alternatives:

- Sort the data by using the SORT procedure with a similar BY statement.

- Specify the NOTSORTED or DESCENDING option in the BY statement for the CORR procedure. The NOTSORTED option does not mean that the data are unsorted but rather that the data are arranged in groups (according to values of the BY variables) and that these groups are not necessarily in alphabetical or increasing numeric order.

- Create an index on the BY variables by using the DATASETS procedure (in Base SAS software).

For more information about BY-group processing, see the discussion in *SAS Language Reference: Concepts*. For more information about the DATASETS procedure, see the discussion in the *SAS Visual Data Management and Utility Procedures Guide*.

FREQ Statement

FREQ *variable* ;

The FREQ statement lists a numeric variable whose value represents the frequency of the observation. If you use the FREQ statement, the procedure assumes that each observation represents n observations, where n is the value of the FREQ variable. If n is not an integer, SAS truncates it. If n is less than 1 or is missing, the observation is excluded from the analysis. The sum of the frequency variable represents the total number of observations.

The effects of the FREQ and WEIGHT statements are similar except when calculating degrees of freedom.

ID Statement

ID *variables* ;

The ID statement specifies one or more additional tip variables to identify observations in scatter plots and scatter plot matrix. For each plot, the tip variables include the X-axis variable, the Y-axis variable, and the variable for observation numbers.

PARTIAL Statement

PARTIAL *variables* ;

The PARTIAL statement lists variables to use in the calculation of partial correlation statistics. Only the Pearson partial correlation, Spearman partial rank-order correlation, and Kendall's partial tau-b can be computed. When you use the PARTIAL statement, observations with missing values are excluded.

With a PARTIAL statement, PROC CORR also displays the partial variance and standard deviation for each analysis variable if the PEARSON option is specified.

VAR Statement

VAR *variables* ;

The VAR statement lists variables for which to compute correlation coefficients. If the VAR statement is not specified, PROC CORR computes correlations for all numeric variables not listed in other statements.

WEIGHT Statement

WEIGHT *variable* ;

The WEIGHT statement lists weights to use in the calculation of Pearson weighted product-moment correlation. The HOEFFDING, KENDALL, and SPEARMAN options are not valid with the WEIGHT statement.

The observations with missing weights are excluded from the analysis. By default, for observations with nonpositive weights, weights are set to zero and the observations are included in the analysis. You can use the EXCLNPWGT option to exclude observations with negative or zero weights from the analysis.

WITH Statement

WITH *variables* ;

The WITH statement lists variables with which correlations of the VAR statement variables are to be computed. The WITH statement requests correlations of the form $r(X_i, Y_j)$, where $X_1, \ldots, X_m$ are analysis variables specified in the VAR statement, and $Y_1, \ldots, Y_n$ are variables specified in the WITH statement. The correlation matrix has a rectangular structure of the form

$$\begin{bmatrix} r(Y_1, X_1) & \cdots & r(Y_1, X_m) \\ \vdots & \ddots & \vdots \\ r(Y_n, X_1) & \cdots & r(Y_n, X_m) \end{bmatrix}$$

For example, the statements

```
proc corr;
   var x1 x2;
   with y1 y2 y3;
run;
```

produce correlations for the following combinations:

$$\begin{bmatrix} r(Y1, X1) & r(Y1, X2) \\ r(Y2, X1) & r(Y2, X2) \\ r(Y3, X1) & r(Y3, X2) \end{bmatrix}$$

Details: CORR Procedure

Pearson Product-Moment Correlation

The Pearson product-moment correlation is a parametric measure of association for two variables. It measures both the strength and the direction of a linear relationship. If one variable X is an exact linear function of another variable Y, a positive relationship exists if the correlation is 1 and a negative relationship exists if the correlation is –1. If there is no linear predictability between the two variables, the correlation is 0. If the two variables are normal with a correlation 0, the two variables are independent. However, correlation does not imply causality because, in some cases, an underlying causal relationship might not exist.

The scatter plot matrix in Figure 2.4 displays the relationship between two numeric random variables in various situations.

Figure 2.4 Correlations between Two Variables

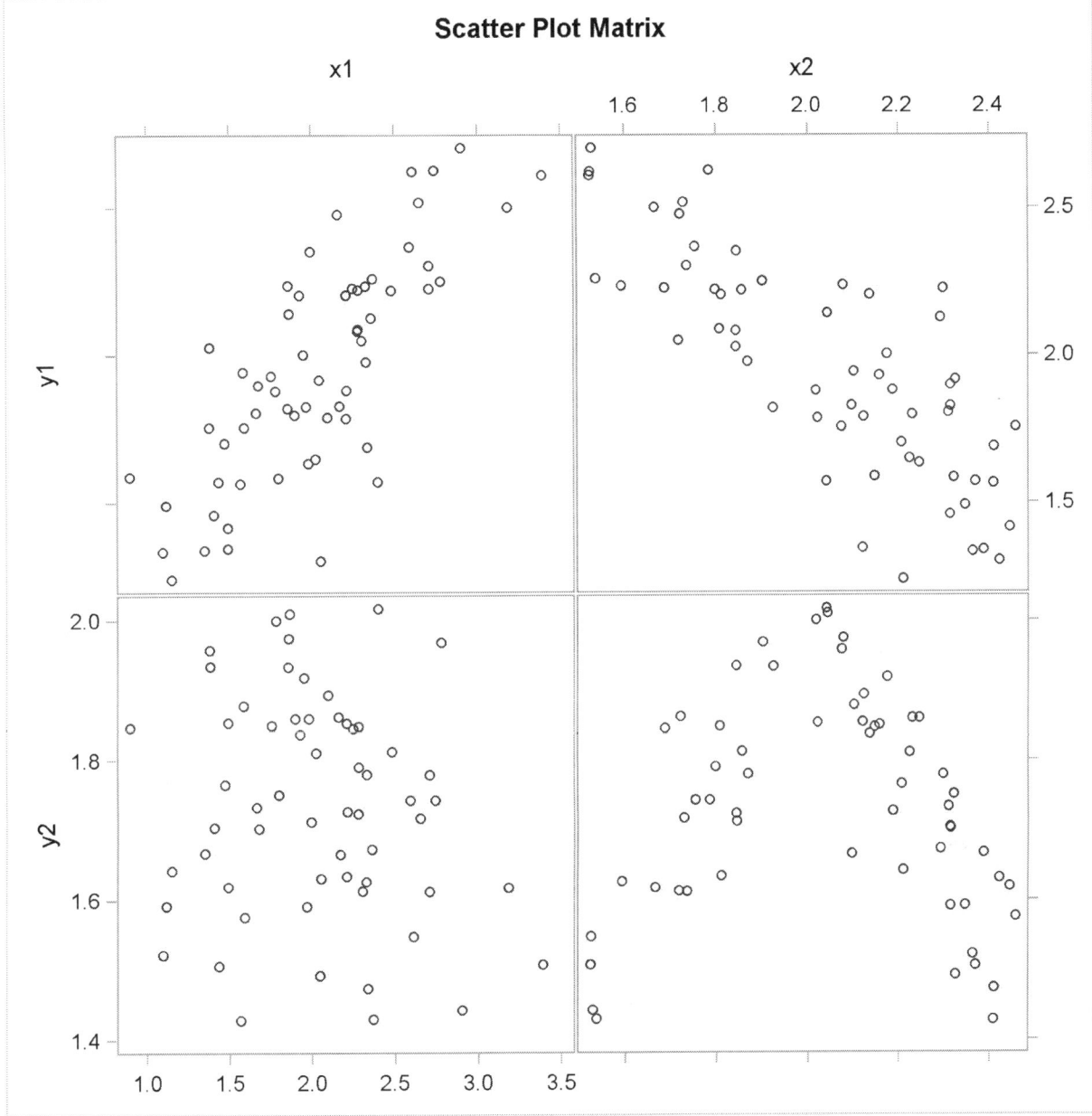

The scatter plot matrix shows a positive correlation between variables Y1 and X1, a negative correlation between Y1 and X2, and no clear correlation between Y2 and X1. The plot also shows no clear linear correlation between Y2 and X2, even though Y2 is dependent on X2.

The formula for the population Pearson product-moment correlation, denoted ρ_{xy}, is

$$\rho_{xy} = \frac{\text{Cov}(x, y)}{\sqrt{\text{V}(x)\text{V}(y)}} = \frac{\text{E}((x - \text{E}(x))(y - \text{E}(y)))}{\sqrt{\text{E}(x - \text{E}(x))^2 \, \text{E}(y - \text{E}(y))^2}}$$

The sample correlation, such as a Pearson product-moment correlation or weighted product-moment correlation, estimates the population correlation. The formula for the sample Pearson product-moment correlation is

$$r_{xy} = \frac{\sum_i ((x_i - \bar{x})(y_i - \bar{y}))}{\sqrt{\sum_i (x_i - \bar{x})^2 \sum_i (y_i - \bar{y})^2}}$$

where $\bar{x}$ is the sample mean of x and $\bar{y}$ is the sample mean of y. The formula for a weighted Pearson product-moment correlation is

$$r_{xy} = \frac{\sum_i w_i (x_i - \bar{x}_w)(y_i - \bar{y}_w)}{\sqrt{\sum_i w_i (x_i - \bar{x}_w)^2 \sum_i w_i (y_i - \bar{y}_w)^2}}$$

where w_i is the weight, $\bar{x}_w$ is the weighted mean of x, and $\bar{y}_w$ is the weighted mean of y.

Probability Values

Probability values for the Pearson correlation are computed by treating

$$t = (n-2)^{1/2} \left(\frac{r^2}{1 - r^2} \right)^{1/2}$$

as coming from a t distribution with $(n - 2)$ degrees of freedom, where r is the sample correlation.

Spearman Rank-Order Correlation

Spearman rank-order correlation is a nonparametric measure of association based on the ranks of the data values. The formula is

$$\theta = \frac{\sum_i ((R_i - \bar{R})(S_i - \bar{S}))}{\sqrt{\sum_i (R_i - \bar{R})^2 \sum (S_i - \bar{S})^2}}$$

where R_i is the rank of x_i, S_i is the rank of y_i, $\bar{R}$ is the mean of the R_i values, and $\bar{S}$ is the mean of the S_i values.

PROC CORR computes the Spearman correlation by ranking the data and using the ranks in the Pearson product-moment correlation formula. In case of ties, the averaged ranks are used.

Probability Values

Probability values for the Spearman correlation are computed by treating

$$t = (n-2)^{1/2} \left(\frac{r^2}{1 - r^2} \right)^{1/2}$$

as coming from a t distribution with $(n - 2)$ degrees of freedom, where r is the sample Spearman correlation.

Kendall's Tau-b Correlation Coefficient

Kendall's tau-b is a nonparametric measure of association based on the number of concordances and discordances in paired observations. Concordance occurs when paired observations vary together, and discordance occurs when paired observations vary differently. The formula for Kendall's tau-b is

$$\tau = \frac{\sum_{i<j} (\text{sgn}(x_i - x_j)\text{sgn}(y_i - y_j))}{\sqrt{(T_0 - T_1)(T_0 - T_2)}}$$

where $T_0 = n(n-1)/2$, $T_1 = \sum_k t_k(t_k - 1)/2$, and $T_2 = \sum_l u_l(u_l - 1)/2$. The t_k is the number of tied x values in the kth group of tied x values, u_l is the number of tied y values in the lth group of tied y values, n is the number of observations, and $\text{sgn}(z)$ is defined as

$$\text{sgn}(z) = \begin{cases} 1 & \text{if } z > 0 \\ 0 & \text{if } z = 0 \\ -1 & \text{if } z < 0 \end{cases}$$

PROC CORR computes Kendall's tau-b by ranking the data and using a method similar to Knight (1966). The data are double sorted by ranking observations according to values of the first variable and reranking the observations according to values of the second variable. PROC CORR computes Kendall's tau-b from the number of interchanges of the first variable and corrects for tied pairs (pairs of observations with equal values of X or equal values of Y).

Probability Values

Probability values for Kendall's tau-b are computed by treating

$$\frac{s}{\sqrt{V(s)}}$$

as coming from a standard normal distribution where

$$s = \sum_{i<j} (\text{sgn}(x_i - x_j)\text{sgn}(y_i - y_j))$$

and $V(s)$, the variance of s, is computed as

$$V(s) = \frac{v_0 - v_t - v_u}{18} + \frac{v_1}{2n(n-1)} + \frac{v_2}{9n(n-1)(n-2)}$$

where

$v_0 = n(n-1)(2n+5)$

$v_t = \sum_k t_k(t_k - 1)(2t_k + 5)$

$v_u = \sum_l u_l(u_l - 1)(2u_l + 5)$

$v_1 = \left(\sum_k t_k(t_k - 1)\right)\left(\sum u_i(u_l - 1)\right)$

$v_2 = \left(\sum_l t_i(t_k - 1)(t_k - 2)\right)\left(\sum u_l(u_l - 1)(u_l - 2)\right)$

The sums are over tied groups of values where t_i is the number of tied x values and u_i is the number of tied y values (Noether 1967). The sampling distribution of Kendall's partial tau-b is unknown; therefore, the probability values are not available.

Hoeffding Dependence Coefficient

Hoeffding's measure of dependence, D, is a nonparametric measure of association that detects more general departures from independence. The statistic approximates a weighted sum over observations of chi-square statistics for two-by-two classification tables (Hoeffding 1948). Each set of (x, y) values are cut points for the classification. The formula for Hoeffding's D is

$$D = 30 \frac{(n-2)(n-3)D_1 + D_2 - 2(n-2)D_3}{n(n-1)(n-2)(n-3)(n-4)}$$

where $D_1 = \sum_i (Q_i - 1)(Q_i - 2)$, $D_2 = \sum_i (R_i - 1)(R_i - 2)(S_i - 1)(S_i - 2)$, and $D_3 = \sum_i (R_i - 2)(S_i - 2)(Q_i - 1)$. R_i is the rank of x_i, S_i is the rank of y_i, and Q_i (also called the bivariate rank) is 1 plus the number of points with both x and y values less than the ith point.

A point that is tied on only the x value or y value contributes 1/2 to Q_i if the other value is less than the corresponding value for the ith point.

A point that is tied on both x and y contributes 1/4 to Q_i. PROC CORR obtains the Q_i values by first ranking the data. The data are then double sorted by ranking observations according to values of the first variable and reranking the observations according to values of the second variable. Hoeffding's D statistic is computed using the number of interchanges of the first variable. When no ties occur among data set observations, the D statistic values are between –0.5 and 1, with 1 indicating complete dependence. However, when ties occur, the D statistic might result in a smaller value. That is, for a pair of variables with identical values, the Hoeffding's D statistic might be less than 1. With a large number of ties in a small data set, the D statistic might be less than –0.5. For more information about Hoeffding's D, see Hollander and Wolfe (1999).

Probability Values

The probability values for Hoeffding's D statistic are computed using the asymptotic distribution computed by Blum, Kiefer, and Rosenblatt (1961). The formula is

$$\frac{(n-1)\pi^4}{60} D + \frac{\pi^4}{72}$$

which comes from the asymptotic distribution. If the sample size is less than 10, refer to the tables for the distribution of D in Hollander and Wolfe (1999).

Partial Correlation

A partial correlation measures the strength of a relationship between two variables, while controlling the effect of other variables. The Pearson partial correlation between two variables, after controlling for variables in the PARTIAL statement, is equivalent to the Pearson correlation between the residuals of the two variables after regression on the controlling variables.

Let $\mathbf{y} = (y_1, y_2, \ldots, y_v)$ be the set of variables to correlate and $\mathbf{z} = (z_1, z_2, \ldots, z_p)$ be the set of controlling variables. The population Pearson partial correlation between the ith and the jth variables of $\mathbf{y}$ given $\mathbf{z}$ is the correlation between errors $(y_i - \mathrm{E}(y_i))$ and $(y_j - \mathrm{E}(y_j))$, where

$$\mathrm{E}(y_i) = \alpha_i + \mathbf{z}\boldsymbol{\beta}_i \quad \text{and} \quad \mathrm{E}(y_j) = \alpha_j + \mathbf{z}\boldsymbol{\beta}_j$$

are the regression models for variables y_i and y_j given the set of controlling variables $\mathbf{z}$, respectively.

For a given sample of observations, a sample Pearson partial correlation between y_i and y_j given z is derived from the residuals $y_i - \hat{y}_i$ and $y_j - \hat{y}_j$, where

$$\hat{y}_i = \hat{\alpha}_i + \mathbf{z}\hat{\boldsymbol{\beta}}_i \quad \text{and} \quad \hat{y}_j = \hat{\alpha}_j + \mathbf{z}\hat{\boldsymbol{\beta}}_j$$

are fitted values from regression models for variables y_i and y_j given z.

The partial corrected sums of squares and crossproducts (CSSCP) of **y** given z are the corrected sums of squares and crossproducts of the residuals $\mathbf{y} - \hat{\mathbf{y}}$. Using these partial corrected sums of squares and crossproducts, you can calculate the partial covariances and partial correlations.

PROC CORR derives the partial corrected sums of squares and crossproducts matrix by applying the Cholesky decomposition algorithm to the CSSCP matrix. For Pearson partial correlations, let S be the partitioned CSSCP matrix between two sets of variables, **z** and **y**:

$$\mathbf{S} = \begin{bmatrix} \mathbf{S}_{zz} & \mathbf{S}_{zy} \\ \mathbf{S}'_{zy} & \mathbf{S}_{yy} \end{bmatrix}$$

PROC CORR calculates $S_{yy.z}$, the partial CSSCP matrix of **y** after controlling for **z**, by applying the Cholesky decomposition algorithm sequentially on the rows associated with **z**, the variables being partialled out.

After applying the Cholesky decomposition algorithm to each row associated with variables **z**, PROC CORR checks all higher-numbered diagonal elements associated with **z** for singularity. A variable is considered singular if the value of the corresponding diagonal element is less than ε times the original unpartialled corrected sum of squares of that variable. You can specify the singularity criterion ε by using the SINGULAR= option. For Pearson partial correlations, a controlling variable **z** is considered singular if the R^2 for predicting this variable from the variables that are already partialled out exceeds $1 - \varepsilon$. When this happens, PROC CORR excludes the variable from the analysis. Similarly, a variable is considered singular if the R^2 for predicting this variable from the controlling variables exceeds $1 - \varepsilon$. When this happens, its associated diagonal element and all higher-numbered elements in this row or column are set to zero.

After the Cholesky decomposition algorithm is applied to all rows associated with **z**, the resulting matrix has the form

$$T = \begin{bmatrix} \mathbf{T}_{zz} & \mathbf{T}_{zy} \\ 0 & \mathbf{S}_{yy.z} \end{bmatrix}$$

where T_{zz} is an upper triangular matrix with $T'_{zz}T_{zz} = S'_{zz}$, $T'_{zz}T_{zy} = S'_{zy}$, and $S_{yy.z} = S_{yy} - T'_{zy}T_{zy}$.

If S_{zz} is positive definite, then $T_{zy} = T'_{zz}{}^{-1}S'_{zy}$ and the partial CSSCP matrix $S_{yy.z}$ is identical to the matrix derived from the formula

$$S_{yy.z} = S_{yy} - S'_{zy}S'^{-1}_{zz}S_{zy}$$

The partial variance-covariance matrix is calculated with the variance divisor (VARDEF= option). PROC CORR then uses the standard Pearson correlation formula on the partial variance-covariance matrix to calculate the Pearson partial correlation matrix.

When a correlation matrix is positive definite, the resulting partial correlation between variables x and y after adjusting for a single variable z is identical to that obtained from the first-order partial correlation formula

$$r_{xy.z} = \frac{r_{xy} - r_{xz}r_{yz}}{\sqrt{(1 - r_{xz}^2)(1 - r_{yz}^2)}}$$

where r_{xy}, r_{xz}, and r_{yz} are the appropriate correlations.

The formula for higher-order partial correlations is a straightforward extension of the preceding first-order formula. For example, when the correlation matrix is positive definite, the partial correlation between x and y controlling for both z_1 and z_2 is identical to the second-order partial correlation formula

$$r_{xy.z_1 z_2} = \frac{r_{xy.z_1} - r_{xz_2.z_1} r_{yz_2.z_1}}{\sqrt{(1 - r^2_{xz_2.z_1})(1 - r^2_{yz_2.z_1})}}$$

where $r_{xy.z_1}$, $r_{xz_2.z_1}$, and $r_{yz_2.z_1}$ are first-order partial correlations among variables x, y, and z_2 given z_1.

To derive the corresponding Spearman partial rank-order correlations and Kendall partial tau-b correlations, PROC CORR applies the Cholesky decomposition algorithm to the Spearman rank-order correlation matrix and Kendall's tau-b correlation matrix and uses the correlation formula. That is, the Spearman partial correlation is equivalent to the Pearson correlation between the residuals of the linear regression of the ranks of the two variables on the ranks of the partialled variables. Thus, if a PARTIAL statement is specified with the CORR=SPEARMAN option, the residuals of the ranks of the two variables are displayed in the plot. The partial tau-b correlations range from −1 to 1. However, the sampling distribution of this partial tau-b is unknown; therefore, the probability values are not available.

Probability Values

Probability values for the Pearson and Spearman partial correlations are computed by treating

$$\frac{(n - k - 2)^{1/2} r}{(1 - r^2)^{1/2}}$$

as coming from a t distribution with $(n - k - 2)$ degrees of freedom, where r is the partial correlation and k is the number of variables being partialled out.

Fisher's z Transformation

For a sample correlation r that uses a sample from a bivariate normal distribution with correlation $\rho = 0$, the statistic

$$t_r = (n - 2)^{1/2} \left(\frac{r^2}{1 - r^2} \right)^{1/2}$$

has a Student's t distribution with $(n-2)$ degrees of freedom.

With the monotone transformation of the correlation r (Fisher 1921)

$$z_r = \tanh^{-1}(r) = \frac{1}{2} \log \left(\frac{1 + r}{1 - r} \right)$$

the statistic z_r has an approximate normal distribution with mean and variance

$$E(z_r) = \zeta + \frac{\rho}{2(n - 1)}$$

$$V(z_r) = \frac{1}{n - 3}$$

where $\zeta = \tanh^{-1}(\rho)$.

For the transformed z_r, the approximate variance $V(z_r) = 1/(n-3)$ is independent of the correlation ρ. Furthermore, even the distribution of z_r is not strictly normal, it tends to normality rapidly as the sample size increases for any values of ρ (Fisher 1973, pp. 200–201).

For the null hypothesis $H_0: \rho = \rho_0$, the p-values are computed by treating

$$z_r - \zeta_0 - \frac{\rho_0}{2(n-1)}$$

as a normal random variable with mean zero and variance $1/(n-3)$, where $\zeta_0 = \tanh^{-1}(\rho_0)$ (Fisher 1973, p. 207; Anderson 1984, p. 123).

Note that the bias adjustment, $\rho_0/(2(n-1))$, is always used when computing p-values under the null hypothesis $H_0: \rho = \rho_0$ in the CORR procedure.

The ALPHA= option in the FISHER option specifies the value α for the confidence level $1 - \alpha$, the RHO0= option specifies the value ρ_0 in the hypothesis $H_0: \rho = \rho_0$, and the BIASADJ= option specifies whether the bias adjustment is to be used for the confidence limits.

The TYPE= option specifies the type of confidence limits. The TYPE=TWOSIDED option requests two-sided confidence limits and a p-value under the hypothesis $H_0: \rho = \rho_0$. For a one-sided confidence limit, the TYPE=LOWER option requests a lower confidence limit and a p-value under the hypothesis $H_0: \rho <= \rho_0$, and the TYPE=UPPER option requests an upper confidence limit and a p-value under the hypothesis $H_0: \rho >= \rho_0$.

Confidence Limits for the Correlation

The confidence limits for the correlation ρ are derived through the confidence limits for the parameter ζ, with or without the bias adjustment.

Without a bias adjustment, confidence limits for ζ are computed by treating

$$z_r - \zeta$$

as having a normal distribution with mean zero and variance $1/(n-3)$.

That is, the two-sided confidence limits for ζ are computed as

$$\zeta_l = z_r - z_{(1-\alpha/2)} \sqrt{\frac{1}{n-3}}$$

$$\zeta_u = z_r + z_{(1-\alpha/2)} \sqrt{\frac{1}{n-3}}$$

where $z_{(1-\alpha/2)}$ is the $100(1 - \alpha/2)$ percentage point of the standard normal distribution.

With a bias adjustment, confidence limits for ζ are computed by treating

$$z_r - \zeta - \text{bias}(r)$$

as having a normal distribution with mean zero and variance $1/(n-3)$, where the bias adjustment function (Keeping 1962, p. 308) is

$$\text{bias}(r) = \frac{r}{2(n-1)}$$

That is, the two-sided confidence limits for ζ are computed as

$$\zeta_l = z_r - \mathrm{bias}(r) - z_{(1-\alpha/2)}\sqrt{\frac{1}{n-3}}$$

$$\zeta_u = z_r - \mathrm{bias}(r) + z_{(1-\alpha/2)}\sqrt{\frac{1}{n-3}}$$

These computed confidence limits of ζ_l and ζ_u are then transformed back to derive the confidence limits for the correlation ρ:

$$r_l = \tanh(\zeta_l) = \frac{\exp(2\zeta_l) - 1}{\exp(2\zeta_l) + 1}$$

$$r_u = \tanh(\zeta_u) = \frac{\exp(2\zeta_u) - 1}{\exp(2\zeta_u) + 1}$$

Note that with a bias adjustment, the CORR procedure also displays the following correlation estimate:

$$r_{adj} = \tanh(z_r - \mathrm{bias}(r))$$

Applications of Fisher's z Transformation

Fisher (1973, p. 199) describes the following practical applications of the z transformation:

- testing whether a population correlation is equal to a given value
- testing for equality of two population correlations
- combining correlation estimates from different samples

To test if a population correlation ρ_1 from a sample of n_1 observations with sample correlation r_1 is equal to a given ρ_0, first apply the z transformation to r_1 and ρ_0: $z_1 = \tanh^{-1}(r_1)$ and $\zeta_0 = \tanh^{-1}(\rho_0)$.

The p-value is then computed by treating

$$z_1 - \zeta_0 - \frac{\rho_0}{2(n_1 - 1)}$$

as a normal random variable with mean zero and variance $1/(n_1 - 3)$.

Assume that sample correlations r_1 and r_2 are computed from two independent samples of n_1 and n_2 observations, respectively. To test whether the two corresponding population correlations, ρ_1 and ρ_2, are equal, first apply the z transformation to the two sample correlations: $z_1 = \tanh^{-1}(r_1)$ and $z_2 = \tanh^{-1}(r_2)$.

The p-value is derived under the null hypothesis of equal correlation. That is, the difference $z_1 - z_2$ is distributed as a normal random variable with mean zero and variance $1/(n_1 - 3) + 1/(n_2 - 3)$.

Assuming further that the two samples are from populations with identical correlation, a combined correlation estimate can be computed. The weighted average of the corresponding z values is

$$\bar{z} = \frac{(n_1 - 3)z_1 + (n_2 - 3)z_2}{n_1 + n_2 - 6}$$

where the weights are inversely proportional to their variances.

Thus, a combined correlation estimate is $\bar{r} = \tanh(\bar{z})$ and $V(\bar{z}) = 1/(n_1 + n_2 - 6)$. See Example 2.4 for further illustrations of these applications.

Note that this approach can be extended to include more than two samples.

Polychoric Correlation

Polychoric correlation measures the correlation between two unobserved, continuous variables that have a bivariate normal distribution. Information about each unobserved variable is obtained through an observed ordinal variable that is derived from the unobserved variable by classifying its values into a finite set of discrete, ordered values (Olsson 1979; Drasgow 1986). Polychoric correlation between two observed binary variables is also known as tetrachoric correlation.

The polychoric correlation coefficient is the maximum likelihood estimate of the product-moment correlation between the underlying normal variables. The range of the polychoric correlation is from –1 to 1. Olsson (1979) gives the likelihood equations and the asymptotic standard errors for estimating the polychoric correlation. The underlying continuous variables relate to the observed ordinal variables through thresholds, which define a range of numeric values that correspond to each categorical level. PROC CORR uses Olsson's maximum likelihood method for simultaneous estimation of the polychoric correlation and the thresholds.

PROC CORR iteratively solves the likelihood equations by using a Newton-Raphson algorithm. The initial estimates of the thresholds are computed from the inverse of the normal distribution function at the cumulative marginal proportions of the table. Iterative computation of the polychoric correlation stops when the convergence measure falls below the convergence criterion or when the maximum number of iterations is reached, whichever occurs first.

Probability Values

The CORR procedure computes two types of testing for the zero polychoric correlation: the Wald test and the likelihood ratio (LR) test.

Given the maximum likelihood estimate of the polychoric correlation $\hat{\rho}$ and its asymptotic standard error $\text{StdErr}(\hat{\rho})$, the Wald chi-square test statistic is computed as

$$\left(\frac{\hat{\rho}}{\text{StdErr}(\hat{\rho})} \right)^2$$

The Wald statistic has an asymptotic chi-square distribution with one degree of freedom.

For the LR test, the maximum likelihood function assuming zero polychoric correlation is also needed. The LR test statistic is computed as

$$-2 \log \left(\frac{L_0}{L_1} \right)$$

where L_1 is the likelihood function with the maximum likelihood estimates for all parameters, and L_0 is the likelihood function with the maximum likelihood estimates for all parameters except the polychoric correlation, which is set to 0. The LR statistic also has an asymptotic chi-square distribution with one degree of freedom.

Polyserial Correlation

Polyserial correlation measures the correlation between two continuous variables with a bivariate normal distribution, where one variable is observed directly, and the other is unobserved. Information about the unobserved variable is obtained through an observed ordinal variable that is derived from the unobserved variable by classifying its values into a finite set of discrete, ordered values (Olsson, Drasgow, and Dorans 1982).

Let X be the observed continuous variable from a normal distribution with mean μ and variance σ^2, let Y be the unobserved continuous variable, and let ρ be the Pearson correlation between X and Y. Furthermore, assume that an observed ordinal variable D is derived from Y as follows:

$$D = \begin{cases} d_{(1)} & \text{if } Y < \tau_1 \\ d_{(k)} & \text{if } \tau_{k-1} \leq Y < \tau_k, \ k = 2, 3, \ldots, K-1 \\ d_{(K)} & \text{if } Y \geq \tau_{K-1} \end{cases}$$

where $d_{(1)} < d_{(2)} < \ldots < d_{(K)}$ are ordered observed values, and $\tau_1 < \tau_2 < \ldots < \tau_{K-1}$ are ordered unknown threshold values.

The likelihood function for the joint distribution (X, D) from a sample of N observations (x_j, d_j) is

$$L = \prod_{j=1}^{N} f(x_j, d_j) = \prod_{j=1}^{N} f(x_j) \, P(D = d_j \mid x_j)$$

where $f(x_j)$ is the normal density function with mean μ and standard deviation σ (Drasgow 1986).

The conditional distribution of Y given $X = x_j$ is normal with mean ρz_j and variance $1 - \rho^2$, where $z_j = (x_j - \mu)/\sigma$ is a standard normal variate. Without loss of generality, assume the variable Y has a standard normal distribution. Then if $d_j = d_{(k)}$, the kth ordered value in D, the resulting conditional density is

$$P(D = d_{(k)} \mid x_j) = \begin{cases} \Phi\left(\frac{\tau_1 - \rho z_j}{\sqrt{1-\rho^2}}\right) & \text{if } k = 1 \\ \Phi\left(\frac{\tau_k - \rho z_j}{\sqrt{1-\rho^2}}\right) - \Phi\left(\frac{\tau_{k-1} - \rho z_j}{\sqrt{1-\rho^2}}\right) & \text{if } k = 2, 3, \ldots, K-1 \\ 1 - \Phi\left(\frac{\tau_{K-1} - \rho z_j}{\sqrt{1-\rho^2}}\right) & \text{if } k = K \end{cases}$$

where Φ is the cumulative normal distribution function.

Cox (1974) derives the maximum likelihood estimates for all parameters μ, σ, ρ and $\tau_1, \ldots, \tau_{k-1}$. The maximum likelihood estimates for μ and σ^2 can be derived explicitly. The maximum likelihood estimate for μ is the sample mean and the maximum likelihood estimate for σ^2 is the sample variance

$$\frac{\sum_{j=1}^{N}(x_j - \bar{x})^2}{N}$$

The maximum likelihood estimates for the remaining parameters, including the polyserial correlation ρ and thresholds $\tau_1, \ldots, \tau_{k-1}$, can be computed by an iterative process, as described by Cox (1974). The asymptotic standard error of the maximum likelihood estimate of ρ can also be computed after this process.

For a vector of parameters, the information matrix is the negative of the Hessian matrix (the matrix of second partial derivatives of the log likelihood function), and is used in the computation of the maximum likelihood estimates of these parameters. The CORR procedure uses the observed information matrix (the information matrix evaluated at the current parameter estimates) in the computation. After the maximum likelihood estimates are derived, the asymptotic covariance matrix for these parameter estimates is computed as the inverse of the observed information matrix (the information matrix evaluated at the maximum likelihood estimates).

Probability Values

The CORR procedure computes two types of testing for the zero polyserial correlation: the Wald test and the likelihood ratio (LR) test.

Given the maximum likelihood estimate of the polyserial correlation $\hat{\rho}$ and its asymptotic standard error $\text{StdErr}(\hat{\rho})$, the Wald chi-square test statistic is computed as

$$\left(\frac{\hat{\rho}}{\text{StdErr}(\hat{\rho})}\right)^2$$

The Wald statistic has an asymptotic chi-square distribution with one degree of freedom.

For the LR test, the maximum likelihood function assuming zero polyserial correlation is also needed. If $\rho = 0$, the likelihood function is reduced to

$$L = \prod_{j=1}^{N} f(x_j, d_j) = \prod_{j=1}^{N} f(x_j) \prod_{j=1}^{N} P(D = d_j)$$

In this case, the maximum likelihood estimates for all parameters can be derived explicitly. The maximum likelihood estimates for μ is the sample mean and the maximum likelihood estimate for σ^2 is the sample variance

$$\frac{\sum_{j=1}^{N}(x_j - \bar{x})^2}{N}$$

In addition, the maximum likelihood estimate for the threshold τ_k, $k=1, \ldots, K\text{-}1$, is

$$\Phi^{-1}\left(\frac{\sum_{g=1}^{k} n_g}{N}\right)$$

where n_g is the number of observations in the gth ordered group of the ordinal variable D, and $N = \sum_{g=1}^{K} n_g$ is the total number of observations.

The LR test statistic is computed as

$$-2 \log\left(\frac{L_0}{L_1}\right)$$

where L_1 is the likelihood function with the maximum likelihood estimates for all parameters, and L_0 is the likelihood function with the maximum likelihood estimates for all parameters except the polyserial correlation, which is set to 0. The LR statistic also has an asymptotic chi-square distribution with one degree of freedom.

Cronbach's Coefficient Alpha

Analyzing latent constructs such as job satisfaction, motor ability, sensory recognition, or customer satisfaction requires instruments to accurately measure the constructs. Interrelated items can be summed to obtain an overall score for each participant. Cronbach's coefficient alpha estimates the reliability of this type of scale by determining the internal consistency of the test or the average correlation of items within the test (Cronbach 1951).

When a value is recorded, the observed value contains some degree of measurement error. Two sets of measurements on the same variable for the same individual might not have identical values. However, repeated measurements for a series of individuals will show some consistency. Reliability measures internal consistency from one set of measurements to another. The observed value Y is divided into two components, a true value T and a measurement error E. The measurement error is assumed to be independent of the true value; that is,

$$Y = T + E \quad \text{Cov}(T, E) = 0$$

The reliability coefficient of a measurement test is defined as the squared correlation between the observed value Y and the true value T; that is,

$$r^2(Y, T) = \frac{\text{Cov}(Y, T)^2}{V(Y)V(T)} = \frac{V(T)^2}{V(Y)V(T)} = \frac{V(T)}{V(Y)}$$

which is the proportion of the observed variance due to true differences among individuals in the sample. If Y is the sum of several observed variables measuring the same feature, you can estimate $V(T)$. Cronbach's coefficient alpha, based on a lower bound for $V(T)$, is an estimate of the reliability coefficient.

Suppose p variables are used with $Y_j = T_j + E_j$ for $j = 1, 2, \ldots, p$, where Y_j is the observed value, T_j is the true value, and E_j is the measurement error. The measurement errors (E_j) are independent of the true values (T_j) and are also independent of each other. Let $Y_0 = \sum_j Y_j$ be the total observed score and let $T_0 = \sum_j T_j$ be the total true score. Because

$$(p-1) \sum_j V(T_j) \geq \sum_{i \neq j} \text{Cov}(T_i, T_j)$$

a lower bound for $V(T_0)$ is given by

$$\frac{p}{p-1} \sum_{i \neq j} \text{Cov}(T_i, T_j)$$

With $\text{Cov}(Y_i, Y_j) = \text{Cov}(T_i, T_j)$ for $i \neq j$, a lower bound for the reliability coefficient, $V(T_0)/V(Y_0)$, is then given by the Cronbach's coefficient alpha:

$$\alpha = \left(\frac{p}{p-1}\right) \frac{\sum_{i \neq j} \text{Cov}(Y_i, Y_j)}{V(Y_0)} = \left(\frac{p}{p-1}\right) \left(1 - \frac{\sum_j V(Y_j)}{V(Y_0)}\right)$$

If the variances of the items vary widely, you can standardize the items to a standard deviation of 1 before computing the coefficient alpha. If the variables are dichotomous (0,1), the coefficient alpha is equivalent to the Kuder-Richardson 20 (KR-20) reliability measure.

When the correlation between each pair of variables is 1, the coefficient alpha has a maximum value of 1. With negative correlations between some variables, the coefficient alpha can have a value less than zero. The larger the overall alpha coefficient, the more likely that items contribute to a reliable scale. Nunnally and Bernstein (1994) suggests 0.70 as an acceptable reliability coefficient; smaller reliability coefficients are seen as inadequate. However, this varies by discipline.

To determine how each item reflects the reliability of the scale, you calculate a coefficient alpha after deleting each variable independently from the scale. Cronbach's coefficient alpha from all variables except the kth variable is given by

$$\alpha_k = \left(\frac{p-1}{p-2}\right)\left(1 - \frac{\sum_{i \neq k} V(Y_i)}{V(\sum_{i \neq k} Y_i)}\right)$$

If the reliability coefficient increases after an item is deleted from the scale, you can assume that the item is not correlated highly with other items in the scale. Conversely, if the reliability coefficient decreases, you can assume that the item is highly correlated with other items in the scale. Refer to Yu (2001) for more information about how to interpret Cronbach's coefficient alpha.

Listwise deletion of observations with missing values is necessary to correctly calculate Cronbach's coefficient alpha. PROC CORR does not automatically use listwise deletion if you specify the ALPHA option. Therefore, you should use the NOMISS option if the data set contains missing values. Otherwise, PROC CORR prints a warning message indicating the need to use the NOMISS option with the ALPHA option.

Confidence and Prediction Ellipses

When the relationship between two variables is nonlinear or when outliers are present, the correlation coefficient might incorrectly estimate the strength of the relationship. Plotting the data enables you to verify the linear relationship and to identify the potential outliers.

The partial correlation between two variables, after controlling for variables in the PARTIAL statement, is the correlation between the residuals of the linear regression of the two variables on the partialled variables. Thus, if a PARTIAL statement is also specified, the residuals of the analysis variables are displayed in the scatter plot matrix and scatter plots.

The CORR procedure optionally provides two types of ellipses for each pair of variables in a scatter plot. One is a confidence ellipse for the population mean, and the other is a prediction ellipse for a new observation. Both assume a bivariate normal distribution.

Let $\bar{\mathbf{Z}}$ and $\mathbf{S}$ be the sample mean and sample covariance matrix of a random sample of size n from a bivariate normal distribution with mean $\boldsymbol{\mu}$ and covariance matrix $\boldsymbol{\Sigma}$. The variable $\bar{\mathbf{Z}} - \boldsymbol{\mu}$ is distributed as a bivariate normal variate with mean zero and covariance $(1/n)\boldsymbol{\Sigma}$, and it is independent of $\mathbf{S}$. Using Hotelling's T^2 statistic, which is defined as

$$T^2 = n(\bar{\mathbf{Z}} - \boldsymbol{\mu})'\mathbf{S}^{-1}(\bar{\mathbf{Z}} - \boldsymbol{\mu})$$

a $100(1-\alpha)\%$ confidence ellipse for $\boldsymbol{\mu}$ is computed from the equation

$$\frac{n}{n-1}(\bar{\mathbf{Z}} - \boldsymbol{\mu})'\mathbf{S}^{-1}(\bar{\mathbf{Z}} - \boldsymbol{\mu}) = \frac{2}{n-2}F_{2,n-2}(1-\alpha)$$

where $F_{2,n-2}(1-\alpha)$ is the $(1-\alpha)$ critical value of an F distribution with degrees of freedom 2 and n-2.

A prediction ellipse is a region for predicting a new observation in the population. It also approximates a region that contains a specified percentage of the population.

Denote a new observation as the bivariate random variable $\mathbf{Z}_{\text{new}}$. The variable

$$\mathbf{Z}_{\text{new}} - \bar{\mathbf{Z}} = (\mathbf{Z}_{\text{new}} - \boldsymbol{\mu}) - (\bar{\mathbf{Z}} - \boldsymbol{\mu})$$

is distributed as a bivariate normal variate with mean zero (the zero vector) and covariance $(1 + 1/n)\Sigma$, and it is independent of **S**. A $100(1 - \alpha)\%$ prediction ellipse is then given by the equation

$$\frac{n}{n-1}(\bar{\mathbf{Z}} - \boldsymbol{\mu})'\mathbf{S}^{-1}(\bar{\mathbf{Z}} - \boldsymbol{\mu}) = \frac{2(n+1)}{n-2} F_{2,n-2}(1-\alpha)$$

The family of ellipses generated by different critical values of the F distribution has a common center (the sample mean) and common major and minor axis directions.

The shape of an ellipse depends on the aspect ratio of the plot. The ellipse indicates the correlation between the two variables if the variables are standardized (by dividing the variables by their respective standard deviations). In this situation, the ratio between the major and minor axis lengths is

$$\sqrt{\frac{1+|r|}{1-|r|}}$$

In particular, if $r=0$, the ratio is 1, which corresponds to a circular confidence contour and indicates that the variables are uncorrelated. A larger value of the ratio indicates a larger positive or negative correlation between the variables.

Missing Values

PROC CORR excludes observations with missing values in the WEIGHT and FREQ variables. By default, PROC CORR uses *pairwise deletion* when observations contain missing values. PROC CORR includes all nonmissing pairs of values for each pair of variables in the statistical computations. Therefore, the correlation statistics might be based on different numbers of observations.

If you specify the NOMISS option, PROC CORR uses *listwise deletion* when a value of the VAR or WITH statement variable is missing. PROC CORR excludes all observations with missing values from the analysis. Therefore, the number of observations for each pair of variables is identical.

The PARTIAL statement always excludes the observations with missing values by automatically invoking the NOMISS option. With the NOMISS option, the data are processed more efficiently because fewer resources are needed. Also, the resulting correlation matrix is nonnegative definite.

In contrast, if the data set contains missing values for the analysis variables and the NOMISS option is not specified, the resulting correlation matrix might not be nonnegative definite. This leads to several statistical difficulties if you use the correlations as input to regression or other statistical procedures.

In-Database Computation

The CORR procedure can use in-database computation to compute univariate statistics and the SSCP matrix if the DATA= input data set is stored as a table in a database management system (DBMS). When the CORR procedure performs in-database computation for the DATA= data set, the procedure generates an SQL query that computes summary tables of univariate statistics and the SSCP matrix. The query is passed to the DBMS and executed in-database. The results of the query are then passed back to the SAS System and transmitted to PROC CORR. The CORR procedure then uses these summary tables to perform the remaining tasks (such as producing the correlation and covariance matrices) in the usual way (out of the database).

In-database computation can provide the advantages of faster processing and reduced data transfer between the database and SAS software. For information about in-database computation, see the section "In-Database Procedures" in *SAS/ACCESS for Relational Databases: Reference* Instead of transferring the entire data set over the network between the database and SAS software, the in-database method transfers only the summary tables. This can substantially reduce processing time when the dimensions of the summary tables (in terms of rows and columns) are much smaller than the dimensions of the entire database table (in terms of individual observations). Additionally, in-database summarization uses efficient parallel processing, which can also provide performance advantages.

By default, PROC CORR uses in-database computation when possible. If in-database computation is used, the EXCLNPWGT option is activated to exclude observations with nonpositive weights. The ID statement requires row-level access and therefore cannot be used in-database. In addition, the HOEFFDING, KENDALL, SPEARMAN, OUTH=, OUTK=, OUTS=, and PLOTS= options also require row-level access and cannot be used in-database.

In-database computation is controlled by the SQLGENERATION option, which you can specify in either a LIBNAME statement or an OPTIONS statement. See the section "In-Database Procedures" in *SAS/ACCESS for Relational Databases: Reference* for details about the SQLGENERATION option and other options that affect in-database computation. There are no CORR procedure options that control in-database computation.

The order of observations is not inherently defined for DBMS tables. The following options relate to the order of observations and therefore should not be specified for PROC CORR in-database computation:

- If you specify the FIRSTOBS= or OBS= data set option, PROC CORR does not perform in-database computation.
- If you specify the NOTSORTED option in the BY statement, PROC CORR in-database computation ignores it and uses the default ASCENDING order for BY variables.

NOTE: In-database computing in the CORR procedure requires installation of the SAS Analytics Accelerator.

Output Tables

By default, PROC CORR prints a report that includes descriptive statistics and correlation statistics for each variable. The descriptive statistics include the number of observations with nonmissing values, the mean, the standard deviation, the minimum, and the maximum.

If a nonparametric measure of association is requested, the descriptive statistics include the median. Otherwise, the sample sum is included. If a Pearson partial correlation is requested, the descriptive statistics also include the partial variance and partial standard deviation.

If variable labels are available, PROC CORR labels the variables. If you specify the CSSCP, SSCP, or COV option, the appropriate sums of squares and crossproducts and covariance matrix appear at the top of the correlation report. If the data set contains missing values, PROC CORR prints additional statistics for each pair of variables. These statistics, calculated from the observations with nonmissing row and column variable values, might include the following:

- SSCP('W','V'), uncorrected sums of squares and crossproducts
- USS('W'), uncorrected sums of squares for the row variable
- USS('V'), uncorrected sums of squares for the column variable

- CSSCP('W','V'), corrected sums of squares and crossproducts
- CSS('W'), corrected sums of squares for the row variable
- CSS('V'), corrected sums of squares for the column variable
- COV('W','V'), covariance
- VAR('W'), variance for the row variable
- VAR('V'), variance for the column variable
- DF('W','V'), divisor for calculating covariance and variances

For each pair of variables, PROC CORR prints the correlation coefficients, the number of observations used to calculate the coefficient, and the *p*-value.

If you specify the ALPHA option, PROC CORR prints Cronbach's coefficient alpha, the correlation between the variable and the total of the remaining variables, and Cronbach's coefficient alpha by using the remaining variables for the raw variables and the standardized variables.

Output Data Sets

If you specify the OUTP=, OUTS=, OUTK=, or OUTH= option, PROC CORR creates an output data set that contains statistics for Pearson correlation, Spearman correlation, Kendall's tau-b, or Hoeffding's *D*, respectively. By default, the output data set is a special data set type (TYPE=CORR) that many SAS/STAT procedures recognize, including PROC REG and PROC FACTOR. When you specify the NOCORR option and the COV, CSSCP, or SSCP option, use the TYPE= data set option to change the data set type to COV, CSSCP, or SSCP.

The output data set includes the following variables:

- BY variables, which identify the BY group when using a BY statement
- _TYPE_ variable, which identifies the type of observation
- _NAME_ variable, which identifies the variable that corresponds to a given row of the correlation matrix
- INTERCEPT variable, which identifies variable sums when specifying the SSCP option
- VAR variables, which identify the variables listed in the VAR statement

You can use a combination of the _TYPE_ and _NAME_ variables to identify the contents of an observation. The _NAME_ variable indicates which row of the correlation matrix the observation corresponds to. The values of the _TYPE_ variable are as follows:

- SSCP, uncorrected sums of squares and crossproducts
- CSSCP, corrected sums of squares and crossproducts

- COV, covariances
- MEAN, mean of each variable
- STD, standard deviation of each variable
- N, number of nonmissing observations for each variable
- SUMWGT, sum of the weights for each variable when using a WEIGHT statement
- CORR, correlation statistics for each variable

If you specify the SSCP option, the OUTP= data set includes an additional observation that contains intercept values. If you specify the ALPHA option, the OUTP= data set also includes observations with the following _TYPE_ values:

- RAWALPHA, Cronbach's coefficient alpha for raw variables
- STDALPHA, Cronbach's coefficient alpha for standardized variables
- RAWALDEL, Cronbach's coefficient alpha for raw variables after deleting one variable
- STDALDEL, Cronbach's coefficient alpha for standardized variables after deleting one variable
- RAWCTDEL, the correlation between a raw variable and the total of the remaining raw variables
- STDCTDEL, the correlation between a standardized variable and the total of the remaining standardized variables

If you use a PARTIAL statement, the statistics are calculated after the variables are partialled. If PROC CORR computes Pearson correlation statistics, MEAN equals zero and STD equals the partial standard deviation associated with the partial variance for the OUTP=, OUTK=, and OUTS= data sets. Otherwise, PROC CORR assigns missing values to MEAN and STD.

ODS Table Names

PROC CORR assigns a name to each table it creates. You must use these names to reference tables when using the Output Delivery System (ODS). These names are listed in Table 2.3 and Table 2.4. For more information about ODS, see Chapter 20, "Using the Output Delivery System" (*SAS/STAT User's Guide*).

Table 2.3 ODS Tables Produced by PROC CORR

ODS Table Name	Description	Option
Cov	Covariances	COV
CronbachAlpha	Coefficient alpha	ALPHA
CronbachAlphaDel	Coefficient alpha with deleted variable	ALPHA
Csscp	Corrected sums of squares and crossproducts	CSSCP
FisherPearsonCorr	Pearson correlation statistics using Fisher's z transformation	FISHER

Table 2.3 *continued*

ODS Table Name	Description	Option
FisherSpearmanCorr	Spearman correlation statistics using Fisher's z transformation	FISHER SPEARMAN
HoeffdingCorr	Hoeffding's D statistics	HOEFFDING
KendallCorr	Kendall's tau-b coefficients	KENDALL
PearsonCorr	Pearson correlations	PEARSON
PolychoricCorr	Polychoric correlations	POLYCHORIC
PolyserialCorr	Polyserial correlations	POLYSERIAL
SimpleStats	Simple descriptive statistics	
SpearmanCorr	Spearman correlations	SPEARMAN
Sscp	Sums of squares and crossproducts	SSCP
VarInformation	Variable information	

Table 2.4 ODS Tables Produced with the PARTIAL Statement

ODS Table Name	Description	Option
FisherPearsonPartialCorr	Pearson partial correlation statistics using Fisher's z transformation	FISHER
FisherSpearmanPartialCorr	Spearman partial correlation statistics using Fisher's z transformation	FISHER SPEARMAN
PartialCsscp	Partial corrected sums of squares and crossproducts	CSSCP
PartialCov	Partial covariances	COV
PartialKendallCorr	Partial Kendall tau-b coefficients	KENDALL
PartialPearsonCorr	Partial Pearson correlations	
PartialSpearmanCorr	Partial Spearman correlations	SPEARMAN

ODS Graphics

Statistical procedures use ODS Graphics to create graphs as part of their output. ODS Graphics is described in detail in Chapter 21, "Statistical Graphics Using ODS" (*SAS/STAT User's Guide*).

Before you create graphs, ODS Graphics must be enabled (for example, with the ODS GRAPHICS ON statement). For more information about enabling and disabling ODS Graphics, see the section "Enabling and Disabling ODS Graphics" in that chapter.

The overall appearance of graphs is controlled by ODS styles. Styles and other aspects of using ODS Graphics are discussed in the section "A Primer on ODS Statistical Graphics" in that chapter.

PROC CORR assigns a name to each graph it creates using ODS. You can use these names to reference the graphs when using ODS. To request these graphs, ODS Graphics must be enabled and you must specify the options indicated in Table 2.5.

Table 2.5 Graphs Produced by PROC CORR

ODS Graph Name	Plot Description	Option
ScatterPlot	Scatter plot	PLOTS=SCATTER
MatrixPlot	Scatter plot matrix	PLOTS=MATRIX

Examples: CORR Procedure

Example 2.1: Computing Four Measures of Association

This example produces a correlation analysis with descriptive statistics and four measures of association: the Pearson product-moment correlation, the Spearman rank-order correlation, Kendall's tau-b coefficients, and Hoeffding's measure of dependence, D.

The Fitness data set created in the section "Getting Started: CORR Procedure" on page 5 contains measurements from a study of physical fitness of 31 participants. The following statements request all four measures of association for the variables Weight, Oxygen, and Runtime:

```
ods graphics on;
title 'Measures of Association for a Physical Fitness Study';
proc corr data=Fitness pearson spearman kendall hoeffding
          plots=matrix(histogram);
   var Weight Oxygen RunTime;
run;
```

Note that Pearson correlations are computed by default only if all three nonparametric correlations (SPEARMAN, KENDALL, and HOEFFDING) are not specified. Otherwise, you need to specify the PEARSON option explicitly to compute Pearson correlations.

The "Simple Statistics" table in Output 2.1.1 displays univariate descriptive statistics for analysis variables. By default, observations with nonmissing values for each variable are used to derive the univariate statistics for that variable. When nonparametric measures of association are specified, the procedure displays the median instead of the sum as an additional descriptive measure.

Output 2.1.1 Simple Statistics

Measures of Association for a Physical Fitness Study

The CORR Procedure

3 Variables: Weight Oxygen RunTime

Simple Statistics

Variable	N	Mean	Std Dev	Median	Minimum	Maximum
Weight	31	77.44452	8.32857	77.45000	59.08000	91.63000
Oxygen	29	47.22721	5.47718	46.67200	37.38800	60.05500
RunTime	29	10.67414	1.39194	10.50000	8.17000	14.03000

The "Pearson Correlation Coefficients" table in Output 2.1.2 displays Pearson correlation statistics for pairs of analysis variables. The Pearson correlation is a parametric measure of association for two continuous random variables. When there are missing data, the number of observations used to calculate the correlation can vary.

Output 2.1.2 Pearson Correlation Coefficients

Pearson Correlation Coefficients Prob > |r| under H0: Rho=0 Number of Observations	Weight	Oxygen	RunTime
Weight	1.00000 31	-0.15358 0.4264 29	0.20072 0.2965 29
Oxygen	-0.15358 0.4264 29	1.00000 29	-0.86843 <.0001 28
RunTime	0.20072 0.2965 29	-0.86843 <.0001 28	1.00000 29

The table shows that the Pearson correlation between Runtime and Oxygen is –0.86843, which is significant with a *p*-value less than 0.0001. This indicates a strong negative linear relationship between these two variables. As Runtime increases, Oxygen decreases linearly.

The Spearman rank-order correlation is a nonparametric measure of association based on the ranks of the data values. The "Spearman Correlation Coefficients" table in Output 2.1.3 displays results similar to those of the "Pearson Correlation Coefficients" table in Output 2.1.2.

Output 2.1.3 Spearman Correlation Coefficients

Spearman Correlation Coefficients Prob > |r| under H0: Rho=0 Number of Observations	Weight	Oxygen	RunTime
Weight	1.00000 31	-0.06824 0.7250 29	0.13749 0.4769 29
Oxygen	-0.06824 0.7250 29	1.00000 29	-0.80131 <.0001 28
RunTime	0.13749 0.4769 29	-0.80131 <.0001 28	1.00000 29

Kendall's tau-b is a nonparametric measure of association based on the number of concordances and discordances in paired observations. The "Kendall Tau b Correlation Coefficients" table in Output 2.1.4 displays results similar to those of the "Pearson Correlation Coefficients" table in Output 2.1.2.

Output 2.1.4 Kendall's Tau-b Correlation Coefficients

Kendall Tau b Correlation Coefficients Prob > |tau| under H0: Tau=0 Number of Observations			
	Weight	Oxygen	RunTime
Weight	1.00000 31	-0.00988 0.9402 29	0.06675 0.6123 29
Oxygen	-0.00988 0.9402 29	1.00000 29	-0.62434 <.0001 28
RunTime	0.06675 0.6123 29	-0.62434 <.0001 28	1.00000 29

Hoeffding's measure of dependence, D, is a nonparametric measure of association that detects more general departures from independence. Without ties in the variables, the values of the D statistic can vary between -0.5 and 1, with 1 indicating complete dependence. Otherwise, the D statistic can result in a smaller value. The "Hoeffding Dependence Coefficients" table in Output 2.1.5 displays Hoeffding dependence coefficients. Since ties occur in the variable Weight, the D statistic for the Weight variable is less than 1.

Output 2.1.5 Hoeffding's Dependence Coefficients

Hoeffding Dependence Coefficients Prob > D under H0: D=0 Number of Observations			
	Weight	Oxygen	RunTime
Weight	0.97690 <.0001 31	-0.00497 0.5101 29	-0.02355 1.0000 29
Oxygen	-0.00497 0.5101 29	1.00000 29	0.23449 <.0001 28
RunTime	-0.02355 1.0000 29	0.23449 <.0001 28	1.00000 29

When you use the PLOTS=MATRIX(HISTOGRAM) option, the CORR procedure displays a symmetric matrix plot for the analysis variables listed in the VAR statement (Output 2.1.6).

Output 2.1.6 Symmetric Scatter Plot Matrix

The strong negative linear relationship between Oxygen and Runtime is evident in Output 2.1.6.

Note that this graphical display is requested by enabling ODS Graphics and by specifying the PLOTS= option. For more information about ODS Graphics, see Chapter 21, "Statistical Graphics Using ODS" (*SAS/STAT User's Guide*).

Example 2.2: Computing Correlations between Two Sets of Variables

The following statements create the data set Setosa, which contains measurements for four iris parts from Fisher's iris data (1936): sepal length, sepal width, petal length, and petal width. The data set has been altered to contain some missing values.

```
*-------------------- Data on Iris Setosa --------------------*
| The data set contains 50 iris specimens from the species   |
| Iris Setosa with the following four measurements:          |
| SepalLength  (sepal length)                                |
| SepalWidth   (sepal width)                                 |
| PetalLength  (petal length)                                |
| PetalWidth   (petal width)                                 |
| Certain values were changed to missing for the analysis.   |
*------------------------------------------------------------*;
data Setosa;
   input SepalLength SepalWidth PetalLength PetalWidth @@;
   label sepallength='Sepal Length in mm.'
         sepalwidth='Sepal Width in mm.'
         petallength='Petal Length in mm.'
         petalwidth='Petal Width in mm.';
   datalines;
50 33 14 02   46 34 14 03   46 36  . 02
51 33 17 05   55 35 13 02   48 31 16 02
52 34 14 02   49 36 14 01   44 32 13 02
50 35 16 06   44 30 13 02   47 32 16 02
48 30 14 03   51 38 16 02   48 34 19 02
50 30 16 02   50 32 12 02   43 30 11  .
58 40 12 02   51 38 19 04   49 30 14 02
51 35 14 02   50 34 16 04   46 32 14 02
57 44 15 04   50 36 14 02   54 34 15 04
52 41 15  .   55 42 14 02   49 31 15 02
54 39 17 04   50 34 15 02   44 29 14 02
47 32 13 02   46 31 15 02   51 34 15 02
50 35 13 03   49 31 15 01   54 37 15 02
54 39 13 04   51 35 14 03   48 34 16 02
48 30 14 01   45 23 13 03   57 38 17 03
51 38 15 03   54 34 17 02   51 37 15 04
52 35 15 02   53 37 15 02
;
```

The following statements request a correlation analysis between two sets of variables, the sepal measurements (length and width) and the petal measurements (length and width):

```
ods graphics on;
title 'Fisher (1936) Iris Setosa Data';
proc corr data=Setosa sscp cov plots=matrix;
   var  sepallength sepalwidth;
   with petallength petalwidth;
run;
```

The "Simple Statistics" table in Output 2.2.1 displays univariate statistics for variables in the VAR and WITH statements.

Output 2.2.1 Simple Statistics

Fisher (1936) Iris Setosa Data

The CORR Procedure

2 With Variables:	PetalLength PetalWidth
2 Variables:	SepalLength SepalWidth

Simple Statistics

Variable	N	Mean	Std Dev	Sum	Minimum	Maximum	Label
PetalLength	49	14.71429	1.62019	721.00000	11.00000	19.00000	Petal Length in mm.
PetalWidth	48	2.52083	1.03121	121.00000	1.00000	6.00000	Petal Width in mm.
SepalLength	50	50.06000	3.52490	2503	43.00000	58.00000	Sepal Length in mm.
SepalWidth	50	34.28000	3.79064	1714	23.00000	44.00000	Sepal Width in mm.

When the WITH statement is specified together with the VAR statement, the CORR procedure produces rectangular matrices for statistics such as covariances and correlations. The matrix rows correspond to the WITH variables (PetalLength and PetalWidth), while the matrix columns correspond to the VAR variables (SepalLength and SepalWidth). The CORR procedure uses the WITH variable labels to label the matrix rows.

The SSCP option requests a table of the uncorrected sum-of-squares and crossproducts matrix, and the COV option requests a table of the covariance matrix. The SSCP and COV options also produce a table of the Pearson correlations.

The sum-of-squares and crossproducts statistics for each pair of variables are computed by using observations with nonmissing row and column variable values. The "Sums of Squares and Crossproducts" table in Output 2.2.2 displays the crossproduct, sum of squares for the row variable, and sum of squares for the column variable for each pair of variables.

Output 2.2.2 Sums of Squares and Crossproducts

Sums of Squares and Crossproducts
SSCP / Row Var SS / Col Var SS

	SepalLength	SepalWidth
PetalLength	36214.00000	24756.00000
Petal Length in mm.	10735.00000	10735.00000
	123793.0000	58164.0000
PetalWidth	6113.00000	4191.00000
Petal Width in mm.	355.00000	355.00000
	121356.0000	56879.0000

The variances are computed by using observations with nonmissing row and column variable values. The "Variances and Covariances" table in Output 2.2.3 displays the covariance, variance for the row variable, variance for the column variable, and associated degrees of freedom for each pair of variables.

Output 2.2.3 Variances and Covariances

Variances and Covariances Covariance / Row Var Variance / Col Var Variance / DF		
	SepalLength	SepalWidth
PetalLength Petal Length in mm.	1.270833333 2.625000000 12.33333333 48	1.363095238 2.625000000 14.60544218 48
PetalWidth Petal Width in mm.	0.911347518 1.063386525 11.80141844 47	1.048315603 1.063386525 13.62721631 47

When there are missing values in the analysis variables, the "Pearson Correlation Coefficients" table in Output 2.2.4 displays the correlation, the p-value under the null hypothesis of zero correlation, and the number of observations for each pair of variables. Only the correlation between PetalWidth and SepalLength and the correlation between PetalWidth and SepalWidth are slightly positive.

Output 2.2.4 Pearson Correlation Coefficients

Pearson Correlation Coefficients Prob > |r| under H0: Rho=0 Number of Observations		
	SepalLength	SepalWidth
PetalLength Petal Length in mm.	0.22335 0.1229 49	0.22014 0.1285 49
PetalWidth Petal Width in mm.	0.25726 0.0775 48	0.27539 0.0582 48

When ODS Graphics is enabled, the PLOTS= option displays a scatter matrix plot by default. Output 2.2.5 displays a rectangular scatter plot matrix for the two sets of variables: the VAR variables SepalLength and SepalWidth are listed across the top of the matrix, and the WITH variables PetalLength and PetalWidth are listed down the side of the matrix. As measured in Output 2.2.4, the plot for PetalWidth and SepalLength and the plot for PetalWidth and SepalWidth also show slight positive correlations.

Output 2.2.5 Rectangular Matrix Plot

Note that this graphical display is requested by enabling ODS Graphics and by specifying the PLOTS= option. For more information about ODS Graphics, see Chapter 21, "Statistical Graphics Using ODS" (*SAS/STAT User's Guide*).

Example 2.3: Analysis Using Fisher's z Transformation

The following statements request Pearson correlation statistics by using Fisher's z transformation for the data set Fitness:

```
proc corr data=Fitness nosimple fisher;
   var weight oxygen runtime;
run;
```

The NOSIMPLE option suppresses the table of univariate descriptive statistics. By default, PROC CORR displays the "Pearson Correlation Coefficients" table in Output 2.3.1.

Output 2.3.1 Pearson Correlations

The CORR Procedure

Pearson Correlation Coefficients
Prob > |r| under H0: Rho=0
Number of Observations

	Weight	Oxygen	RunTime
Weight	1.00000	-0.15358	0.20072
		0.4264	0.2965
	31	29	29
Oxygen	-0.15358	1.00000	-0.86843
	0.4264		<.0001
	29	29	28
RunTime	0.20072	-0.86843	1.00000
	0.2965	<.0001	
	29	28	29

Using the FISHER option, the CORR procedure displays correlation statistics by using Fisher's z transformation in Output 2.3.2.

Output 2.3.2 Correlation Statistics Using Fisher's z Transformation

Pearson Correlation Statistics (Fisher's z Transformation)

Variable	With Variable	N	Sample Correlation	Fisher's z	Bias Adjustment	Correlation Estimate	95% Confidence Limits	p Value for H0:Rho=0
Weight	Oxygen	29	-0.15358	-0.15480	-0.00274	-0.15090	-0.490289 0.228229	0.4299
Weight	RunTime	29	0.20072	0.20348	0.00358	0.19727	-0.182422 0.525765	0.2995
Oxygen	RunTime	28	-0.86843	-1.32665	-0.01608	-0.86442	-0.935728 -0.725221	<.0001

The table also displays confidence limits and a p-value for the default null hypothesis $H_0: \rho = \rho_0$. See the section "Fisher's z Transformation" on page 24 for details on Fisher's z transformation.

The following statements request one-sided hypothesis tests and confidence limits for the correlations using Fisher's z transformation:

```
proc corr data=Fitness nosimple nocorr fisher (type=lower);
   var weight oxygen runtime;
run;
```

The NOSIMPLE option suppresses the "Simple Statistics" table, and the NOCORR option suppresses the "Pearson Correlation Coefficients" table.

Output 2.3.3 displays correlation statistics by using Fisher's z transformation.

Output 2.3.3 One-Sided Correlation Analysis Using Fisher's z Transformation

The CORR Procedure

			Pearson Correlation Statistics (Fisher's z Transformation)					
Variable	With Variable	N	Sample Correlation	Fisher's z	Bias Adjustment	Correlation Estimate	Lower 95% CL	p Value for H0:Rho<=0
Weight	Oxygen	29	-0.15358	-0.15480	-0.00274	-0.15090	-0.441943	0.7850
Weight	RunTime	29	0.20072	0.20348	0.00358	0.19727	-0.122077	0.1497
Oxygen	RunTime	28	-0.86843	-1.32665	-0.01608	-0.86442	-0.927408	1.0000

The FISHER(TYPE=LOWER) option requests a lower confidence limit and a *p*-value for the test of the one-sided hypothesis $H_0: \rho \leq 0$ against the alternative hypothesis $H_1: \rho > 0$. Here Fisher's *z*, the bias adjustment, and the estimate of the correlation are the same as for the two-sided alternative. However, because TYPE=LOWER is specified, only a lower confidence limit is computed for each correlation, and one-sided *p*-values are computed.

Example 2.4: Applications of Fisher's *z* Transformation

This example illustrates some applications of Fisher's *z* transformation. For details, see the section "Fisher's *z* Transformation" on page 24.

The following statements simulate independent samples of variables X and Y from a bivariate normal distribution. The first batch of 150 observations is sampled using a known correlation of 0.3, the second batch of 150 observations is sampled using a known correlation of 0.25, and the third batch of 100 observations is sampled using a known correlation of 0.3.

```
data Sim (drop=i);
do i=1 to 400;
  X = rannor(135791);
  Batch = 1 + (i>150) + (i>300);
  if Batch = 1 then Y = 0.3*X + 0.9*rannor(246791);
  if Batch = 2 then Y = 0.25*X + sqrt(.8375)*rannor(246791);
  if Batch = 3 then Y = 0.3*X + 0.9*rannor(246791);
  output;
end;
run;
```

This data set will be used to illustrate the following applications of Fisher's *z* transformation:

- testing whether a population correlation is equal to a given value
- testing for equality of two population correlations
- combining correlation estimates from different samples

Testing Whether a Population Correlation Is Equal to a Given Value ρ_0

You can use the following statements to test the null hypothesis H_0: $\rho = 0.5$ against a two-sided alternative H_1: $\rho \neq 0.5$. The test is requested with the option FISHER(RHO0=0.5).

```
title 'Analysis for Batch 1';
proc corr data=Sim (where=(Batch=1)) fisher(rho0=.5);
   var X Y;
run;
```

Output 2.4.1 displays the results based on Fisher's transformation. The null hypothesis is rejected since the p-value is less than 0.0001.

Output 2.4.1 Fisher's Test for $H_0 : \rho = \rho_0$

Analysis for Batch 1

The CORR Procedure

Pearson Correlation Statistics (Fisher's z Transformation)

With Variable	Variable	N	Sample Correlation	Fisher's z	Bias Adjustment	Correlation Estimate	95% Confidence Limits	H0:Rho=Rho0 Rho0	p Value
X	Y	150	0.22081	0.22451	0.0007410	0.22011	0.062034 0.367409	0.50000	<.0001

Testing for Equality of Two Population Correlations

You can use the following statements to test for equality of two population correlations, ρ_1 and ρ_2. Here, the null hypothesis H_0: $\rho_1 = \rho_2$ is tested against the alternative H_1: $\rho_1 \neq \rho_2$.

```
ods output FisherPearsonCorr=SimCorr;
title 'Testing Equality of Population Correlations';
proc corr data=Sim (where=(Batch=1 or Batch=2)) fisher;
   var X Y;
   by Batch;
run;
```

The ODS OUTPUT statement saves the "FisherPearsonCorr" table into an output data set in the CORR procedure. The output data set SimCorr contains Fisher's z statistics for both batches.

The following statements display (in Figure 2.4.2) the output data set SimCorr:

```
proc print data=SimCorr;
run;
```

Output 2.4.2 Fisher's Correlation Statistics

Obs	Batch	Var	WithVar	NObs	Corr	ZVal	BiasAdj	CorrEst	Lcl	Ucl	pValue
1	1	X	Y	150	0.22081	0.22451	0.0007410	0.22011	0.062034	0.367409	0.0065
2	2	X	Y	150	0.33694	0.35064	0.00113	0.33594	0.185676	0.470853	<.0001

The p-value for testing H_0 is derived by treating the difference $z_1 - z_2$ as a normal random variable with mean zero and variance $1/(n_1 - 3) + 1/(n_2 - 3)$, where z_1 and z_2 are Fisher's z transformation of the sample correlations r_1 and r_2, respectively, and where n_1 and n_2 are the corresponding sample sizes.

48 ✦ *Chapter 2: The CORR Procedure*

The following statements compute the *p*-value in Output 2.4.3:

```
data SimTest (drop=Batch);
   merge SimCorr (where=(Batch=1) keep=Nobs ZVal Batch
                  rename=(Nobs=n1 ZVal=z1))
         SimCorr (where=(Batch=2) keep=Nobs ZVal Batch
                  rename=(Nobs=n2 ZVal=z2));
   variance = 1/(n1-3) + 1/(n2-3);
   z = (z1 - z2) / sqrt( variance );
   pval = probnorm(z);
   if (pval > 0.5) then pval = 1 - pval;
   pval = 2*pval;
run;

proc print data=SimTest noobs;
run;
```

Output 2.4.3 Test of Equality of Observed Correlations

n1	z1	n2	z2	variance	z	pval
150	0.22451	150	0.35064	0.013605	-1.08135	0.27954

In Output 2.4.3, the *p*-value of 0.2795 does not provide evidence to reject the null hypothesis that $\rho_1 = \rho_2$. The sample sizes $n_1 = 150$ and $n_2 = 150$ are not large enough to detect the difference $\rho_1 - \rho_2 = 0.05$ at a significance level of $\alpha = 0.05$.

Combining Correlation Estimates from Different Samples

Assume that sample correlations r_1 and r_2 are computed from two independent samples of n_1 and n_2 observations, respectively. A combined correlation estimate is given by $\bar{r} = \tanh(\bar{z})$, where $\bar{z}$ is the weighted average of the *z* transformations of r_1 and r_2:

$$\bar{z} = \frac{(n_1 - 3)z_1 + (n_2 - 3)z_2}{n_1 + n_2 - 6}$$

The following statements compute a combined estimate of ρ by using Batch 1 and Batch 3:

```
ods output FisherPearsonCorr=SimCorr2;
proc corr data=Sim (where=(Batch=1 or Batch=3)) fisher;
   var X Y;
   by Batch;
run;

data SimComb (drop=Batch);
   merge SimCorr2 (where=(Batch=1) keep=Nobs ZVal Batch
                   rename=(Nobs=n1 ZVal=z1))
         SimCorr2 (where=(Batch=3) keep=Nobs ZVal Batch
                   rename=(Nobs=n2 ZVal=z2));
   z = ((n1-3)*z1 + (n2-3)*z2) / (n1+n2-6);
   corr = tanh(z);
   var = 1/(n1+n2-6);
   zlcl = z - probit(0.975)*sqrt(var);
   zucl = z + probit(0.975)*sqrt(var);
```

```
   lcl= tanh(zlcl);
   ucl= tanh(zucl);
   pval= probnorm( z/sqrt(var));
   if (pval > .5) then pval= 1 - pval;
   pval= 2*pval;
run;

proc print data=SimComb noobs;
   var n1 z1 n2 z2 corr lcl ucl pval;
run;
```

Output 2.4.4 displays the combined estimate of ρ. The table shows that a correlation estimate from the combined samples is *r=0.2264*. The 95% confidence interval is (0.10453,0.34156), using the variance of the combined estimate. Note that this interval contains the population correlation 0.3.

Output 2.4.4 Combined Correlation Estimate

n1	z1	n2	z2	corr	lcl	ucl	pval
150	0.22451	100	0.23929	0.22640	0.10453	0.34156	.000319748

Example 2.5: Computing Polyserial Correlations

The following statements create the data set Fitness1. This data set contains an ordinal variable Oxygen that is derived from a continuous measurement of oxygen intake which is not directly observed.

```
*---------------- Data on Physical Fitness ----------------*
| These measurements were made on men involved in a physical |
| fitness course at N.C. State University.                   |
| The variables are Age (years), Weight (kg),                |
| Runtime (time to run 1.5 miles in minutes), and            |
| Oxygen (an ordinal variable based on oxygen intake,        |
|         ml per kg body weight per minute)                  |
| Certain values were changed to missing for the analysis.   |
*----------------------------------------------------------*;
data Fitness1;
   input Age Weight RunTime Oxygen @@;
   datalines;
44 89.47 11.37  8     40 75.07 10.07  9
44 85.84  8.65 10     42 68.15  8.17 11
38 89.02   .    9     47 77.45 11.63  8
40 75.98 11.95  9     43 81.19 10.85  9
44 81.42 13.08  7     38 81.87  8.63 12
44 73.03 10.13 10     45 87.66 14.03  7
45 66.45 11.12  8     47 79.15 10.60  9
54 83.12 10.33 10     49 81.42  8.95  9
51 69.63 10.95  8     51 77.91 10.00  9
48 91.63 10.25  9     49 73.37 10.08   .
57 73.37 12.63  7     54 79.38 11.17  9
52 76.32  9.63  9     50 70.87  8.92 10
51 67.25 11.08  9     54 91.63 12.88  7
51 73.71 10.47  9     57 59.08  9.93 10
49 76.32   .    .     48 61.24 11.50  9
52 82.78 10.50  9
;
```

The following statements compute Pearson correlations and polyserial correlations:

```
proc corr data=Fitness1 pearson polyserial;
   with Oxygen;
   var  Age Weight RunTime;
run;
```

For the purpose of computing Pearson correlations, the variables in the WITH and VAR statements are treated as continuous variables. For the purpose of computing polyserial correlations, the variables in the WITH statement are treated as ordinal variables by default, and the variables in the VAR statement are treated as continuous variables.

The "Simple Statistics" table in Output 2.5.1 displays univariate descriptive statistics for each analysis variable.

Output 2.5.1 Simple Statistics

The CORR Procedure

1 With Variables:	Oxygen
3 Variables:	Age Weight RunTime

Simple Statistics

Variable	N	Mean	Std Dev	Median	Minimum	Maximum
Oxygen	29	8.93103	1.16285	9.00000	7.00000	12.00000
Age	31	47.67742	5.21144	48.00000	38.00000	57.00000
Weight	31	77.44452	8.32857	77.45000	59.08000	91.63000
RunTime	29	10.67414	1.39194	10.50000	8.17000	14.03000

The "Pearson Correlation Coefficients" table in Output 2.5.2 displays Pearson correlation statistics between Oxygen and the other three variables. The table shows a strong correlation between variables Oxygen and RunTime.

Output 2.5.2 Pearson Correlation Coefficients

Pearson Correlation Coefficients
Prob > |r| under H0: Rho=0
Number of Observations

	Age	Weight	RunTime
Oxygen	-0.25581	-0.22211	-0.85750
	0.1804	0.2469	<.0001
	29	29	28

The "Polyserial Correlations" table in Output 2.5.3 displays polyserial correlation statistics between Oxygen and the three continuous variables. The variable Oxygen is treated as an ordinal variable derived from oxygen intake (the underlying continuous variable), assuming a bivariate normal distribution for oxygen intake and each of the three continuous variables Age, Weight, and RunTime. The CORR procedure provides two tests for a zero polyserial correlation: the Wald test and the likelihood ratio test. The table shows a strong polyserial correlation between RunTime and the underlying continuous variable of Oxygen from both tests.

Output 2.5.3 Polyserial Correlation Coefficients

					Wald Test		LR Test	
Continuous Variable	Ordinal Variable	N	Correlation	Standard Error	Chi-Square	Pr > ChiSq	Chi-Square	Pr > ChiSq
Age	Oxygen	29	-0.23586	0.18813	1.5717	0.2100	1.4466	0.2291
Weight	Oxygen	29	-0.24514	0.18421	1.7709	0.1833	1.6185	0.2033
RunTime	Oxygen	28	-0.91042	0.04071	500.0345	<.0001	38.6963	<.0001

Example 2.6: Computing Cronbach's Coefficient Alpha

The following statements create the data set Fish1 from the Fish data set used in Chapter 111, "The STEPDISC Procedure" (*SAS/STAT User's Guide*). The cubic root of the weight (Weight3) is computed as a one-dimensional measure of the size of a fish.

```
/*-------------------- Fish Measurement Data ---------------------*
 | The data set contains 35 fish from the species Bream caught in |
 | Finland's lake Laengelmavesi with the following measurements:  |
 | Weight    (in grams)                                           |
 | Length3   (length from the nose to the end of its tail, in cm) |
 | HtPct     (max height, as percentage of Length3)               |
 | WidthPct  (max width,  as percentage of Length3)               |
 *----------------------------------------------------------------*;
data Fish1 (drop=HtPct WidthPct);
   title 'Fish Measurement Data';
   input Weight Length3 HtPct WidthPct @@;
   Weight3= Weight**(1/3);
   Height=HtPct*Length3/100;
   Width=WidthPct*Length3/100;
   datalines;
242.0 30.0 38.4 13.4      290.0 31.2 40.0 13.8
340.0 31.1 39.8 15.1      363.0 33.5 38.0 13.3
430.0 34.0 36.6 15.1      450.0 34.7 39.2 14.2
500.0 34.5 41.1 15.3      390.0 35.0 36.2 13.4
450.0 35.1 39.9 13.8      500.0 36.2 39.3 13.7
475.0 36.2 39.4 14.1      500.0 36.2 39.7 13.3
500.0 36.4 37.8 12.0        .   37.3 37.3 13.6
600.0 37.2 40.2 13.9      600.0 37.2 41.5 15.0
700.0 38.3 38.8 13.8      700.0 38.5 38.8 13.5
610.0 38.6 40.5 13.3      650.0 38.7 37.4 14.8
575.0 39.5 38.3 14.1      685.0 39.2 40.8 13.7
620.0 39.7 39.1 13.3      680.0 40.6 38.1 15.1
700.0 40.5 40.1 13.8      725.0 40.9 40.0 14.8
720.0 40.6 40.3 15.0      714.0 41.5 39.8 14.1
850.0 41.6 40.6 14.9     1000.0 42.6 44.5 15.5
920.0 44.1 40.9 14.3      955.0 44.0 41.1 14.3
925.0 45.3 41.4 14.9      975.0 45.9 40.6 14.7
950.0 46.5 37.9 13.7
;
```

The following statements request a correlation analysis and compute Cronbach's coefficient alpha for the variables Weight3, Length3, Height, and Width:

```
ods graphics on;
title 'Fish Measurement Data';
proc corr data=fish1 nomiss alpha plots=matrix;
   var Weight3 Length3 Height Width;
run;
```

The ALPHA option computes Cronbach's coefficient alpha for the analysis variables.

The "Simple Statistics" table in Output 2.6.1 displays univariate descriptive statistics for each analysis variable.

Output 2.6.1 Simple Statistics

Fish Measurement Data

The CORR Procedure

4 **Variables:** Weight3 Length3 Height Width

Simple Statistics

Variable	N	Mean	Std Dev	Sum	Minimum	Maximum
Weight3	34	8.44751	0.97574	287.21524	6.23168	10.00000
Length3	34	38.38529	4.21628	1305	30.00000	46.50000
Height	34	15.22057	1.98159	517.49950	11.52000	18.95700
Width	34	5.43805	0.72967	184.89370	4.02000	6.74970

The "Pearson Correlation Coefficients" table in Output 2.6.2 displays Pearson correlation statistics for pairs of analysis variables.

Output 2.6.2 Pearson Correlation Coefficients

Pearson Correlation Coefficients, N = 34
Prob > |r| under H0: Rho=0

	Weight3	Length3	Height	Width
Weight3	1.00000	0.96523 <.0001	0.96261 <.0001	0.92789 <.0001
Length3	0.96523 <.0001	1.00000	0.95492 <.0001	0.92171 <.0001
Height	0.96261 <.0001	0.95492 <.0001	1.00000	0.92632 <.0001
Width	0.92789 <.0001	0.92171 <.0001	0.92632 <.0001	1.00000

Since the data set contains only one species of fish, all the variables are highly correlated. Using the ALPHA option, the CORR procedure computes Cronbach's coefficient alpha in Output 2.6.3. The Cronbach's coefficient alpha is a lower bound for the reliability coefficient for the raw variables and the standardized variables. Positive correlation is needed for the alpha coefficient because variables measure a common entity.

Output 2.6.3 Cronbach's Coefficient Alpha

Cronbach Coefficient Alpha	
Variables	Alpha
Raw	0.822134
Standardized	0.985145

Because the variances of some variables vary widely, you should use the standardized score to estimate reliability. The overall standardized Cronbach's coefficient alpha of 0.985145 provides an acceptable lower bound for the reliability coefficient. This is much greater than the suggested value of 0.70 given by Nunnally and Bernstein (1994).

The standardized alpha coefficient provides information about how each variable reflects the reliability of the scale with standardized variables. If the standardized alpha decreases after removing a variable from the construct, then this variable is strongly correlated with other variables in the scale. On the other hand, if the standardized alpha increases after removing a variable from the construct, then removing this variable from the scale makes the construct more reliable. The "Cronbach Coefficient Alpha with Deleted Variables" table in Output 2.6.4 does not show significant increase or decrease in the standardized alpha coefficients. See the section "Cronbach's Coefficient Alpha" on page 29 for more information about Cronbach's alpha.

Output 2.6.4 Cronbach's Coefficient Alpha with Deleted Variables

	Cronbach Coefficient Alpha with Deleted Variable			
	Raw Variables		Standardized Variables	
Deleted Variable	Correlation with Total	Alpha	Correlation with Total	Alpha
Weight3	0.975379	0.783365	0.973464	0.977103
Length3	0.967602	0.881987	0.967177	0.978783
Height	0.964715	0.655098	0.968079	0.978542
Width	0.934635	0.824069	0.937599	0.986626

Example 2.7: Saving Correlations in an Output Data Set

The following statements compute Pearson correlations:

```
title 'Correlations for a Fitness and Exercise Study';
proc corr data=Fitness nomiss outp=CorrOutp;
   var weight oxygen runtime;
run;
```

The NOMISS option excludes observations with missing values of the VAR statement variables from the analysis—that is, the same set of 28 observations is used to compute the correlation for each pair of variables. The OUTP= option creates an output data set named CorrOutp that contains the Pearson correlation statistics.

The "Pearson Correlation Coefficients" table in Output 2.7.1 displays the correlation and the *p*-value under the null hypothesis of zero correlation.

Output 2.7.1 Pearson Correlation Coefficients

Correlations for a Fitness and Exercise Study

The CORR Procedure

Pearson Correlation Coefficients, N = 28 Prob > |r| under H0: Rho=0			
	Weight	Oxygen	RunTime
Weight	1.00000	-0.18419 0.3481	0.19505 0.3199
Oxygen	-0.18419 0.3481	1.00000	-0.86843 <.0001
RunTime	0.19505 0.3199	-0.86843 <.0001	1.00000

The following statements display (in Output 2.7.2) the output data set:

```
title 'Output Data Set from PROC CORR';
proc print data=CorrOutp noobs;
run;
```

Output 2.7.2 OUTP= Data Set with Pearson Correlations

Output Data Set from PROC CORR

TYPE	_NAME_	Weight	Oxygen	RunTime
MEAN		77.2168	47.1327	10.6954
STD		8.4495	5.5535	1.4127
N		28.0000	28.0000	28.0000
CORR	Weight	1.0000	-0.1842	0.1950
CORR	Oxygen	-0.1842	1.0000	-0.8684
CORR	RunTime	0.1950	-0.8684	1.0000

The output data set has the default type CORR and can be used as an input data set for regression or other statistical procedures. For example, the following statements request a regression analysis using CorrOutp, without reading the original data in the REG procedure:

```
title 'Input Type CORR Data Set from PROC REG';
proc reg data=CorrOutp;
   model runtime= weight oxygen;
run;
```

The following statements generate the same results as the preceding statements:

```
proc reg data=Fitness;
   model runtime= weight oxygen;
run;
```

Example 2.8: Creating Scatter Plots

The following statements request a correlation analysis and a scatter plot matrix for the variables in the data set Fish1, which was created in Example 2.6.

```
ods graphics on;
title 'Fish Measurement Data';
proc corr data=fish1 nomiss plots=matrix(histogram);
   var Height Width Length3 Weight3;
run;
```

The "Simple Statistics" table in Output 2.8.1 displays univariate descriptive statistics for analysis variables.

Output 2.8.1 Simple Statistics

Fish Measurement Data

The CORR Procedure

4 **Variables:** Height Width Length3 Weight3

Variable	N	Mean	Std Dev	Sum	Minimum	Maximum
Height	34	15.22057	1.98159	517.49950	11.52000	18.95700
Width	34	5.43805	0.72967	184.89370	4.02000	6.74970
Length3	34	38.38529	4.21628	1305	30.00000	46.50000
Weight3	34	8.44751	0.97574	287.21524	6.23168	10.00000

The "Pearson Correlation Coefficients" table in Output 2.8.2 displays Pearson correlation statistics for pairs of analysis variables.

Output 2.8.2 Pearson Correlation Coefficients

Pearson Correlation Coefficients, N = 34
Prob > |r| under H0: Rho=0

	Height	Width	Length3	Weight3
Height	1.00000	0.92632 <.0001	0.95492 <.0001	0.96261 <.0001
Width	0.92632 <.0001	1.00000	0.92171 <.0001	0.92789 <.0001
Length3	0.95492 <.0001	0.92171 <.0001	1.00000	0.96523 <.0001
Weight3	0.96261 <.0001	0.92789 <.0001	0.96523 <.0001	1.00000

The variables are highly correlated. For example, the correlation between Height and Width is 0.92632.

The PLOTS=MATRIX(HISTOGRAM) option requests a scatter plot matrix for the VAR statement variables in Output 2.8.3.

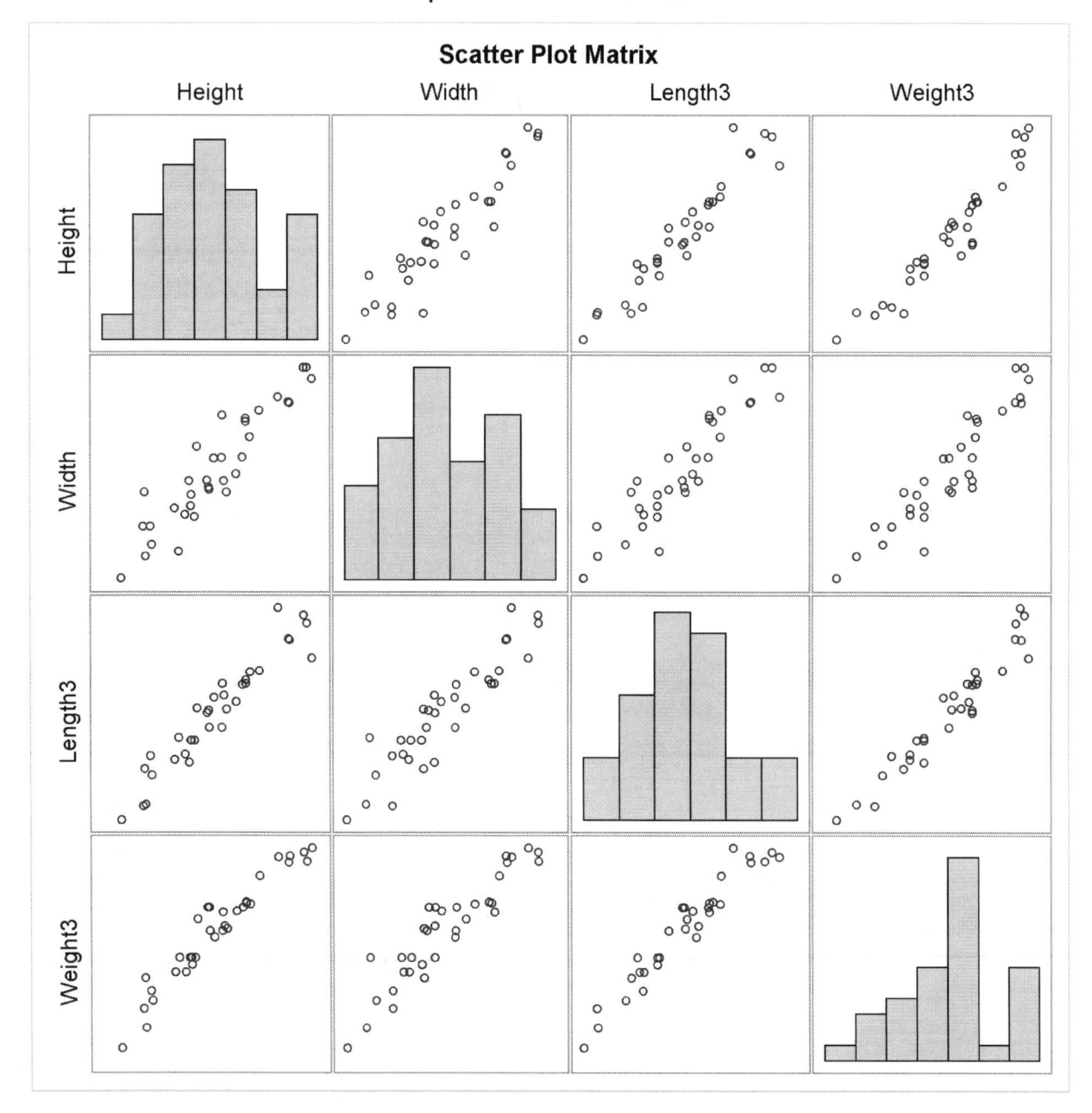

Output 2.8.3 Scatter Plot Matrix

Note that this graphical display is requested by enabling ODS Graphics and by specifying the PLOTS= option. For more information about ODS Graphics, see Chapter 21, "Statistical Graphics Using ODS" (*SAS/STAT User's Guide*).

To explore the correlation between Height and Width, the following statements display (in Output 2.8.4) a scatter plot with prediction ellipses for the two variables:

```
ods graphics on;
proc corr data=fish1 nomiss
          plots=scatter(nvar=2 alpha=.20 .30);
   var Height Width Length3 Weight3;
run;
```

The PLOTS=SCATTER(NVAR=2) option requests a scatter plot for the first two variables in the VAR list. The ALPHA=.20 .30 suboption requests 80% and 70% prediction ellipses, respectively.

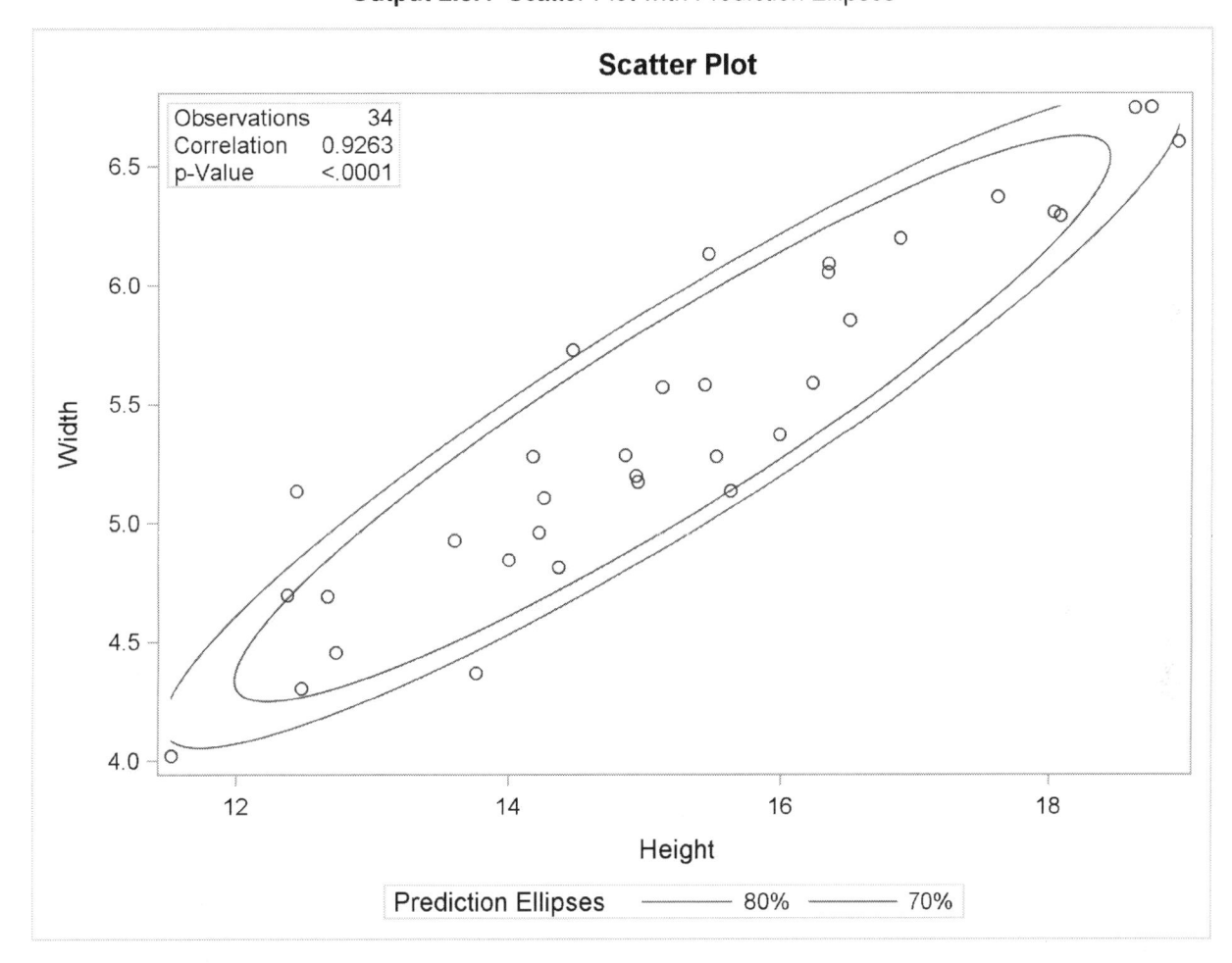

Output 2.8.4 Scatter Plot with Prediction Ellipses

A prediction ellipse is a region for predicting a new observation from the population, assuming bivariate normality. It also approximates a region that contains a specified percentage of the population. The displayed prediction ellipse is centered at the means $(\bar{x}, \bar{y})$. For further details, see the section "Confidence and Prediction Ellipses" on page 31.

58 ✦ *Chapter 2: The CORR Procedure*

Note that the following statements also display (in Output 2.8.5) a scatter plot for Height and Width:

```
ods graphics on;
proc corr data=fish1
          plots=scatter(alpha=.20 .30);
   var Height Width;
run;
```

Output 2.8.5 Scatter Plot with Prediction Ellipses

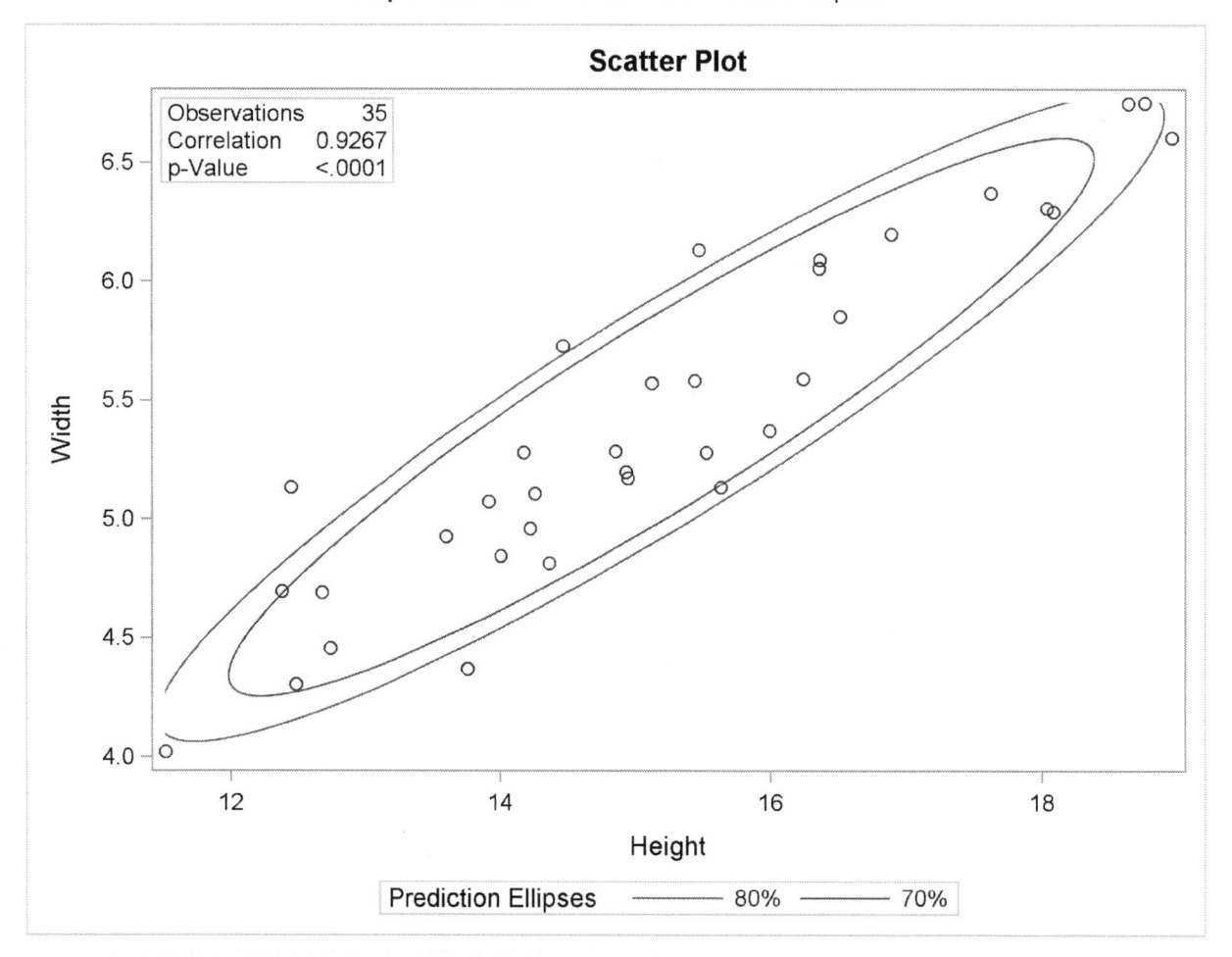

Scatter Plot

Observations	35
Correlation	0.9267
p-Value	<.0001

Prediction Ellipses ——— 80% ——— 70%

Output 2.8.5 includes the point (13.9, 5.1), which was excluded from Output 2.8.4 because the observation had a missing value for Weight3. The prediction ellipses in Output 2.8.5 also reflect the inclusion of this observation.

The following statements display (in Output 2.8.6) a scatter plot with confidence ellipses for the mean:

```
ods graphics on;
title 'Fish Measurement Data';
proc corr data=fish1 nomiss
          plots=scatter(ellipse=confidence nvar=2 alpha=.05 .01);
   var Height Width Length3 Weight3;
run;
```

The NVAR=2 suboption within the PLOTS= option restricts the number of plots created to the first two variables in the VAR statement, and the ELLIPSE=CONFIDENCE suboption requests confidence ellipses for the mean. The ALPHA=.05 .01 suboption requests 95% and 99% confidence ellipses, respectively.

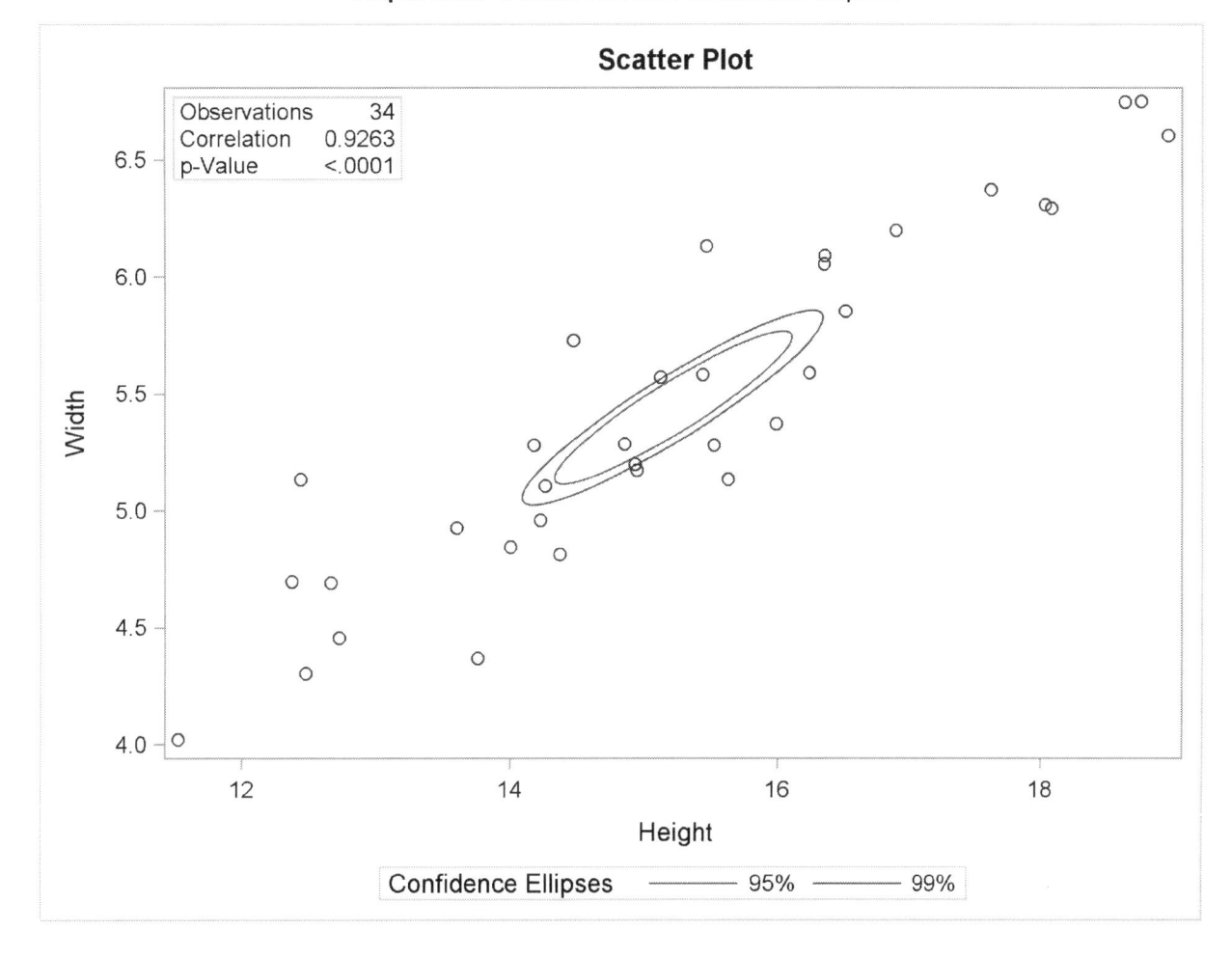

Output 2.8.6 Scatter Plot with Confidence Ellipses

Scatter Plot

Observations	34
Correlation	0.9263
p-Value	<.0001

Confidence Ellipses ——— 95% ——— 99%

The confidence ellipse for the mean is centered at the means $(\bar{x}, \bar{y})$. For further details, see the section "Confidence and Prediction Ellipses" on page 31.

Example 2.9: Computing Partial Correlations

A partial correlation measures the strength of the linear relationship between two variables, while adjusting for the effect of other variables.

The following statements request a partial correlation analysis of variables Height and Width while adjusting for the variables Length3 and Weight. The latter variables, which are said to be "partialled out" of the analysis, are specified with the PARTIAL statement.

```
ods graphics on;
title 'Fish Measurement Data';
proc corr data=fish1 plots=scatter(alpha=.20 .30);
   var Height Width;
   partial Length3 Weight3;
run;
```

Output 2.9.1 displays descriptive statistics for all the variables. The partial variance and partial standard deviation for the variables in the VAR statement are also displayed.

Output 2.9.1 Descriptive Statistics

Fish Measurement Data

The CORR Procedure

2 Partial Variables:	Length3 Weight3
2 Variables:	Height Width

Simple Statistics

Variable	N	Mean	Std Dev	Sum	Minimum	Maximum	Partial Variance	Partial Std Dev
Length3	34	38.38529	4.21628	1305	30.00000	46.50000		
Weight3	34	8.44751	0.97574	287.21524	6.23168	10.00000		
Height	34	15.22057	1.98159	517.49950	11.52000	18.95700	0.26607	0.51582
Width	34	5.43805	0.72967	184.89370	4.02000	6.74970	0.07315	0.27047

When you specify a PARTIAL statement, observations with missing values are excluded from the analysis. Output 2.9.2 displays partial correlations for the variables in the VAR statement.

Output 2.9.2 Pearson Partial Correlation Coefficients

Pearson Partial Correlation Coefficients, N = 34
Prob > |r| under H0: Partial Rho=0

	Height	Width
Height	1.00000	0.25692
		0.1558
Width	0.25692	1.00000
	0.1558	

The partial correlation between the variables Height and Width is 0.25692, which is much less than the unpartialled correlation, 0.92632 (in Output 2.9.2). The *p*-value for the partial correlation is 0.1558.

The PLOTS=SCATTER option displays (in Output 2.9.3) a scatter plot of the residuals for the variables Height and Width after controlling for the effect of variables Length3 and Weight. The ALPHA=.20 .30 suboption requests 80% and 70% prediction ellipses, respectively.

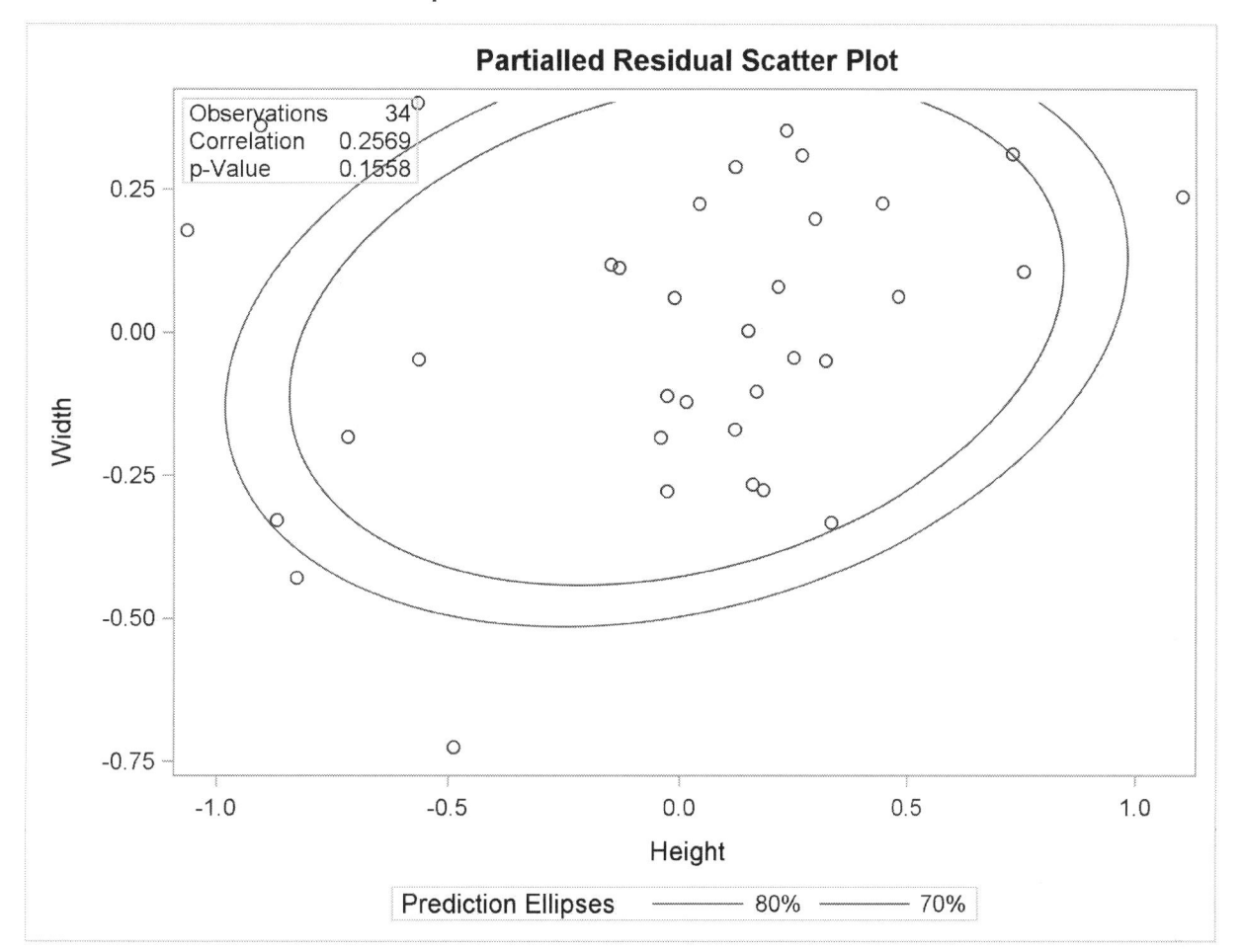

Output 2.9.3 Partial Residual Scatter Plot

Partialled Residual Scatter Plot

Observations 34
Correlation 0.2569
p-Value 0.1558

Prediction Ellipses —— 80% —— 70%

In Output 2.9.3, a standard deviation of Height has roughly the same length on the X axis as a standard deviation of Width on the Y axis. The major axis length is not significantly larger than the minor axis length, indicating a weak partial correlation between Height and Width.

References

Anderson, T. W. (1984). *An Introduction to Multivariate Statistical Analysis*. 2nd ed. New York: John Wiley & Sons.

Blum, J. R., Kiefer, J., and Rosenblatt, M. (1961). "Distribution Free Tests of Independence Based on the Sample Distribution Function." *Annals of Mathematical Statistics* 32:485–498.

Cox, N. R. (1974). "Estimation of the Correlation between a Continuous and a Discrete Variable." *Biometrics* 30:171–178.

Cronbach, L. J. (1951). "Coefficient Alpha and the Internal Structure of Tests." *Psychometrika* 16:297–334.

Drasgow, F. (1986). "Polychoric and Polyserial Correlations." In *Encyclopedia of Statistical Sciences*, vol. 7, edited by S. Kotz, N. L. Johnson, and C. B. Read. New York: John Wiley & Sons.

Fisher, R. A. (1921). "On the 'Probable Error' of a Coefficient of Correlation Deduced from a Small Sample." *Metron* 1:3–32.

Fisher, R. A. (1936). "The Use of Multiple Measurements in Taxonomic Problems." *Annals of Eugenics* 7:179–188.

Fisher, R. A. (1973). *Statistical Methods for Research Workers*. 14th ed. New York: Hafner Publishing.

Hoeffding, W. (1948). "A Non-parametric Test of Independence." *Annals of Mathematical Statistics* 19:546–557.

Hollander, M., and Wolfe, D. A. (1999). *Nonparametric Statistical Methods*. 2nd ed. New York: John Wiley & Sons.

Keeping, E. S. (1962). *Introduction to Statistical Inference*. New York: D. Van Nostrand.

Knight, W. E. (1966). "A Computer Method for Calculating Kendall's Tau with Ungrouped Data." *Journal of the American Statistical Association* 61:436–439.

Noether, G. E. (1967). *Elements of Nonparametric Statistics*. New York: John Wiley & Sons.

Nunnally, J. C., and Bernstein, I. H. (1994). *Psychometric Theory*. 3rd ed. New York: McGraw-Hill.

Olsson, U. (1979). "Maximum Likelihood Estimation of the Polychoric Correlation Coefficient." *Psychometrika* 12:443–460.

Olsson, U., Drasgow, F., and Dorans, N. J. (1982). "The Polyserial Correlation Coefficient." *Biometrika* 47:337–347.

Yu, C. H. (2001). "An Introduction to Computing and Interpreting Cronbach Coefficient Alpha in SAS." In *Proceedings of the Twenty-Sixth Annual SAS Users Group International Conference*. Cary, NC: SAS Institute Inc. http://www2.sas.com/proceedings/sugi26/p246-26.pdf.

Chapter 3
The FREQ Procedure

Contents

Overview: FREQ Procedure	**64**
Getting Started: FREQ Procedure	**65**
Frequency Tables and Statistics	65
Agreement Study	72
Syntax: FREQ Procedure	**75**
PROC FREQ Statement	75
BY Statement	77
EXACT Statement	78
OUTPUT Statement	87
TABLES Statement	99
TEST Statement	144
WEIGHT Statement	147
Details: FREQ Procedure	**148**
Inputting Frequency Counts	148
Grouping with Formats	149
Missing Values	150
In-Database Computation	152
Statistical Computations	153
Definitions and Notation	153
Chi-Square Tests and Statistics	155
Measures of Association	160
Binomial Proportion	170
Risks and Risk Differences	178
Common Risk Difference	188
Odds Ratio and Relative Risks for 2×2 Tables	192
Cochran-Armitage Test for Trend	202
Jonckheere-Terpstra Test	203
Tests and Measures of Agreement	205
Cochran-Mantel-Haenszel Statistics	212
Gail-Simon Test for Qualitative Interactions	221
Exact Statistics	221
Computational Resources	226
Output Data Sets	226
Displayed Output	229
ODS Table Names	239
ODS Graphics	243

Examples: FREQ Procedure **244**
    Example 3.1: Output Data Set of Frequencies 244
    Example 3.2: Frequency Dot Plots 247
    Example 3.3: Chi-Square Goodness-of-Fit Tests 250
    Example 3.4: Binomial Proportions 253
    Example 3.5: Analysis of a 2x2 Contingency Table 256
    Example 3.6: Output Data Set of Chi-Square Statistics 258
    Example 3.7: Cochran-Mantel-Haenszel Statistics 260
    Example 3.8: Cochran-Armitage Trend Test 262
    Example 3.9: Friedman's Chi-Square Test 266
    Example 3.10: Cochran's Q Test 267
References **269**

Overview: FREQ Procedure

The FREQ procedure produces one-way to n-way frequency and contingency (crosstabulation) tables. For two-way tables, PROC FREQ computes tests and measures of association. For n-way tables, PROC FREQ provides stratified analysis by computing statistics within strata and across strata.

For one-way frequency tables, PROC FREQ provides goodness-of-fit tests for equal proportions or specified null proportions. For one-way tables, PROC FREQ also provides confidence limits and tests for binomial proportions, including tests for noninferiority and equivalence.

For contingency tables, PROC FREQ can compute various statistics to examine the relationships between two classification variables. For some pairs of variables, you might want to examine the existence or strength of any association between the variables. To determine if an association exists, PROC FREQ computes chi-square tests. To estimate the strength of an association, PROC FREQ computes measures of association that tend to be close to zero when there is no association and close to the maximum (or minimum) value when there is perfect association. The statistics for contingency tables include the following:

- chi-square tests and measures
- measures of association
- risks (binomial proportions) and risk differences for 2×2 tables
- odds ratios and relative risks for 2×2 tables
- tests for trend
- tests and measures of agreement
- Cochran-Mantel-Haenszel statistics

PROC FREQ computes asymptotic standard errors, confidence intervals, and tests for measures of association and measures of agreement. Exact p-values and confidence intervals are available for many test statistics and measures. PROC FREQ also performs analyses that adjust for stratification variables by computing statistics within and across strata for n-way tables. These statistics include Cochran-Mantel-Haenszel statistics and measures of agreement.

In choosing measures of association to use in analyzing a two-way table, you should consider the study design (which indicates whether the row and column variables are dependent or independent), the measurement scale of the variables (nominal, ordinal, or interval), the type of association that each measure is designed to detect, and any assumptions required for valid interpretation of a measure. You should exercise care in selecting measures that are appropriate for your data.

Similar comments apply to the choice and interpretation of test statistics. For example, the Mantel-Haenszel chi-square statistic requires an ordinal scale for both variables and is designed to detect a linear association. The Pearson chi-square, on the other hand, is appropriate for all variables and can detect any kind of association, but it is less powerful for detecting a linear association because its power is dispersed over a greater number of degrees of freedom (except for 2×2 tables).

For more information about selecting the appropriate statistical analyses, see Agresti (2007) and Stokes, Davis, and Koch (2012).

Several SAS procedures produce frequency counts; only PROC FREQ computes chi-square tests for one-way to n-way tables and measures of association and agreement for contingency tables. Other procedures to consider for counting include the TABULATE and UNIVARIATE procedures. When you want to produce contingency tables and tests of association for sample survey data, you can use PROC SURVEYFREQ. For more information, see Chapter 14, "Introduction to Survey Procedures" (*SAS/STAT User's Guide*). When you want to fit models to categorical data, you can use a procedure such as CATMOD, GENMOD, GLIMMIX, LOGISTIC, PROBIT, or SURVEYLOGISTIC. For more information, see Chapter 8, "Introduction to Categorical Data Analysis Procedures" (*SAS/STAT User's Guide*).

PROC FREQ uses the Output Delivery System (ODS), a SAS subsystem that provides capabilities for displaying and controlling the output from SAS procedures. ODS enables you to convert any of the output from PROC FREQ into a SAS data set. See the section "ODS Table Names" on page 239 for more information.

PROC FREQ uses ODS Graphics to create graphs as part of its output. For general information about ODS Graphics, see Chapter 21, "Statistical Graphics Using ODS" (*SAS/STAT User's Guide*). For information about the statistical graphics that PROC FREQ produces, see the PLOTS= option in the TABLES statement and the section "ODS Graphics" on page 243.

Getting Started: FREQ Procedure

Frequency Tables and Statistics

The FREQ procedure provides easy access to statistics for testing for association in a crosstabulation table.

In this example, high school students applied for courses in a summer enrichment program; these courses included journalism, art history, statistics, graphic arts, and computer programming. The students accepted

66 ✦ Chapter 3: The FREQ Procedure

were randomly assigned to classes with and without internships in local companies. Table 3.1 contains counts of the students who enrolled in the summer program by gender and whether they were assigned an internship slot.

Table 3.1 Summer Enrichment Data

Gender	Internship	Enrollment Yes	No	Total
boys	yes	35	29	64
boys	no	14	27	41
girls	yes	32	10	42
girls	no	53	23	76

The SAS data set SummerSchool is created by inputting the summer enrichment data as cell count data, or providing the frequency count for each combination of variable values. The following DATA step statements create the SAS data set SummerSchool:

```
data SummerSchool;
   input Gender $ Internship $ Enrollment $ Count @@;
   datalines;
boys  yes yes 35   boys  yes no 29
boys  no  yes 14   boys  no  no 27
girls yes yes 32   girls yes no 10
girls no  yes 53   girls no  no 23
;
```

The variable Gender takes the values 'boys' or 'girls,' the variable Internship takes the values 'yes' and 'no,' and the variable Enrollment takes the values 'yes' and 'no.' The variable Count contains the number of students that correspond to each combination of data values. The double at sign (@@) indicates that more than one observation is included on a single data line. In this DATA step, two observations are included on each line.

Researchers are interested in whether there is an association between internship status and summer program enrollment. The Pearson chi-square statistic is an appropriate statistic to assess the association in the corresponding 2 × 2 table. The following PROC FREQ statements specify this analysis.

You specify the table for which you want to compute statistics with the TABLES statement. You specify the statistics you want to compute with options after a slash (/) in the TABLES statement.

```
proc freq data=SummerSchool order=data;
   tables Internship*Enrollment / chisq;
   weight Count;
run;
```

The ORDER= option controls the order in which variable values are displayed in the rows and columns of the table. By default, the values are arranged according to the alphanumeric order of their unformatted values. If you specify ORDER=DATA, the data are displayed in the same order as they occur in the input data set. Here, because 'yes' appears before 'no' in the data, 'yes' appears first in any table. Other options for controlling order include ORDER=FORMATTED, which orders according to the formatted values, and ORDER=FREQ, which orders by descending frequency count.

In the TABLES statement, Internship*Enrollment specifies a table where the rows are internship status and the columns are program enrollment. The CHISQ option requests chi-square statistics for assessing association between these two variables. Because the input data are in cell count form, the WEIGHT statement is required. The WEIGHT statement names the variable Count, which provides the frequency of each combination of data values.

Figure 3.1 presents the crosstabulation of Internship and Enrollment. In each cell, the values printed under the cell count are the table percentage, row percentage, and column percentage, respectively. For example, in the first cell, 63.21 percent of the students offered courses with internships accepted them and 36.79 percent did not.

Figure 3.1 Crosstabulation Table

The FREQ Procedure

Frequency Percent Row Pct Col Pct	Table of Internship by Enrollment			
		Enrollment		
	Internship	yes	no	Total
	yes	67 30.04 63.21 50.00	39 17.49 36.79 43.82	106 47.53
	no	67 30.04 57.26 50.00	50 22.42 42.74 56.18	117 52.47
	Total	134 60.09	89 39.91	223 100.00

Figure 3.2 displays the statistics produced by the CHISQ option. The Pearson chi-square statistic is labeled 'Chi-Square' and has a value of 0.8189 with 1 degree of freedom. The associated *p*-value is 0.3655, which means that there is no significant evidence of an association between internship status and program enrollment. The other chi-square statistics have similar values and are asymptotically equivalent. The other statistics (phi coefficient, contingency coefficient, and Cramér's *V*) are measures of association derived from the Pearson chi-square. For Fisher's exact test, the two-sided *p*-value is 0.4122, which also shows no association between internship status and program enrollment.

Figure 3.2 Statistics Produced with the CHISQ Option

Statistic	DF	Value	Prob
Chi-Square	1	0.8189	0.3655
Likelihood Ratio Chi-Square	1	0.8202	0.3651
Continuity Adj. Chi-Square	1	0.5899	0.4425
Mantel-Haenszel Chi-Square	1	0.8153	0.3666
Phi Coefficient		0.0606	
Contingency Coefficient		0.0605	
Cramer's V		0.0606	

Fisher's Exact Test	
Cell (1,1) Frequency (F)	67
Left-sided Pr <= F	0.8513
Right-sided Pr >= F	0.2213
Table Probability (P)	0.0726
Two-sided Pr <= P	0.4122

The analysis, so far, has ignored gender. However, it might be of interest to ask whether program enrollment is associated with internship status after adjusting for gender. You can address this question by doing an analysis of a set of tables (in this case, by analyzing the set consisting of one for boys and one for girls). The Cochran-Mantel-Haenszel (CMH) statistic is appropriate for this situation: it addresses whether rows and columns are associated after controlling for the stratification variable. In this case, you would be stratifying by gender.

The PROC FREQ statements for this analysis are very similar to those for the first analysis, except that there is a third variable, Gender, in the TABLES statement. When you cross more than two variables, the two rightmost variables construct the rows and columns of the table, respectively, and the leftmost variables determine the stratification.

The following PROC FREQ statements also request frequency plots for the crosstabulation tables. PROC FREQ produces these plots by using ODS Graphics to create graphs as part of the procedure output. ODS Graphics must be enabled before producing plots. The PLOTS(ONLY)=FREQPLOT option requests frequency plots. The TWOWAY=CLUSTER *plot-option* specifies a cluster layout for the two-way frequency plots.

```
ods graphics on;
proc freq data=SummerSchool;
   tables Gender*Internship*Enrollment /
          chisq cmh plots(only)=freqplot(twoway=cluster);
   weight Count;
run;
ods graphics off;
```

This execution of PROC FREQ first produces two individual crosstabulation tables of Internship by Enrollment: one for boys and one for girls. Frequency plots and chi-square statistics are produced for each individual table. Figure 3.3, Figure 3.4, and Figure 3.5 show the results for boys. Note that the chi-square statistic for boys is significant at the $\alpha = 0.05$ level of significance. Boys offered a course with an internship are more likely to enroll than boys who are not.

Figure 3.4 displays the frequency plot of Internship by Enrollment for boys. By default, frequency plots are displayed as bar charts. You can use PLOTS= options to request dot plots instead of bar charts, to change the orientation of the bars from vertical to horizontal, and to change the scale from frequencies to percents. You can also use PLOTS= options to specify other two-way layouts (stacked, vertical groups, or horizontal groups) and to change the primary grouping from column levels to row levels.

Figure 3.6, Figure 3.7, and Figure 3.8 display the crosstabulation table, frequency plot, and chi-square statistics for girls. You can see that there is no evidence of association between internship offers and program enrollment for girls.

Figure 3.3 Crosstabulation Table for Boys

The FREQ Procedure

Frequency Percent Row Pct Col Pct	Table 1 of Internship by Enrollment Controlling for Gender=boys		
		Enrollment	
Internship	no	yes	Total
no	27 25.71 65.85 48.21	14 13.33 34.15 28.57	41 39.05
yes	29 27.62 45.31 51.79	35 33.33 54.69 71.43	64 60.95
Total	56 53.33	49 46.67	105 100.00

Figure 3.4 Frequency Plot for Boys

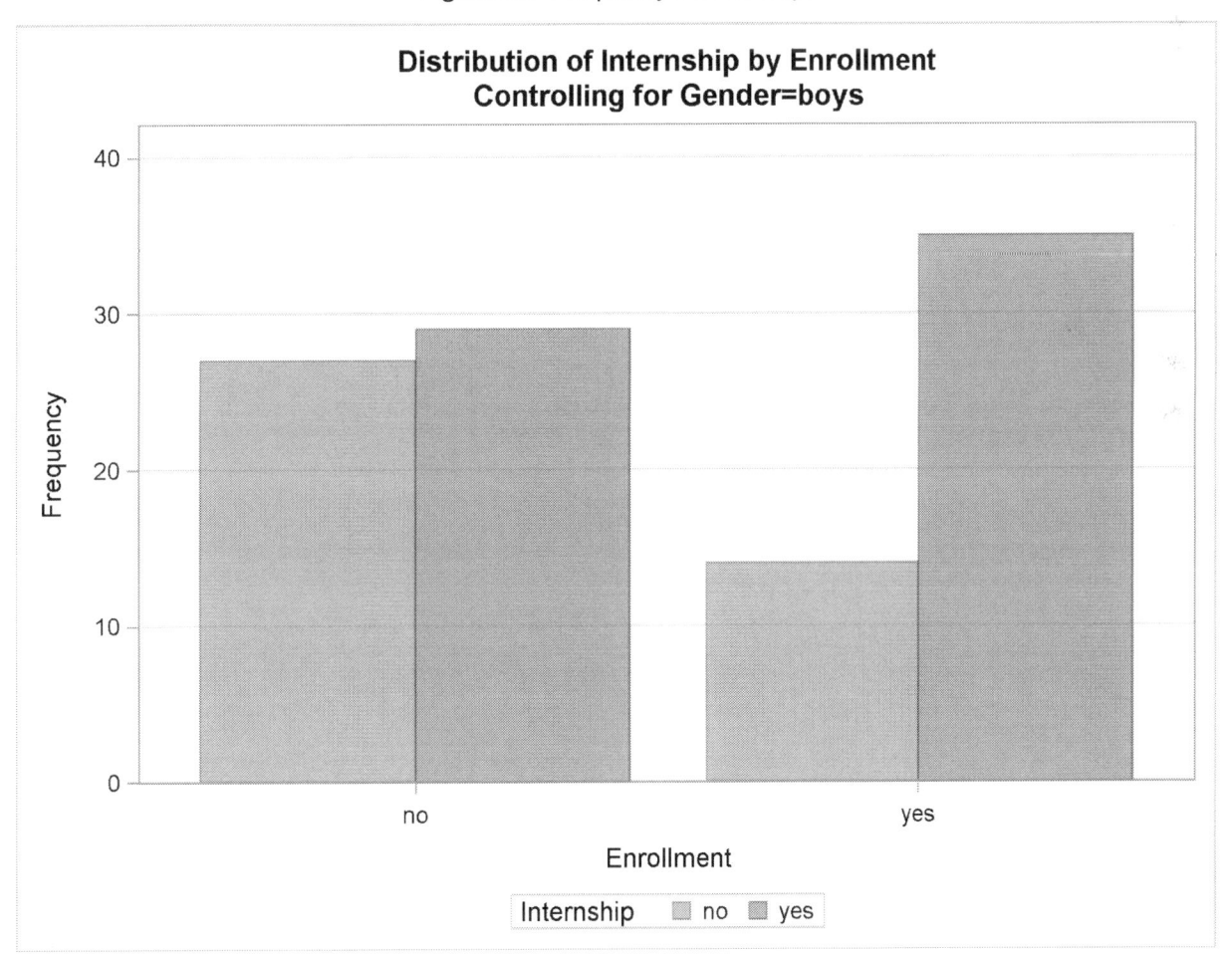

Distribution of Internship by Enrollment
Controlling for Gender=boys

Figure 3.5 Chi-Square Statistics for Boys

Statistic	DF	Value	Prob
Chi-Square	1	4.2366	0.0396
Likelihood Ratio Chi-Square	1	4.2903	0.0383
Continuity Adj. Chi-Square	1	3.4515	0.0632
Mantel-Haenszel Chi-Square	1	4.1963	0.0405
Phi Coefficient		0.2009	
Contingency Coefficient		0.1969	
Cramer's V		0.2009	

Fisher's Exact Test	
Cell (1,1) Frequency (F)	27
Left-sided Pr <= F	0.9885
Right-sided Pr >= F	0.0311
Table Probability (P)	0.0196
Two-sided Pr <= P	0.0467

Figure 3.6 Crosstabulation Table for Girls

Frequency Percent Row Pct Col Pct	Table 2 of Internship by Enrollment Controlling for Gender=girls		
	Enrollment		
Internship	no	yes	Total
no	23 19.49 30.26 69.70	53 44.92 69.74 62.35	76 64.41
yes	10 8.47 23.81 30.30	32 27.12 76.19 37.65	42 35.59
Total	33 27.97	85 72.03	118 100.00

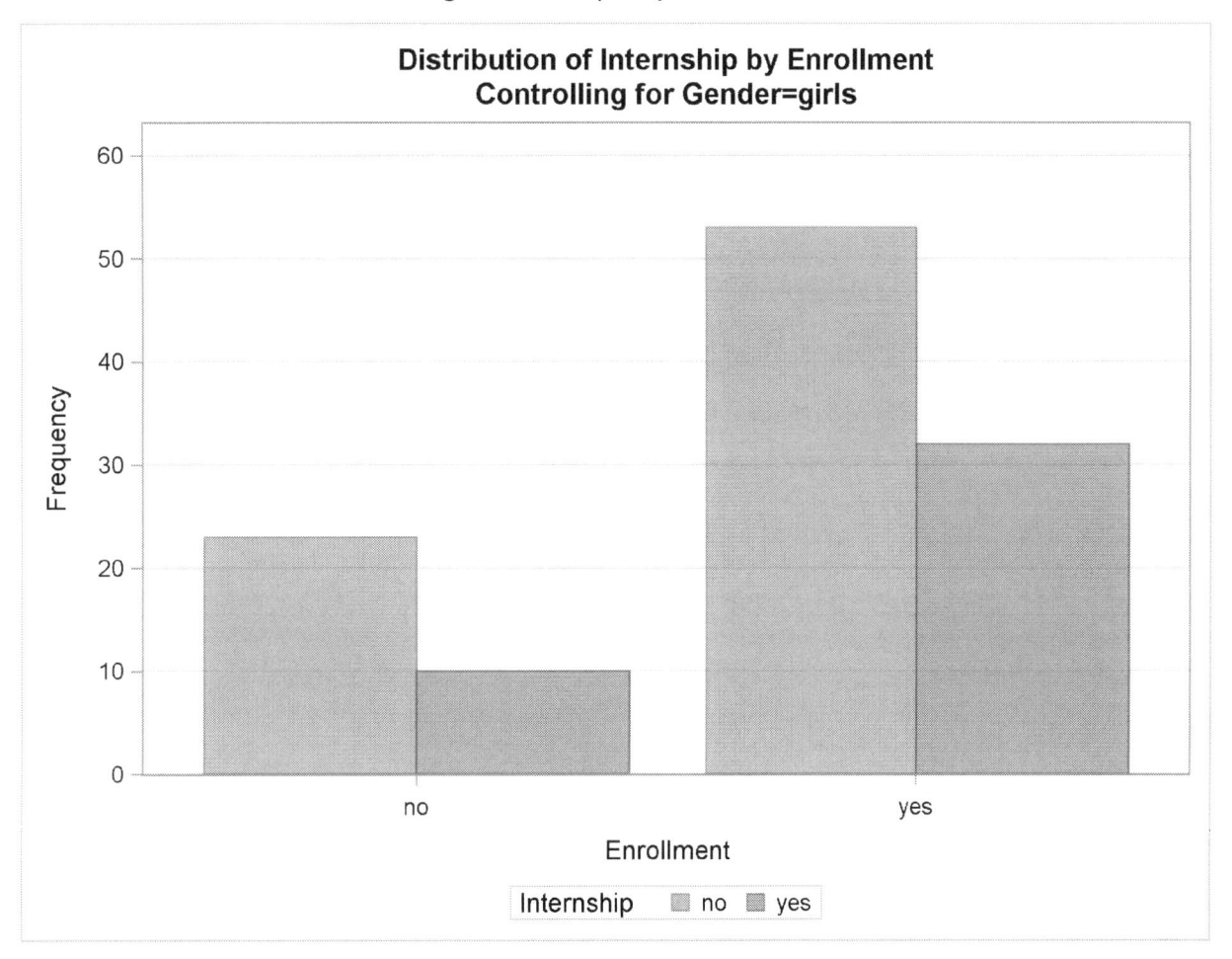

Figure 3.7 Frequency Plot for Girls

**Distribution of Internship by Enrollment
Controlling for Gender=girls**

Figure 3.8 Chi-Square Statistics for Girls

Statistic	DF	Value	Prob
Chi-Square	1	0.5593	0.4546
Likelihood Ratio Chi-Square	1	0.5681	0.4510
Continuity Adj. Chi-Square	1	0.2848	0.5936
Mantel-Haenszel Chi-Square	1	0.5545	0.4565
Phi Coefficient		0.0688	
Contingency Coefficient		0.0687	
Cramer's V		0.0688	

Fisher's Exact Test	
Cell (1,1) Frequency (F)	23
Left-sided Pr <= F	0.8317
Right-sided Pr >= F	0.2994
Table Probability (P)	0.1311
Two-sided Pr <= P	0.5245

These individual table results demonstrate the occasional problems with combining information into one table and not accounting for information in other variables such as Gender. Figure 3.9 contains the CMH results. There are three summary (CMH) statistics; which one you use depends on whether your rows and/or columns have an order in $r \times c$ tables. However, in the case of 2×2 tables, ordering does not matter and all three statistics take the same value. The CMH statistic follows the chi-square distribution under the hypothesis of no association, and here, it takes the value 4.0186 with 1 degree of freedom. The associated p-value is 0.0450, which indicates a significant association at the $\alpha = 0.05$ level.

Thus, when you adjust for the effect of gender in these data, there is an association between internship and program enrollment. But, if you ignore gender, no association is found. Note that the CMH option also produces other statistics, including estimates and confidence limits for relative risk and odds ratios for 2×2 tables and the Breslow-Day Test. These results are not displayed here.

Figure 3.9 Test for the Hypothesis of No Association

Cochran-Mantel-Haenszel Statistics (Based on Table Scores)

Statistic	Alternative Hypothesis	DF	Value	Prob
1	Nonzero Correlation	1	4.0186	0.0450
2	Row Mean Scores Differ	1	4.0186	0.0450
3	General Association	1	4.0186	0.0450

Agreement Study

Medical researchers are interested in evaluating the efficacy of a new treatment for a skin condition. Dermatologists from participating clinics were trained to conduct the study and to evaluate the condition. After the training, two dermatologists examined patients with the skin condition from a pilot study and rated the same patients. The possible evaluations are terrible, poor, marginal, and clear. Table 3.2 contains the data.

Table 3.2 Skin Condition Data

| Dermatologist 1 | Dermatologist 2 ||||
	Terrible	Poor	Marginal	Clear
Terrible	10	4	1	0
Poor	5	10	12	2
Marginal	2	4	12	5
Clear	0	2	6	13

The following DATA step statements create the SAS dataset SkinCondition. The dermatologists' evaluations of the patients are contained in the variables Derm1 and Derm2; the variable Count is the number of patients given a particular pair of ratings.

```
data SkinCondition;
   input Derm1 $ Derm2 $ Count;
   datalines;
terrible terrible 10
terrible     poor 4
terrible marginal 1
terrible    clear 0
```

```
poor     terrible 5
poor     poor     10
poor     marginal 12
poor     clear    2
marginal terrible 2
marginal poor     4
marginal marginal 12
marginal clear    5
clear    terrible 0
clear    poor     2
clear    marginal 6
clear    clear    13
;
```

The following PROC FREQ statements request an agreement analysis of the skin condition data. In order to evaluate the agreement of the diagnoses (a possible contribution to measurement error in the study), the *kappa coefficient* is computed.

The TABLES statement requests a crosstabulation of the variables Derm1 and Derm2. The AGREE option in the TABLES statement requests the kappa coefficient, together with its standard error and confidence limits. The KAPPA option in the TEST statement requests a test for the null hypothesis that kappa is 0, which indicates that the agreement is purely by chance. The NOPRINT option in the TABLES statement suppresses the display of the two-way table. The PLOTS= option requests an agreement plot for the two dermatologists. ODS Graphics must be enabled before producing plots.

```
ods graphics on;
proc freq data=SkinCondition order=data;
   tables Derm1*Derm2 /
          agree noprint plots=agreeplot;
   test kappa;
   weight Count;
run;
ods graphics off;
```

Figure 3.10 and Figure 3.11 show the results. The kappa coefficient has the value 0.3449, which indicates some agreement between the dermatologists, and the hypothesis test confirms that you can reject the null hypothesis of no agreement. This conclusion is further supported by the confidence interval of (0.2030, 0.4868), which suggests that the true kappa is greater than 0. The AGREE option also produces Bowker's symmetry test and the weighted kappa coefficient, but that output is not shown here. Figure 3.11 displays the agreement plot for the ratings of the two dermatologists.

Figure 3.10 Agreement Study

The FREQ Procedure

Statistics for Table of Derm1 by Derm2

Kappa Statistics

Statistic	Estimate	Standard Error	95% Confidence Limits	
Simple Kappa	0.3449	0.0724	0.2030	0.4868
Weighted Kappa	0.5082	0.0655	0.3798	0.6366

74 ✦ *Chapter 3: The FREQ Procedure*

Figure 3.10 *continued*

Test of H0: Kappa = 0				
Estimate	H0 Std Err	Z	Pr > Z	Pr > |Z|
0.3449	0.0612	5.6366	<.0001	<.0001

Figure 3.11 Agreement Plot

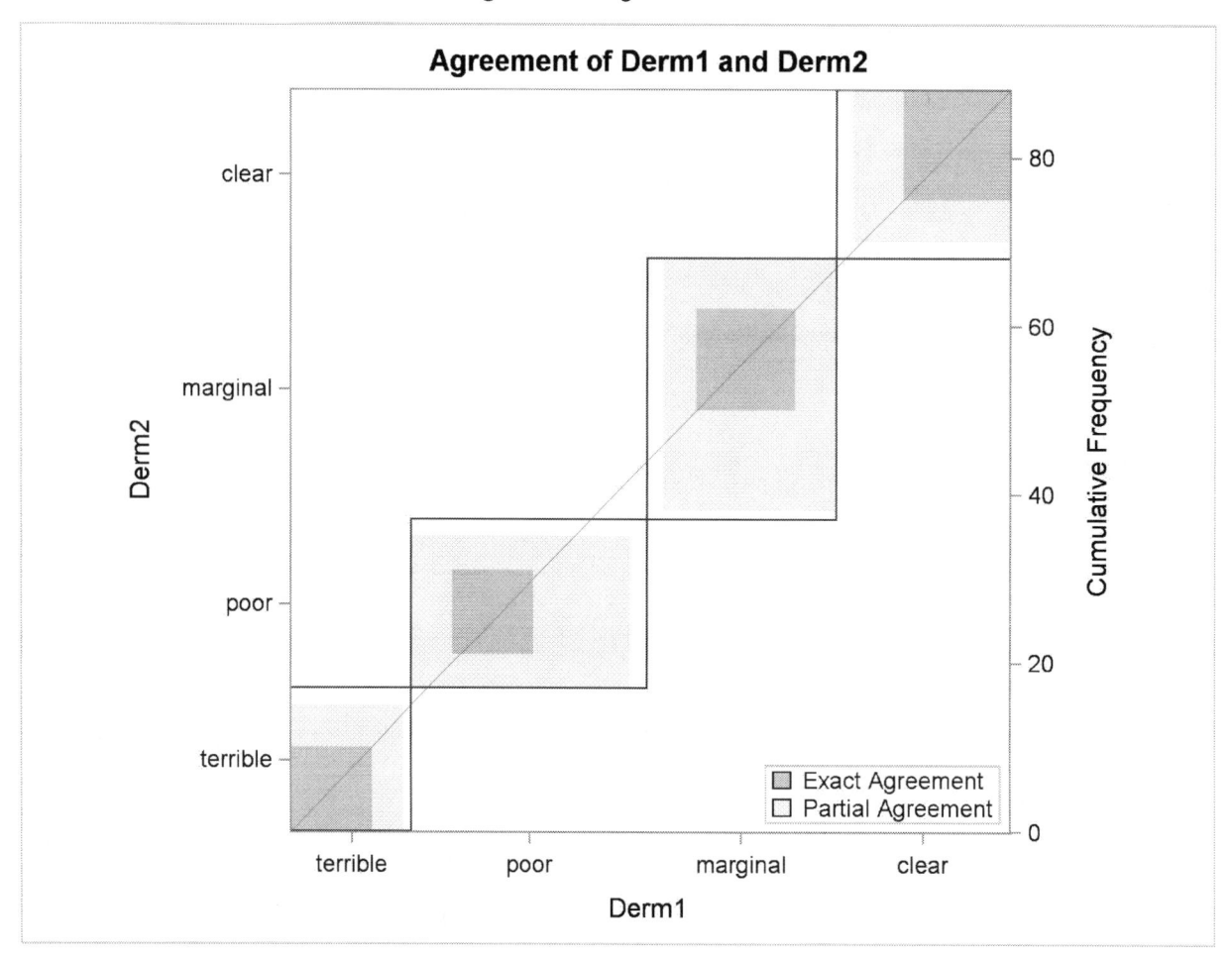

Syntax: FREQ Procedure

The following statements are available in the FREQ procedure:

PROC FREQ <*options*> ;
   BY *variables* ;
   EXACT *statistic-options* < / *computation-options*> ;
   OUTPUT <**OUT**=*SAS-data-set*> *output-options* ;
   TABLES *requests* < / *options*> ;
   TEST *options* ;
   WEIGHT *variable* < / *option*> ;

The PROC FREQ statement is the only required statement for the FREQ procedure. If you specify the following statements, PROC FREQ produces a one-way frequency table for each variable in the most recently created data set.

```
proc freq;
run;
```

Table 3.3 summarizes the basic functions of the procedure statements. The following sections provide detailed syntax information for the BY, EXACT, OUTPUT, TABLES, TEST, and WEIGHT statements in alphabetical order after the description of the PROC FREQ statement.

Table 3.3 Summary of PROC FREQ Statements

Statement	Description
BY	Provides separate analyses for each BY group
EXACT	Requests exact tests
OUTPUT	Requests an output data set
TABLES	Specifies tables and requests analyses
TEST	Requests tests for measures of association and agreement
WEIGHT	Identifies a weight variable

PROC FREQ Statement

PROC FREQ <*options*> ;

The PROC FREQ statement invokes the FREQ procedure. Optionally, it also identifies the input data set. By default, the procedure uses the most recently created SAS data set.

Table 3.4 lists the *options* available in the PROC FREQ statement. Descriptions of the *options* follow in alphabetical order.

Table 3.4 PROC FREQ Statement Options

Option	Description
COMPRESS	Begins the next one-way table on the current page
DATA=	Names the input data set
FORMCHAR=	Specifies the outline and cell divider characters for crosstabulation tables
NLEVELS	Displays the number of levels for all TABLES variables
NOPRINT	Suppresses all displayed output
ORDER=	Specifies the order for reporting variable values
PAGE	Displays one table per page

You can specify the following *options* in the PROC FREQ statement.

COMPRESS

begins display of the next one-way frequency table on the same page as the preceding one-way table if there is enough space to begin the table. By default, the next one-way table begins on the current page only if the entire table fits on that page. The COMPRESS option is not valid with the PAGE option.

DATA=*SAS-data-set*

names the *SAS-data-set* to be analyzed by PROC FREQ. If you omit the DATA= option, the procedure uses the most recently created SAS data set.

FORMCHAR(1,2,7)='*formchar-string*'

defines the characters to be used for constructing the outlines and dividers for the cells of crosstabulation table displays. The *formchar-string* should be three characters long. The characters are used to draw the vertical separators (1), the horizontal separators (2), and the vertical-horizontal intersections (7). If you do not specify the FORMCHAR= option, PROC FREQ uses FORMCHAR(1,2,7)='|-+' by default. Table 3.5 summarizes the formatting characters used by PROC FREQ.

Table 3.5 Formatting Characters Used by PROC FREQ

Position	Default	Used to Draw
1	|	Vertical separators
2	-	Horizontal separators
7	+	Intersections of vertical and horizontal separators

The FORMCHAR= option can specify 20 different SAS formatting characters used to display output; however, PROC FREQ uses only the first, second, and seventh formatting characters. Therefore, the proper specification for PROC FREQ is FORMCHAR(1,2,7)= '*formchar-string*'.

Specifying all blanks for *formchar-string* produces crosstabulation tables with no outlines or dividers— for example, FORMCHAR(1,2,7)=' '. You can use any character in *formchar-string*, including hexadecimal characters. If you use hexadecimal characters, you must put an *x* after the closing quote. For information about which hexadecimal codes to use for which characters, see the documentation for your hardware.

See the CALENDAR, PLOT, and TABULATE procedures in the *SAS Visual Data Management and Utility Procedures Guide* for more information about form characters.

NLEVELS
> displays the "Number of Variable Levels" table, which provides the number of levels for each variable named in the TABLES statements. For more information, see the section "Number of Variable Levels Table" on page 229. PROC FREQ determines the variable levels from the formatted variable values, as described in the section "Grouping with Formats" on page 149.

NOPRINT
> suppresses the display of all output. You can use the NOPRINT option when you only want to create an output data set. See the section "Output Data Sets" on page 226 for information about the output data sets produced by PROC FREQ. Note that the NOPRINT option temporarily disables the Output Delivery System (ODS). For more information, see Chapter 20, "Using the Output Delivery System" (*SAS/STAT User's Guide*).
>
> **NOTE:** A NOPRINT option is also available in the TABLES statement. It suppresses display of the crosstabulation tables but allows display of the requested statistics.

ORDER=DATA | FORMATTED | FREQ | INTERNAL
> specifies the order of the variable levels in the frequency and crosstabulation tables, which you request in the TABLES statement.
>
> The ORDER= option can take the following values:

Value of ORDER=	Levels Ordered By
DATA	Order of appearance in the input data set
FORMATTED	External formatted value, except for numeric variables with no explicit format, which are sorted by their unformatted (internal) value
FREQ	Descending frequency count; levels with the most observations come first in the order
INTERNAL	Unformatted value

> By default, ORDER=INTERNAL. The FORMATTED and INTERNAL orders are machine-dependent. The ORDER= option does not apply to missing values, which are always ordered first.
>
> For more information about sort order, see the chapter on the SORT procedure in the *SAS Visual Data Management and Utility Procedures Guide* and the discussion of BY-group processing in *SAS Language Reference: Concepts*.

PAGE
> displays only one table per page. Otherwise, PROC FREQ displays multiple tables per page as space permits. The PAGE option is not valid with the COMPRESS option.

BY Statement

> **BY** *variables* ;

You can specify a BY statement with PROC FREQ to obtain separate analyses of observations in groups that are defined by the BY variables. When a BY statement appears, the procedure expects the input data set to be sorted in order of the BY variables. If you specify more than one BY statement, only the last one specified is used.

If your input data set is not sorted in ascending order, use one of the following alternatives:

- Sort the data by using the SORT procedure with a similar BY statement.

- Specify the NOTSORTED or DESCENDING option in the BY statement for the FREQ procedure. The NOTSORTED option does not mean that the data are unsorted but rather that the data are arranged in groups (according to values of the BY variables) and that these groups are not necessarily in alphabetical or increasing numeric order.

- Create an index on the BY variables by using the DATASETS procedure (in Base SAS software).

For more information about BY-group processing, see the discussion in *SAS Language Reference: Concepts*. For more information about the DATASETS procedure, see the discussion in the *SAS Visual Data Management and Utility Procedures Guide*.

EXACT Statement

EXACT *statistic-options* < / *computation-options* > ;

The EXACT statement requests exact tests and confidence limits for selected statistics. The *statistic-options* identify which statistics to compute, and the *computation-options* specify options for computing exact statistics. For more information, see the section "Exact Statistics" on page 221.

NOTE: PROC FREQ computes exact tests by using fast and efficient algorithms that are superior to direct enumeration. Exact tests are appropriate when a data set is small, sparse, skewed, or heavily tied. For some large problems, computation of exact tests might require a considerable amount of time and memory. Consider using asymptotic tests for such problems. Alternatively, when asymptotic methods might not be sufficient for such large problems, consider using Monte Carlo estimation of exact *p*-values. You can request Monte Carlo estimation by specifying the MC *computation-option* in the EXACT statement. See the section "Computational Resources" on page 224 for more information.

Statistic Options

The *statistic-options* specify which exact tests and confidence limits to compute. Table 3.6 lists the available *statistic-options* and the exact statistics that are computed. Descriptions of the *statistic-options* follow the table in alphabetical order.

For one-way tables, exact *p*-values are available for binomial proportion tests, the chi-square goodness-of-fit test, and the likelihood ratio chi-square test. Exact (Clopper-Pearson) confidence limits are available for the binomial proportion.

For two-way tables, exact *p*-values are available for the following tests: Pearson chi-square test, likelihood ratio chi-square test, Mantel-Haenszel chi-square test, Fisher's exact test, Jonckheere-Terpstra test, Cochran-Armitage test for trend, and Bowker's symmetry test. Exact *p*-values are also available for tests of the following statistics: Pearson correlation coefficient, Spearman correlation coefficient, Kendall's tau-*b*, Stuart's tau-*c*, Somers' $D(C|R)$, Somers' $D(R|C)$, simple kappa coefficient, and weighted kappa coefficient.

For 2 × 2 tables, PROC FREQ provides the exact McNemar's test, exact confidence limits for the odds ratio, and Barnard's unconditional exact test for the risk (proportion) difference. PROC FREQ also provides exact unconditional confidence limits for the risk difference and for the relative risk (ratio of proportions). For stratified 2 × 2 tables, PROC FREQ provides Zelen's exact test for equal odds ratios, exact confidence limits for the common odds ratio, and an exact test for the common odds ratio.

Most of the *statistic-option* names listed in Table 3.6 are identical to the corresponding option names in the TABLES and OUTPUT statements. You can request exact computations for groups of statistics by using *statistic-options* that are identical to the TABLES statement options CHISQ, MEASURES, and AGREE. For example, when you specify the CHISQ *statistic-option* in the EXACT statement, PROC FREQ computes exact *p*-values for the Pearson chi-square, likelihood ratio chi-square, and Mantel-Haenszel chi-square tests for two-way tables. You can request an exact test for an individual statistic by specifying the corresponding *statistic-option* from the list in Table 3.6.

Using the EXACT Statement with the TABLES Statement

You must use a TABLES statement with the EXACT statement. If you use only one TABLES statement, you do not need to specify the same options in both the TABLES and EXACT statements; when you specify a *statistic-option* in the EXACT statement, PROC FREQ automatically invokes the corresponding TABLES statement option. However, when you use an EXACT statement with multiple TABLES statements, you must specify options in the TABLES statements to request statistics. PROC FREQ then provides exact tests or confidence limits for those statistics that you also specify in the EXACT statement.

Table 3.6 EXACT Statement Statistic Options

Statistic Option	Exact Statistics
AGREE	McNemar's test (for 2 × 2 tables), simple kappa test, weighted kappa test
BARNARD	Barnard's test (for 2 × 2 tables)
BINOMIAL | BIN	Binomial proportion tests for one-way tables
CHISQ	Chi-square goodness-of-fit test for one-way tables; Pearson chi-square, likelihood ratio chi-square, and Mantel-Haenszel chi-square tests for two-way tables
COMOR	Confidence limits for the common odds ratio, common odds ratio test (for $h \times 2 \times 2$ tables)
EQOR | ZELEN	Zelen's test for equal odds ratios (for $h \times 2 \times 2$ tables)
FISHER	Fisher's exact test
JT	Jonckheere-Terpstra test
KAPPA	Test for the simple kappa coefficient
KENTB | TAUB	Test for Kendall's tau-*b*
LRCHI	Likelihood ratio chi-square test (one-way and two-way tables)
MCNEM	McNemar's test (for 2 × 2 tables)
MEASURES	Tests for the Pearson correlation and Spearman correlation, confidence limits for the odds ratio (for 2 × 2 tables)
MHCHI	Mantel-Haenszel chi-square test
OR | ODDSRATIO	Confidence limits for the odds ratio (for 2 × 2 tables)
PCHI	Pearson chi-square test (one-way and two-way tables)
PCORR	Test for the Pearson correlation coefficient
RELRISK	Confidence limits for the relative risk (for 2 × 2 tables)

Table 3.6 *continued*

Statistic Option	Exact Statistics
RISKDIFF	Confidence limits for the risk difference (for 2 × 2 tables)
SCORR	Test for the Spearman correlation coefficient
SMDCR	Test for Somers' $D(C\|R)$
SMDRC	Test for Somers' $D(R\|C)$
STUTC | TAUC	Test for Stuart's tau-c
SYMMETRY | BOWKER	Symmetry test
TREND	Cochran-Armitage test for trend
WTKAPPA | WTKAP	Test for the weighted kappa coefficient

You can specify the following *statistic-options*:

AGREE

requests McNemar's exact test, an exact test for the simple kappa coefficient, and an exact test for the weighted kappa coefficient. For more information, see the sections "Tests and Measures of Agreement" on page 205 and "Exact Statistics" on page 221.

For McNemar's test, you can specify the null hypothesis ratio of discordant proportions by using the AGREE(MNULLRATIO=) option in the TABLES statement; by default, MNULLRATIO=1. For the weighted kappa coefficient, you can request Fleiss-Cohen weights by specifying the AGREE(WT=FC) option in the TABLES statement; by default, PROC FREQ computes the weighted kappa coefficient by using Cicchetti-Allison agreement weights.

McNemar's test is available for 2 × 2 tables. Kappa coefficients are defined only for square two-way tables, where the number of rows equals the number of columns. If your table is not square because some observations have weights of 0, you can specify the ZEROS option in the WEIGHT statement to include these observations in the analysis. For more information, see the section "Tables with Zero-Weight Rows or Columns" on page 212.

For 2 × 2 tables, the weighted kappa coefficient is equivalent to the simple kappa coefficient, and PROC FREQ displays only analyses for the simple kappa coefficient.

BARNARD

requests Barnard's exact unconditional test for the risk (proportion) difference for 2 × 2 tables. For more information, see the section "Barnard's Unconditional Exact Test" on page 187.

To request exact unconditional confidence limits for the risk difference. you can specify the RISKDIFF option in the EXACT statement. The RISKDIFF option in the TABLES statement provides asymptotic tests and several types of confidence limits for the risk difference. For more information, see the section "Risks and Risk Differences" on page 178.

BINOMIAL
BIN

requests an exact test for the binomial proportion (for one-way tables). For more information, see the section "Binomial Tests" on page 173. You can specify the null hypothesis proportion by using the BINOMIAL(P=) option in the TABLES statement; by default, P=0.5.

The BINOMIAL option in the TABLES statement provides exact (Clopper-Pearson) confidence limits for the binomial proportion by default. You can specify the BINOMIAL(CL=MIDP) option in the TABLES statement to request exact mid-p confidence limits for the binomial proportion. The BINOMIAL option in the TABLES statement also provides asymptotic (Wald) tests and several other confidence limit types for the binomial proportion. For more information, see the section "Binomial Proportion" on page 170.

CHISQ

requests the following exact chi-square tests for two-way tables: Pearson chi-square, likelihood ratio chi-square, and Mantel-Haenszel chi-square. For more information, see the section "Chi-Square Tests and Statistics" on page 155. The CHISQ option in the TABLES statement provides asymptotic tests for these statistics.

For one-way tables, the CHISQ option requests an exact chi-square goodness-of-fit test. You can specify null hypothesis proportions for this test by using the CHISQ(TESTP=) option in the TABLES statement. By default, the one-way chi-square test is based on the null hypothesis of equal proportions. For more information, see the section "Chi-Square Test for One-Way Tables" on page 155.

COMOR

requests an exact test and exact confidence limits for the common odds ratio for multiway 2×2 tables. For more information, see the section "Exact Confidence Limits for the Common Odds Ratio" on page 219. The CMH option in the TABLES statement provides Mantel-Haenszel and logit estimates of the common odds ratio along with their asymptotic confidence limits.

EQOR
ZELEN

requests Zelen's exact test for equal odds ratios for multiway 2×2 tables. For more information, see the section "Zelen's Exact Test for Equal Odds Ratios" on page 218. The CMH option in the TABLES statement provides an (asymptotic) Breslow-Day test for homogeneity of odds ratios.

FISHER

requests Fisher's exact test. For more information, see the sections "Fisher's Exact Test" on page 158 and "Exact Statistics" on page 221. For 2×2 tables, the CHISQ option in the TABLES statement provides Fisher's exact test. For general $R \times C$ tables, Fisher's exact test is also known as the Freeman-Halton test.

JT

requests an exact Jonckheere-Terpstra test. For more information, see the sections "Jonckheere-Terpstra Test" on page 203 and "Exact Statistics" on page 221. The JT option in the TABLES statement provides an asymptotic Jonckheere-Terpstra test.

KAPPA

requests an exact test for the simple kappa coefficient. For more information, see the sections "Simple Kappa Coefficient" on page 206 and "Exact Statistics" on page 221. The AGREE option in the TABLES statement provides the simple kappa estimate, standard error, and confidence limits. The KAPPA option in the TEST statement provides an asymptotic test for the simple kappa coefficient.

Kappa coefficients are defined only for square two-way tables, where the number of rows equals the number of columns. If your table is not square because some observations have weights of 0, you can specify the ZEROS option in the WEIGHT statement to include these observations in the analysis. For more information, see the section "Tables with Zero-Weight Rows or Columns" on page 212.

KENTB

TAUB

requests an exact test for Kendall's tau-*b*. For more information, see the sections "Kendall's Tau-b" on page 162 and "Exact Statistics" on page 221. The MEASURES option in the TABLES statement provides an estimate and standard error of Kendall's tau-*b*. The KENTB option in the TEST statement provides an asymptotic test for Kendall's tau-*b*.

LRCHI

requests an exact test for the likelihood ratio chi-square for two-way tables. For more information, see the sections "Likelihood Ratio Chi-Square Test" on page 157 and "Exact Statistics" on page 221. The CHISQ option in the TABLES statement provides an asymptotic likelihood ratio chi-square test for two-way tables.

For one-way tables, the LRCHI option requests an exact likelihood ratio goodness-of-fit test. You can specify null hypothesis proportions by using the CHISQ(TESTP=) option in the TABLES statement. By default, the one-way test is based on the null hypothesis of equal proportions. For more information, see the section "Likelihood Ratio Chi-Square Test for One-Way Tables" on page 157.

MCNEM

requests an exact McNemar's test. For more information, see the sections "McNemar's Test" on page 205 and "Exact Statistics" on page 221. You can specify the null hypothesis ratio of discordant proportions by using the AGREE(MNULLRATIO=) option in the TABLES statement; by default, MNULLRATIO=1. The AGREE option in the TABLES statement provides an asymptotic McNemar's test.

MEASURES

requests exact tests for the Pearson and Spearman correlations. For more information, see the sections "Pearson Correlation Coefficient" on page 164, "Spearman Rank Correlation Coefficient" on page 165, and "Exact Statistics" on page 221. The PCORR and SCORR options in the TEST statement provide asymptotic tests for the Pearson and Spearman correlations, respectively.

The MEASURES option also requests exact confidence limits for the odds ratio for 2 × 2 tables. For more information, see the subsection Exact Confidence Limits in the section "Confidence Limits for the Odds Ratio" on page 193. You can also request exact confidence limits for the odds ratio by specifying the OR option in the EXACT statement.

MHCHI

requests an exact test for the Mantel-Haenszel chi-square. For more information, see the sections "Mantel-Haenszel Chi-Square Test" on page 158 and "Exact Statistics" on page 221. The CHISQ option in the TABLES statement provides an asymptotic Mantel-Haenszel chi-square test.

OR

ODDSRATIO

requests exact confidence limits for the odds ratio for 2 × 2 tables. For more information, see the subsection "Exact Confidence Limits" in the section "Confidence Limits for the Odds Ratio" on page 193.

You can request exact mid-*p* confidence limits for the odds ratio by specifying the OR(CL=MIDP) option in the TABLES statement. The OR(CL=) option in the TABLES statement also provides other types of confidence limits for the odds ratio. For more information, see the section "Confidence Limits for the Odds Ratio" on page 193.

The ALPHA= option in the TABLES statement determines the confidence level of the exact confidence limits; by default, ALPHA=0.05, which produces 95% confidence limits for the odds ratio.

PCHI

requests an exact test for the Pearson chi-square for two-way tables. For more information, see the sections "Pearson Chi-Square Test for Two-Way Tables" on page 156 and "Exact Statistics" on page 221. The CHISQ option in the TABLES statement provides an asymptotic Pearson chi-square test.

For one-way tables, the PCHI option requests an exact chi-square goodness-of-fit test. You can specify null hypothesis proportions by using the CHISQ(TESTP=) option in the TABLES statement. By default, the goodness-of-fit test is based on the null hypothesis of equal proportions. For more information, see the section "Chi-Square Test for One-Way Tables" on page 155.

PCORR

requests an exact test for the Pearson correlation coefficient. For more information, see the sections "Pearson Correlation Coefficient" on page 164 and "Exact Statistics" on page 221. The MEASURES option in the TABLES statement provides the estimate and standard error of the Pearson correlation. The PCORR option in the TEST statement provides an asymptotic test for the Pearson correlation.

RELRISK < (options) >

requests exact unconditional confidence limits for the relative risk for 2×2 tables. By default (beginning in SAS/STAT 14.3), the exact confidence limits are computed by inverting two separate one-sided exact tests that are based on the score statistic (Chan and Zhang 1999). For more information, see the subsection "Exact Unconditional Confidence Limits" in the section "Confidence Limits for the Relative Risk" on page 196.

The RELRISK(CL=) option in the TABLES statement provides additional types of confidence limits for the relative risk. For more information, see the section "Confidence Limits for the Risk Difference" on page 179.

The ALPHA= option in the TABLES statement determines the confidence level; by default, ALPHA=0.05, which produces 95% confidence limits for the relative risk.

You can specify the following *options*:

COLUMN=1 | 2 | BOTH

specifies the table column of the relative risk. By default, COLUMN=1, which provides exact confidence limits for the column 1 relative risk. COLUMN=BOTH provides exact confidence limits for both column 1 and column 2 relative risks.

METHOD=NOSCORE | SCORE | SCORE2

specifies the computation method for the exact confidence limits. By default, METHOD=SCORE.

You can specify one of the following methods:

NOSCORE

computes the exact confidence limits by inverting two separate one-sided exact tests that are based on the unstandardized relative risk (Santner and Snell 1980). For more information, see the subsection "Exact Unconditional Confidence Limits" in the section "Confidence Limits for the Relative Risk" on page 196. This method is the default in releases before SAS/STAT 14.3.

SCORE

computes the exact confidence limits by inverting two separate one-sided exact tests that are based on the score statistic (Chan and Zhang 1999). For more information, see the subsection "Exact Unconditional Confidence Limits" in the section "Confidence Limits for the Relative Risk" on page 196. This method is the default beginning in SAS/STAT 14.3.

SCORE2

computes the exact confidence limits by inverting a single two-sided exact test that is based on the score statistic (Agresti and Min 2001). For more information, see the subsection "Exact Unconditional Confidence Limits" in the section "Confidence Limits for the Relative Risk" on page 196.

RISKDIFF < (*options*) >

requests exact unconditional confidence limits for the risk difference for 2 × 2 tables. By default (beginning in SAS/STAT 14.3), the exact confidence limits are computed by inverting two separate one-sided exact tests that are based on the score statistic (Chan and Zhang 1999). For more information, see the subsection "Exact Unconditional Confidence Limits" in the section "Confidence Limits for the Risk Difference" on page 179.

The RISKDIFF(CL=) option in the TABLES statement provides additional types of confidence limits for the risk difference. For more information, see the section "Confidence Limits for the Risk Difference" on page 179.

The ALPHA= option in the TABLES statement determines the confidence level; by default, ALPHA=0.05, which produces 95% confidence limits for the risk difference.

You can specify the following *options*:

COLUMN=1 | 2 | BOTH

specifies the table column of the risk difference. By default, COLUMN=BOTH and the exact confidence limits are displayed in the 'Risk Estimates' tables. If you specify the RISKDIFF(NORISKS) option in the TABLES statement to suppress the 'Risk Estimates' tables, COLUMN=1 by default and the exact confidence limits are displayed in the 'Risk Difference Confidence Limits' table.

METHOD=NOSCORE | SCORE | SCORE2

specifies the computation method for the exact confidence limits. By default, METHOD=SCORE.

You can specify one of the following methods:

NOSCORE

computes the exact confidence limits by inverting two separate one-sided exact tests that are based on the unstandardized risk difference (Santner and Snell 1980). For more information, see the subsection "Exact Unconditional Confidence Limits" in the section "Confidence Limits for the Risk Difference" on page 179. This method is the default in releases before SAS/STAT 14.3.

SCORE

computes the exact confidence limits by inverting two separate one-sided exact tests that are based on the score statistic (Chan and Zhang 1999). For more information, see the subsection "Exact Unconditional Confidence Limits" in the section "Confidence Limits for the Risk Difference" on page 179. This method is the default beginning in SAS/STAT 14.3.

SCORE2

computes the exact confidence limits by inverting a single two-sided exact test that is based on the score statistic (Agresti and Min 2001). For more information, see the subsection "Exact Unconditional Confidence Limits" in the section "Confidence Limits for the Risk Difference" on page 179.

SCORR

requests an exact test for the Spearman correlation coefficient. For more information, see the sections "Spearman Rank Correlation Coefficient" on page 165 and "Exact Statistics" on page 221. The MEASURES option in the TABLES statement provides the estimate and standard error of the Spearman correlation. The SCORR option in the TEST statement provides an asymptotic test for the Spearman correlation.

SMDCR

requests an exact test for Somers' $D(C|R)$. For more information, see the sections "Somers' D" on page 163 and "Exact Statistics" on page 221. The MEASURES option in the TABLES statement provides the estimate and standard error of Somers' $D(C|R)$. The SMDCR option in the TEST statement provides an asymptotic test for Somers' $D(C|R)$.

SMDRC

requests an exact test for Somers' $D(R|C)$. For more information, see the sections "Somers' D" on page 163 and "Exact Statistics" on page 221. The MEASURES option in the TABLES statement provides the estimate and standard error of Somers' $D(R|C)$. The SMDRC option in the TEST statement provides an asymptotic test for Somers' $D(C|R)$.

STUTC
TAUC

requests an exact test for Stuart's tau-c. For more information, see the sections "Stuart's Tau-c" on page 163 and "Exact Statistics" on page 221. The MEASURES option in the TABLES statement provides the estimate and standard error of Stuart's tau-c. The STUTC option in the TEST statement provides an asymptotic test for Stuart's tau-c.

SYMMETRY
BOWKER

requests an exact symmetry test. This test is available for square $R \times R$ two-way tables where the table dimension R is greater than 2. For more information, see the section "Exact Symmetry Test" on page 206. The AGREE option in the TABLES statement provides an asymptotic symmetry test.

TREND

requests the exact Cochran-Armitage test for trend. For more information, see the sections "Cochran-Armitage Test for Trend" on page 202 and "Exact Statistics" on page 221. The TREND option in the TABLES statement provides an asymptotic Cochran-Armitage test for trend. This test is available for tables of dimensions $2 \times C$ or $R \times 2$.

WTKAPPA
WTKAP

requests an exact test for the weighted kappa coefficient. For more information, see the sections "Weighted Kappa Coefficient" on page 208 and "Exact Statistics" on page 221. By default, PROC FREQ computes the weighted kappa coefficient by using Cicchetti-Allison agreement weights. You can request Fleiss-Cohen agreement weights by specifying the AGREE(WT=FC) option in the TABLES statement.

Kappa coefficients are defined only for square two-way tables, where the number of rows equals the number of columns. If your table is not square because some observations have weights of 0, you can specify the ZEROS option in the WEIGHT statement to include these observations in the analysis. For more information, see the section "Tables with Zero-Weight Rows or Columns" on page 212.

For 2×2 tables, the weighted kappa coefficient is equivalent to the simple kappa coefficient, and PROC FREQ displays only analyses for the simple kappa coefficient.

Computation Options

The *computation-options* specify options for computing exact statistics. You can specify the following *computation-options* in the EXACT statement after a slash (/).

ALPHA=α

specifies the level of the confidence limits for Monte Carlo p-value estimates. The value of α must be between 0 and 1; a confidence level of α produces $100(1 - \alpha)\%$ confidence limits. By default ALPHA=0.01, which produces 99% confidence limits for the Monte Carlo estimates.

The ALPHA= option invokes the MC option.

MAXTIME=*value*

specifies the maximum clock time (in seconds) that PROC FREQ can use to compute an exact p-value. If the procedure does not complete the computation within the specified time, the computation terminates. The MAXTIME= *value* must be a positive number. This option is available for Monte Carlo estimation of exact p-values, in addition to direct exact p-value computation. For more information, see the section "Computational Resources" on page 224.

MC

requests Monte Carlo estimation of exact p-values instead of direct exact p-value computation. Monte Carlo estimation can be useful for large problems that require a considerable amount of time and memory for exact computations but for which asymptotic approximations might not be sufficient. For more information, see the section "Monte Carlo Estimation" on page 225.

This option is available for all EXACT *statistic-options* except the BINOMIAL option and the following options that apply only to 2×2 or $h \times 2 \times 2$ tables: BARNARD, COMOR, EQOR, MCNEM, OR, RELRISK, and RISKDIFF. PROC FREQ always computes exact tests or confidence limits (not Monte Carlo estimates) for these statistics.

The ALPHA=, N=, and SEED= options invoke the MC option.

MIDP

requests exact mid p-values for the exact tests. The exact mid p-value is defined as the exact p-value minus half the exact point probability. For more information, see the section "Definition of p-Values" on page 223.

The MIDP option is available for all EXACT statement *statistic-options* except the following: BARNARD, EQOR, OR, RELRISK, and RISKDIFF. You cannot specify both the MIDP option and the MC option.

N=n

specifies the number of samples for Monte Carlo estimation. The value of n must be a positive integer, and the default is 10,000. Larger values of n produce more precise estimates of exact p-values. Because larger values of n generate more samples, the computation time increases.

The N= option invokes the MC option.

PFORMAT=format-name | **EXACT**

specifies the display format for exact p-values. PROC FREQ applies this format to one- and two-sided exact p-values, exact point probabilities, and exact mid p-values. By default, PROC FREQ displays exact p-values in the PVALUE6.4 format.

You can provide a *format-name* or you can specify PFORMAT=EXACT to control the format of exact p-values. The value of *format-name* can be any standard SAS numeric format or a user-defined format. The format length must not exceed 24. For information about formats, see the FORMAT procedure in the *SAS Visual Data Management and Utility Procedures Guide* and the FORMAT statement and SAS format in *SAS Formats and Informats: Reference*.

If you specify PFORMAT=EXACT, PROC FREQ uses the 6.4 format to display exact p-values that are greater than or equal to 0.001; the procedure uses the E10.3 format to display values that are between 0.000 and 0.001. This is the format that PROC FREQ uses to display exact p-values in releases before SAS/STAT 12.3. Beginning in SAS/STAT 12.3, by default PROC FREQ uses the PVALUE6.4 format to display exact p-values.

POINT

requests exact point probabilities for the exact tests. The exact point probability is the exact probability that the test statistic equals the observed value. For more information, see the section "Definition of p-Values" on page 223.

The POINT option is available for all EXACT statement *statistic-options* except the following: BARNARD, EQOR, OR, RELRISK, and RISKDIFF. You cannot specify both the POINT option and the MC option.

SEED=number

specifies the initial seed for random number generation for Monte Carlo estimation. The value of the SEED= option must be an integer. If you do not specify the SEED= option or if the SEED= value is negative or 0, PROC FREQ uses the time of day from the computer's clock to obtain the initial seed.

The SEED= option invokes the MC option.

OUTPUT Statement

OUTPUT < **OUT=**SAS-data-set> output-options ;

The OUTPUT statement creates a SAS data set that contains statistics that are computed by PROC FREQ. Table 3.7 lists the statistics that can be stored in the output data set. You identify which statistics to include by specifying *output-options*.

You must use a TABLES statement with the OUTPUT statement. The OUTPUT statement stores statistics for only one table request. If you use multiple TABLES statements, the contents of the output data set correspond to the last TABLES statement. If you use multiple table requests in a single TABLES statement, the contents of the output data set correspond to the last table request. Only one OUTPUT statement is allowed in a single invocation of the procedure.

For a one-way or two-way table, the output data set contains one observation that stores the requested statistics for the table. For a multiway table, the output data set contains an observation for each two-way table (stratum) of the multiway crosstabulation. If you request summary statistics for the multiway table, the output data set also contains an observation that stores the across-strata summary statistics. If you use a BY statement, the output data set contains an observation or set of observations for each BY group. For more information about the contents of the output data set, see the section "Contents of the OUTPUT Statement Output Data Set" on page 228.

The output data set that is created by the OUTPUT statement is not the same as the output data set that is created by the OUT= option in the TABLES statement. The OUTPUT statement creates a data set that contains statistics (such as the Pearson chi-square and its *p*-value), and the OUT= option in the TABLES statement creates a data set that contains frequency table counts and percentages. See the section "Output Data Sets" on page 226 for more information.

As an alternative to the OUTPUT statement, you can use the Output Delivery System (ODS) to store statistics that PROC FREQ computes. ODS can create a SAS data set from any table that PROC FREQ produces. See the section "ODS Table Names" on page 239 for more information.

You can specify the following *options* in the OUTPUT statement:

OUT=SAS-data-set
: specifies the name of the output data set. When you use an OUTPUT statement but do not use the OUT= option, PROC FREQ creates a data set and names it by using the DATA*n* convention.

output-options
: specify the statistics to include in the output data set. Table 3.7 lists the *output-options* that are available in the OUTPUT statement, together with the TABLES statement options that are required to produce the statistics. Descriptions of the *output-options* follow the table in alphabetical order.

You can specify *output-options* to request individual statistics, or you can request groups of statistics by using *output-options* that are identical to the group options in the TABLES statement (for example, the CHISQ, MEASURES, CMH, AGREE, and ALL options).

When you specify an *output-option*, the output data set includes statistics from the corresponding analysis. In addition to the estimate or test statistic, the output data set includes associated values such as standard errors, confidence limits, *p*-values, and degrees of freedom. For more information, see the section "Contents of the OUTPUT Statement Output Data Set" on page 228.

To store a statistic in the output data set, you must also request computation of that statistic with the appropriate TABLES, EXACT, or TEST statement option. For example, the PCHI *output-option* includes the Pearson chi-square in the output data set. You must also request computation of the Pearson chi-square by specifying the CHISQ option in the TABLES statement. Or, if you use only one TABLES statement, you can request computation of the Pearson chi-square by specifying the PCHI or CHISQ option in the EXACT statement. Table 3.7 lists the TABLES statement options that are required to produce the OUTPUT data set statistics.

Table 3.7 OUTPUT Statement Output Options

Output Option	Output Data Set Statistics	Required TABLES Statement Option
AGREE	McNemar's test (2 × 2 tables), Bowker's test, simple and weighted kappas; for multiple strata, overall simple and weighted kappas, tests for equal kappas, and Cochran's Q ($h \times 2 \times 2$ tables)	AGREE
AJCHI	Continuity-adjusted chi-square (2 × 2 tables)	CHISQ
ALL	CHISQ, MEASURES, and CMH statistics; N (number of nonmissing observations)	ALL
BDCHI	Breslow-Day test ($h \times 2 \times 2$ tables)	CMH, CMH1, or CMH2
BINOMIAL | BIN	Binomial statistics (one-way tables)	BINOMIAL
CHISQ	For one-way tables, goodness-of-fit test; for two-way tables, Pearson, likelihood ratio, continuity-adjusted, and Mantel-Haenszel chi-squares, Fisher's exact test (2 × 2 tables), phi and contingency coefficients, Cramér's V	CHISQ
CMH	Cochran-Mantel-Haenszel (CMH) correlation, row mean scores (ANOVA), and general association statistics; for 2 × 2 tables, logit and Mantel-Haenszel common odds ratios and relative risks, Breslow-Day test	CMH
CMH1	CMH statistics, except row mean scores (ANOVA) and general association statistics	CMH or CMH1
CMH2	CMH statistics, except general association statistic	CMH or CMH2
CMHCOR	CMH correlation statistic	CMH, CMH1, or CMH2
CMHGA	CMH general association statistic	CMH
CMHRMS	CMH row mean scores (ANOVA) statistic	CMH or CMH2
COCHQ	Cochran's Q ($h \times 2 \times 2$ tables)	AGREE
CONTGY	Contingency coefficient	CHISQ
CRAMV	Cramér's V	CHISQ
EQKAP	Test for equal simple kappas	AGREE
EQOR | ZELEN	Zelen's test for equal odds ratios ($h \times 2 \times 2$ tables)	CMH and EXACT EQOR
EQWKP	Test for equal weighted kappas	AGREE
FISHER	Fisher's exact test	CHISQ or FISHER [1]
GAMMA	Gamma	MEASURES
GS | GAILSIMON	Gail-Simon test	CMH(GAILSIMON)
JT	Jonckheere-Terpstra test	JT
KAPPA	Simple kappa coefficient	AGREE
KENTB | TAUB	Kendall's tau-b	MEASURES
LAMCR	Lambda asymmetric ($C \vert R$)	MEASURES
LAMDAS	Lambda symmetric	MEASURES
LAMRC	Lambda asymmetric ($R \vert C$)	MEASURES

[1] CHISQ computes Fisher's exact test for 2 × 2 tables. Use the FISHER option to compute Fisher's exact test for general $r \times c$ tables.

Table 3.7 *continued*

Output Option	Output Data Set Statistics	Required TABLES Statement Option
LGOR	Logit common odds ratio	CMH, CMH1, or CMH2
LGRRC1	Logit common relative risk, column 1	CMH, CMH1, or CMH2
LGRRC2	Logit common relative risk, column 2	CMH, CMH1, or CMH2
LRCHI	Likelihood ratio chi-square	CHISQ
MCNEM	McNemar's test (2×2 tables)	AGREE
MEASURES	Gamma, Kendall's tau-b, Stuart's tau-c, Somers' $D(C|R)$ and $D(R|C)$, Pearson and Spearman correlations, lambda asymmetric $(C|R)$ and $(R|C)$, lambda symmetric, uncertainty coefficients $(C|R)$ and $(R|C)$, symmetric uncertainty coefficient; odds ratio and relative risks (2×2 tables)	MEASURES
MHCHI	Mantel-Haenszel chi-square	CHISQ
MHOR | COMOR	Mantel-Haenszel common odds ratio	CMH, CMH1, or CMH2
MHRRC1	Mantel-Haenszel common relative risk, column 1	CMH, CMH1, or CMH2
MHRRC2	Mantel-Haenszel common relative risk, column 2	CMH, CMH1, or CMH2
N	Number of nonmissing observations	
NMISS	Number of missing observations	
OR | ODDSRATIO	Odds ratio (2×2 tables)	MEASURES, OR, or RELRISK
PCHI	Chi-square goodness-of-fit test (one-way tables), Pearson chi-square (two-way tables)	CHISQ
PCORR	Pearson correlation coefficient	MEASURES
PHI	Phi coefficient	CHISQ
PLCORR	Polychoric correlation coefficient	PLCORR
RDIF1	Column 1 risk difference (row 1 – row 2)	RISKDIFF
RDIF2	Column 2 risk difference (row 1 – row 2)	RISKDIFF
RELRISK	Odds ratio and relative risks (2×2 tables)	MEASURES or RELRISK
RISKDIFF	Risks and risk differences (2×2 tables)	RISKDIFF
RISKDIFF1	Risks and risk difference, column 1	RISKDIFF
RISKDIFF2	Risks and risk difference, column 2	RISKDIFF
RRC1 | RELRISK1	Relative risk, column 1	MEASURES or RELRISK
RRC2 | RELRISK2	Relative risk, column 2	MEASURES or RELRISK
RSK1 | RISK1	Column 1 overall risk	RISKDIFF
RSK11 | RISK11	Column 1 risk for row 1	RISKDIFF
RSK12 | RISK12	Column 2 risk for row 1	RISKDIFF
RSK2 | RISK2	Column 2 overall risk	RISKDIFF
RSK21 | RISK21	Column 1 risk for row 2	RISKDIFF
RSK22 | RISK22	Column 2 risk for row 2	RISKDIFF
SCORR	Spearman correlation coefficient	MEASURES
SMDCR	Somers' $D(C|R)$	MEASURES
SMDRC	Somers' $D(R|C)$	MEASURES
STUTC | TAUC	Stuart's tau-c	MEASURES

Table 3.7 continued

Output Option	Output Data Set Statistics	Required TABLES Statement Option
TREND	Cochran-Armitage test for trend	TREND
TSYMM | BOWKER	Bowker's symmetry test	AGREE
U	Symmetric uncertainty coefficient	MEASURES
UCR	Uncertainty coefficient $(C\|R)$	MEASURES
URC	Uncertainty coefficient $(R\|C)$	MEASURES
WTKAPPA | WTKAP	Weighted kappa coefficient	AGREE

You can specify the following *output-options* in the OUTPUT statement.

AGREE

includes the following tests and measures of agreement in the output data set: McNemar's test (for 2×2 tables), Bowker's symmetry test, the simple kappa coefficient, and the weighted kappa coefficient. For multiway tables, the AGREE option also includes the following statistics in the output data set: overall simple and weighted kappa coefficients, tests for equal simple and weighted kappa coefficients, and Cochran's Q test.

The AGREE option in the TABLES statement requests computation of tests and measures of agreement. For more information, see the section "Tests and Measures of Agreement" on page 205.

AGREE statistics are computed only for square tables, where the number of rows equals the number of columns. PROC FREQ provides Bowker's symmetry test and weighted kappa coefficients only for tables larger than 2×2. (For 2×2 tables, Bowker's test is identical to McNemar's test, and the weighted kappa coefficient equals the simple kappa coefficient.) Cochran's Q is available for multiway 2×2 tables.

AJCHI

includes the continuity-adjusted chi-square in the output data set. The continuity-adjusted chi-square is available for 2×2 tables and is provided by the CHISQ option in the TABLES statement. For more information, see the section "Continuity-Adjusted Chi-Square Test" on page 158.

ALL

includes all statistics that are requested by the CHISQ, MEASURES, and CMH *output-options* in the output data set. ALL also includes the number of nonmissing observations, which you can request individually by specifying the N *output-option*.

BDCHI

includes the Breslow-Day test in the output data set. The Breslow-Day test for homogeneity of odds ratios is computed for multiway 2×2 tables and is provided by the CMH, CMH1, and CMH2 options in the TABLES statement. For more information, see the section "Breslow-Day Test for Homogeneity of the Odds Ratios" on page 218.

BINOMIAL

BIN

includes the binomial proportion estimate, confidence limits, and tests in the output data set. The BINOMIAL option in the TABLES statement requests computation of binomial statistics, which are available for one-way tables. For more information, see the section "Binomial Proportion" on page 170.

CHISQ

includes the following chi-square tests and measures in the output data set for two-way tables: Pearson chi-square, likelihood ratio chi-square, Mantel-Haenszel chi-square, phi coefficient, contingency coefficient, and Cramér's V. For 2×2 tables, CHISQ also includes Fisher's exact test and the continuity-adjusted chi-square in the output data set. For more information, see the section "Chi-Square Tests and Statistics" on page 155. For one-way tables, CHISQ includes the chi-square goodness-of-fit test in the output data set. For more information, see the section "Chi-Square Test for One-Way Tables" on page 155. The CHISQ option in the TABLES statement requests computation of these statistics.

If you specify the CHISQ(WARN=OUTPUT) option in the TABLES statement, the CHISQ option also includes the variable WARN_PCHI in the output data set. This variable indicates the validity warning for the asymptotic Pearson chi-square test.

CMH

includes the following Cochran-Mantel-Haenszel statistics in the output data set: correlation, row mean scores (ANOVA), and general association. For 2×2 tables, the CMH option also includes the Mantel-Haenszel and logit estimates of the common odds ratio and relative risks. For multiway (stratified) 2×2 tables, the CMH option includes the Breslow-Day test for homogeneity of odds ratios. The CMH option in the TABLES statement requests computation of these statistics. For more information, see the section "Cochran-Mantel-Haenszel Statistics" on page 212.

If you specify the CMH(MANTELFLEISS) option in the TABLES statement, the CMH option includes the Mantel-Fleiss analysis in the output data set. The variables MF_CMH and WARN_CMH contain the Mantel-Fleiss criterion and the warning indicator, respectively.

CMH1

includes the CMH statistics in the output data set, with the exception of the row mean scores (ANOVA) statistic and the general association statistic. The CMH1 option in the TABLES statement requests computation of these statistics. For more information, see the section "Cochran-Mantel-Haenszel Statistics" on page 212.

CMH2

includes the CMH statistics in the output data set, with the exception of the general association statistic. The CMH2 option in the TABLES statement requests computation of these statistics. For more information, see the section "Cochran-Mantel-Haenszel Statistics" on page 212.

CMHCOR

includes the Cochran-Mantel-Haenszel correlation statistic in the output data set. The CMH option in the TABLES statement requests computation of this statistic. For more information, see the section "Correlation Statistic" on page 214.

CMHGA

includes the Cochran-Mantel-Haenszel general association statistic in the output data set. The CMH option in the TABLES statement requests computation of this statistic. For more information, see the section "General Association Statistic" on page 215.

CMHRMS

includes the Cochran-Mantel-Haenszel row mean scores (ANOVA) statistic in the output data set. The CMH option in the TABLES statement requests computation of this statistic. For more information, see the section "ANOVA (Row Mean Scores) Statistic" on page 214.

COCHQ

includes Cochran's Q test in the output data set. The AGREE option in the TABLES statement requests computation of this test, which is available for multiway 2×2 tables. For more information, see the section "Cochran's Q Test" on page 212.

CONTGY

includes the contingency coefficient in the output data set. The CHISQ option in the TABLES statement requests computation of the contingency coefficient. For more information, see the section "Contingency Coefficient" on page 160.

CRAMV

includes Cramér's V in the output data set. The CHISQ option in the TABLES statement requests computation of Cramér's V. For more information, see the section "Cramér's V" on page 160.

EQKAP

includes the test for equal simple kappa coefficients in the output data set. The AGREE option in the TABLES statement requests computation of this test, which is available for multiway, square ($h \times r \times r$) tables. For more information, see the section "Tests for Equal Kappa Coefficients" on page 211.

EQOR

ZELEN

includes Zelen's exact test for equal odds ratios in the output data set. The EQOR option in the EXACT statement requests computation of this test, which is available for multiway 2×2 tables. For more information, see the section "Zelen's Exact Test for Equal Odds Ratios" on page 218.

EQWKP

includes the test for equal weighted kappa coefficients in the output data set. The AGREE option in the TABLES statement requests computation of this test. The test for equal weighted kappas is available for multiway, square ($h \times r \times r$) tables where $r > 2$. For more information, see the section "Tests for Equal Kappa Coefficients" on page 211.

FISHER

includes Fisher's exact test in the output data set. For 2×2 tables, the CHISQ option in the TABLES statement provides Fisher's exact test. For tables larger than 2×2, the FISHER option in the EXACT statement provides Fisher's exact test. For more information, see the section "Fisher's Exact Test" on page 158.

GAMMA

includes the gamma statistic in the output data set. The MEASURES option in the TABLES statement requests computation of the gamma statistic. For more information, see the section "Gamma" on page 162.

GS

GAILSIMON

includes the Gail-Simon test for qualitative interaction in the output data set. The CMH(GAILSIMON) option in the TABLES statement requests computation of this test. For more information, see the section "Gail-Simon Test for Qualitative Interactions" on page 221.

JT

includes the Jonckheere-Terpstra test in the output data set. The JT option in the TABLES statement requests the Jonckheere-Terpstra test. For more information, see the section "Jonckheere-Terpstra Test" on page 203.

KAPPA

includes the simple kappa coefficient in the output data set. The AGREE option in the TABLES statement requests computation of kappa, which is available for square tables (where the number of rows equals the number of columns). For multiway square tables, the KAPPA option also includes the overall kappa coefficient in the output data set. For more information, see the sections "Simple Kappa Coefficient" on page 206 and "Overall Kappa Coefficient" on page 211.

KENTB

TAUB

includes Kendall's tau-b in the output data set. The MEASURES option in the TABLES statement requests computation of Kendall's tau-b. For more information, see the section "Kendall's Tau-b" on page 162.

LAMCR

includes the asymmetric lambda $\lambda(C|R)$ in the output data set. The MEASURES option in the TABLES statement requests computation of lambda. For more information, see the section "Lambda (Asymmetric)" on page 167.

LAMDAS

includes the symmetric lambda in the output data set. The MEASURES option in the TABLES statement requests computation of lambda. For more information, see the section "Lambda (Symmetric)" on page 168.

LAMRC

includes the asymmetric lambda $\lambda(R|C)$ in the output data set. The MEASURES option in the TABLES statement requests computation of lambda. For more information, see the section "Lambda (Asymmetric)" on page 167.

LGOR

includes the logit estimate of the common odds ratio in the output data set. The CMH option in the TABLES statement requests computation of this statistic, which is available for 2×2 tables. For more information, see the section "Adjusted Odds Ratio and Relative Risk Estimates" on page 216.

LGRRC1

includes the logit estimate of the common relative risk (column 1) in the output data set. The CMH option in the TABLES statement requests computation of this statistic, which is available for 2×2 tables. For more information, see the section "Adjusted Odds Ratio and Relative Risk Estimates" on page 216.

LGRRC2

includes the logit estimate of the common relative risk (column 2) in the output data set. The CMH option in the TABLES statement requests computation of this statistic, which is available for 2×2 tables. For more information, see the section "Adjusted Odds Ratio and Relative Risk Estimates" on page 216.

LRCHI

includes the likelihood ratio chi-square in the output data set. The CHISQ option in the TABLES statement requests computation of the likelihood ratio chi-square. For more information, see the section "Likelihood Ratio Chi-Square Test" on page 157.

MCNEM

includes McNemar's test (for 2 × 2 tables) in the output data set. The AGREE option in the TABLES statement requests computation of McNemar's test. For more information, see the section "McNemar's Test" on page 205.

MEASURES

includes the following measures of association in the output data set: gamma, Kendall's tau-b, Stuart's tau-c, Somers' $D(C|R)$, Somers' $D(R|C)$, Pearson and Spearman correlation coefficients, lambda (symmetric and asymmetric), and uncertainty coefficients (symmetric and asymmetric). For 2×2 tables, the MEASURES option also includes the odds ratio, column 1 relative risk, and column 2 relative risk. The MEASURES option in the TABLES statement requests computation of these statistics. For more information, see the section "Measures of Association" on page 160.

MHCHI

includes the Mantel-Haenszel chi-square in the output data set. The CHISQ option in the TABLES statement requests computation of the Mantel-Haenszel chi-square. For more information, see the section "Mantel-Haenszel Chi-Square Test" on page 158.

MHOR
COMOR

includes the Mantel-Haenszel estimate of the common odds ratio in the output data set. The CMH option in the TABLES statement requests computation of this statistic, which is available for 2 × 2 tables. For more information, see the section "Adjusted Odds Ratio and Relative Risk Estimates" on page 216.

MHRRC1

includes the Mantel-Haenszel estimate of the common relative risk (column 1) in the output data set. The CMH option in the TABLES statement requests computation of this statistic, which is available for 2 × 2 tables. For more information, see the section "Adjusted Odds Ratio and Relative Risk Estimates" on page 216.

MHRRC2

includes the Mantel-Haenszel estimate of the common relative risk (column 2) in the output data set. The CMH option in the TABLES statement requests computation of this statistic, which is available for 2 × 2 tables. For more information, see the section "Adjusted Odds Ratio and Relative Risk Estimates" on page 216.

N

includes the number of nonmissing observations in the output data set.

NMISS

includes the number of missing observations in the output data set. For more information, see the section "Missing Values" on page 150.

OR
ODDSRATIO
RROR

    includes the odds ratio (for 2 × 2 tables) in the output data set. The MEASURES, OR, and RELRISK options in the TABLES statement request this statistic. For more information, see the section "Odds Ratio" on page 192.

PCHI

    includes the Pearson chi-square in the output data set for two-way tables. For more information, see the section "Pearson Chi-Square Test for Two-Way Tables" on page 156. For one-way tables, the PCHI option includes the chi-square goodness-of-fit test in the output data set. For more information, see the section "Chi-Square Test for One-Way Tables" on page 155. The CHISQ option in the TABLES statement requests computation of these statistics.

    If you specify the CHISQ(WARN=OUTPUT) option in the TABLES statement, the PCHI option also includes the variable WARN_PCHI in the output data set. This variable indicates the validity warning for the asymptotic Pearson chi-square test.

PCORR

    includes the Pearson correlation coefficient in the output data set. The MEASURES option in the TABLES statement requests computation of the Pearson correlation. For more information, see the section "Pearson Correlation Coefficient" on page 164.

PHI

    includes the phi coefficient in the output data set. The CHISQ option in the TABLES statement requests computation of the phi coefficient. For more information, see the section "Phi Coefficient" on page 159.

PLCORR

    includes the polychoric correlation coefficient in the output data set. For 2 × 2 tables, this statistic is known as the tetrachoric correlation coefficient. The PLCORR option in the TABLES statement requests computation of the polychoric correlation. For more information, see the section "Polychoric Correlation" on page 166.

RDIF1

    includes the column 1 risk difference (row 1 − row 2) in the output data set. The RISKDIFF option in the TABLES statement requests computation of risks and risk differences, which are available for 2 × 2 tables. For more information, see the section "Risks and Risk Differences" on page 178.

RDIF2

    includes the column 2 risk difference (row 1 − row 2) in the output data set. The RISKDIFF option in the TABLES statement requests computation of risks and risk differences, which are available for 2 × 2 tables. For more information, see the section "Risks and Risk Differences" on page 178.

RELRISK

    includes the column 1 and column 2 relative risks (for 2 × 2 tables) in the output data set. The MEASURES and RELRISK options in the TABLES statement request these statistics. For more information, see the section "Relative Risks" on page 195.

RISKDIFF

includes risks (binomial proportions) and risk differences for 2 × 2 tables in the output data set. These statistics include the row 1 risk, row 2 risk, total (overall) risk, and risk difference (row 1 – row 2) for column 1 and column 2. The RISKDIFF option in the TABLES statement requests computation of these statistics. For more information, see the section "Risks and Risk Differences" on page 178.

RISKDIFF1

includes column 1 risks (binomial proportions) and risk differences for 2 × 2 tables in the output data set. These statistics include the row 1 risk, row 2 risk, total (overall) risk, and risk difference (row 1 – row 2). The RISKDIFF option in the TABLES statement requests computation of these statistics. For more information, see the section "Risks and Risk Differences" on page 178.

RISKDIFF2

includes column 2 risks (binomial proportions) and risk differences for 2 × 2 tables in the output data set. These statistics include the row 1 risk, row 2 risk, total (overall) risk, and risk difference (row 1 – row 2). The RISKDIFF option in the TABLES statement requests computation of these statistics. For more information, see the section "Risks and Risk Differences" on page 178.

RRC1

RELRISK1

includes the column 1 relative risk in the output data set. The MEASURES and RELRISK options in the TABLES statement request relative risks, which are available for 2 × 2 tables. For more information, see the section "Odds Ratio and Relative Risks for 2 × 2 Tables" on page 192.

RRC2

RELRISK2

includes the column 2 relative risk in the output data set. The MEASURES and RELRISK options in the TABLES statement request relative risks, which are available for 2 × 2 tables. For more information, see the section "Odds Ratio and Relative Risks for 2 × 2 Tables" on page 192.

RSK1

RISK1

includes the overall column 1 risk in the output data set. The RISKDIFF option in the TABLES statement requests computation of risks and risk differences, which are available for 2 × 2 tables. For more information, see the section "Risks and Risk Differences" on page 178.

RSK11

RISK11

includes the column 1 risk for row 1 in the output data set. The RISKDIFF option in the TABLES statement requests computation of risks and risk differences, which are available for 2 × 2 tables. For more information, see the section "Risks and Risk Differences" on page 178.

RSK12

RISK12

includes the column 2 risk for row 1 in the output data set. The RISKDIFF option in the TABLES statement requests computation of risks and risk differences, which are available for 2 × 2 tables. For more information, see the section "Risks and Risk Differences" on page 178.

RSK2

RISK2

includes the overall column 2 risk in the output data set. The RISKDIFF option in the TABLES statement requests computation of risks and risk differences. For more information, see the section "Risks and Risk Differences" on page 178.

RSK21

RISK21

includes the column 1 risk for row 2 in the output data set. The RISKDIFF option in the TABLES statement requests computation of risks and risk differences, which are available for 2×2 tables. For more information, see the section "Risks and Risk Differences" on page 178.

RSK22

RISK22

includes the column 2 risk for row 2 in the output data set. The RISKDIFF option in the TABLES statement requests computation of risks and risk differences, which are available for 2×2 tables. For more information, see the section "Risks and Risk Differences" on page 178.

SCORR

includes the Spearman correlation coefficient in the output data set. The MEASURES option in the TABLES statement requests computation of the Spearman correlation. For more information, see the section "Spearman Rank Correlation Coefficient" on page 165.

SMDCR

includes Somers' $D(C|R)$ in the output data set. The MEASURES option in the TABLES statement requests computation of Somers' D. For more information, see the section "Somers' D" on page 163.

SMDRC

includes Somers' $D(R|C)$ in the output data set. The MEASURES option in the TABLES statement requests computation of Somers' D. For more information, see the section "Somers' D" on page 163.

STUTC

TAUC

includes Stuart's tau-c in the output data set. The MEASURES option in the TABLES statement requests computation of tau-c. For more information, see the section "Stuart's Tau-c" on page 163.

TREND

includes the Cochran-Armitage test for trend in the output data set. The TREND option in the TABLES statement requests computation of the trend test. This test is available for tables of dimension $2 \times C$ or $R \times 2$. For more information, see the section "Cochran-Armitage Test for Trend" on page 202.

TSYMM

BOWKER

includes Bowker's symmetry test in the output data set. The AGREE option in the TABLES statement requests computation of Bowker's test. For more information, see the section "Bowker's Symmetry Test" on page 206.

U

includes the uncertainty coefficient (symmetric) in the output data set. The MEASURES option in the TABLES statement requests computation of the uncertainty coefficient. For more information, see the section "Uncertainty Coefficient (Symmetric)" on page 169.

UCR

includes the asymmetric uncertainty coefficient $U(C|R)$ in the output data set. The MEASURES option in the TABLES statement requests computation of the uncertainty coefficient. For more information, see the section "Uncertainty Coefficients (Asymmetric)" on page 169.

URC

includes the asymmetric uncertainty coefficient $U(R|C)$ in the output data set. The MEASURES option in the TABLES statement requests computation of the uncertainty coefficient. For more information, see the section "Uncertainty Coefficients (Asymmetric)" on page 169.

WTKAPPA

WTKAP

includes the weighted kappa coefficient in the output data set. The AGREE option in the TABLES statement requests computation of weighted kappa, which is available for square tables larger than 2×2. For multiway tables, the WTKAPPA option also includes the overall weighted kappa coefficient in the output data set. For more information, see the sections "Weighted Kappa Coefficient" on page 208 and "Overall Kappa Coefficient" on page 211.

TABLES Statement

TABLES *requests* < / *options* > ;

The TABLES statement requests one-way to *n*-way frequency and crosstabulation tables and statistics for those tables.

If you omit the TABLES statement, PROC FREQ generates one-way frequency tables for all data set variables that are not listed in the other statements.

The following argument is required in the TABLES statement.

requests

specify the frequency and crosstabulation tables to produce. A request is composed of one variable name or several variable names separated by asterisks. To request a one-way frequency table, use a single variable. To request a two-way crosstabulation table, use an asterisk between two variables. To request a multiway table (an *n*-way table, where $n > 2$), separate the variables with asterisks. The unique values of these variables form the rows, columns, and strata of the table. You can include up to 50 variables in a single multiway table request.

For two-way to multiway tables, the values of the last variable form the crosstabulation table columns, and the values of the next-to-last variable form the rows. Each level (or combination of levels) of the other variables forms one stratum. PROC FREQ produces a separate crosstabulation table for each stratum. For example, a specification of A*B*C*D in a TABLES statement produces *k* tables, where *k* is the number of different combinations of values for A and B. Each table lists the values for C down the side and the values for D across the top.

You can use multiple TABLES statements in the PROC FREQ step. PROC FREQ builds all the table requests in one pass of the data, so that there is essentially no loss of efficiency. You can also specify any number of table requests in a single TABLES statement. To specify multiple table requests quickly, use a grouping syntax by placing parentheses around several variables and joining other variables or variable combinations. For example, the statements shown in Table 3.8 illustrate grouping syntax.

Table 3.8 Grouping Syntax

TABLES Request	Equivalent to
A*(B C)	A*B A*C
(A B)*(C D)	A*C B*C A*D B*D
(A B C)*D	A*D B*D C*D
A - - C	A B C
(A - - C)*D	A*D B*D C*D

The TABLES statement variables are one or more variables from the DATA= input data set. These variables can be either character or numeric, but the procedure treats them as categorical variables. PROC FREQ uses the formatted values of the TABLES variable to determine the categorical variable levels. So if you assign a format to a variable with a FORMAT statement, PROC FREQ formats the values before dividing observations into the levels of a frequency or crosstabulation table. See the FORMAT procedure in the *SAS Visual Data Management and Utility Procedures Guide* and the FORMAT statement and SAS formats in *SAS Formats and Informats: Reference*.

If you use PROC FORMAT to create a user-written format that combines missing and nonmissing values into one category, PROC FREQ treats the entire category of formatted values as missing. See the discussion in the section "Grouping with Formats" on page 149 for more information.

By default, the frequency or crosstabulation table lists the values of both character and numeric variables in ascending order based on internal (unformatted) variable values. You can change the order of the values in the table by specifying the ORDER= option in the PROC FREQ statement. To list the values in ascending order by formatted value, use ORDER=FORMATTED.

Without Options

If you request a one-way frequency table for a variable without specifying options, PROC FREQ produces frequencies, cumulative frequencies, percentages of the total frequency, and cumulative percentages for each value of the variable. If you request a two-way or an *n*-way crosstabulation table without specifying any options, PROC FREQ produces crosstabulation tables that include cell frequencies, cell percentages of the total frequency, cell percentages of row frequencies, and cell percentages of column frequencies. The procedure excludes observations with missing values from the table but displays the total frequency of missing observations following each table.

Options

Table 3.9 lists the *options* available in the TABLES statement. Descriptions of the *options* follow in alphabetical order.

Table 3.9 TABLES Statement Options

Option	Description
Control Statistical Analysis	
AGREE	Requests tests and measures of classification agreement
ALL	Requests tests and measures of association produced by the CHISQ, MEASURES, and CMH options
ALPHA=	Sets confidence level for confidence limits
BINOMIAL | BIN	Requests binomial proportions, confidence limits, and tests for one-way tables
CHISQ	Requests chi-square tests and measures based on chi-square
CL	Requests confidence limits for MEASURES statistics
CMH	Requests all Cochran-Mantel-Haenszel statistics
CMH1	Requests CMH correlation statistic, adjusted odds ratios, and adjusted relative risks
CMH2	Requests CMH correlation and row mean scores (ANOVA) statistics, adjusted odds ratios, and adjusted relative risks
COMMONRISKDIFF	Requests common risk difference for $h \times 2 \times 2$ tables
FISHER	Requests Fisher's exact test for tables larger than 2×2
GAILSIMON	Requests Gail-Simon test for qualitative interactions
JT	Requests Jonckheere-Terpstra test
MEASURES	Requests measures of association
MISSING	Treats missing values as nonmissing
OR	Requests the odds ratio for 2×2 tables
PLCORR	Requests polychoric correlation
RELRISK	Requests relative risks for 2×2 tables
RISKDIFF	Requests risks and risk differences for 2×2 tables
SCORES=	Specifies type of row and column scores
TREND	Requests Cochran-Armitage test for trend
Control Additional Table Information	
CELLCHI2	Displays cell contributions to the Pearson chi-square statistic
CUMCOL	Displays cumulative column percentages
DEVIATION	Displays deviations of cell frequencies from expected values
EXPECTED	Displays expected cell frequencies
MISSPRINT	Displays missing value frequencies
PEARSONRES	Displays Pearson residuals in the CROSSLIST table
PRINTKWTS	Displays kappa coefficient weights
SCOROUT	Displays row and column scores
SPARSE	Includes all possible combinations of variable levels in the LIST table and OUT= data set
STDRES	Displays standardized residuals in the CROSSLIST table
TOTPCT	Displays percentages of total frequency for n-way tables ($n > 2$)
Control Displayed Output	
CONTENTS=	Specifies contents label for crosstabulation tables
CROSSLIST	Displays crosstabulation tables in ODS column format

Table 3.9 *continued*

Option	Description
FORMAT=	Formats frequencies in crosstabulation tables
LIST	Displays two-way to *n*-way tables in list format
MAXLEVELS=	Specifies maximum number of levels to display in one-way tables
NOCOL	Suppresses display of column percentages
NOCUM	Suppresses display of cumulative frequencies and percentages
NOFREQ	Suppresses display of frequencies
NOPERCENT	Suppresses display of percentages
NOPRINT	Suppresses display of crosstabulation tables but displays statistics
NOROW	Suppresses display of row percentages
NOSPARSE	Suppresses zero-frequency levels in the CROSSLIST table, LIST table, and OUT= data set
NOWARN	Suppresses log warning message for the chi-square test
Produce Statistical Graphics	
PLOTS=	Requests plots from ODS Graphics
Create an Output Data Set	
OUT=	Names an output data set to contain frequency counts
OUTCUM	Includes cumulative frequencies and percentages in the output data set for one-way tables
OUTEXPECT	Includes expected frequencies in the output data set
OUTPCT	Includes row, column, and two-way table percentages in the output data set

You can specify the following *options* in a TABLES statement.

AGREE < (*agree-options*) >

requests tests and measures of classification agreement for square tables. This option provides the simple and weighted kappa coefficients along with their standard errors and confidence limits. For multiway tables, this option also produces the overall simple and weighted kappa coefficients (along with their standard errors and confidence limits) and tests for equal kappas among strata. For 2×2 tables, this option provides McNemar's test; for square tables that have more than two response categories (levels), this option provides Bowker's symmetry test. For multiway tables that have two response categories, this option also produces Cochran's Q test. For more information, see the section "Tests and Measures of Agreement" on page 205.

Measures of agreement can be computed only for square tables, where the number of rows equals the number of columns. If your table is not square because some observations have weights of 0, you can specify the ZEROS option in the WEIGHT statement to include these observations in the analysis. For more information, see the section "Tables with Zero-Weight Rows or Columns" on page 212.

For 2×2 tables, the weighted kappa coefficient is equivalent to the simple kappa coefficient, and PROC FREQ displays only analyses for the simple kappa coefficient.

You can specify the confidence level in the ALPHA= option. By default, ALPHA=0.05, which produces 95% confidence limits.

You can specify the EXACT statement to request McNemar's exact test (for 2 × 2 tables), an exact symmetry test, and exact tests for the simple and weighted kappa coefficients. For more information, see the section "Exact Statistics" on page 221.

You can specify the following *agree-options*:

AC1
> requests the AC1 agreement coefficient. For more information, see the section "AC1 Agreement Coefficient" on page 211.

DFSYM=*df* **| ADJUST**
> controls the degrees of freedom for Bowker's symmetry test. You can specify the value of the degrees of freedom (*df*), or you can specify DFSYM=ADJUST to adjust the degrees of freedom for empty table cells. The value of *df* must be a positive number. By default, *df* is $R(R-1)/2$, where R is the dimension of the two-way table.
>
> When you specify DFSYM=ADJUST, the degrees of freedom are reduced by the number of off-diagonal table-cell pairs that have a total frequency of 0. By default, the degrees of freedom count all off-diagonal table-cell pairs. For more information, see the section "Bowker's Symmetry Test" on page 206.

KAPPADETAILS
DETAILS
> displays the "Kappa Details" table, which includes the following statistics for the simple kappa coefficient: observed agreement, chance-expected agreement, maximum kappa, and the B_n measure. If the two-way table is 2 × 2, the "Kappa Details" table also includes the prevalence index and the bias index. For more information, see the section "Simple Kappa Coefficient" on page 206.
>
> If the two-way table is larger than 2 × 2, this option also displays the "Weighted Kappa Details" table, which includes the observed agreement and chance-expected agreement components of the weighted kappa coefficient. For more information, see the section "Weighted Kappa Coefficient" on page 208.

MNULLRATIO=*value*
> specifies the null *value* of the ratio of discordant proportions for McNemar's test. By default, MNULLRATIO=1. For more information, see the section "McNemar's Test" on page 205.

NULLKAPPA=*value*
> requests the simple kappa coefficient test and specifies the null *value* for the test. The null value must be between –1 and 1. By default, NULLKAPPA=0. For more information, see the section "Simple Kappa Coefficient" on page 206.
>
> This option is not available when you specify the KAPPA option in the EXACT statement, which requests an exact test for the kappa coefficient.

NULLWTKAPPA=*value*
> requests the weighted kappa coefficient test and specifies the null *value* for the test. The null value must be between –1 and 1. By default, NULLWTKAPPA=0. For more information, see the section "Weighted Kappa Coefficient" on page 208.
>
> This option is not available when you specify the WTKAPPA option in the EXACT statement, which requests an exact test for the weighted kappa coefficient.

PABAK
> requests the prevalence-adjusted bias-adjusted kappa coefficient. For more information, see the section "Prevalence-Adjusted Bias-Adjusted Kappa" on page 210.

PRINTKWTS
> displays the agreement weights that PROC FREQ uses to compute the weighted kappa coefficient. Agreement weights reflect the relative agreement between pairs of variable levels. By default, PROC FREQ uses Cicchetti-Allison agreement weights. If you specify the WT=FC option, the procedure uses Fleiss-Cohen agreement weights. For more information, see the section "Weighted Kappa Coefficient" on page 208.

TABLES=RESTORE
> displays agreement tables (which are produced by the AGREE option) in factoid (label-value) format, which is the format of these tables in releases before SAS/STAT 14.2. Beginning in SAS/STAT 14.3, PROC FREQ displays all agreement tables in tabular format (instead of factoid format) by default.
>
> In SAS/STAT 14.2, PROC FREQ displays agreement tables in tabular format (instead of factoid format) by default when you specify any of the following *agree-options*: AC1, KAPPADETAILS, NULLKAPPA=, NULLWTKAPPA=, PABAK, or WTKAPPADETAILS.

WT=FC
> specifies Fleiss-Cohen agreement weights in the computation of the weighted kappa coefficient. Agreement weights reflect the relative agreement between pairs of variable levels. By default, PROC FREQ uses Cicchetti-Allison agreement weights to compute the weighted kappa coefficient. For more information, see the section "Weighted Kappa Coefficient" on page 208.

WTKAPPADETAILS
> displays the "Weighted Kappa Details" table, which includes the observed agreement and chance-expected agreement components of the weighted kappa coefficient. This information is available for two-way tables that are larger than 2×2. For more information, see the section "Weighted Kappa Coefficient" on page 208.

ALL
> requests all tests and measures that are produced by the CHISQ, MEASURES, and CMH options. You can control the number of CMH statistics to compute by specifying the CMH1 or CMH2 option.

ALPHA=α

specifies the level of confidence limits. The value of α must be between 0 and 1; a confidence level of α produces $100(1-\alpha)\%$ confidence limits. By default ALPHA=0.05, which produces 95% confidence limits.

This option applies to confidence limits that you request in the TABLES statement. The ALPHA= option in the EXACT statement applies to confidence limits for Monte Carlo estimates of exact *p*-values, which you request by specifying the MC option in the EXACT statement.

BINOMIAL < (*binomial-options*) >

BIN < (*binomial-options*) >

requests the binomial proportion for one-way tables. When you specify this option, by default PROC FREQ provides the asymptotic standard error, asymptotic Wald and exact (Clopper-Pearson) confidence limits, and the asymptotic equality test for the binomial proportion.

You can specify *binomial-options* in parentheses after the BINOMIAL option. The LEVEL= *binomial-option* identifies the variable level for which to compute the proportion. If you do not specify this option, PROC FREQ computes the proportion for the first level that appears in the one-way frequency table. The P= *binomial-option* specifies the null proportion for the binomial tests. If you do not specify this option, PROC FREQ uses 0.5 as the null proportion for the binomial tests.

You can also specify *binomial-options* to request additional tests and confidence limits for the binomial proportion. The EQUIV, NONINF, and SUP *binomial-options* request tests of equivalence, noninferiority, and superiority, respectively. The CL= *binomial-option* requests confidence limits for the binomial proportion.

You can specify the level for the binomial confidence limits in the ALPHA= option. By default, ALPHA=0.05, which produces 95% confidence limits. As part of the noninferiority, superiority, and equivalence analyses, PROC FREQ provides null-based equivalence limits that have a confidence coefficient of $100(1-2\alpha)\%$ (Schuirmann 1999). In these analyses, the default of ALPHA=0.05 produces 90% equivalence limits. For more information, see the sections "Noninferiority Test" on page 174 and "Equivalence Test" on page 176.

To request exact tests for the binomial proportion, you can specify the BINOMIAL option in the EXACT statement. PROC FREQ computes exact *p*-values for all binomial tests that you request, which can include noninferiority, superiority, and equivalence tests, in addition to the equality test that the BINOMIAL option produces by default.

For more information, see the section "Binomial Proportion" on page 170.

Table 3.10 summarizes the *binomial-options*.

Table 3.10 BINOMIAL Options

Option	Description
CORRECT	Requests continuity correction
LEVEL=	Specifies the variable level
OUTLEVEL	Includes the level in the output data sets
P=	Specifies the null proportion
Request Confidence Limits	
CL=AGRESTICOULL | AC	Requests Agresti-Coull confidence limits
CL=BLAKER	Requests Blaker confidence limits
CL=EXACT | CLOPPERPEARSON	Requests exact (Clopper-Pearson) confidence limits
CL=JEFFREYS	Requests Jeffreys confidence limits
CL=LIKELIHOODRATIO | LR	Requests likelihood ratio confidence limits
CL=LOGIT	Requests logit confidence limits
CL=MIDP	Requests exact mid-p confidence limits
CL=WALD	Requests Wald confidence limits
CL=WILSON | SCORE	Requests Wilson (score) confidence limits
Request Tests	
EQUIV | EQUIVALENCE	Requests an equivalence test
MARGIN=	Specifies the test margin
NONINF | NONINFERIORITY	Requests a noninferiority test
SUP | SUPERIORITY	Requests a superiority test
VAR=NULL | SAMPLE	Specifies the test variance

You can specify the following *binomial-options*:

CL=*type* | (*types*)

requests confidence limits for the binomial proportion. You can specify one or more *types* of confidence limits. When you specify only one *type*, you can omit the parentheses around the request. PROC FREQ displays the confidence limits in the "Binomial Confidence Limits" table.

The ALPHA= option determines the level of the confidence limits that the CL= *binomial-option* provides. By default, ALPHA=0.05, which produces 95% confidence limits for the binomial proportion.

You can specify the CL= *binomial-option* with or without requests for binomial tests. The confidence limits that CL= produces do not depend on the tests that you request and do not use the value of the test margin (which you can specify in the MARGIN= *binomial-option*).

If you do not specify the CL= *binomial-option*, the BINOMIAL option displays Wald and exact (Clopper-Pearson) confidence limits in the "Binomial Proportion" table.

You can specify the following *types*:

AGRESTICOULL

AC

requests Agresti-Coull confidence limits for the binomial proportion. For more information, see the section "Agresti-Coull Confidence Limits" on page 171.

BLAKER

requests Blaker confidence limits for the binomial proportion. For more information, see the section "Blaker Confidence Limits" on page 171.

EXACT
CLOPPERPEARSON

requests exact (Clopper-Pearson) confidence limits for the binomial proportion. For more information, see the section "Exact (Clopper-Pearson) Confidence Limits" on page 170.

If you do not specify the CL= *binomial-option*, PROC FREQ displays Wald and exact (Clopper-Pearson) confidence limits in the "Binomial Proportion" table. To request exact tests for the binomial proportion, you can specify the BINOMIAL option in the EXACT statement.

JEFFREYS

requests Jeffreys confidence limits for the binomial proportion. For more information, see the section "Jeffreys Confidence Limits" on page 172.

LIKELIHOODRATIO
LR

requests likelihood ratio confidence limits for the binomial proportion. For more information, see the section "Likelihood Ratio Confidence Limits" on page 172.

LOGIT

requests logit confidence limits for the binomial proportion. For more information, see the section "Logit Confidence Limits" on page 172.

MIDP

requests exact mid-p confidence limits for the binomial proportion. For more information, see the section "Mid-p Confidence Limits" on page 172.

WALD < (CORRECT) >

requests Wald confidence limits for the binomial proportion. For more information, see the section "Wald Confidence Limits" on page 170.

If you specify CL=WALD(CORRECT), the Wald confidence limits include a continuity correction. If you specify the CORRECT *binomial-option*, both the Wald confidence limits and the Wald tests include continuity corrections.

If you do not specify the CL= *binomial-option*, PROC FREQ displays Wald and exact (Clopper-Pearson) confidence limits in the "Binomial Proportion" table.

WILSON < (CORRECT) >
SCORE < (CORRECT) >

requests Wilson confidence limits for the binomial proportion. These are also known as *score* confidence limits. For more information, see the section "Wilson (Score) Confidence Limits" on page 173.

If you specify CL=WILSON(CORRECT) or the CORRECT *binomial-option*, the Wilson confidence limits include a continuity correction.

CORRECT
> includes a continuity correction in the Wald confidence limits, Wald tests, and Wilson confidence limits.
>
> You can request continuity corrections individually for Wald or Wilson confidence limits by specifying the CL=WALD(CORRECT) or CL=WILSON(CORRECT) *binomial-option*, respectively.

EQUIV
EQUIVALENCE
> requests a test of equivalence for the binomial proportion. For more information, see the section "Equivalence Test" on page 176. You can specify the equivalence test margins, the null proportion, and the variance type in the MARGIN=, P=, and VAR= *binomial-options*, respectively. To request an exact equivalence test, you can specify the BINOMIAL option in the EXACT statement.

LEVEL=ature*level-number* | *'level-value'*
> specifies the variable level for the binomial proportion. You can specify the *level-number*, which is the order in which the level appears in the one-way frequency table. Or you can specify the *level-value*, which is the formatted value of the variable level. The *level-number* must be a positive integer. You must enclose the *level-value* in single quotes.
>
> By default, PROC FREQ computes the binomial proportion for the first variable level that appears in the one-way frequency table.

MARGIN=*value* | (*lower*, *upper*)
> specifies the margin for the noninferiority, superiority, and equivalence tests, which you can request by specifying the NONINF, SUP, and EQUIV *binomial-options*, respectively. By default, MARGIN=0.2.
>
> For noninferiority and superiority tests, specify a single *value* in the MARGIN= option. The MARGIN= *value* must be a positive number. You can specify *value* as a number between 0 and 1. Or you can specify *value* in percentage form as a number between 1 and 100, and PROC FREQ converts that number to a proportion. PROC FREQ treats the value 1 as 1%.
>
> For noninferiority and superiority tests, the test limits must be between 0 and 1. The limits are determined by the null proportion value (which you can specify in the P= *binomial-option*) and by the margin value. The noninferiority limit is the null proportion minus the margin. By default, the null proportion is 0.5 and the margin is 0.2, which produces a noninferiority limit of 0.3. The superiority limit is the null proportion plus the margin, which is 0.7 by default.
>
> For an equivalence test, you can specify a single MARGIN= *value*, or you can specify both *lower* and *upper* values. If you specify a single MARGIN= *value*, it must be a positive number, as described previously. If you specify a single MARGIN= *value* for an equivalence test, PROC FREQ uses –*value* as the lower margin and *value* as the upper margin for the test. If you specify both *lower* and *upper* values for an equivalence test, you can specify them in proportion form as numbers between –1 and 1. Or you can specify them in percentage form as numbers between –100 and 100, and PROC FREQ converts the numbers to proportions. The value of *lower* must be less than the value of *upper*.
>
> The equivalence limits must be between 0 and 1. The equivalence limits are determined by the null proportion value (which you can specify in the P= *binomial-option*) and by the margin values. The lower equivalence limit is the null proportion plus the lower margin. By default, the null

proportion is 0.5 and the lower margin is –0.2, which produces a lower equivalence limit of 0.3. The upper equivalence limit is the null proportion plus the upper margin, which is 0.7 by default.

For more information, see the sections "Noninferiority Test" on page 174 and "Equivalence Test" on page 176.

NONINF
NONINFERIORITY

    requests a test of noninferiority for the binomial proportion. For more information, see the section "Noninferiority Test" on page 174. You can specify the noninferiority test margin, the null proportion, and the variance type in the MARGIN=, P=, and VAR= *binomial-options*, respectively. To request an exact noninferiority test, you can specify the BINOMIAL option in the EXACT statement.

OUTLEVEL

    includes the variables LevelNumber and LevelValue in all ODS output data sets that PROC FREQ produces when you specify the BINOMIAL option in the TABLES statement. The OUTLEVEL option also includes the variables LevelNumber and LevelValue in the statistics output data set that PROC FREQ produces when you specify the BINOMIAL option in the OUTPUT statement.

    The LevelNumber and LevelValue variables identify the analysis variable level for which PROC FREQ computes the binomial proportion. The value of LevelNumber is the order of the level in the one-way frequency table. The value of LevelValue is the formatted value of the level. You can specify the OUTLEVEL *binomial-option* with or without the LEVEL= *binomial-option*.

P=*value*

    specifies the null hypothesis proportion for the binomial tests. The null proportion *value* must be a positive number. You can specify *value* as a number between 0 and 1. Or you can specify *value* in percentage form (as a number between 1 and 100), and PROC FREQ converts that number to a proportion. PROC FREQ treats the value 1 as 1%. By default, P=0.5.

SUP
SUPERIORITY

    requests a test of superiority for the binomial proportion. For more information, see the section "Superiority Test" on page 176. You can specify the superiority test margin, the null proportion, and the variance type in the MARGIN=, P=, and VAR= *binomial-options*, respectively. To request an exact superiority test, you can specify the BINOMIAL option in the EXACT statement.

VAR=NULL | SAMPLE

    specifies the type of variance to use in the Wald tests of noninferiority, superiority, and equivalence. If you specify VAR=SAMPLE, PROC FREQ computes the variance estimate by using the sample proportion. If you specify VAR=NULL, PROC FREQ computes a test-based variance by using the null hypothesis proportion (which you can specify in the P= *binomial-option*). For more information, see the sections "Noninferiority Test" on page 174 and "Equivalence Test" on page 176. The default is VAR=SAMPLE.

CELLCHI2

displays each table cell's contribution to the Pearson chi-square statistic in the crosstabulation table. The cell chi-square is computed as $(frequency - expected)^2 / expected$, where *frequency* is the table cell frequency (count) and *expected* is the expected cell frequency, which is computed under the null

110 ✦ *Chapter 3: The FREQ Procedure*

hypothesis that the row and column variables are independent. For more information, see the section "Pearson Chi-Square Test for Two-Way Tables" on page 156. This option has no effect for one-way tables or for tables that are displayed in list format (which you can request by specifying the LIST option).

CHISQ < (*chisq-options*) >

requests chi-square tests of homogeneity or independence and measures of association that are based on the chi-square statistic. For two-way tables, the chi-square tests include the Pearson chi-square, likelihood ratio chi-square, and Mantel-Haenszel chi-square tests. The chi-square measures include the phi coefficient, contingency coefficient, and Cramér's V. For 2×2 tables, the CHISQ option also provides Fisher's exact test and the continuity-adjusted chi-square test. For more information, see the section "Chi-Square Tests and Statistics" on page 155.

For one-way tables, the CHISQ option provides the Pearson chi-square goodness-of-fit test. You can also request the likelihood ratio goodness-of-fit test for one-way tables by specifying the LRCHI *chisq-option* in parentheses after the CHISQ option. By default, the one-way chi-square tests are based on the null hypothesis of equal proportions. Alternatively, you can provide null hypothesis proportions or frequencies by specifying the TESTP= or TESTF= *chisq-option*, respectively. See the section "Chi-Square Test for One-Way Tables" on page 155 for more information.

To request Fisher's exact test for tables larger than 2×2, specify the FISHER option in the EXACT statement. Exact *p*-values are also available for the Pearson, likelihood ratio, and Mantel-Haenszel chi-square tests. See the description of the EXACT statement for more information.

You can specify the following *chisq-options*:

DF=*df*

specifies the degrees of freedom for the chi-square tests. The value of *df* must not be 0. If the value of *df* is positive, PROC FREQ uses *df* as the degrees of freedom for the chi-square tests. If the value of *df* is negative, PROC FREQ uses *df* to adjust the default degrees of freedom for the chi-square tests.

By default for one-way tables, the value of *df* is $(n-1)$, where n is the number of variable levels in the table. By default for two-way tables, the value of *df* is $(r-1)(c-1)$, where r is the number of rows in the table and c is the number of columns. See the sections "Chi-Square Test for One-Way Tables" on page 155 and "Chi-Square Tests and Statistics" on page 155 for more information.

If you specify a negative value of *df*, PROC FREQ adjusts the default degrees of freedom by adding the (negative) value of *df* to the default value to produce the adjusted degrees of freedom. The adjusted degrees of freedom must be positive.

The DF= *chisq-option* specifies or adjusts the degrees of freedom for the following chi-square tests: the Pearson and likelihood ratio goodness-of-fit tests for one-way tables; and the Pearson, likelihood ratio, and Mantel-Haenszel chi-square tests for two-way tables.

LRCHI

requests the likelihood ratio goodness-of-fit test for one-way tables. See the section "Likelihood Ratio Chi-Square Test for One-Way Tables" on page 157 for more information.

By default, this test is based on the null hypothesis of equal proportions. You can provide null hypothesis proportions or frequencies by specifying the TESTP= or TESTF= *chisq-option*, respectively. You can request an exact likelihood ratio goodness-of-fit test by specifying the LRCHI option in the EXACT statement.

TESTF=(*values*)| *SAS-data-set*

specifies null hypothesis frequencies for the one-way chi-square goodness-of-fit tests. For more information, see the section "Chi-Square Test for One-Way Tables" on page 155. You can list the null frequencies as *values* in parentheses after TESTF=. Or you can provide the null frequencies in a secondary input data set by specifying TESTF=*SAS-data-set*. The TESTF=*SAS-data-set* cannot be the same data set that you specify in the DATA= option. You can specify only one TESTF= or TESTP= data set in a single invocation of the procedure.

If you list the null frequencies as *values*, you can separate the *values* with blanks or commas. The *values* must be positive numbers. The number of *values* must equal the number of variable levels in the one-way table. The sum of the *values* must equal the total frequency for the one-way table. Order the *values* to match the order in which the corresponding variable levels appear in the one-way frequency table.

If you provide the null frequencies in a secondary input data set (TESTF=*SAS-data-set*), the variable that contains the null frequencies should be named _TESTF_, TestFrequency, or Frequency. The null frequencies must be positive numbers. The number of frequencies must equal the number of levels in the one-way frequency table, and the sum of the frequencies must equal the total frequency for the one-way table. Order the null frequencies in the data set to match the order in which the corresponding variable levels appear in the one-way frequency table.

TESTP=(*values*)| *SAS-data-set*

specifies null hypothesis proportions for the one-way chi-square goodness-of-fit tests. For more information, see the section "Chi-Square Test for One-Way Tables" on page 155. You can list the null proportions as *values* in parentheses after TESTP=. Or you can provide the null proportions in a secondary input data set by specifying TESTP=*SAS-data-set*. The TESTP=*SAS-data-set* cannot be the same data set that you specify in the DATA= option. You can specify only one TESTF= or TESTP= data set in a single invocation of the procedure.

If you list the null proportions as *values*, you can separate the *values* with blanks or commas. The *values* must be positive numbers. The number of *values* must equal the number of variable levels in the one-way table. Order the *values* to match the order in which the corresponding variable levels appear in the one-way frequency table. You can specify *values* in probability form as numbers between 0 and 1, where the proportions sum to 1. Or you can specify *values* in percentage form as numbers between 0 and 100, where the percentages sum to 100.

If you provide the null proportions in a secondary input data set (TESTP=*SAS-data-set*), the variable that contains the null proportions should be named _TESTP_, TestPercent, or Percent. The null proportions must be positive numbers. The number of proportions must equal the number of levels in the one-way frequency table. You can provide the proportions in probability form as numbers between 0 and 1, where the proportions sum to 1. Or you can provide the proportions in percentage form as numbers between 0 and 100, where the percentages sum to 100. Order the null proportions in the data set to match the order in which the corresponding variable levels appear in the one-way frequency table.

WARN=*type* | (*types*)

controls the warning message for the validity of the asymptotic Pearson chi-square test. By default, PROC FREQ displays a warning message when more than 20% of the table cells have expected frequencies that are less than 5. If you specify the NOPRINT option in the PROC FREQ statement, the procedure displays the warning in the log; otherwise, the procedure displays the warning as a footnote in the chi-square table. You can use the WARN= option to suppress the warning and to include a warning indicator in the output data set.

You can specify one or more of the following *types* in the WARN= option. If you specify more than one *type* value, enclose the values in parentheses after WARN=. For example, `warn = (output noprint)`.

Value of WARN=	Description
OUTPUT	Adds a warning indicator variable to the output data set
NOLOG	Suppresses the chi-square warning message in the log
NOPRINT	Suppresses the chi-square warning message in the display
NONE	Suppresses the chi-square warning message entirely

If you specify the WARN=OUTPUT option, the ODS output data set ChiSq contains a variable named Warning that equals 1 for the Pearson chi-square observation when more than 20% of the table cells have expected frequencies that are less than 5 and equals 0 otherwise. If you specify WARN=OUTPUT and also specify the CHISQ option in the OUTPUT statement, the statistics output data set contains a variable named WARN_PCHI that indicates the warning.

The WARN=NOLOG option has the same effect as the NOWARN option in the TABLES statement.

CL

requests confidence limits for the measures of association, which you can request by specifying the MEASURES option. For more information, see the sections "Measures of Association" on page 160 and "Confidence Limits" on page 160. You can set the level of the confidence limits by using the ALPHA= option; by default, ALPHA=0.05, which produces 95% confidence limits.

If you omit the MEASURES option, the CL option invokes MEASURES. The CL option is equivalent to the MEASURES(CL) option.

CMH < (*cmh-options*) >

requests Cochran-Mantel-Haenszel statistics, which test for association between the row and column variables after adjusting for the remaining variables in a multiway table. The Cochran-Mantel-Haenszel statistics include the nonzero correlation statistic, the row mean scores (ANOVA) statistic, and the general association statistic. In addition, for 2 × 2 tables, the CMH option provides the adjusted Mantel-Haenszel and logit estimates of the odds ratio and relative risks, together with their confidence limits. For stratified 2 × 2 tables, the CMH option provides the Breslow-Day test for homogeneity of odds ratios. (To request Tarone's adjustment for the Breslow-Day test, specify the BDT *cmh-option*.) For more information, see the section "Cochran-Mantel-Haenszel Statistics" on page 212.

You can use the CMH1 or CMH2 option to control the number of CMH statistics that PROC FREQ computes.

For stratified 2 × 2 tables, you can request Zelen's exact test for equal odds ratios by specifying the EQOR option in the EXACT statement. For more information, see the section "Zelen's Exact Test for Equal Odds Ratios" on page 218. You can request exact confidence limits for the common odds ratio by specifying the COMOR option in the EXACT statement. This option also provides a common odds ratio test. For more information, see the section "Exact Confidence Limits for the Common Odds Ratio" on page 219.

You can specify the following *cmh-options* in parentheses after the CMH option. These *cmh-options*, which apply to stratified 2 × 2 tables, are also available with the CMH1 or CMH2 option.

BDT

requests Tarone's adjustment in the Breslow-Day test for homogeneity of odds ratios. For more information, see the section "Breslow-Day Test for Homogeneity of the Odds Ratios" on page 218.

GAILSIMON <(COLUMN=1 | 2)>

GS <(COLUMN=1 | 2)>

requests the Gail-Simon test for qualitative interaction, which applies to stratified 2 × 2 tables. For more information, see the section "Gail-Simon Test for Qualitative Interactions" on page 221.

The COLUMN= option specifies the column of the risk differences to use to compute the Gail-Simon test. By default, PROC FREQ uses column 1 risk differences. If you specify COLUMN=2, PROC FREQ uses column 2 risk differences.

The GAILSIMON *cmh-option* has the same effect as the GAILSIMON option in the TABLES statement.

MANTELFLEISS

MF

requests the Mantel-Fleiss criterion for the Mantel-Haenszel statistic for stratified 2 × 2 tables. For more information, see the section "Mantel-Fleiss Criterion" on page 215.

CMH1 <(*cmh-options*)>

requests the Cochran-Mantel-Haenszel correlation statistic. This option does not provide the CMH row mean scores (ANOVA) statistic or the general association statistic, which are provided by the CMH option. For tables larger than 2 × 2, the CMH1 option requires less memory than the CMH option, which can require an enormous amount of memory for large tables.

For 2 × 2 tables, the CMH1 option also provides the adjusted Mantel-Haenszel and logit estimates of the odds ratio and relative risks, together with their confidence limits. For stratified 2 × 2 tables, the CMH1 option provides the Breslow-Day test for homogeneity of odds ratios.

The *cmh-options* for CMH1 are the same as the *cmh-options* that are available with the CMH option. For more information, see the description of the CMH option.

CMH2 <(*cmh-options*)>

requests the Cochran-Mantel-Haenszel correlation statistic and the row mean scores (ANOVA) statistic. This option does not provide the CMH general association statistic, which is provided by the CMH option. For tables larger than 2 × 2, the CMH2 option requires less memory than the CMH option, which can require an enormous amount of memory for large tables.

For 2 × 2 tables, the CMH1 option also provides the adjusted Mantel-Haenszel and logit estimates of the odds ratio and relative risks, together with their confidence limits. For stratified 2 × 2 tables, the CMH1 option provides the Breslow-Day test for homogeneity of odds ratios.

The *cmh-options* for CMH2 are the same as the *cmh-options* that are available with the CMH option. For more information, see the description of the CMH option.

COMMONRISKDIFF < *options* >

requests the common (stratified) risk difference for multiway 2 × 2 tables, where the risk difference is the difference between the row 1 proportion and the row 2 proportion in a 2 × 2 table. By default, this option provides Mantel-Haenszel and summary score estimates of the common risk difference, together with their confidence limits. For more information, see the section "Common Risk Difference" on page 188.

You can specify the following *options* to request confidence limit types and tests for the common risk difference:

CL=*type* | (*types*)

requests confidence limits for the common risk difference. You can specify one or more *types* of confidence limits. When you specify only one *type*, you can omit the parentheses. You can specify CL=NONE to suppress the "Confidence Limits for the Common Risk Difference" table.

You can specify the confidence level in the ALPHA= option. By default, ALPHA=0.05, which produces 95% confidence limits for the common risk difference.

You can specify one or more of the following *types*:

MH

requests Mantel-Haenszel confidence limits, which are computed by using Mantel-Haenszel stratum weights and the Sato variance estimator (Sato 1989). For more information, see the section "Mantel-Haenszel Confidence Limits and Test" on page 188.

MR
MINRISK

requests minimum risk confidence limits, which are computed by using minimum risk weights. For more information, see the section "Minimum Risk Confidence Limits and Test" on page 189.

NEWCOMBE

requests stratified Newcombe confidence limits that use Mantel-Haenszel weights to combine the stratum components. For more information, see the section "Stratified Newcombe Confidence Limits" on page 191.

NEWCOMBEMR

requests stratified Newcombe confidence limits that use minimum risk weights to combine the stratum components. For more information, see the section "Stratified Newcombe Confidence Limits" on page 191.

NONE

suppresses the "Confidence Limits for the Common Risk Difference" table.

SCORE

requests summary score confidence limits. For more information, see the section "Summary Score Confidence Limits" on page 191.

COLUMN=1 | 2

specifies the table column for which to compute the common risk difference statistics. If you do not specify this option but you do specify the RISKDIFF(COLUMN=) option, PROC FREQ provides the common risk difference statistics for the column that you specify in the RISKDIFF(COLUMN=) option. If you do not specify either of these options, COLUMN=1 by default.

CORRECT=NO

removes the continuity correction in the minimum risk confidence limits and in the minimum risk test, which you can request by specifying the CL=MR and TEST=MR options, respectively. For more information, see the section "Minimum Risk Confidence Limits and Test" on page 189.

PRINTWTS < =(MH | MR) >

displays the stratum weights together with the stratum risk differences and frequencies. By default, this option displays the weight type or types for the confidence limits and tests that you request. Optionally, you can specify the weight type to display; the PRINTWTS=MH option displays Mantel-Haenszel weights and the PRINTWTS=MR option displays minimum risk weights. You can display both weight types by specifying PRINTWTS=(MH MR).

TEST < =type | (types) >

requests common risk difference tests. You can specify one or more *types*. When you specify only one *type*, you can omit the parentheses. If you do not specify *types*, this option provides tests that correspond to the confidence limit *types* that you specify in the CL= option.

You can specify one or more of the following *types*:

MH

requests a Mantel-Haenszel test, which is computed by using Mantel-Haenszel stratum weights and the Sato variance estimator (Sato 1989). For more information, see the section "Mantel-Haenszel Confidence Limits and Test" on page 188.

MR < (VAR=SAMPLE) >

MINRISK < (VAR=SAMPLE) >

requests the minimum risk test, which is computed by using minimum risk weights. If you specify VAR=SAMPLE, PROC FREQ uses the sample (observed) variance estimate instead of a null variance estimate to compute the minimum risk test statistic. For more information, see the section "Minimum Risk Confidence Limits and Test" on page 189.

SCORE

requests the summary score test. For more information, see the section "Summary Score Confidence Limits" on page 191.

CONTENTS= 'string'

specifies the label to use for crosstabulation tables in the contents file, the Results window, and the trace record. For information about output presentation, see the *SAS Output Delivery System: User's Guide*.

If you omit the CONTENTS= option, the contents label for crosstabulation tables is "Cross-Tabular Freq Table" by default.

Note that contents labels for all crosstabulation tables that are produced by a single TABLES statement use the same text. To specify different contents labels for different crosstabulation tables, request the tables in separate TABLES statements and use the CONTENTS= option in each TABLES statement.

To remove the crosstabulation table entry from the contents file, you can specify a null label with CONTENTS=''.

The CONTENTS= option affects only contents labels for crosstabulation tables. It does not affect contents labels for other PROC FREQ tables.

To specify the contents label for any PROC FREQ table, you can use PROC TEMPLATE to create a customized table template. The CONTENTS_LABEL attribute in the DEFINE TABLE statement of PROC TEMPLATE specifies the contents label for the table. See the chapter "The TEMPLATE Procedure" in the *SAS Output Delivery System: User's Guide* for more information.

CROSSLIST < (*options*) >

displays crosstabulation tables by using an ODS column format instead of the default crosstabulation cell format. In the CROSSLIST table display, the rows correspond to the crosstabulation table cells, and the columns correspond to descriptive statistics such as frequencies and percentages. The CROSSLIST table displays the same information as the default crosstabulation table (but it uses an ODS column format). For more information about the contents of the CROSSLIST table, See the section "Two-Way and Multiway Tables" on page 231.

You can control the contents of a CROSSLIST table by specifying the same options available for the default crosstabulation table. These include the NOFREQ, NOPERCENT, NOROW, and NOCOL options. You can request additional information in a CROSSLIST table by specifying the CELLCHI2, DEVIATION, EXPECTED, MISSPRINT, and TOTPCT options. You can also display standardized residuals or Pearson residuals in a CROSSLIST table by specifying the CROSSLIST(STDRES) or CROSSLIST(PEARSONRES) option, respectively; these options are not available for the default crosstabulation table. The FORMAT= and CUMCOL options have no effect on CROSSLIST tables. You cannot specify both the LIST option and the CROSSLIST option in the same TABLES statement.

You can specify the NOSPARSE option along with the CROSSLIST option to suppress variable levels that have frequencies of 0. By default for CROSSLIST tables, PROC FREQ displays all levels of the column variable within each level of the row variable, including any levels that have frequencies of 0. By default for multiway CROSSLIST tables, PROC FREQ displays all levels of the row variable within each stratum of the table, including any row levels that have frequencies of 0 in the stratum.

You can specify the following *options*:

STDRES

displays the standardized residuals of the table cells in the CROSSLIST table. The standardized residual is the ratio of (*frequency − expected*) to its standard error, where *frequency* is the table cell frequency (count) and *expected* is the expected table cell frequency, which is computed under the null hypothesis that the row and column variables are independent. For more information, see the section "Standardized Residuals" on page 156. You can display the expected values and deviations by specifying the EXPECTED and DEVIATION options, respectively.

PEARSONRES

displays the Pearson residuals of the table cells in the CROSSLIST table. The Pearson residual is the square root of the table cell's contribution to the Pearson chi-square statistic. The Pearson residual is computed as $(frequency - expected)/\sqrt{expected}$, where *frequency* is the table cell frequency (count) and *expected* is the expected table cell frequency, which is computed under the null hypothesis that the row and column variables are independent. For more information, see the section "Pearson Chi-Square Test for Two-Way Tables" on page 156. You can display the expected values, deviations, and cell chi-squares by specifying the EXPECTED, DEVIATION, and CELLCHI2 options, respectively.

CUMCOL

displays the cumulative column percentages in the cells of the crosstabulation table. The CUMCOL option does not apply to crosstabulation tables produced with the LIST or CROSSLIST option.

DEVIATION

displays the deviations of the frequencies from the expected frequencies (*frequency − expected*) in the crosstabulation table. The expected frequencies are computed under the null hypothesis that the row and column variables are independent. For more information, see the section "Pearson Chi-Square Test for Two-Way Tables" on page 156. You can display the expected values by specifying the EXPECTED option. This option has no effect for one-way tables or for tables that are displayed in list format (which you can request by specifying the LIST option).

EXPECTED

displays the expected cell frequencies in the crosstabulation table. The expected frequencies are computed under the null hypothesis that the row and column variables are independent. For more information, see the section "Pearson Chi-Square Test for Two-Way Tables" on page 156. This option has no effect for one-way tables or for tables that are displayed in list format (which you can request by specifying the LIST option).

FISHER

requests Fisher's exact test for tables that are larger than 2×2. (For 2×2 tables, the CHISQ option provides Fisher's exact test.) This test is also known as the Freeman-Halton test. See the sections "Fisher's Exact Test" on page 158 and "Exact Statistics" on page 221 for more information.

If you omit the CHISQ option in the TABLES statement, the FISHER option invokes CHISQ. You can also request Fisher's exact test by specifying the FISHER option in the EXACT statement.

NOTE: PROC FREQ computes exact tests by using fast and efficient algorithms that are superior to direct enumeration. Exact tests are appropriate when a data set is small, sparse, skewed, or heavily tied. For some large problems, computation of exact tests might require a substantial amount of time and memory. Consider using asymptotic tests for such problems. Alternatively, when asymptotic methods might not be sufficient for such large problems, consider using Monte Carlo estimation of exact *p*-values. You can request Monte Carlo estimation by specifying the MC *computation-option* in the EXACT statement. See the section "Computational Resources" on page 224 for more information.

FORMAT=*format-name*

specifies a format for the following crosstabulation table cell values: frequency, expected frequency, and deviation. PROC FREQ also uses the specified format to display the row and column total frequencies and the overall total frequency in crosstabulation tables.

You can specify any standard SAS numeric format or a numeric format defined with the FORMAT procedure. The format length must not exceed 24. If you omit the FORMAT= option, by default PROC FREQ uses the BEST6. format to display frequencies less than 1E6, and the BEST7. format otherwise.

The FORMAT= option applies only to crosstabulation tables displayed in the default format. It does not apply to crosstabulation tables produced with the LIST or CROSSLIST option.

To change display formats in any FREQ table, you can use PROC TEMPLATE. See the chapter "The TEMPLATE Procedure" in the *SAS Output Delivery System: User's Guide* for more information.

GAILSIMON < (COLUMN=1 | 2) >

GS < (COLUMN=1 | 2) >

requests the Gail-Simon test for qualitative interaction, which applies to stratified 2×2 tables. For more information, see the section "Gail-Simon Test for Qualitative Interactions" on page 221.

The COLUMN= option specifies the column of the risk differences to use to compute the Gail-Simon test. By default, PROC FREQ uses column 1 risk differences. If you specify COLUMN=2, PROC FREQ uses column 2 risk differences.

JT

requests the Jonckheere-Terpstra test. For more information, see the section "Jonckheere-Terpstra Test" on page 203. To request exact *p*-values for the Jonckheere-Terpstra test, specify the JT option in the EXACT statement. See the section "Exact Statistics" on page 221 for more information.

LIST

displays two-way and multiway tables by using a list format instead of the default crosstabulation cell format. This option displays an entire multiway table in one table, instead of displaying a separate two-way table for each stratum. For more information, see the section "Two-Way and Multiway Tables" on page 231.

The LIST option is not available when you request tests and statistics; you must use the standard crosstabulation table display or the CROSSLIST display when you request tests and statistics.

MAXLEVELS=*n*

specifies the maximum number of variable levels to display in one-way frequency tables. The value of *n* must be a positive integer. PROC FREQ displays the first *n* variable levels, matching the order in which the levels appear in the one-way frequency table. (The ORDER= option controls the order of the variable levels. By default, ORDER=INTERNAL, which orders the variable levels by unformatted value.)

The MAXLEVELS= option also applies to one-way frequency plots, which you can request by specifying the PLOTS=FREQPLOT option when ODS Graphics is enabled.

If you specify the MISSPRINT option to display missing levels in the frequency table, the MAXLEVELS= option displays the first *n* nonmissing levels.

The MAXLEVELS= option does not apply to the OUT= output data set, which includes all variable levels. The MAXLEVELS= option does not affect the computation of percentages, statistics, or tests for the one-way table; these values are based on the complete table.

MEASURES < (CL) >

requests measures of association and their asymptotic standard errors. This option provides the following measures: gamma, Kendall's tau-*b*, Stuart's tau-*c*, Somers' $D(C|R)$, Somers' $D(R|C)$, Pearson and Spearman correlation coefficients, lambda (symmetric and asymmetric), and uncertainty coefficients (symmetric and asymmetric). If you specify the CL option in parentheses after the MEASURES option, PROC FREQ provides confidence limits for the measures of association. For more information, see the section "Measures of Association" on page 160.

For 2×2 tables, the MEASURES option also provides the odds ratio, column 1 relative risk, column 2 relative risk, and their asymptotic Wald confidence limits. You can request the odds ratio and relative risks separately (without the other measures of association) by specifying the RELRISK option. You can request confidence limits for the odds ratio by specifying the OR(CL=) option.

You can use the TEST statement to request asymptotic tests for the following measures of association: gamma, Kendall's tau-b, Stuart's tau-c, Somers' $D(C|R)$, Somers' $D(R|C)$, and Pearson and Spearman correlation coefficients. You can use the EXACT statement to request exact confidence limits for the odds ratio, exact unconditional confidence limits for the relative risks, and exact tests for the following measures of association: Kendall's tau-b, Stuart's tau-c, Somers' $D(C|R)$ and $D(R|C)$, and Pearson and Spearman correlation coefficients. For more information, see the descriptions of the TEST and EXACT statements and the section "Exact Statistics" on page 221.

MISSING

treats missing values as a valid nonmissing level for all TABLES variables. The MISSING option displays the missing levels in frequency and crosstabulation tables and includes them in all calculations of percentages, tests, and measures.

By default, if you do not specify the MISSING or MISSPRINT option, an observation is excluded from a table if it has a missing value for any of the variables in the TABLES request. When PROC FREQ excludes observations with missing values, it displays the total frequency of missing observations following the table. See the section "Missing Values" on page 150 for more information.

MISSPRINT

displays missing value frequencies in frequency and crosstabulation tables but does not include the missing value frequencies in any computations of percentages, tests, or measures.

By default, if you do not specify the MISSING or MISSPRINT option, an observation is excluded from a table if it has a missing value for any of the variables in the TABLES request. When PROC FREQ excludes observations with missing values, it displays the total frequency of missing observations following the table. See the section "Missing Values" on page 150 for more information.

NOCOL

suppresses the display of column percentages in crosstabulation table cells.

NOCUM

suppresses the display of cumulative frequencies and percentages in one-way frequency tables. The NOCUM option also suppresses the display of cumulative frequencies and percentages in crosstabulation tables in list format, which you request with the LIST option.

NOFREQ

suppresses the display of cell frequencies in crosstabulation tables. The NOFREQ option also suppresses row total frequencies. This option has no effect for one-way tables or for crosstabulation tables in list format, which you request with the LIST option.

NOPERCENT

suppresses the display of overall percentages in crosstabulation tables. These percentages include the cell percentages of the total (two-way) table frequency, and the row and column percentages of the total table frequency. To suppress the display of cell percentages of row or column totals, use the NOROW or NOCOL option, respectively.

For one-way frequency tables and crosstabulation tables in list format, the NOPERCENT option suppresses the display of percentages and cumulative percentages.

NOPRINT

suppresses the display of frequency and crosstabulation tables but displays all requested tests and statistics. To suppress the display of all output, including tests and statistics, use the NOPRINT option in the PROC FREQ statement.

NOROW

suppresses the display of row percentages in crosstabulation table cells.

NOSPARSE

suppresses zero-frequency cells in the LIST table, CROSSLIST table, and OUT= data set.

The NOSPARSE option is available when you specify the ZEROS option in the WEIGHT statement, which include observations that have weights of 0. By default, the ZEROS option invokes the SPARSE option, which displays zero-frequency cells in the LIST table and includes them in the OUT= data set; the NOSPARSE option suppresses the zero-frequency cells. For more information, see the description of the ZEROS option.

The NOSPARSE option is also available when you specify the CROSSLIST option. By default for CROSSLIST tables, PROC FREQ displays all levels of the column variable within each level of the row variable, including any levels that have frequencies of 0. By default for multiway CROSSLIST tables, PROC FREQ displays all levels of the row variable within each stratum of the table, including any row levels that have 0 frequencies in the stratum. The NOSPARSE option suppresses the zero-frequency levels in the CROSSLIST table.

NOWARN

suppresses the log warning message for the validity of the asymptotic Pearson chi-square test. By default, PROC FREQ provides a validity warning for the asymptotic Pearson chi-square test when more than 20cells have expected frequencies that are less than 5. This warning message appears in the log if you specify the NOPRINT option in the PROC FREQ statement,

The NOWARN option is equivalent to the CHISQ(WARN=NOLOG) option. You can also use the CHISQ(WARN=) option to suppress the warning message in the display and to request a warning variable in the chi-square ODS output data set or in the OUTPUT data set.

OR <(CL=*type* | (*types*)**)>**

ODDSRATIO <(CL=*type* | (*types*)**)>**

requests the odds ratio and confidence limits for 2 × 2 tables. For more information, see the section "Odds Ratio" on page 192.

You can specify one or more *types* of confidence limits. When you specify only one confidence limit *type*, you can omit the parentheses around the request. PROC FREQ displays the confidence limits in the "Confidence Limits for the Odds Ratio" table.

Specifying the OR option without the CL= option is equivalent to specifying the RELRISK option, which produces the "Odds Ratio and Relative Risks" table. For more information, see the description of the RELRISK option. When you specify the OR(CL=) option, PROC FREQ does not produce the "Odds Ratio and Relative Risks" table unless you also specify the RELRISK or MEASURES option.

The ALPHA= option determines the confidence level; by default, ALPHA=0.05, which produces 95% confidence limits for the odds ratio.

You can specify the following *types*:

EXACT
> displays exact confidence limits for the odds ratio in the "Confidence Limits for the Odds Ratio" table. (By default, PROC FREQ displays the exact confidence limits in a separate table.) You must also request computation of the exact confidence limits by specifying the OR option in the EXACT statement. For more information, see the subsection "Exact Confidence Limits" in the section "Confidence Limits for the Odds Ratio" on page 193.

LR
LIKELIHOODRATIO
> requests likelihood ratio confidence limits for the odds ratio. For more information, see the subsection "Likelihood Ratio Confidence Limits" in the section "Confidence Limits for the Odds Ratio" on page 193.

MIDP
> requests exact mid-p confidence limits for the odds ratio. For more information, see the subsection "Exact Mid-p Confidence Limits" in the section "Confidence Limits for the Odds Ratio" on page 193.

SCORE < (CORRECT=NO) >
> requests score confidence limits for the odds ratio. For more information, see the subsection "Score Confidence Limits" in the section "Confidence Limits for the Odds Ratio" on page 193. If you specify CORRECT=NO, PROC FREQ provides the uncorrected form of the score confidence limits.

WALD
> requests asymptotic Wald confidence limits, which are based on a log transformation of the odds ratio. For more information, see the subsection "Wald Confidence Limits" in the section "Confidence Limits for the Odds Ratio" on page 193.

WALDMODIFIED
> requests Wald modified confidence limits for the odds ratio. For more information, see the subsection "Wald Modified Confidence Limits" in the section "Confidence Limits for the Odds Ratio" on page 193.

OUT=SAS-data-set
> names an output data set that contains frequency or crosstabulation table counts and percentages. If more than one table request appears in the TABLES statement, the contents of the OUT= data set correspond to the last table request in the TABLES statement. The OUT= data set variable COUNT contains the frequencies and the variable PERCENT contains the percentages. For more information, see the section "Output Data Sets" on page 226. You can specify the following options to include additional information in the OUT= data set: OUTCUM, OUTEXPECT, and OUTPCT.

OUTCUM
> includes cumulative frequencies and cumulative percentages in the OUT= data set for one-way tables. The variable CUM_FREQ contains the cumulative frequencies, and the variable CUM_PCT contains the cumulative percentages. For more information, see the section "Output Data Sets" on page 226. The OUTCUM option has no effect for two-way or multiway tables.

OUTEXPECT

includes expected cell frequencies in the OUT= data set for crosstabulation tables. The variable EXPECTED contains the expected cell frequencies. For more information, see the section "Output Data Sets" on page 226. The EXPECTED option has no effect for one-way tables.

OUTPCT

includes the following additional variables in the OUT= data set for crosstabulation tables:

PCT_COL	percentage of column frequency
PCT_ROW	percentage of row frequency
PCT_TABL	percentage of stratum (two-way table) frequency, for n-way tables where $n > 2$

For more information, see the section "Output Data Sets" on page 226. The OUTPCT option has no effect for one-way tables.

PLCORR <(*options*)>

POLYCHORIC <(*options*)>

requests the polychoric correlation coefficient and its asymptotic standard error. For 2×2 tables, this statistic is more commonly known as the tetrachoric correlation coefficient and is labeled as such in the displayed output. For more information, see the section "Polychoric Correlation" on page 166.

If you also specify the CL or MEASURES(CL) option, PROC FREQ provides confidence limits for the polychoric correlation. If you specify the PLCORR option in the TEST statement, the procedure provides Wald and likelihood ratio tests for the polychoric correlation. The PLCORR option invokes the MEASURES option.

You can specify the following *options*:

ADJUST

replaces a 2×2 table cell frequency of 0 by 0.5 before computing the tetrachoric correlation (Brown and Benedetti 1977a, p. 353). To maintain the row and column marginal frequencies, adjacent cell frequencies are decreased by 0.5 and the opposite cell frequency is increased by 0.5.

This option is available for 2×2 tables and is applied only when a single cell frequency is 0. It has no effect when both off-diagonal cell frequencies are 0 (and therefore the correlation is 1) or when both diagonal cell frequencies are 1 (and therefore the correlation is –1).

CONVERGE=*value*

specifies the convergence criterion. The *value* must be a positive number. By default, CONVERGE=0.0001. Iterative computation of the polychoric correlation stops when the convergence measure falls below *value* or when the number of iterations exceeds the MAXITER= *number*, whichever happens first. For parameter values that are less than 0.01, PROC FREQ evaluates convergence by using the absolute difference instead of the relative difference. For more information, see the section "Polychoric Correlation" on page 166.

MAXITER=*number*

specifies the maximum *number* of iterations. The value of *number* must be a positive integer. By default, MAXITER=50. Iterative computation of the polychoric correlation stops when the number of iterations exceeds the maximum *number* or when the convergence measure falls below the CONVERGE= *value*, whichever happens first. For more information, see the section "Polychoric Correlation" on page 166.

PLOTS < *(global-plot-options)* > < =*plot-request* <*(plot-options)*> >
PLOTS < *(global-plot-options)* > < =(*plot-request* <*(plot-options)*> <... *plot-request* <*(plot-options)*> >) >

controls the plots that are produced through ODS Graphics. *Plot-requests* identify the plots, and *plot-options* control the appearance and content of the plots. You can specify *plot-options* in parentheses after a *plot-request*. A *global-plot-option* applies to all plots for which it is available unless it is altered by a specific *plot-option*. You can specify *global-plot-options* in parentheses after the PLOTS option.

When you specify only one *plot-request*, you can omit the parentheses around the request. For example:

```
plots=all
plots=freqplot
plots=(freqplot oddsratioplot)
plots(only)=(cumfreqplot deviationplot)
```

ODS Graphics must be enabled before plots can be requested. For example:

```
ods graphics on;
proc freq;
   tables treatment*response / chisq plots=freqplot;
   weight wt;
run;
ods graphics off;
```

For more information about enabling and disabling ODS Graphics, see the section "Enabling and Disabling ODS Graphics" in Chapter 21, "Statistical Graphics Using ODS" (*SAS/STAT User's Guide*).

If ODS Graphics is enabled but you do not specify the PLOTS= option, PROC FREQ produces all plots that are associated with the analyses that you request, with the exception of the frequency, cumulative frequency, and mosaic plots. To produce a frequency plot or cumulative frequency plot when ODS Graphics is enabled, you must specify the FREQPLOT or CUMFREQPLOT *plot-request*, respectively, in the PLOTS= option, or you must specify the PLOTS=ALL option. To produce a mosaic plot when ODS Graphics is enabled, you must specify the MOSAICPLOT *plot-request* in the PLOTS= option, or you must specify the PLOTS=ALL option.

PROC FREQ produces the remaining plots (listed in Table 3.11) by default when you request the corresponding TABLES statement options. You can suppress default plots and request specific plots by using the PLOTS(ONLY)= option; PLOTS(ONLY)=(*plot-requests*) produces only the plots that are specified as *plot-requests*. You can suppress all plots by specifying the PLOTS=NONE option. The PLOTS option has no effect when you specify the NOPRINT option in the PROC FREQ statement.

Plot Requests

Table 3.11 lists the available *plot-requests* together with their required TABLES statement options. Descriptions of the *plot-requests* follow the table in alphabetical order.

Table 3.11 Plot Requests

Plot Request	Description	Required TABLES Statement Option
AGREEPLOT	Agreement plot	AGREE ($r \times r$ table)
ALL	All plots	None
CUMFREQPLOT	Cumulative frequency plot	One-way table request
DEVIATIONPLOT	Deviation plot	CHISQ (one-way table)
FREQPLOT	Frequency plot	Any table request
KAPPAPLOT	Kappa plot	AGREE ($h \times r \times r$ table)
MOSAICPLOT	Mosaic plot	Two-way or multiway table request
NONE	No plots	None
ODDSRATIOPLOT	Odds ratio plot	MEASURES, OR, or RELRISK ($h \times 2 \times 2$ table)
RELRISKPLOT	Relative risk plot	MEASURES or RELRISK ($h \times 2 \times 2$ table)
RISKDIFFPLOT	Risk difference plot	RISKDIFF ($h \times 2 \times 2$ table)
WTKAPPAPLOT	Weighted kappa plot	AGREE ($h \times r \times r$ table, $r > 2$)

You can specify the following *plot-requests*:

AGREEPLOT < (*plot-options*) >

requests an agreement plot (Bangdiwala and Bryan 1987). An agreement plot displays the strength of agreement in a two-way table, where the row and column variables represent two independent ratings of *n* subjects. For information about agreement plots, see Bangdiwala (1988), Bangdiwala et al. (2008), and Friendly (2000, Section 3.7.2).

To produce an agreement plot, you must also specify the AGREE option in the TABLES statement. Agreement statistics and plots are available for two-way square tables, where the number of rows equals the number of columns.

Table 3.12 lists the *plot-options* that are available for agreement plots. For descriptions of the *plot-options*, see the subsection "Plot Options" in this section.

Table 3.12 Plot Options for AGREEPLOT

Plot Option	Description	Values
LEGEND=	Legend	NO or YES*
PARTIAL=	Partial agreement	NO or YES*
SHOWSCALE=	Frequency scale	NO or YES*
STATS	Statistics	None

*Default

If you specify the STATS *plot-option*, the agreement plot displays the values of the kappa coefficient, the weighted kappa coefficient, the B_n measure (Bangdiwala and Bryan 1987), and the sample size. PROC FREQ stores these statistics in an ODS table named BnMeasure, which is not displayed. For more information, see the section "ODS Table Names" on page 239.

ALL

requests all plots that are associated with the specified analyses. Table 3.11 lists the available *plot-requests* and the corresponding analysis options. If you specify the PLOTS=ALL option, PROC FREQ produces the frequency, cumulative frequency, and mosaic plots that are associated with the tables that you request. (These plots are not produced by default when ODS Graphics is enabled.)

CUMFREQPLOT < (*plot-options*) >

requests a plot of cumulative frequencies. Cumulative frequency plots are available for one-way frequency tables.

To produce a cumulative frequency plot, you must specify the CUMFREQPLOT *plot-request* in the PLOTS= option, or you must specify the PLOTS=ALL option. PROC FREQ does not produce cumulative frequency plots by default when ODS Graphics is enabled.

Table 3.13 lists the *plot-options* that are available for cumulative frequency plots. For descriptions of the *plot-options*, see the subsection "Plot Options" in this section.

Table 3.13 Plot Options for CUMFREQPLOT

Plot Option	Description	Values
ORIENT=	Orientation	HORIZONTAL or VERTICAL*
SCALE=	Scale	FREQ* or PERCENT
TYPE=	Type	BARCHART* or DOTPLOT

*Default

DEVIATIONPLOT < (*plot-options*) >

requests a plot of relative deviations from expected frequencies. Deviation plots are available for chi-square analysis of one-way frequency tables. To produce a deviation plot, you must also specify the CHISQ option in the TABLES statement for a one-way frequency table.

Table 3.14 lists the *plot-options* that are available for deviation plots. For descriptions of the *plot-options*, see the subsection "Plot Options" in this section.

Table 3.14 Plot Options for DEVIATIONPLOT

Plot Option	Description	Values
NOSTAT	No statistic	None
ORIENT=	Orientation	HORIZONTAL or VERTICAL*
TYPE=	Type	BARCHART* or DOTPLOT

*Default

FREQPLOT < (*plot-options*) >

requests a frequency plot. Frequency plots are available for frequency and crosstabulation tables. For multiway crosstabulation tables, PROC FREQ provides a two-way frequency plot for each stratum (two-way table).

To produce a frequency plot, you must specify the FREQPLOT *plot-request* in the PLOTS= option, or you must specify the PLOTS=ALL option. PROC FREQ does not produce frequency plots by default when ODS Graphics is enabled.

Table 3.15 lists the *plot-options* that are available for frequency plots. For descriptions of the *plot-options*, see the subsection "Plot Options" in this section.

Table 3.15 Plot Options for FREQPLOT

Plot Option	Description	Values
GROUPBY=**	Primary group	COLUMN* or ROW
NPANELPOS=**	Sections per panel	Number (4*)
ORIENT=	Orientation	HORIZONTAL or VERTICAL*
SCALE=	Scale	FREQ*, GROUPPERCENT**, LOG, PERCENT, SQRT
TWOWAY=**	Two-way layout	CLUSTER, GROUPHORIZONTAL, GROUPVERTICAL*, or STACKED
TYPE=	Type	BARCHART* or DOTPLOT

*Default
**For two-way tables

You can specify the following *plot-options* for all frequency plots: ORIENT=, SCALE=, and TYPE=. You can specify the following *plot-options* for frequency plots of two-way (and multiway) tables: GROUPBY=, NPANELPOS=, and TWOWAY=. The NPANELPOS= *plot-option* is not available with the TWOWAY=CLUSTER or TWOWAY=STACKED layout, which is always displayed in a single panel.

By default, PROC FREQ displays frequency plots as bar charts. To display frequency plots as dot plots, specify TYPE=DOTPLOT. To plot percentages instead of frequencies, specify SCALE=PERCENT. For two-way tables, there are four frequency plot layouts available, which you can request by specifying the TWOWAY= *plot-option*. For more information, see the subsection "Plot Options" in this section.

By default, graph cells in a two-way layout are first grouped by column variable levels; row variable levels are then displayed within the column variable levels. To group first by row variable levels, specify GROUPBY=ROW.

KAPPAPLOT < (*plot-options*) >

requests a plot of kappa statistics along with confidence limits. Kappa plots are available for multiway square tables and display the kappa statistic (with confidence limits) for each two-way table (stratum). Kappa plots also display the overall kappa statistic unless you specify the COMMON=NO *plot-option*. To produce a kappa plot, you must specify the AGREE option in the TABLES statement to compute kappa statistics.

Table 3.16 lists the *plot-options* that are available for kappa plots. For descriptions of the *plot-options*, see the subsection "Plot Options" in this section.

Table 3.16 Plot Options for KAPPAPLOT and WTKAPPAPLOT

Plot Option	Description	Values
CLDISPLAY=	Error bar type	BAR, LINE, LINEARROW, SERIF*, or SERIFARROW
COMMON=	Overall kappa	NO or YES*
NPANELPOS=	Statistics per graphic	Number (all*)
ORDER=	Order of two-way levels	ASCENDING or DESCENDING
RANGE=	Range to display	Values or CLIP
STATS	Statistic values	None

*Default

MOSAICPLOT < (*plot-options*) >

requests a mosaic plot. Mosaic plots are available for two-way and multiway crosstabulation tables; for multiway tables, PROC FREQ provides a mosaic plot for each two-way table (stratum).

To produce a mosaic plot, you must specify the MOSAICPLOT *plot-request* in the PLOTS= option, or you must specify the PLOTS=ALL option. PROC FREQ does not produce mosaic plots by default when ODS Graphics is enabled.

Mosaic plots display tiles that correspond to the crosstabulation table cells. The areas of the tiles are proportional to the frequencies of the table cells. The column variable is displayed on the X axis, and the tile widths are proportional to the relative frequencies of the column variable levels. The row variable is displayed on the Y axis, and the tile heights are proportional to the relative frequencies of the row levels within column levels. For more information, see Friendly (2000).

By default, the colors of the tiles correspond to the row variable levels. If you specify the COLORSTAT= *plot-option*, the tiles are colored according to the values of the Pearson or standardized residuals.

You can specify the following *plot-options*:

COLORSTAT < =PEARSONRES | STDRES >

colors the mosaic plot tiles according to the values of residuals. If you specify COLORSTAT=PEARSONRES, the tiles are colored according to the Pearson residuals of the corresponding table cells. For more information, see the section "Pearson Chi-Square Test for Two-Way Tables" on page 156. If you specify COLORSTAT=STDRES, the tiles are colored according to the standardized residuals of the corresponding table cells. For more information, see the section "Standardized Residuals" on page 156. You can display the Pearson or standardized residuals in the CROSSLIST table by specifying the CROSSLIST(PEARSONRES) or CROSSLIST(STDRES) option, respectively.

SQUARE

produces a square mosaic plot, where the height of the Y axis equals the width of the X axis. In a square mosaic plot, the scale of the relative frequencies is the same on both axes. By default, PROC FREQ produces a rectangular mosaic plot.

NONE
> suppresses all plots.

ODDSRATIOPLOT < *(plot-options)* >
> requests a plot of odds ratios along with confidence limits. Odds ratio plots are available for multiway 2 × 2 tables and display the odds ratio (with confidence limits) for each 2 × 2 table (stratum). To produce an odds ratio plot, you must also specify the MEASURES, OR, or RELRISK option in the TABLES statement to compute the odds ratios.
>
> Table 3.17 lists the *plot-options* that are available for odds ratio plots. For descriptions of the *plot-options*, see the subsection "Plot Options" in this section.
>
> **Table 3.17** Plot Options for ODDSRATIOPLOT, RELRISKPLOT, and RISKDIFFPLOT
>
> | Plot Option | Description | Values |
> |---|---|---|
> | CL= | Confidence limit type | Type |
> | CLDISPLAY= | Error bar type | BAR, LINE, LINEARROW, SERIF*, or SERIFARROW |
> | COLUMN=** | Risk column | 1* or 2 |
> | COMMON= | Common value | NO or YES* |
> | LOGBASE=*** | Axis scale | 2, E, or 10 |
> | NPANELPOS= | Statistics per graphic | Number (all*) |
> | ORDER= | Order of two-way levels | ASCENDING or DESCENDING |
> | RANGE= | Range to display | Values or CLIP |
> | STATS | Statistic values | None |
>
> *Default
> **Available for RELRISKPLOT and RISKDIFFPLOT
> ***Available for ODDSRATIOPLOT and RELRISKPLOT
>
> You can specify one of the following confidence limit types for the odds ratio plot: exact (CL=EXACT), likelihood ratio (CL=LR), exact mid-*p* (CL=MIDP), score (CL=SCORE), Wald (CL=WALD), or Wald modified (CL=WALDMODIFIED). By default, the odds ratio plot displays Wald confidence limits. For more information, see the descriptions of the CL= *plot-option* and the OR(CL=) option.
>
> To display exact confidence limits in the odds ratio plot, you must also request their computation by specifying the OR option in the EXACT statement.
>
> When CL=WALD or CL=EXACT, the odds ratio plot displays the common odds ratio by default when it is available. To compute the common odds ratio along with Wald confidence limits, specify the CMH option in the TABLES statement. To compute the common odds ratio along with exact confidence limits, specify the COMOR option in the EXACT statement. To suppress display of the common odds ratio, specify COMMON=NO.

RELRISKPLOT < *(plot-options)* >
> requests a plot of relative risks along with confidence limits. Relative risk plots are available for multiway 2 × 2 tables and display the relative risk (with confidence limits) for each 2 × 2 table (stratum). To produce a relative risk plot, you must also specify the MEASURES or RELRISK option in the TABLES statement to compute relative risks.

Table 3.17 lists the *plot-options* that are available for relative risk plots. For descriptions of the *plot-options*, see the subsection "Plot Options" in this section.

You can specify one of the following confidence limit types for the relative risk plot: exact (CL=EXACT), likelihood ratio (CL=LR), score (CL=SCORE), Wald (CL=WALD), or Wald modified (CL=WALDMODIFIED). By default, the relative risk plot displays Wald confidence limits. For more information, see the descriptions of the CL= *plot-option* and the RELRISK(CL=) option.

To display exact confidence limits in the relative risk plot, you must also request their computation by specifying the RELRISK option in the EXACT statement. The risk column that you specify for the confidence limits must match the risk column that you specify for the plot.

The relative risk plot displays the common relative risk by default when you specify CL=WALD and the CMH option in the TABLES statement. To suppress display of the common relative risk, specify COMMON=NO.

RISKDIFFPLOT < (*plot-options*) >

requests a plot of risk (proportion) differences along with confidence limits. Risk difference plots are available for multiway 2 × 2 tables and display the risk difference (with confidence limits) for each 2 × 2 table (stratum). To produce a risk difference plot, you must also specify the RISKDIFF option in the TABLES statement to compute risk differences.

Table 3.17 lists the *plot-options* that are available for risk difference plots. For descriptions of the *plot-options*, see the subsection "Plot Options" in this section.

You can specify one of the following confidence limit types for the risk difference plot: Agresti-Caffo (CL=AC), exact (CL=EXACT), Hauck-Anderson (CL=HA), Miettinen-Nurminen (score) (CL=MN), Newcombe (CL=NEWCOMBE), and Wald (CL=WALD). By default, the plot displays Wald confidence limits for the risk difference. For more information, see the descriptions of the CL= *plot-option* and the RISKDIFF(CL=) option.

To display exact confidence limits in the risk difference plot, you must also request their computation by specifying the RISKDIFF option in the EXACT statement. The risk column that you specify for the confidence limits must match the risk column that you specify for the plot.

By default, the risk difference plot displays the common risk difference when you specify the RISKDIFF(COMMON) option and one of the following confidence limit types in the CL= *plot-option*: Miettinen-Nurminen (score) (CL=MN), Newcombe (CL=NEWCOMBE), or Wald (CL=WALD). To suppress display of the common risk difference, specify COMMON=NO.

WTKAPPAPLOT < (*plot-options*) >

requests a plot of weighted kappa coefficients along with confidence limits. Weighted kappa plots are available for multiway square tables and display the weighted kappa coefficient (with confidence limits) for each two-way table (stratum). Weighted kappa plots also display the overall weighted kappa coefficient unless you specify the COMMON=NO *plot-option*.

To produce a weighted kappa plot, you must specify the AGREE option in the TABLES statement to compute weighted kappa coefficients, and the table dimension must be greater than 1.

Table 3.16 lists the *plot-options* that are available for weighted kappa plots. For descriptions of the *plot-options*, see the subsection "Plot Options" in this section.

Global Plot Options

A *global-plot-option* applies to all plots for which the option is available unless it is altered by an individual *plot-option*. You can specify *global-plot-options* in parentheses after the PLOTS option. For example:

```
plots(order=ascending stats)=(riskdiffplot oddsratioplot)
plots(only)=freqplot
```

The following *plot-options* are available as *global-plot-options*: CLDISPLAY=, COLUMN=, COMMON=, EXACT, LOGBASE=, NPANELPOS=, ORDER=, ORIENT=, RANGE=, SCALE=, STATS, and TYPE=. For descriptions of these *plot-options*, see the subsection "Plot Options" in this section.

In addition to these *plot-options*, you can specify the following *global-plot-option*:

ONLY
    suppresses the default plots and requests only the plots that are specified as *plot-requests*.

Plot Options

You can specify the following *plot-options* in parentheses after a *plot-request*:

CL=type
    specifies the *type* of confidence limits to display. You can specify the CL= *plot-option* when you specify any of the following *plot-requests*: ODDSRATIOPLOT, RELRISKPLOT, and RISKDIFFPLOT.

    For odds ratio plots (ODDSRATIOPLOT), the available confidence limit types include the following: exact (CL=EXACT), likelihood ratio (CL=LR), exact mid-p (CL=MIDP), score (CL=SCORE), Wald (CL=WALD), and Wald modified (CL=WALDMODIFIED). For more information, see the description of the OR(CL=) option and the section "Confidence Limits for the Odds Ratio" on page 193. By default, CL=WALD. When you specify CL=EXACT to display exact confidence limits, you must also request computation of exact confidence limits by specifying the OR option in the EXACT statement.

    For relative risk plots (RELRISKPLOT), the available confidence limit types include the following: exact (CL=EXACT), likelihood ratio (CL=LR), score (CL=SCORE), Wald (CL=WALD), and Wald modified (CL=WALDMODIFIED). For more information, see the description of the RELRISK(CL=) option and the section "Confidence Limits for the Relative Risk" on page 196. By default, CL=WALD. When you specify CL=EXACT to display exact confidence limits, you must also request computation of exact confidence limits by specifying the RELRISK option in the EXACT statement.

    For risk difference plots (RISKDIFFPLOT), the available confidence limit types include the following: Agresti-Caffo (CL=AC), exact (CL=EXACT), Hauck-Anderson (CL=HA), Miettinen-Nurminen (score) (CL=MN), Newcombe (CL=NEWCOMBE), and Wald (CL=WALD). For more information, see the description of the RISKDIFF(CL=) option and the section "Confidence Limits for the Risk Difference" on page 179. By default, CL=WALD. When you specify CL=EXACT to display exact confidence limits in the plot, you must also request computation of exact confidence limits by specifying the RISKDIFF option in the EXACT statement.

CLDISPLAY=BAR < width > | **LINE** | **LINEARROW** | **SERIF** | **SERIFARROW**

controls the appearance of the confidence limit error bars. You can specify the CLDISPLAY= *plot-option* when you specify the following *plot-requests*: KAPPAPLOT, ODDSRATIOPLOT, RELRISKPLOT, RISKDIFFPLOT, and WTKAPPAPLOT.

The default is CLDISPLAY=SERIF, which displays the confidence limits as lines with serifs. CLDISPLAY=LINE displays the confidence limits as plain lines without serifs. The CLDISPLAY=SERIFARROW and CLDISPLAY=LINEARROW *plot-options* display arrowheads on any error bars that are clipped by the RANGE= *plot-option*; if an entire error bar is cut from the plot, the plot displays an arrowhead that points toward the statistic.

CLDISPLAY=BAR displays the confidence limits as bars. By default, the width of the bars equals the size of the marker for the estimate. You can control the width of the bars and the size of the marker by specifying the value of *width* as a percentage of the distance between bars, $0 < width \leq 1$. The bar might disappear when the value of *width* is very small.

COLUMN=1 | 2

specifies the table column to use to compute the risks (proportion) for the relative risk plot (RELRISKPLOT) and the risk difference plot (RISKDIFFPLOT). If you specify COLUMN=1, the plot displays the column 1 relative risks or the column 1 risk differences. Similarly, if you specify COLUMN=2, the plot displays the column 2 relative risks or risk differences.

For relative risk plots, the default is COLUMN=1. For risk difference plots, the default is COLUMN=1 if you request computation of both column 1 and column 2 risk differences by specifying the RISKDIFF option. If you request computation of only the column 1 (or column 2) risk differences by specifying the RISKDIFF(COLUMN=1) (or RISKDIFF(COLUMN=2)) option, by default the risk difference plot displays the risk differences for the column that you specify.

COMMON=NO | YES

controls the display of the common (overall) statistic in plots that display stratum (two-way table) statistics for multiway tables. You can specify the COMMON= *plot-option* when you specify the following *plot-requests*: KAPPAPLOT, ODDSRATIOPLOT, RELRISKPLOT, RISKDIFFPLOT, and WTKAPPAPLOT.

COMMON=NO suppresses display of the common statistic and its confidence limits. By default, COMMON=YES, which displays the common statistic and its confidence limits when these values are available. For more information, see the descriptions of the *plot-requests*.

EXACT

requests display of exact confidence limits instead of asymptotic confidence limits. You can specify the EXACT *plot-option* when you specify the following *plot-requests*: ODDSRATIOPLOT, RELRISKPLOT, and RISKDIFFPLOT. The EXACT *plot-option* is equivalent to the CL=EXACT *plot-option*.

When you specify the EXACT *plot-option*, you must also request computation of exact confidence limits by specifying the appropriate *statistic-option* in the EXACT statement.

GROUPBY=COLUMN | ROW

specifies the primary grouping for two-way frequency plots, which you can request by specifying the FREQPLOT *plot=request*. The default is GROUPBY=COLUMN, which groups graph cells first by column variable and displays row variable levels within column variable levels. You can specify GROUPBY=ROW to group first by row variable. In two-way and multiway table requests, the column variable is the last variable specified and forms the columns of the crosstabulation table. The row variable is the next-to-last variable specified and forms the rows of the table.

By default for a bar chart that is displayed in the TWOWAY=STACKED layout, bars correspond to the column variable levels, and row levels are displayed (stacked) within each column bar. By default for a bar chart that is displayed in the TWOWAY=CLUSTER layout, bars are first grouped by column variable levels, and row levels are displayed as adjacent bars within each column-level group. You can reverse the default row and column variable grouping by specifying GROUPBY=ROW.

LOGBASE=2 | E | 10

applies to the odds ratio plot (ODDSRATIOPLOT) and the relative risk plot (RELRISKPLOT). This *plot-option* displays the odds ratio or relative risk axis on the log scale that you specify.

LEGEND=NO | YES

applies to the agreement plot (AGREEPLOT). LEGEND=NO suppresses the legend that identifies the areas of exact and partial agreement. The default is LEGEND=YES.

NOSTAT

applies to the deviation plot (DEVIATIONPLOT). NOSTAT suppresses the chi-square *p*-value that deviation plot displays by default.

NPANELPOS=*n*

divides the plot into multiple panels that display at most |*n*| statistics or sections.

If *n* is positive, the number of statistics or sections per panel is balanced; if *n* is negative, the number of statistics per panel is not balanced. For example, suppose you want to display 21 odds ratios. NPANELPOS=20 displays two panels, the first with 11 odds ratios and the second with 10 odds ratios; NPANELPOS=−20 displays 20 odds ratios in the first panel but only 1 odds ratio in the second panel. This *plot-option* is available for all plots except mosaic plots and one-way weighted frequency plots.

For two-way frequency plots (FREQPLOT), NPANELPOS=*n* requests that panels display at most |*n*| sections, where sections correspond to row or column variable levels, depending on the type of plot and the grouping. By default, *n*=4 and each panel includes at most four sections. This *plot-option* applies to two-way plots that are displayed in the TWOWAY=GROUPVERTICAL or TWOWAY=GROUPHORIZONTAL layout. The NPANELPOS= *plot-option* does not apply to the TWOWAY=CLUSTER and TWOWAY=STACKED layouts, which are always displayed in a single panel.

For plots that display statistics along with confidence limits, NPANELPOS=*n* requests that panels display at most |*n*| statistics. By default, *n*=0 and all statistics are displayed in a single panel. This *plot-option* applies to the following plots: KAPPAPLOT, ODDSRATIOPLOT, RELRISKPLOT, RISKDIFFPLOT, and WTKAPPAPLOT.

ORDER=ASCENDING | DESCENDING

displays the two-way table (strata) statistics in order of the statistic value. You can specify the ORDER= *plot-option* when you specify the following *plot-requests*: KAPPAPLOT, ODDSRATIOPLOT, RELRISKPLOT, RISKDIFFPLOT, and WTKAPPAPLOT.

If you specify ORDER=ASCENDING or ORDER=DESCENDING, the plot displays the statistics in ascending or descending order, respectively. By default, the order of the statistics in the plot matches the order that the two-way table strata appear in the multiway table display.

ORIENT=HORIZONTAL | VERTICAL

controls the orientation of the plot. You can specify the ORIENT= *plot-option* when you specify the following *plot-requests*: CUMFREQPLOT, DEVIATIONPLOT, and FREQPLOT.

ORIENT=HORIZONTAL places the variable levels on the Y axis and the frequencies, percentages, or statistic values on the X axis. ORIENT=VERTICAL places the variable levels on the X axis. The default orientation is ORIENT=VERTICAL for bar charts (TYPE=BARCHART) and ORIENT=HORIZONTAL for dot plots (TYPE=DOTPLOT).

PARTIAL=NO | YES

controls the display of partial agreement in the agreement plot (AGREEPLOT). PARTIAL=NO suppresses the display of partial agreement. When you specify PARTIAL=NO, the agreement plot displays only exact agreement. Exact agreement includes the diagonal cells of the square table, where the row and column variable levels are the same. Partial agreement includes the adjacent off-diagonal table cells, where the row and column values are within one level of exact agreement. The default is PARTIAL=YES.

RANGE=(< *min* > <, *max* >)| CLIP

specifies the range of values to display. You can specify the RANGE= *plot-option* when you specify the following *plot-requests*: KAPPAPLOT, ODDSRATIOPLOT, RELRISKPLOT, RISKDIFFPLOT, and WTKAPPAPLOT.

If you specify RANGE=CLIP, the confidence limits are clipped and the display range is determined by the minimum and maximum values of the statistics. By default, the display range includes all confidence limits.

SCALE=FREQ | GROUPPERCENT | LOG | PERCENT | SQRT

specifies the scale of the frequencies to display. This *plot-option* is available for frequency plots (FREQPLOT) and cumulative frequency plots (CUMFREQPLOT).

The default is SCALE=FREQ, which displays unscaled frequencies. SCALE=PERCENT displays percentages (relative frequencies) of the total frequency. SCALE=LOG displays log (base 10) frequencies. SCALE=SQRT displays square roots of the frequencies, producing a plot known as a *rootogram*.

SCALE=GROUPPERCENT is available for two-way frequency plots. This option displays the row or column percentages instead of the overall percentages (of the table frequency). By default (or when you specify the GROUPBY=COLUMN *plot-option*), SCALE=GROUPPERCENT displays the column percentages. If you specify the GROUPBY=ROW *plot-option*, the primary grouping of graph cells is by row variable level and the plot displays row percentages. For more information, see the description of the GROUPBY= *plot-option*.

SHOWSCALE=NO | YES

controls the display of the cumulative frequency scale on the right side of the agreement plot (AGREEPLOT). SHOWSCALE=NO suppresses the display of the scale. The default is SHOWSCALE=YES.

STATS
: displays statistic values in the plot. For the following *plot-requests*, the STATS *plot-option* displays the statistics and their confidence limits on the right side of the plot: KAPPAPLOT, ODDSRATIOPLOT, RELRISKPLOT, RISKDIFFPLOT, and WTKAPPAPLOT.

  For the agreement plot (AGREEPLOT), the STATS *plot-option* displays the values of the kappa statistic, the weighted kappa statistic, the B_n measure (Bangdiwala and Bryan 1987), and the sample size. PROC FREQ stores these statistics in an ODS table named BnMeasure, which is not displayed. For more information, see the section "ODS Table Names" on page 239.

  If you do not request the STATS *plot-option*, these plots do not display the statistic values.

TWOWAY=CLUSTER | GROUPHORIZONTAL | GROUPVERTICAL | STACKED
: specifies the layout for two-way frequency plots.

  All TWOWAY= layouts are available for bar charts (TYPE=BARCHART). All TWOWAY= layouts except TWOWAY=CLUSTER are available for dot plots (TYPE=DOTPLOT). The ORIENT= and GROUPBY= *plot-options* are available for all TWOWAY= layouts.

  The default two-way layout is TWOWAY=GROUPVERTICAL, which produces a grouped plot that has a vertical common baseline. By default for bar charts (TYPE=BARCHART, ORIENT=VERTICAL), the X axis displays column variable levels, and the Y axis displays frequencies. The plot includes a vertical (Y-axis) block for each row variable level. The relative positions of the graph cells in this plot layout are the same as the relative positions of the table cells in the crosstabulation table. You can reverse the default row and column grouping by specifying the GROUPBY=ROW *plot-option*.

  The TWOWAY=GROUPHORIZONTAL layout produces a grouped plot that has a horizontal common baseline. By default (GROUPBY=COLUMN), the plot displays a block on the X axis for each column variable level. Within each column-level block, the plot displays row variable levels.

  The TWOWAY=STACKED layout produces stacked displays of frequencies. By default (GROUPBY=COLUMN) in a stacked bar chart, the bars correspond to column variable levels, and row levels are stacked within each column level. By default in a stacked dot plot, the dotted lines correspond to column levels, and cell frequencies are plotted as data dots on the corresponding column line. The dot color identifies the row level.

  The TWOWAY=CLUSTER layout, which is available only for bar charts, displays groups of adjacent bars. By default, the primary grouping is by column variable level, and row levels are displayed within each column level.

  You can reverse the default row and column grouping in any layout by specifying the GROUPBY=ROW *plot-option*. The default is GROUPBY=COLUMN, which groups first by column variable.

TYPE=BARCHART | DOTPLOT
: specifies the plot type (format) of the frequency (FREQPLOT), cumulative frequency (CUMFREQPLOT), and deviation plots (DEVIATIONPLOT). TYPE=BARCHART produces a bar chart and TYPE=DOTPLOT produces a dot plot. The default is TYPE=BARCHART.

PRINTKWTS

displays the agreement weights that PROC FREQ uses to compute the weighted kappa coefficient. Agreement weights reflect the relative agreement between pairs of variable levels. By default, PROC FREQ uses the Cicchetti-Allison form of agreement weights. If you specify the AGREE(WT=FC) option, the procedure uses the Fleiss-Cohen form of agreement weights. For more information, see the section "Weighted Kappa Coefficient" on page 208.

This option has no effect unless you also specify the AGREE option to compute the weighted kappa coefficient. The PRINTKWTS option is equivalent to the AGREE(PRINTKWTS) option.

RELRISK < (*relrisk-options*) >

requests relative risk measures and their confidence limits for 2 × 2 tables. These measures include the odds ratio, the column 1 relative risk, and the column 2 relative risk. For more information, see the section "Odds Ratio and Relative Risks for 2 × 2 Tables" on page 192. By default, PROC FREQ displays the relative risk measures and their asymptotic Wald confidence limits in the "Odds Ratio and Relative Risks" table. You can also obtain this table by specifying the MEASURES option, which produces other measures of association in addition to the relative risks.

You can specify *relrisk-options* in parentheses after the RELRISK option to request tests and additional confidence limits for the column 1 or column 2 relative risk. Table 3.18 summarizes the *relrisk-options*.

When you request tests or additional confidence limit types for the relative risk, PROC FREQ does not display the "Odds Ratio and Relative Risks" table unless you also specify the PRINTALL *relrisk-option*.

Table 3.18   RELRISK (Relative Risk) Options

Option	Description
COLUMN=1	2	Specifies the risk column
PRINTALL	Displays "Odds Ratio and Relative Risks" table
Request Confidence Limits	
CL=EXACT	Displays exact confidence limits
CL=LR	Requests likelihood ratio confidence limits
CL=SCORE	Requests score confidence limits
CL=WALD	Requests Wald confidence limits
CL=WALDMODIFIED	Requests Wald modified confidence limits
Request Tests	
EQUAL(NULL=)	Requests an equality test
EQUIV	EQUIVALENCE	Requests an equivalence test
MARGIN=	Specifies the test margin
METHOD=	Specifies the test method
NONINF	NONINFERIORITY	Requests a noninferiority test
SUP	SUPERIORITY	Requests a superiority test

You can specify the following *relrisk-options*:

CL=*type* | (*types*)

specifies confidence limit types for the relative risk. You can specify one or more *types* of confidence limits. When you specify only one *type*, you can omit the parentheses around the request. When you specify the CL= *relrisk-option*, PROC FREQ displays the confidence limits in the "Confidence Limits for the Relative Risk" table.

The ALPHA= option determines the level of the confidence limits that the CL= *relrisk-option* provides. By default, ALPHA=0.05, which produces 95% confidence limits for the relative risk.

You can specify the following *types*:

EXACT

> displays exact unconditional confidence limits for the relative risk in the "Confidence Limits for the Relative Risk" table. (By default, PROC FREQ displays the exact confidence limits in a separate table.) You must also request computation of the exact confidence limits by specifying the RELRISK option in the EXACT statement. For more information, see the subsection "Exact Unconditional Confidence Limits" in the section "Confidence Limits for the Relative Risk" on page 196.

LR
LIKELIHOOD RATIO

> requests likelihood ratio confidence limits for the relative risk. For more information, see the subsection "Likelihood Ratio Confidence Limits" in the section "Confidence Limits for the Relative Risk" on page 196.

SCORE < (CORRECT=NO) >

> requests score confidence limits for the relative risk. For more information, see the subsection "Score Confidence Limits" in the section "Confidence Limits for the Relative Risk" on page 196. If you specify CORRECT=NO, PROC FREQ provides the uncorrected form of the confidence limits.

WALD

> requests asymptotic Wald confidence limits, which are based on a log transformation of the relative risk. For more information, see the subsection "Wald Confidence Limits" in the section "Confidence Limits for the Relative Risk" on page 196.

WALDMODIFIED

> requests Wald modified confidence limits for the odds ratio. For more information, see the subsection "Wald Modified Confidence Limits" in the section "Confidence Limits for the Relative Risk" on page 196.

COLUMN=1 | 2

> specifies the table column for which to compute the relative risk confidence limits (which you request by specifying the CL= *relrisk-option*) and the relative risk tests (EQUAL, EQUIV, NONINF, and SUP). By default, COLUMN=1.
>
> This option has no effect on the "Odds Ratio and Relative Risks" table, which displays both column 1 and column 2 relative risks.

EQUAL < (NULL=*value* **)>**

> requests an equality test for the relative risk. For more information, see the subsection "Equality Test" in the section "Relative Risk Tests" on page 199. You can specify the test in the METHOD= *relrisk-option*, and you can specify the null hypothesis *value* of the relative risk in the NULL= option. The null *value* must be a positive number. By default, METHOD=WALD and NULL=1.

EQUIV
EQUIVALENCE
> requests an equivalence test for the relative risk. For more information, see the subsection "Equivalence Test" in the section "Relative Risk Tests" on page 199. You can specify the test method in the METHOD= *relrisk-option*, and you can specify the test margins in the MARGIN= *relrisk-option*. By default, METHOD=WALD and MARGIN=(0.8,1.25).

MARGIN=*value* | (*lower*, *upper*)
> specifies the margin for the noninferiority, superiority, and equivalence tests, which you request by specifying the NONINF, SUP, and EQUIV *relrisk-options*, respectively. By default, MARGIN=0.8 for noninferiority tests, MARGIN=1.25 for superiority tests, and MARGIN=(0.8,1.25) for equivalence tests.
>
> For noninferiority and superiority tests, specify a single *value* in the MARGIN= option. The *value* must be a positive number. For a noninferiority test, the *value* should be less than 1; for a superiority test, the *value* should be greater than 1.
>
> For an equivalence test, you can specify a single MARGIN= *value*, or you can specify both *lower* and *upper* values. All values must be positive numbers. If you specify a single *value*, PROC FREQ uses *value* as the lower margin and the inverse of *value* as the upper margin. If you specify both *lower* and *upper* values, the value of *lower* must be less than the value of *upper*.

METHOD=*method*
> specifies the method to be used for the equality, equivalence, noninferiority, and superiority tests, which you request by specifying the EQUAL, EQUIV, NONINF, and SUP *relrisk-options*, respectively. By default, METHOD=WALD.
>
> You can specify one of the following *methods*:
>
> **FM**
> **SCORE**
>> requests Farrington-Manning (score) tests for the equality, equivalence, noninferiority, and superiority analyses of the relative risk. For more information, see the subsection "Farrington-Manning (Score) Test" in the section "Relative Risk Tests" on page 199.
>
> **LR**
> **LIKELIHOODRATIO**
>> requests likelihood ratio tests for the equality, equivalence, noninferiority, and superiority analyses of the relative risk. For more information, see the subsection "Likelihood Ratio Test" in the section "Relative Risk Tests" on page 199.
>
> **WALD**
>> requests Wald tests for the equality, equivalence, noninferiority, and superiority analyses of the relative risk. For more information, see the subsection "Wald Test" in the section "Relative Risk Tests" on page 199.
>
> **WALDMODIFIED**
>> requests Wald modified tests for the equality, equivalence, noninferiority, and superiority analyses of the relative risk. For more information, see the subsection "Wald Modified Test" in the section "Relative Risk Tests" on page 199.

NONINF
NONINFERIORITY

    requests a noninferiority test for the relative risk. For more information, see the subsection "Noninferiority Test" in the section "Relative Risk Tests" on page 199. You can specify the test method in the METHOD= *relrisk-option*, and you can specify the margin in the MARGIN= *relrisk-option*. By default, METHOD=WALD and MARGIN=0.8.

PRINTALL

    displays the "Odds Ratio and Relative Risks" table when you request tests or additional confidence limits by specifying *relrisk-options*. By default, PROC FREQ does not display this table when you request tests or additional confidence limits for the relative risk.

SUP
SUPERIORITY

    requests a superiority test for the relative risk. For more information, see the subsection "Superiority Test" in the section "Relative Risk Tests" on page 199. You can specify the test method in the METHOD= *relrisk-option*, and you can specify the margin in the MARGIN= *relrisk-option*. By default, METHOD=WALD and MARGIN=1.25.

RISKDIFF < (*riskdiff-options*) >

requests risks (binomial proportions) and risk differences for 2 × 2 tables. By default, this option provides the row 1 risk, row 2 risk, total (overall) risk, and risk difference (row 1 – row 2), together with their asymptotic standard errors and Wald confidence limits; by default, this option also provides exact (Clopper-Pearson) confidence limits for the row 1, row 2, and total risks. You can request exact unconditional confidence limits for the risk difference by specifying the RISKDIFF option in the EXACT statement. PROC FREQ displays these results in the column 1 and column 2 "Risk Estimates" tables (which you can suppress by specifying the NORISKS *riskdiff-option*).

You can specify *riskdiff-options* in parentheses after the RISKDIFF option to request tests and additional confidence limits for the risk difference, in addition to estimates of the common risk difference for multiway 2 × 2 tables. Table 3.19 summarizes the *riskdiff-options*.

The CL= *riskdiff-option* requests confidence limits for the risk difference. Available confidence limit types include Agresti-Caffo, exact unconditional, Hauck-Anderson, Miettinen-Nurminen (score), Newcombe, and Wald. Continuity-corrected Newcombe and Wald confidence limits are also available. You can request more than one type of confidence limits in the same analysis. PROC FREQ displays the confidence limits in the "Confidence Limits for the Risk Difference" table.

The CL=EXACT *riskdiff-option* displays exact unconditional confidence limits in the "Confidence Limits for the Risk Difference" table. When you specify CL=EXACT, you must also request computation of the exact confidence limits by specifying the RISKDIFF option in the EXACT statement.

The EQUAL, EQUIV, NONINF, and SUP *riskdiff-options* request tests of equality, equivalence, noninferiority, and superiority, respectively, for the risk difference. Available test methods include Farrington-Manning (score), Hauck-Anderson, and Wald. Newcombe (hybrid-score) confidence limits are available for the equivalence, noninferiority, and superiority analyses.

As part of the noninferiority, superiority, and equivalence analyses, PROC FREQ provides null-based equivalence limits that have a confidence coefficient of $100(1 - 2\alpha)\%$ (Schuirmann 1999). The ALPHA= option determines the confidence level; by default, ALPHA=0.05, which produces 90% equivalence limits for these analyses. For more information, see the sections "Noninferiority Tests" on page 184 and "Equivalence Test" on page 187.

Table 3.19 RISKDIFF (Proportion Difference) Options

Option	Description
COLUMN=1 | 2	Specifies the risk column
COMMON	Requests common risk difference
CORRECT	Requests continuity correction
NORISKS	Suppresses default risk tables
Request Confidence Limits	
CL=AC	Requests Agresti-Caffo confidence limits
CL=EXACT	Displays exact confidence limits
CL=HA	Requests Hauck-Anderson confidence limits
CL=MN | SCORE	Requests Miettinen-Nurminen confidence limits
CL=NEWCOMBE	Requests Newcombe confidence limits
CL=WALD	Requests Wald confidence limits
Request Tests	
EQUAL(NULL=)	Requests an equality test
EQUIV | EQUIVALENCE	Requests an equivalence test
MARGIN=	Specifies the test margin
METHOD=	Specifies the test method
NONINF | NONINFERIORITY	Requests a noninferiority test
SUP | SUPERIORITY	Requests a superiority test
VAR=SAMPLE | NULL	Specifies the test variance

You can specify the following *riskdiff-options*:

CL=*type* | (*types*)

requests confidence limits for the risk difference. You can specify one or more *types* of confidence limits. When you specify only one *type*, you can omit the parentheses around the request. PROC FREQ displays the confidence limits in the "Confidence Limits for the Risk Difference" table.

The ALPHA= option determines the level of the confidence limits. By default, ALPHA=0.05, which produces 95% confidence limits for the risk difference.

You can specify the CL= *riskdiff-option* with or without requests for risk difference tests. The confidence limits that CL= produces do not depend on the tests that you request and do not use the value of the test margin (which you can specify in the MARGIN= *riskdiff-option*).

You can specify the following *types*:

AC

AGRESTICAFFO

requests Agresti-Caffo confidence limits for the risk difference. For more information, see the subsection "Agresti-Caffo Confidence Limits" in the section "Confidence Limits for the Risk Difference" on page 179.

EXACT

displays exact unconditional confidence limits for the risk difference in the "Confidence Limits for the Risk Difference" table. You must also request computation of the exact confidence limits by specifying the RISKDIFF option in the EXACT statement.

By default, PROC FREQ computes the exact confidence limits by inverting two separate one-sided exact tests that are based on the score statistic. For more information, see the RISKDIFF option in the EXACT statement and the subsection "Exact Unconditional Confidence Limits" in the section "Confidence Limits for the Risk Difference" on page 179.

By default, PROC FREQ also displays these exact confidence limits in the "Risk Estimates" table. You can suppress this table by specifying the NORISKS *riskdiff-option*.

HA

requests Hauck-Anderson confidence limits for the risk difference. For more information, see the subsection "Hauck-Anderson Confidence Limits" in the section "Confidence Limits for the Risk Difference" on page 179.

MN <(CORRECT=NO | MEE)>
SCORE <(CORRECT=NO | MEE)>

requests Miettinen-Nurminen (score) confidence limits for the risk difference. For more information, see the subsection "Miettinen-Nurminen (Score) Confidence Limits" in the section "Confidence Limits for the Risk Difference" on page 179. By default, the Miettinen-Nurminen confidence limits include a bias correction factor (Miettinen and Nurminen 1985; Newcombe and Nurminen 2011). If you specify CL=MN(CORRECT=NO), PROC FREQ provides the uncorrected form of the confidence limits (Mee 1984).

NEWCOMBE <(CORRECT)>

requests Newcombe hybrid-score confidence limits for the risk difference. If you specify CL=NEWCOMBE(CORRECT) or the CORRECT *riskdiff-option*, the Newcombe confidence limits include a continuity correction. For more information, see the subsection "Newcombe Confidence Limits" in the section "Confidence Limits for the Risk Difference" on page 179.

WALD <(CORRECT)>

requests Wald confidence limits for the risk difference. If you specify CL=WALD(CORRECT) or the CORRECT *riskdiff-option*, the Wald confidence limits include a continuity correction. For more information, see the subsection "Wald Confidence Limits" in the section "Confidence Limits for the Risk Difference" on page 179.

COLUMN=1 | 2 | BOTH

specifies the table column for which to compute the risk difference tests (EQUAL, EQUIV, NONINF, and SUP) and the risk difference confidence limits (which you request by specifying the CL= *riskdiff-option*). By default, COLUMN=1.

This option has no effect on the "Risk Estimates" table, which is produced for both column 1 and column 2. You can suppress the "Risk Estimates" table by specifying the NORISKS *riskdiff-option*.

COMMON

requests estimates of the common (overall) risk difference for multiway 2 × 2 tables. This option provides Mantel-Haenszel and summary score estimates for the common risk difference, together with their confidence limits. If you specify the RISKDIFF(CL=NEWCOMBE) option, the RISKDIFF(COMMON) option also provides Newcombe confidence limits for the common risk difference. For more information, see the section "Common Risk Difference" on page 188.

You can use the COMMONRISKDIFF option to request additional confidence limits and tests for the common risk difference.

If you do not specify the COLUMN= *riskdiff-option*, PROC FREQ provides the common risk difference for column 1 by default. If you specify COLUMN=2, PROC FREQ provides the common risk difference for column 2. COLUMN=BOTH does not apply to the common risk difference.

CORRECT

includes a continuity correction in the Wald confidence limits, Wald tests, and Newcombe confidence limits. For more information, see the section "Risks and Risk Differences" on page 178.

EQUAL < (NULL=value **)>**

requests an equality test for the risk difference. For more information, see the section "Equality Tests" on page 184. You can specify the test method in the METHOD= *riskdiff-option*, and you can specify the null hypothesis *value* of the risk difference in the NULL= option. By default, METHOD=WALD and NULL=0. You can specify the null *value* in proportion form as a number between −1 and 1, or you can specify the null *value* in percentage form as a number between −100 and 100. When the *value* is between −100 and −1 or between 1 and 100, PROC FREQ converts the number to a proportion. PROC FREQ treats the values −1 and 1 as percentages.

EQUIV

EQUIVALENCE

requests an equivalence test for the risk difference. For more information, see the section "Equivalence Test" on page 187. You can specify the test method in the METHOD= *riskdiff-option*, and you can specify the margins in the MARGIN= *riskdiff-option*. By default, METHOD=WALD and MARGIN=0.2.

MARGIN=value | (lower, upper)

specifies the margin for the noninferiority, superiority, and equivalence tests, which you request by specifying the NONINF, SUP, and EQUIV *riskdiff-options*, respectively. By default, MARGIN=0.2.

For noninferiority and superiority tests, specify a single *value* in the MARGIN= option. The *value* must be a positive number. You can specify *value* as a number between 0 and 1. Or you can specify *value* in percentage form as a number between 1 and 100, and PROC FREQ converts that number to a proportion. PROC FREQ treats the value 1 as 1%.

For an equivalence test, you can specify a single MARGIN= *value*, or you can specify both *lower* and *upper* values. If you specify a single *value*, it must be a positive number, as described previously. If you specify a single *value* for an equivalence test, PROC FREQ uses −*value* as the lower margin and *value* as the upper margin for the test. If you specify both *lower* and *upper* values for an equivalence test, you can specify them in proportion form as numbers between

–1 and 1. Or you can specify them in percentage form as numbers between –100 and 100, and PROC FREQ converts the numbers to proportions. The value of *lower* must be less than the value of *upper*.

METHOD=*method*

specifies the method to be used for the equality, equivalence, noninferiority, and superiority tests, which you request by specifying the EQUAL, EQUIV, NONINF, and SUP *riskdiff-options*, respectively. By default, METHOD=WALD.

You can specify the following *methods*:

FM
SCORE

requests Farrington-Manning (score) tests for the equality, equivalence, noninferiority, and superiority analyses. For more information, see the subsection "Farrington-Manning (Score) Test" in the section "Noninferiority Tests" on page 184.

HA

requests Hauck-Anderson tests for the equality, equivalence, noninferiority, and superiority analyses. For more information, see the subsection "Hauck-Anderson Test" in the section "Noninferiority Tests" on page 184.

NEWCOMBE

requests Newcombe (hybrid-score) confidence limits for the equivalence, noninferiority, and superiority analyses. If you specify the CORRECT *riskdiff-option*, the Newcombe confidence limits include a continuity correction. For more information, see the subsection "Newcombe Noninferiority Analysis" in the section "Noninferiority Tests" on page 184.

WALD

requests Wald tests for the equality, equivalence, noninferiority, and superiority analyses. If you specify the CORRECT *riskdiff-option*, the Wald tests and confidence limits include a continuity correction. If you specify the VAR=NULL *riskdiff-option*, the tests use the null (test-based) variance instead of the sample variance. For more information, see the subsection "Wald Test" in the section "Noninferiority Tests" on page 184.

NONINF
NONINFERIORITY

requests a noninferiority test for the risk difference. For more information, see the section "Noninferiority Tests" on page 184. You can specify the test method in the METHOD= *riskdiff-option*, and you can specify the margin in the MARGIN= *riskdiff-option*. By default, METHOD=WALD and MARGIN=0.2.

NORISKS

suppresses display of the "Risk Estimates" tables, which the RISKDIFF option produces by default for column 1 and column 2. The "Risk Estimates" tables contain the risks and risk differences, together with their asymptotic standard errors, Wald confidence limits, and exact confidence limits.

SUP
SUPERIORITY

> requests a superiority test for the risk difference. For more information, see the section "Superiority Test" on page 186. You can specify the test method in the METHOD= *riskdiff-option*, and you can specify the margin in the MARGIN= *riskdiff-option*. By default, METHOD=WALD and MARGIN=0.2.

VAR=NULL | SAMPLE

> specifies the type of variance to use in the Wald tests of equality, equivalence, noninferiority, and superiority. If you specify VAR=SAMPLE, PROC FREQ uses the sample variance. If you specify VAR=NULL, PROC FREQ uses a test-based variance that is computed by using the null hypothesis value of the risk difference. For more information, see the sections "Equality Tests" on page 184 and "Noninferiority Tests" on page 184. The default is VAR=SAMPLE.

SCORES=*type*

specifies the type of row and column scores that PROC FREQ uses to compute the following statistics: Mantel-Haenszel chi-square, Pearson correlation, Cochran-Armitage test for trend, weighted kappa coefficient, and Cochran-Mantel-Haenszel statistics. The value of *type* can be one of the following:

- **MODRIDIT**
- **RANK**
- **RIDIT**
- **TABLE**

See the section "Scores" on page 154 for descriptions of these score types.

If you do not specify the SCORES= option, PROC FREQ uses SCORES=TABLE by default. For character variables, the row and column TABLE scores are the row and column numbers. That is, the TABLE score is 1 for row 1, 2 for row 2, and so on. For numeric variables, the row and column TABLE scores equal the variable values. For more information, see the section "Scores" on page 154. Using MODRIDIT, RANK, or RIDIT scores yields nonparametric analyses.

You can use the SCOROUT option to display the row and column scores.

SCOROUT

> displays the row and column scores that PROC FREQ uses to compute score-based tests and statistics. You can specify the score type by using the SCORES= option. For more information, see the section "Scores" on page 154.
>
> The scores are computed and displayed only when PROC FREQ computes statistics for two-way tables. You can use ODS to store the scores in an output data set. See the section "ODS Table Names" on page 239 for more information.

SPARSE

> reports all possible combinations of the variable values for an *n*-way table when $n > 1$, even if a combination does not occur in the data. The SPARSE option applies only to crosstabulation tables displayed in LIST format and to the OUT= output data set. If you do not use the LIST or OUT= option, the SPARSE option has no effect.
>
> When you specify the SPARSE and LIST options, PROC FREQ displays all combinations of variable values in the table listing, including those that have frequency counts of 0. By default, without the

144 ✦ *Chapter 3: The FREQ Procedure*

SPARSE option, PROC FREQ does not display zero-frequency levels in LIST output. When you use the SPARSE and OUT= options, PROC FREQ includes empty crosstabulation table cells in the output data set. By default, PROC FREQ does not include zero-frequency table cells in the output data set.

See the section "Missing Values" on page 150 for more information.

TOTPCT

displays the percentage of the total multiway table frequency in crosstabulation tables for n-way tables, where $n > 2$. By default, PROC FREQ displays the percentage of the individual two-way table frequency but does not display the percentage of the total frequency for multiway crosstabulation tables. See the section "Two-Way and Multiway Tables" on page 231 for more information.

The percentage of total multiway table frequency is displayed by default when you specify the LIST option. It is also provided by default in the PERCENT variable in the OUT= output data set.

TREND

requests the Cochran-Armitage test for trend. The table must be $2 \times C$ or $R \times 2$ to compute the trend test. For more information, see the section "Cochran-Armitage Test for Trend" on page 202. To request exact p-values for the trend test, specify the TREND option in the EXACT statement. See the section "Exact Statistics" on page 221 for more information.

TEST Statement

TEST *test-options* ;

The TEST statement requests asymptotic tests for measures of association and measures of agreement. The *test-options* identify which tests to compute. Table 3.20 lists the available *test-options*, together with their corresponding TABLES statement options. Descriptions of the *test-options* follow the table in alphabetical order.

For each measure of association or agreement that you request in the TEST statement, PROC FREQ provides an asymptotic test that the measure is 0. The procedure displays the asymptotic standard error under the null hypothesis, the test statistic, and the one-sided and two-sided p-values. PROC FREQ also provides confidence limits for the measure. The ALPHA= option in the TABLES statement determines the confidence level; by default, ALPHA=0.05, which provides 95% confidence limits. For more information, see the sections "Asymptotic Tests" on page 161 and "Confidence Limits" on page 160. For information about the individual measures, see the sections "Measures of Association" on page 160 and "Tests and Measures of Agreement" on page 205.

You can also request exact tests for selected measures of association and agreement by using the EXACT statement. For more information, see the section "Exact Statistics" on page 221.

Using the TEST Statement with the TABLES Statement

You must use a TABLES statement with the TEST statement. If you use only one TABLES statement, you do not need to specify the same options in both the TABLES and TEST statements; when you specify an option in the TEST statement, PROC FREQ automatically invokes the corresponding TABLES statement option. However, when you use the TEST statement with multiple TABLES statements, you must specify options in the TABLES statements to request statistics; PROC FREQ then provides asymptotic tests for those statistics that you specify in the TEST statement.

Table 3.20 TEST Statement Options

Test Option	Asymptotic Tests	Required TABLES Statement Option
AGREE	Simple and weighted kappa coefficients	AGREE
GAMMA	Gamma	ALL or MEASURES
KAPPA	Simple kappa coefficient	AGREE
KENTB | TAUB	Kendall's tau-b	ALL or MEASURES
MEASURES	Gamma, Kendall's tau-b, Stuart's tau-c, Somers' $D(C\|R)$, Somers' $D(R\|C)$, Pearson and Spearman correlations	ALL or MEASURES
PCORR	Pearson correlation coefficient	ALL or MEASURES
PLCORR	Polychoric correlation	PLCORR
SCORR	Spearman correlation coefficient	ALL or MEASURES
SMDCR	Somers' $D(C\|R)$	ALL or MEASURES
SMDRC	Somers' $D(R\|C)$	ALL or MEASURES
STUTC | TAUC	Stuart's tau-c	ALL or MEASURES
WTKAPPA | WTKAP	Weighted kappa coefficient	AGREE

You can specify the following *test-options* in the TEST statement.

AGREE

requests asymptotic tests for the simple kappa coefficient and the weighted kappa coefficient. For more information, see the sections "Simple Kappa Coefficient" on page 206 and "Weighted Kappa Coefficient" on page 208.

By default, these tests are based on null values of 0; you can specify nonzero null values for the simple kappa and weighted kappa tests by using the AGREE(NULLKAPPA=) and AGREE(NULLWTKAPPA=) options, respectively, in the TABLES statement.

The AGREE option in the TABLES statement provides estimates, standard errors, and confidence limits for kappa coefficients. You can request exact tests for kappa coefficients by using the EXACT statement.

Kappa coefficients are defined only for square tables, where the number of rows equals the number of columns. Kappa coefficients are not computed for tables that are not square. For 2 × 2 tables, the weighted kappa coefficient is identical to the simple kappa coefficient, and PROC FREQ presents only the simple kappa coefficient.

GAMMA

requests an asymptotic test for the gamma statistic. For more information, see the section "Gamma" on page 162. The MEASURES option in the TABLES statement provides the gamma statistic and its asymptotic standard error.

KAPPA

requests an asymptotic test for the simple kappa coefficient. For more information, see the section "Simple Kappa Coefficient" on page 206.

By default, the null value of kappa for this test is 0; you can specify a nonzero null value by using the AGREE(NULLKAPPA=) option in the TABLES statement.

The AGREE option in the TABLES statement provides the kappa statistic, its standard error, and its confidence limits. You can request an exact test for the simple kappa coefficient by specifying the KAPPA option in the EXACT statement.

Kappa coefficients are defined only for square tables, where the number of rows equals the number of columns. PROC FREQ does not compute kappa coefficients for tables that are not square.

KENTB

TAUB

requests an asymptotic test for Kendall's tau-*b*. For more information, see the section "Kendall's Tau-b" on page 162.

The MEASURES option in the TABLES statement provides Kendall's tau-*b* and its standard error. You can request an exact test for Kendall's tau-*b* by specifying the KENTB option in the EXACT statement.

MEASURES

requests asymptotic tests for the following measures of association: gamma, Kendall's tau-*b*, Pearson correlation coefficient, Somers' $D(C|R)$, Somers' $D(R|C)$, Spearman correlation coefficient, and Stuart's tau-*c*. For more information, see the section "Measures of Association" on page 160.

The MEASURES option in the TABLES statement provides measures of association and their asymptotic standard errors. You can request exact tests for selected measures by using the EXACT statement.

PCORR

requests an asymptotic test for the Pearson correlation coefficient. For more information, see the section "Pearson Correlation Coefficient" on page 164.

The MEASURES option in the TABLES statement provides the Pearson correlation and its standard error. You can request an exact test for the Pearson correlation by specifying the PCORR option in the EXACT statement.

PLCORR

requests Wald and likelihood ratio tests for the polychoric correlation coefficient. For more information, see the section "Polychoric Correlation" on page 166.

The PLCORR option in the TABLES statement provides the polychoric correlation and its standard error.

SCORR

requests an asymptotic test for the Spearman correlation coefficient. For more information, see the section "Spearman Rank Correlation Coefficient" on page 165.

The MEASURES option in the TABLES statement provides the Spearman correlation and its standard error. You can request an exact test for the Spearman correlation by specifying the SCORR option in the EXACT statement.

SMDCR

requests an asymptotic test for Somers' $D(C|R)$. For more information, see the section "Somers' D" on page 163.

The MEASURES option in the TABLES statement provides Somers' $D(C|R)$ and its standard error. You can request an exact test for Somers' $D(C|R)$ by specifying the SMDCR option in the EXACT statement.

SMDRC

requests an asymptotic test for Somers' $D(R|C)$. For more information, see the section "Somers' D" on page 163.

The MEASURES option in the TABLES statement provides Somers' $D(R|C)$ and its standard error. You can request an exact test for Somers' $D(R|C)$ by specifying the SMDRC option in the EXACT statement.

STUTC

TAUC

requests an asymptotic test for Stuart's tau-c. For more information, see the section "Stuart's Tau-c" on page 163.

The MEASURES option in the TABLES statement provides Stuart's tau-c and its standard error. You can request an exact test for Stuart's tau-c by specifying the STUTC option in the EXACT statement.

WTKAPPA

WTKAP

requests an asymptotic test for the weighted kappa coefficient. For more information, see the section "Weighted Kappa Coefficient" on page 208.

By default, the null value of weighted kappa for this test is 0; you can specify a nonzero null value by using the AGREE(NULLWTKAPPA=) option in the TABLES statement.

The AGREE option in the TABLES statement provides the weighted kappa coefficient, its standard error, and confidence limits. You can request an exact test for the weighted kappa by specifying the WTKAPPA option in the EXACT statement.

Kappa coefficients are defined only for square tables, where the number of rows equals the number of columns. PROC FREQ does not compute kappa coefficients for tables that are not square. For 2×2 tables, the weighted kappa coefficient is identical to the simple kappa coefficient, and PROC FREQ presents only the simple kappa coefficient.

WEIGHT Statement

WEIGHT *variable* < / *option* > ;

The WEIGHT statement names a numeric variable that provides a weight for each observation in the input data set. The WEIGHT statement is most commonly used to input cell count data. See the section "Inputting Frequency Counts" on page 148 for more information. If you use a WEIGHT statement, PROC FREQ assumes that an observation represents n observations, where n is the value of *variable*. The value of the WEIGHT variable is not required to be an integer.

If the value of the WEIGHT variable is missing, PROC FREQ does not use that observation in the analysis. If the value of the WEIGHT variable is 0, PROC FREQ ignores the observation unless you specify the ZEROS option, which includes observations that have weights of 0. If you do not specify a WEIGHT statement, PROC FREQ assigns a weight of 1 to each observation. The sum of the WEIGHT variable values represents the total number of observations.

If any value of the WEIGHT variable is negative, PROC FREQ displays the frequencies computed from the weighted values but does not compute percentages and statistics. If you create an output data set by using the OUT= option in the TABLES statement, PROC FREQ assigns missing values to the PERCENT variable. PROC FREQ also assigns missing values to the variables that the OUTEXPECT and OUTPCT options provide. If any value of the WEIGHT variable is negative, you cannot create an output data set by using the OUTPUT statement because statistics are not computed when there are negative weights.

You can specify the following *option* in the WEIGHT statement:

ZEROS

includes observations that have weights of 0. By default, PROC FREQ ignores observations that have weights of 0.

If you specify the ZEROS option, frequency and crosstabulation tables display levels that contain only zero-weight observations. If you do not specify the ZEROS option, PROC FREQ does not process observations that have weights of 0 and therefore does not display levels that contain only zero-weight observations.

When you specify the ZEROS option, PROC FREQ includes zero-weight levels in chi-square tests and binomial computations for one-way tables. This makes it possible to compute binomial tests and estimates for a reference level that contains no observations with positive weights.

For two-way tables, the ZEROS option enables computation of kappa statistics when there are levels that contain no observations with positive weights. For more information, see the section "Tables with Zero-Weight Rows or Columns" on page 212.

Even when you specify the ZEROS option, PROC FREQ does not compute CHISQ or MEASURES statistics for two-way tables that contain a zero-weight row or column because most of these statistics are undefined in this case.

By default, the ZEROS option invokes the SPARSE option in the TABLES statement, which includes zero-weight table cells in the LIST table and OUT= data set. To suppress zero-weight cells, you can specify the NOSPARSE option in the TABLES statement.

Details: FREQ Procedure

Inputting Frequency Counts

PROC FREQ can use either raw data or cell count data to produce frequency and crosstabulation tables. *Raw data*, also known as case-record data, report the data as one record for each subject or sample member. *Cell count data* report the data as a table, listing all possible combinations of data values along with the frequency counts. This way of presenting data often appears in published results.

The following DATA step statements store raw data in a SAS data set:

```
data Raw;
   input Subject $ R C @@;
   datalines;
01 1 1   02 1 1   03 1 1   04 1 1   05 1 1
06 1 2   07 1 2   08 1 2   09 2 1   10 2 1
11 2 1   12 2 1   13 2 2   14 2 2   14 2 2
;
```

You can store the same data as cell counts by using the following DATA step statements:

```
data CellCounts;
   input R C Count @@;
   datalines;
1 1 5   1 2 3
2 1 4   2 2 3
;
```

The variable R contains the values for the rows, and the variable C contains the values for the columns. The variable Count contains the cell count for each row and column combination.

Both the Raw data set and the CellCounts data set produce identical frequency counts, two-way tables, and statistics. When using the CellCounts data set, you must include a WEIGHT statement to specify that the variable Count contains cell counts. For example, the following PROC FREQ statements create a two-way crosstabulation table by using the CellCounts data set:

```
proc freq data=CellCounts;
   tables R*C;
   weight Count;
run;
```

Grouping with Formats

PROC FREQ groups a variable's values according to its formatted values. If you assign a format to a variable with a FORMAT statement, PROC FREQ formats the variable values before dividing observations into the levels of a frequency or crosstabulation table.

For example, suppose that variable X has the values 1.1, 1.4, 1.7, 2.1, and 2.3. Each of these values appears as a level in the frequency table. If you decide to round each value to a single digit, include the following statement in the PROC FREQ step:

```
format X 1.;
```

Now the table lists the frequency count for formatted level 1 as two and for formatted level 2 as three.

PROC FREQ treats formatted character variables in the same way. The formatted values are used to group the observations into the levels of a frequency table or crosstabulation table. PROC FREQ uses the entire value of a character format to classify an observation.

You can also use the FORMAT statement to assign formats that were created with the FORMAT procedure to the variables. User-written formats determine the number of levels for a variable and provide labels for a table. If you use the same data with different formats, you can produce frequency counts and statistics for different classifications of the variable values.

When you use PROC FORMAT to create a user-written format that combines missing and nonmissing values into one category, PROC FREQ treats the entire category of formatted values as missing. For example, a questionnaire codes 1 as yes, 2 as no, and 8 as a no answer. The following PROC FORMAT statements create a user-written format:

```
proc format;
   value Questfmt 1   ='Yes'
                  2   ='No'
                  8,. ='Missing';
run;
```

When you use a FORMAT statement to assign Questfmt. to a variable, the variable's frequency table no longer includes a frequency count for the response of 8. You must use the MISSING or MISSPRINT option in the TABLES statement to list the frequency for no answer. The frequency count for this level includes observations with either a value of 8 or a missing value (.).

The frequency or crosstabulation table lists the values of both character and numeric variables in ascending order based on internal (unformatted) variable values unless you change the order with the ORDER= option. To list the values in ascending order by formatted values, use ORDER=FORMATTED in the PROC FREQ statement.

For more information about the FORMAT statement, see *SAS Formats and Informats: Reference*.

Missing Values

When the value of the WEIGHT variable is missing, PROC FREQ does not include that observation in the analysis.

PROC FREQ treats missing BY variable values like any other BY variable value. The missing values form a separate BY group.

If an observation has a missing value for a variable in a TABLES request, by default PROC FREQ does not include that observation in the frequency or crosstabulation table. Also by default, PROC FREQ does not include observations with missing values in the computation of percentages and statistics. The procedure displays the number of missing observations following each table.

PROC FREQ also reports the number of missing values in output data sets. The TABLES statement OUT= data set includes an observation that contains the missing value frequency. The NMISS option in the OUTPUT statement provides an output data set variable that contains the missing value frequency.

The following options change the way in which PROC FREQ handles missing values of TABLES variables:

MISSPRINT   displays missing value frequencies in frequency or crosstabulation tables but does not include them in computations of percentages or statistics.

MISSING   treats missing values as a valid nonmissing level for all TABLES variables. Displays missing levels in frequency and crosstabulation tables and includes them in computations of percentages and statistics.

This example shows the three ways that PROC FREQ can handle missing values of TABLES variables. The following DATA step statements create a data set with a missing value for the variable A:

```
data one;
   input A Freq;
   datalines;
1 2
```

```
2 2
. 2
;
```

The following PROC FREQ statements request a one-way frequency table for the variable A. The first request does not specify a missing value option. The second request specifies the MISSPRINT option in the TABLES statement. The third request specifies the MISSING option in the TABLES statement.

```
proc freq data=one;
   tables A;
   weight Freq;
   title 'Default';
run;
proc freq data=one;
   tables A / missprint;
   weight Freq;
   title 'MISSPRINT Option';
run;
proc freq data=one;
   tables A / missing;
   weight Freq;
   title 'MISSING Option';
run;
```

Figure 3.12 displays the frequency tables produced by this example. The first table shows PROC FREQ's default behavior for handling missing values. The observation with a missing value of the TABLES variable A is not included in the table, and the frequency of missing values is displayed following the table. The second table, for which the MISSPRINT option is specified, displays the missing observation but does not include its frequency when computing the total frequency and percentages. The third table shows that PROC FREQ treats the missing level as a valid nonmissing level when the MISSING option is specified. The table displays the missing level, and PROC FREQ includes this level when computing frequencies and percentages.

Figure 3.12 Missing Values in Frequency Tables

Default

The FREQ Procedure

A	Frequency	Percent	Cumulative Frequency	Cumulative Percent
1	2	50.00	2	50.00
2	2	50.00	4	100.00

Frequency Missing = 2

MISSPRINT Option

The FREQ Procedure

A	Frequency	Percent	Cumulative Frequency	Cumulative Percent
.	2	.	.	.
1	2	50.00	2	50.00
2	2	50.00	4	100.00

Frequency Missing = 2

Figure 3.12 *continued*

MISSING Option

The FREQ Procedure

A	Frequency	Percent	Cumulative Frequency	Cumulative Percent
.	2	33.33	2	33.33
1	2	33.33	4	66.67
2	2	33.33	6	100.00

When a combination of variable values in a two-way table is missing, PROC FREQ assigns 0 to the frequency count of the corresponding table cell. By default, PROC FREQ does not include missing combinations in the LIST display or the OUT= output data set. To include missing combinations in the LIST display and the OUT= output data set, you can specify the SPARSE option in the TABLES statement.

In-Database Computation

The FREQ procedure can use in-database computation to construct frequency and crosstabulation tables when the DATA= input data set is stored as a table in a supported database management system (DBMS). PROC FREQ supports the following database management systems: Aster, DB2, Greenplum, Hadoop, HAWQ, Impala, Netazza, Oracle, SAP HANA, and Teradata. In-database computation can provide the advantages of faster processing and reduced data transfer between the database and SAS software. For information about in-database computation, see the section "In-Database Procedures" in *SAS/ACCESS for Relational Databases: Reference*.

PROC FREQ performs in-database computation by using SQL implicit pass-through. The procedure generates SQL queries that are based on the tables that you request in the TABLES statement. The database executes these SQL queries to construct initial summary tables, which are then transmitted to PROC FREQ. The procedure uses this summary information to perform the remaining analyses and tasks in the usual way (out of the database). Instead of transferring the entire data set over the network between the database and SAS software, in-database computation transfers only the summary tables. This can substantially reduce processing time when the dimensions of the summary tables (in terms of rows and columns) are much smaller than the dimensions of the entire database table (in terms of individual observations). In addition, in-database summarization uses efficient parallel processing, which can also provide performance advantages.

In-database computation is controlled by the SQLGENERATION option, which you can specify in either a LIBNAME statement or an OPTIONS statement. For information about the SQLGENERATION option and other options that affect in-database computation, see the section "In-Database Procedures" in *SAS/ACCESS for Relational Databases: Reference*. By default, PROC FREQ uses in-database computation when possible. PROC FREQ has no procedure options that control in-database computation.

PROC FREQ uses formatted values to group observations into the levels of frequency and crosstabulation tables. For more information, see the section "Grouping with Formats" on page 149. If formats are available in the database, in-database summarization uses the formats. If formats are not available in the database, the in-database summarization uses the raw data values, and PROC FREQ performs the final, formatted classification (out of the database). For more information, see the section "Deploying and Using SAS Formats in Teradata" in *SAS/ACCESS for Relational Databases: Reference*.

The order of observations is not inherently defined for DBMS tables. The following options relate to the order of observations and therefore should not be specified for PROC FREQ in-database computation:

- If you specify the FIRSTOBS= or OBS= data set option, PROC FREQ does not perform in-database computation.

- If you specify the NOTSORTED option in the BY statement, PROC FREQ in-database computation ignores it and uses the default ASCENDING order for BY variables.

- If you specify the ORDER=DATA option for input data in a DBMS table, PROC FREQ computation might produce different results for separate runs of the same analysis. In addition to determining the order of variable levels in crosstabulation table displays, the ORDER= option can also affect the values of many of the test statistics and measures that PROC FREQ computes.

Statistical Computations

Definitions and Notation

A two-way table represents the crosstabulation of row variable X and column variable Y. Let the table row values or levels be denoted by X_i, $i = 1, 2, \ldots, R$, and the column values by Y_j, $j = 1, 2, \ldots, C$. Let n_{ij} denote the frequency of the table cell in the ith row and jth column and define the following notation:

$$n_{i.} = \sum_j n_{ij} \quad \text{(row totals)}$$

$$n_{.j} = \sum_i n_{ij} \quad \text{(column totals)}$$

$$n = \sum_i \sum_j n_{ij} \quad \text{(overall total)}$$

$$p_{ij} = n_{ij}/n \quad \text{(cell percentages)}$$

$$p_{i.} = n_{i.}/n \quad \text{(row percentages of total)}$$

$$p_{.j} = n_{.j}/n \quad \text{(column percentages of total)}$$

$$R_i = \text{score for row } i$$

$$C_j = \text{score for column } j$$

$$\bar{R} = \sum_i n_{i.} R_i / n \quad \text{(average row score)}$$

$$\bar{C} = \sum_j n_{.j} C_j / n \quad \text{(average column score)}$$

$$A_{ij} = \sum_{k>i}\sum_{l>j} n_{kl} + \sum_{k<i}\sum_{l<j} n_{kl}$$

$$D_{ij} = \sum_{k>i}\sum_{l<j} n_{kl} + \sum_{k<i}\sum_{l>j} n_{kl}$$

$$P = \sum_i \sum_j n_{ij} A_{ij} \quad \text{(twice the number of concordances)}$$

$$Q = \sum_i \sum_j n_{ij} D_{ij} \quad \text{(twice the number of discordances)}$$

Scores

PROC FREQ uses scores of the variable values to compute the Mantel-Haenszel chi-square, Pearson correlation, Cochran-Armitage test for trend, weighted kappa coefficient, and Cochran-Mantel-Haenszel statistics. The SCORES= option in the TABLES statement specifies the score type that PROC FREQ uses. The available score types are TABLE, RANK, RIDIT, and MODRIDIT scores. The default score type is TABLE. Using MODRIDIT, RANK, or RIDIT scores yields nonparametric analyses.

For numeric variables, table scores are the values of the row and column levels. If the row or column variable is formatted, then the table score is the internal numeric value corresponding to that level. If two or more numeric values are classified into the same formatted level, then the internal numeric value for that level is the smallest of these values. For character variables, table scores are defined as the row numbers and column numbers (that is, 1 for the first row, 2 for the second row, and so on).

Rank scores, which you request with the SCORES=RANK option, are defined as

$$R_i^1 = \sum_{k<i} n_{k\cdot} + (n_{i\cdot} + 1)/2 \quad i = 1, 2, \ldots, R$$

$$C_j^1 = \sum_{l<j} n_{\cdot l} + (n_{\cdot j} + 1)/2 \quad j = 1, 2, \ldots, C$$

where R_i^1 is the rank score of row i, and C_j^1 is the rank score of column j. Note that rank scores yield midranks for tied values.

Ridit scores, which you request with the SCORES=RIDIT option, are defined as rank scores standardized by the sample size (Bross 1958; Mack and Skillings 1980). Ridit scores are derived from the rank scores as

$$R_i^2 = R_i^1/n \quad i = 1, 2, \ldots, R$$

$$C_j^2 = C_j^1/n \quad j = 1, 2, \ldots, C$$

Modified ridit scores (SCORES=MODRIDIT) represent the expected values of the order statistics of the uniform distribution on (0,1) (Van Elteren 1960; Lehmann and D'Abrera 2006). Modified ridit scores are derived from rank scores as

$$R_i^3 = R_i^1/(n+1) \quad i = 1, 2, \ldots, R$$

$$C_j^3 = C_j^1/(n+1) \quad j = 1, 2, \ldots, C$$

Chi-Square Tests and Statistics

The CHISQ option provides chi-square tests of homogeneity or independence and measures of association that are based on the chi-square statistic. When you specify the CHISQ option in the TABLES statement, PROC FREQ computes the following chi-square tests for each two-way table: Pearson chi-square, likelihood ratio chi-square, and Mantel-Haenszel chi-square tests. PROC FREQ provides the following measures of association that are based on the Pearson chi-square statistic: phi coefficient, contingency coefficient, and Cramér's *V*. For 2 × 2 tables, the CHISQ option also provides Fisher's exact test and the continuity-adjusted chi-square statistic. You can request Fisher's exact test for general $R \times C$ tables by specifying the FISHER option in the TABLES or EXACT statement.

If you specify the CHISQ option for one-way tables, PROC FREQ provides a one-way Pearson chi-square goodness-of-fit test. If you specify the CHISQ(LRCHI) option for one-way tables, PROC FREQ also provides a one-way likelihood ratio chi-square test. The other tests and statistics that the CHISQ option produces are available only for two-way tables.

For two-way tables, the null hypothesis for the chi-square tests is no association between the row variable and the column variable. When the sample size n is large, the test statistics have asymptotic chi-square distributions under the null hypothesis. When the sample size is not large, or when the data set is sparse or heavily tied, exact tests might be more appropriate than asymptotic tests. PROC FREQ provides exact *p*-values for the Pearson chi-square, likelihood ratio chi-square, and Mantel-Haenszel chi-square tests, in addition to Fisher's exact test. For one-way tables, PROC FREQ provides exact *p*-values for the Pearson and likelihood ratio chi-square goodness-of-fit tests. You can request these exact tests by specifying the corresponding options in the EXACT statement. See the section "Exact Statistics" on page 221 for more information.

The Mantel-Haenszel chi-square statistic is appropriate only when both variables lie on an ordinal scale. The other chi-square tests and statistics in this section are appropriate for either nominal or ordinal variables. The following sections give the formulas that PROC FREQ uses to compute the chi-square tests and statistics. For more information about these statistics, see Agresti (2007) and Stokes, Davis, and Koch (2012), and the other references cited.

Chi-Square Test for One-Way Tables

For one-way frequency tables, the CHISQ option in the TABLES statement provides a chi-square goodness-of-fit test. Let *C* denote the number of classes, or levels, in the one-way table. Let f_i denote the frequency of class *i* (or the number of observations in class *i*) for $i = 1, 2, \ldots, C$. Then PROC FREQ computes the one-way chi-square statistic as

$$Q_P = \sum_{i=1}^{C} (f_i - e_i)^2 / e_i$$

where e_i is the expected frequency for class *i* under the null hypothesis.

In the test for equal proportions, which is the default for the CHISQ option, the null hypothesis specifies equal proportions of the total sample size for each class. Under this null hypothesis, the expected frequency for each class equals the total sample size divided by the number of classes,

$$e_i = n / C \quad \text{for} \quad i = 1, 2, \ldots, C$$

In the test for specified frequencies, which PROC FREQ computes when you input null hypothesis frequencies by using the TESTF= option, the expected frequencies are the TESTF= values that you specify. In the test for specified proportions, which PROC FREQ computes when you input null hypothesis proportions by using the TESTP= option, the expected frequencies are determined from the specified TESTP= proportions p_i as

$$e_i = p_i \times n \quad \text{for } i = 1, 2, \ldots, C$$

Under the null hypothesis (of equal proportions, specified frequencies, or specified proportions), Q_P has an asymptotic chi-square distribution with C–1 degrees of freedom.

In addition to the asymptotic test, you can request an exact one-way chi-square test by specifying the CHISQ option in the EXACT statement. See the section "Exact Statistics" on page 221 for more information.

Pearson Chi-Square Test for Two-Way Tables

The Pearson chi-square for two-way tables involves the differences between the observed and expected frequencies, where the expected frequencies are computed under the null hypothesis of independence. The Pearson chi-square statistic is computed as

$$Q_P = \sum_i \sum_j (n_{ij} - e_{ij})^2 / e_{ij}$$

where n_{ij} is the observed frequency in table cell (i, j) and e_{ij} is the expected frequency for table cell (i, j). The expected frequency is computed under the null hypothesis that the row and column variables are independent,

$$e_{ij} = (n_{i.} \, n_{.j}) / n$$

When the row and column variables are independent, Q_P has an asymptotic chi-square distribution with $(R$–$1)(C$–$1)$ degrees of freedom. For large values of Q_P, this test rejects the null hypothesis in favor of the alternative hypothesis of general association.

In addition to the asymptotic test, you can request an exact Pearson chi-square test by specifying the PCHI or CHISQ option in the EXACT statement. See the section "Exact Statistics" on page 221 for more information.

For 2×2 tables, the Pearson chi-square is also appropriate for testing the equality of two binomial proportions. For $R \times 2$ and $2 \times C$ tables, the Pearson chi-square tests the homogeneity of proportions. For more information, see Fienberg (1980).

Standardized Residuals

When you specify the CROSSLIST(STDRES) option in the TABLES statement for two-way or multiway tables, PROC FREQ displays the standardized residuals in the CROSSLIST table.

The standardized residual of a crosstabulation table cell is the ratio of (*frequency – expected*) to its standard error, where *frequency* is the table cell frequency and *expected* is the estimated expected cell frequency. The expected frequency is computed under the null hypothesis that the row and column variables are independent. See the section "Pearson Chi-Square Test for Two-Way Tables" on page 156 for more information.

PROC FREQ computes the standardized residual of table cell (i, j) as

$$(n_{ij} - e_{ij}) / \sqrt{e_{ij}(1 - p_{i.})(1 - p_{.j})}$$

where n_{ij} is the observed frequency of table cell (i, j), e_{ij} is the expected frequency of the table cell, $p_{i.}$ is the proportion in row i $(n_{i.}/n)$, and $p_{.j}$ is the proportion in column j $(n_{.j}/n)$. The expected frequency of table cell (i, j) is computed as

$$e_{ij} = (n_{i.} \, n_{.j}) / n$$

Under the null hypothesis of independence, each standardized residual has an asymptotic standard normal distribution. See section 2.4.5 of Agresti (2007) for more information.

Likelihood Ratio Chi-Square Test for One-Way Tables

For one-way frequency tables, the CHISQ(LRCHI) option in the TABLES statement provides a likelihood ratio chi-square goodness-of-fit test. By default, the likelihood ratio test is based on the null hypothesis of equal proportions in the C classes (levels) of the one-way table. If you specify null hypothesis proportions or frequencies by using the CHISQ(TESTP=) or CHISQ(TESTF=) option, respectively, the likelihood ratio test is based on the null hypothesis values that you specify.

PROC FREQ computes the one-way likelihood ratio test as

$$G^2 = 2 \sum_{i=1}^{C} f_i \ln(f_i/e_i)$$

where f_i is the observed frequency of class i, and e_i is the expected frequency of class i under the null hypothesis.

For the null hypothesis of equal proportions, the expected frequency of each class equals the total sample size divided by the number of classes,

$$e_i = n / C \quad \text{for} \quad i = 1, 2, \ldots, C$$

If you provide null hypothesis frequencies by specifying the CHISQ(TESTF=) option in the TABLES statement, the expected frequencies are the TESTF= values that you specify. If you provide null hypothesis proportions by specifying the CHISQ(TESTP=) option in the TABLES statement, PROC FREQ computes the expected frequencies as

$$e_i = p_i \times n \quad \text{for} \quad i = 1, 2, \ldots, C$$

where the proportions p_i are the TESTP= values that you specify.

Under the null hypothesis (of equal proportions, specified frequencies, or specified proportions), the likelihood ratio statistic G^2 has an asymptotic chi-square distribution with $C-1$ degrees of freedom.

In addition to the asymptotic test, you can request an exact one-way likelihood ratio chi-square test by specifying the LRCHI option in the EXACT statement. See the section "Exact Statistics" on page 221 for more information.

Likelihood Ratio Chi-Square Test

The likelihood ratio chi-square involves the ratios between the observed and expected frequencies. The likelihood ratio chi-square statistic is computed as

$$G^2 = 2 \sum_i \sum_j n_{ij} \ln(n_{ij}/e_{ij})$$

where n_{ij} is the observed frequency in table cell (i, j) and e_{ij} is the expected frequency for table cell (i, j).

When the row and column variables are independent, G^2 has an asymptotic chi-square distribution with $(R-1)(C-1)$ degrees of freedom.

In addition to the asymptotic test, you can request an exact likelihood ratio chi-square test by specifying the LRCHI or CHISQ option in the EXACT statement. See the section "Exact Statistics" on page 221 for more information.

Continuity-Adjusted Chi-Square Test

The continuity-adjusted chi-square for 2 × 2 tables is similar to the Pearson chi-square, but it is adjusted for the continuity of the chi-square distribution. The continuity-adjusted chi-square is most useful for small sample sizes. The use of the continuity adjustment is somewhat controversial; this chi-square test is more conservative (and more like Fisher's exact test) when the sample size is small. As the sample size increases, the continuity-adjusted chi-square becomes more like the Pearson chi-square.

The continuity-adjusted chi-square statistic is computed as

$$Q_C = \sum_i \sum_j \left(\max(0, |n_{ij} - e_{ij}| - 0.5) \right)^2 / e_{ij}$$

Under the null hypothesis of independence, Q_C has an asymptotic chi-square distribution with $(R-1)(C-1)$ degrees of freedom.

Mantel-Haenszel Chi-Square Test

The Mantel-Haenszel chi-square statistic tests the alternative hypothesis that there is a linear association between the row variable and the column variable. Both variables must lie on an ordinal scale. The Mantel-Haenszel chi-square statistic is computed as

$$Q_{MH} = (n-1)r^2$$

where r is the Pearson correlation between the row variable and the column variable. For a description of the Pearson correlation, see the "Pearson Correlation Coefficient" on page 164. The Pearson correlation and thus the Mantel-Haenszel chi-square statistic use the scores that you specify in the SCORES= option in the TABLES statement. See Mantel and Haenszel (1959) and Landis, Heyman, and Koch (1978) for more information.

Under the null hypothesis of no association, Q_{MH} has an asymptotic chi-square distribution with 1 degree of freedom.

In addition to the asymptotic test, you can request an exact Mantel-Haenszel chi-square test by specifying the MHCHI or CHISQ option in the EXACT statement. See the section "Exact Statistics" on page 221 for more information.

Fisher's Exact Test

Fisher's exact test is another test of association between the row and column variables. This test assumes that the row and column totals are fixed and uses the hypergeometric distribution to compute probabilities of possible tables conditional on the observed row and column totals. Fisher's exact test does not depend on any large-sample distribution assumptions, and so it is appropriate even for small sample sizes and for sparse tables.

2 × 2 Tables For 2 × 2 tables, PROC FREQ gives the following information for Fisher's exact test: table probability, two-sided *p*-value, left-sided *p*-value, and right-sided *p*-value. The table probability equals the hypergeometric probability of the observed table, and is in fact the value of the test statistic for Fisher's exact test.

Where *p* is the hypergeometric probability of a specific table with the observed row and column totals, Fisher's exact *p*-values are computed by summing probabilities *p* over defined sets of tables,

$$\text{Prob} = \sum_A p$$

The two-sided *p*-value is the sum of all possible table probabilities (conditional on the observed row and column totals) that are less than or equal to the observed table probability. For the two-sided *p*-value, the set *A* includes all possible tables with hypergeometric probabilities less than or equal to the probability of the observed table. A small two-sided *p*-value supports the alternative hypothesis of association between the row and column variables.

For 2 × 2 tables, one-sided *p*-values for Fisher's exact test are defined in terms of the frequency of the cell in the first row and first column of the table, the (1,1) cell. Denoting the observed (1,1) cell frequency by n_{11}, the left-sided *p*-value for Fisher's exact test is the probability that the (1,1) cell frequency is less than or equal to n_{11}. For the left-sided *p*-value, the set *A* includes those tables with a (1,1) cell frequency less than or equal to n_{11}. A small left-sided *p*-value supports the alternative hypothesis that the probability of an observation being in the first cell is actually less than expected under the null hypothesis of independent row and column variables.

Similarly, for a right-sided alternative hypothesis, *A* is the set of tables where the frequency of the (1,1) cell is greater than or equal to that in the observed table. A small right-sided *p*-value supports the alternative that the probability of the first cell is actually greater than that expected under the null hypothesis.

Because the (1,1) cell frequency completely determines the 2 × 2 table when the marginal row and column sums are fixed, these one-sided alternatives can be stated equivalently in terms of other cell probabilities or ratios of cell probabilities. The left-sided alternative is equivalent to an odds ratio less than 1, where the odds ratio equals $(n_{11}n_{22}/n_{12}n_{21})$. The left-sided alternative is also equivalent to the column 1 risk for row 1 being less than the column 1 risk for row 2, $p_{1|1} < p_{1|2}$. Similarly, the right-sided alternative is equivalent to the column 1 risk for row 1 being greater than the column 1 risk for row 2, $p_{1|1} > p_{1|2}$. For more information, see Agresti (2007).

R × C Tables Fisher's exact test was extended to general *R* × *C* tables by Freeman and Halton (1951), and this test is also known as the Freeman-Halton test. For *R* × *C* tables, the two-sided *p*-value definition is the same as for 2 × 2 tables. The set *A* contains all tables with *p* less than or equal to the probability of the observed table. A small *p*-value supports the alternative hypothesis of association between the row and column variables. For *R* × *C* tables, Fisher's exact test is inherently two-sided. The alternative hypothesis is defined only in terms of general, and not linear, association. Therefore, Fisher's exact test does not have right-sided or left-sided *p*-values for general *R* × *C* tables.

For *R* × *C* tables, PROC FREQ computes Fisher's exact test by using the network algorithm of Mehta and Patel (1983), which provides a faster and more efficient solution than direct enumeration. See the section "Exact Statistics" on page 221 for more details.

Phi Coefficient

The phi coefficient is a measure of association derived from the Pearson chi-square. The range of the phi coefficient is $-1 \leq \phi \leq 1$ for 2 × 2 tables. For tables larger than 2 × 2, the range is $0 \leq \phi \leq \min(\sqrt{R-1}, \sqrt{C-1})$ (Liebetrau 1983). The phi coefficient is computed as

$$\phi = (n_{11}n_{22} - n_{12}n_{21}) / \sqrt{n_{1\cdot}n_{2\cdot}n_{\cdot 1}n_{\cdot 2}} \quad \text{for 2} \times \text{2 tables}$$

$$\phi = \sqrt{Q_P/n} \quad \text{otherwise}$$

See Fleiss, Levin, and Paik (2003, pp. 98–99) for more information.

Contingency Coefficient

The contingency coefficient is a measure of association derived from the Pearson chi-square. The range of the contingency coefficient is $0 \leq P \leq \sqrt{(m-1)/m}$, where $m = \min(R, C)$ (Liebetrau 1983). The contingency coefficient is computed as

$$P = \sqrt{Q_P / (Q_P + n)}$$

See Kendall and Stuart (1979, pp. 587–588) for more information.

Cramér's V

Cramér's V is a measure of association derived from the Pearson chi-square. It is designed so that the attainable upper bound is always 1. The range of Cramér's V is $-1 \leq V \leq 1$ for 2×2 tables; for tables larger than 2×2, the range is $0 \leq V \leq 1$. Cramér's V is computed as

$$V = \phi \quad \text{for } 2 \times 2 \text{ tables}$$

$$V = \sqrt{\frac{Q_P/n}{\min(R-1, C-1)}} \quad \text{otherwise}$$

See Kendall and Stuart (1979, p. 588) for more information.

Measures of Association

When you specify the MEASURES option in the TABLES statement, PROC FREQ computes several statistics that describe the association between the row and column variables of the contingency table. The following are measures of ordinal association that consider whether the column variable Y tends to increase as the row variable X increases: gamma, Kendall's tau-*b*, Stuart's tau-*c*, and Somers' *D*. These measures are appropriate for ordinal variables, and they classify pairs of observations as *concordant* or *discordant*. A pair is concordant if the observation with the larger value of X also has the larger value of Y. A pair is discordant if the observation with the larger value of X has the smaller value of Y. See Agresti (2007) and the other references cited for the individual measures of association.

The Pearson correlation coefficient and the Spearman rank correlation coefficient are also appropriate for ordinal variables. The Pearson correlation describes the strength of the linear association between the row and column variables, and it is computed by using the row and column scores specified by the SCORES= option in the TABLES statement. The Spearman correlation is computed with rank scores. The polychoric correlation (requested by the PLCORR option) also requires ordinal variables and assumes that the variables have an underlying bivariate normal distribution. The following measures of association do not require ordinal variables and are appropriate for nominal variables: lambda asymmetric, lambda symmetric, and the uncertainty coefficients.

PROC FREQ computes estimates of the measures according to the formulas given in the following sections. For each measure, PROC FREQ computes an asymptotic standard error (ASE), which is the square root of the asymptotic variance denoted by Var in the following sections.

Confidence Limits

If you specify the CL option in the TABLES statement, PROC FREQ computes asymptotic confidence limits for all MEASURES statistics. The confidence coefficient is determined according to the value of the ALPHA= option, which, by default, is 0.05 and produces 95% confidence limits.

The confidence limits are computed as

$$\text{Est} \pm (z_{\alpha/2} \times \text{ASE})$$

where Est is the estimate of the measure, $z_{\alpha/2}$ is the $100(1 - \alpha/2)$th percentile of the standard normal distribution, and ASE is the asymptotic standard error of the estimate.

Asymptotic Tests

For each measure that you specify in the TEST statement, PROC FREQ computes an asymptotic test of the null hypothesis that the measure is 0. Asymptotic tests are available for the following measures of association: gamma, Kendall's tau-*b*, Stuart's tau-*c*, Somers' $D(C|R)$, Somers' $D(R|C)$, the Pearson correlation coefficient, and the Spearman rank correlation coefficient. To compute an asymptotic test, PROC FREQ uses a standardized test statistic z, which has an asymptotic standard normal distribution under the null hypothesis. The test statistic is computed as

$$z = \text{Est} / \sqrt{\text{Var}_0(\text{Est})}$$

where Est is the estimate of the measure and $\text{Var}_0(\text{Est})$ is the variance of the estimate under the null hypothesis. Formulas for $\text{Var}_0(\text{Est})$ for the individual measures of association are given in the following sections.

Note that the ratio of Est to $\sqrt{\text{Var}_0(\text{Est})}$ is the same for the following measures: gamma, Kendall's tau-*b*, Stuart's tau-*c*, Somers' $D(C|R)$, and Somers' $D(R|C)$. Therefore, the tests for these measures are identical. For example, the *p*-values for the test of H_0: gamma $= 0$ equal the *p*-values for the test of H_0: tau $- b = 0$.

PROC FREQ computes one-sided and two-sided *p*-values for each of these tests. When the test statistic z is greater than its null hypothesis expected value of 0, PROC FREQ displays the right-sided *p*-value, which is the probability of a larger value of the statistic occurring under the null hypothesis. A small right-sided *p*-value supports the alternative hypothesis that the true value of the measure is greater than 0. When the test statistic is less than or equal to 0, PROC FREQ displays the left-sided *p*-value, which is the probability of a smaller value of the statistic occurring under the null hypothesis. A small left-sided *p*-value supports the alternative hypothesis that the true value of the measure is less than 0. The one-sided *p*-value P_1 can be expressed as

$$P_1 = \begin{cases} \text{Prob}(Z > z) & \text{if } z > 0 \\ \text{Prob}(Z < z) & \text{if } z \leq 0 \end{cases}$$

where Z has a standard normal distribution. The two-sided *p*-value P_2 is computed as

$$P_2 = \text{Prob}(|Z| > |z|)$$

Exact Tests

Exact tests are available for the following measures of association: Kendall's tau-*b*, Stuart's tau-*c*, Somers' $D(C|R)$ and $(R|C)$, the Pearson correlation coefficient, and the Spearman rank correlation coefficient. If you request an exact test for a measure of association in the EXACT statement, PROC FREQ computes the exact test of the hypothesis that the measure is 0. For more information, see the section "Exact Statistics" on page 221.

Gamma

The gamma (Γ) statistic is based only on the number of concordant and discordant pairs of observations. It ignores tied pairs (that is, pairs of observations that have equal values of X or equal values of Y). Gamma is appropriate only when both variables lie on an ordinal scale. The range of gamma is $-1 \leq \Gamma \leq 1$. If the row and column variables are independent, gamma tends to be close to 0. Gamma is computed as

$$G = (P - Q) / (P + Q)$$

and the asymptotic variance is

$$\text{Var}(G) = \frac{16}{(P+Q)^4} \sum_i \sum_j n_{ij} (QA_{ij} - PD_{ij})^2$$

For 2×2 tables, gamma is equivalent to Yule's Q. See Goodman and Kruskal (1979) and Agresti (2002) for more information.

The variance under the null hypothesis that gamma equals 0 is computed as

$$\text{Var}_0(G) = \frac{4}{(P+Q)^2} \left(\sum_i \sum_j n_{ij} (A_{ij} - D_{ij})^2 - (P-Q)^2/n \right)$$

For more information, see Brown and Benedetti (1977b).

Kendall's Tau-b

Kendall's tau-b (τ_b) is similar to gamma except that tau-b uses a correction for ties. Tau-b is appropriate only when both variables lie on an ordinal scale. The range of tau-b is $-1 \leq \tau_b \leq 1$. Kendall's tau-b is computed as

$$t_b = (P - Q) / \sqrt{w_r w_c}$$

and the asymptotic variance is

$$\text{Var}(t_b) = \frac{1}{w^4} \left(\sum_i \sum_j n_{ij} (2w d_{ij} + t_b v_{ij})^2 - n^3 t_b^2 (w_r + w_c)^2 \right)$$

where

$$w = \sqrt{w_r w_c}$$

$$w_r = n^2 - \sum_i n_{i\cdot}^2$$

$$w_c = n^2 - \sum_j n_{\cdot j}^2$$

$$d_{ij} = A_{ij} - D_{ij}$$

$$v_{ij} = n_{i\cdot} w_c + n_{\cdot j} w_r$$

See Kendall (1955) for more information.

The variance under the null hypothesis that tau-b equals 0 is computed as

$$\text{Var}_0(t_b) = \frac{4}{w_r w_c} \left(\sum_i \sum_j n_{ij}(A_{ij} - D_{ij})^2 - (P-Q)^2/n \right)$$

For more information, see Brown and Benedetti (1977b).

PROC FREQ also provides an exact test for the Kendall's tau-*b*. You can request this test by specifying the KENTB option in the EXACT statement. See the section "Exact Statistics" on page 221 for more information.

Stuart's Tau-c

Stuart's tau-*c* (τ_c) makes an adjustment for table size in addition to a correction for ties. Tau-*c* is appropriate only when both variables lie on an ordinal scale. The range of tau-*c* is $-1 \leq \tau_c \leq 1$. Stuart's tau-*c* is computed as

$$t_c = m(P-Q) / n^2(m-1)$$

and the asymptotic variance is

$$\text{Var}(t_c) = \frac{4m^2}{(m-1)^2 n^4} \left(\sum_i \sum_j n_{ij} d_{ij}^2 - (P-Q)^2/n \right)$$

where $m = \min(R, C)$ and $d_{ij} = A_{ij} - D_{ij}$. The variance under the null hypothesis that tau-*c* equals 0 is the same as the asymptotic variance

$$\text{Var}_0(t_c) = \text{Var}(t_c)$$

For more information, see Brown and Benedetti (1977b).

PROC FREQ also provides an exact test for the Stuart's tau-*c*. You can request this test by specifying the STUTC option in the EXACT statement. See the section "Exact Statistics" on page 221 for more information.

Somers' D

Somers' $D(C|R)$ and Somers' $D(R|C)$ are asymmetric modifications of tau-*b*. $C|R$ indicates that the row variable X is regarded as the independent variable and the column variable Y is regarded as dependent. Similarly, $R|C$ indicates that the column variable Y is regarded as the independent variable and the row variable X is regarded as dependent. Somers' *D* differs from tau-*b* in that it uses a correction only for pairs that are tied on the independent variable. Somers' *D* is appropriate only when both variables lie on an ordinal scale. The range of Somers' *D* is $-1 \leq D \leq 1$. Somers' $D(C|R)$ is computed as

$$D(C|R) = (P-Q) / w_r$$

and its asymptotic variance is

$$\text{Var}(D(C|R)) = \frac{4}{w_r^4} \sum_i \sum_j n_{ij} \left(w_r d_{ij} - (P-Q)(n - n_{i.}) \right)^2$$

where $d_{ij} = A_{ij} - D_{ij}$ and

$$w_r = n^2 - \sum_i n_{i.}^2$$

For more information, see Somers (1962); Goodman and Kruskal (1979); Liebetrau (1983).

The variance under the null hypothesis that $D(C|R)$ equals 0 is computed as

$$\text{Var}_0(D(C|R)) = \frac{4}{w_r^2} \left(\sum_i \sum_j n_{ij}(A_{ij} - D_{ij})^2 - (P-Q)^2/n \right)$$

For more information, see Brown and Benedetti (1977b).

Formulas for Somers' $D(R|C)$ are obtained by interchanging the indices.

PROC FREQ also provides exact tests for Somers' $D(C|R)$ and $(R|C)$. You can request these tests by specifying the SMDCR and SMDCR options in the EXACT statement. See the section "Exact Statistics" on page 221 for more information.

Pearson Correlation Coefficient

The Pearson correlation coefficient (ρ) is computed by using the scores specified in the SCORES= option. This measure is appropriate only when both variables lie on an ordinal scale. The range of the Pearson correlation is $-1 \leq \rho \leq 1$. The Pearson correlation coefficient is computed as

$$r = v/w = ss_{rc}/\sqrt{ss_r ss_c}$$

and its asymptotic variance is

$$\text{Var}(r) = \frac{1}{w^4} \sum_i \sum_j n_{ij} \left(w(R_i - \bar{R})(C_j - \bar{C}) - \frac{b_{ij}v}{2w} \right)^2$$

where R_i and C_j are the row and column scores and

$$ss_r = \sum_i \sum_j n_{ij}(R_i - \bar{R})^2$$

$$ss_c = \sum_i \sum_j n_{ij}(C_j - \bar{C})^2$$

$$ss_{rc} = \sum_i \sum_j n_{ij}(R_i - \bar{R})(C_j - \bar{C})$$

$$b_{ij} = (R_i - \bar{R})^2 ss_c + (C_j - \bar{C})^2 ss_r$$

$$v = ss_{rc}$$

$$w = \sqrt{ss_r ss_c}$$

See Snedecor and Cochran (1989) for more information.

The SCORES= option in the TABLES statement determines the type of row and column scores used to compute the Pearson correlation (and other score-based statistics). The default is SCORES=TABLE. See the section "Scores" on page 154 for details about the available score types and how they are computed.

The variance under the null hypothesis that the correlation equals 0 is computed as

$$\text{Var}_0(r) = \left(\sum_i \sum_j n_{ij}(R_i - \bar{R})^2(C_j - \bar{C})^2 - ss_{rc}^2/n \right) / ss_r ss_c$$

This expression for the variance is derived for multinomial sampling in a contingency table framework, and it differs from the form obtained under the assumption that both variables are continuous and normally distributed. For more information, see Brown and Benedetti (1977b).

PROC FREQ also provides an exact test for the Pearson correlation coefficient. You can request this test by specifying the PCORR option in the EXACT statement. See the section "Exact Statistics" on page 221 for more information.

Spearman Rank Correlation Coefficient

The Spearman correlation coefficient (ρ_s) is computed by using rank scores, which are defined in the section "Scores" on page 154. This measure is appropriate only when both variables lie on an ordinal scale. The range of the Spearman correlation is $-1 \leq \rho_s \leq 1$. The Spearman correlation coefficient is computed as

$$r_s = v / w$$

and its asymptotic variance is

$$\text{Var}(r_s) = \frac{1}{n^2 w^4} \sum_i \sum_j n_{ij}(z_{ij} - \bar{z})^2$$

where R_i^1 and C_j^1 are the row and column rank scores and

$$v = \sum_i \sum_j n_{ij} R(i) C(j)$$

$$w = \frac{1}{12} \sqrt{FG}$$

$$F = n^3 - \sum_i n_{i\cdot}^3$$

$$G = n^3 - \sum_j n_{\cdot j}^3$$

$$R(i) = R_i^1 - n/2$$

$$C(j) = C_j^1 - n/2$$

$$\bar{z} = \frac{1}{n} \sum_i \sum_j n_{ij} z_{ij}$$

$$z_{ij} = w v_{ij} - v w_{ij}$$

$$v_{ij} = n\left(R(i)C(j) + \frac{1}{2}\sum_l n_{il}C(l) + \frac{1}{2}\sum_k n_{kj}R(k) + \sum_l \sum_{k>i} n_{kl}C(l) + \sum_k \sum_{l>j} n_{kl}R(k)\right)$$

$$w_{ij} = \frac{-n}{96w}\left(Fn_{\cdot j}^2 + Gn_{i\cdot}^2\right)$$

See Snedecor and Cochran (1989) for more information.

The variance under the null hypothesis that the correlation equals 0 is computed as

$$\text{Var}_0(r_s) = \frac{1}{n^2 w^2}\sum_i\sum_j n_{ij}(v_{ij} - \bar{v})^2$$

where

$$\bar{v} = \sum_i\sum_j n_{ij}v_{ij}/n$$

This expression for the variance is derived for multinomial sampling in a contingency table framework, and it differs from the form obtained under the assumption that both variables are continuous and normally distributed. For more information, see Brown and Benedetti (1977b).

PROC FREQ also provides an exact test for the Spearman correlation coefficient. You can request this test by specifying the SCORR option in the EXACT statement. See the section "Exact Statistics" on page 221 for more information.

Polychoric Correlation

When you specify the PLCORR option in the TABLES statement, PROC FREQ computes the polychoric correlation and its standard error. The polychoric correlation is based on the assumption that the two ordinal, categorical variables of the frequency table have an underlying bivariate normal distribution. The polychoric correlation coefficient is the maximum likelihood estimate of the product-moment correlation between the underlying normal variables. The range of the polychoric correlation is from −1 to 1. For 2 × 2 tables, the polychoric correlation is also known as the tetrachoric correlation (and it is labeled as such in the displayed output). See Drasgow (1986) for an overview of polychoric correlation coefficient.

Olsson (1979) gives the likelihood equations and the asymptotic standard errors for estimating the polychoric correlation. The underlying continuous variables relate to the observed crosstabulation table through thresholds, which define a range of numeric values that correspond to each categorical (table) level. PROC FREQ uses Olsson's maximum likelihood method for simultaneous estimation of the polychoric correlation and the thresholds. (Olsson also presents a two-step method that estimates the thresholds first.)

PROC FREQ iteratively solves the likelihood equations by using a Newton-Raphson algorithm. The initial estimates of the thresholds are computed from the inverse of the normal distribution function at the cumulative marginal proportions of the table. Iterative computation of the polychoric correlation stops when the convergence measure falls below the convergence criterion or when the maximum number of iterations is reached, whichever occurs first. For parameter values that are less than 0.01, the procedure evaluates convergence by using the absolute difference instead of the relative difference. The PLCORR(CONVERGE=)

option specifies the convergence criterion, which is 0.0001 by default. The PLCORR(MAXITER=) option specifies the maximum number of iterations, which is 20 by default.

If you specify the CL option in the TABLES statement, PROC FREQ provides confidence limits for the polychoric correlation. The confidence limits are computed as

$$\hat{\rho} \pm (z_{\alpha/2} \times \text{SE}(\hat{\rho}))$$

where $\hat{\rho}$ is the estimate of the polychoric correlation, $z_{\alpha/2}$ is the $100(1 - \alpha/2)$th percentile of the standard normal distribution, and $\text{SE}(\hat{\rho})$ is the standard error of the polychoric correlation estimate.

If you specify the PLCORR option in the TEST statement, PROC FREQ provides Wald and likelihood ratio tests of the null hypothesis that the polychoric correlation is 0. The Wald test statistic is computed as

$$z = \hat{\rho} / \text{SE}(\hat{\rho})$$

which has a standard normal distribution under the null hypothesis. PROC FREQ computes one-sided and two-sided p-values for the Wald test. When the test statistic z is greater than its null expected value of 0, PROC FREQ displays the right-sided p-value. When the test statistic is less than or equal to 0, PROC FREQ displays the left-sided p-value.

The likelihood ratio statistic for the polychoric correlation is computed as

$$G^2 = -2 \ln(L_0/L_1)$$

where L_0 is the value of the likelihood function (Olsson 1979) when the polychoric correlation is 0, and L_1 is the value of the likelihood function at the maximum (where all parameters are replaced by their maximum likelihood estimates). Under the null hypothesis, the likelihood ratio statistic has an asymptotic chi-square distribution with 1 degree of freedom.

Lambda (Asymmetric)

Asymmetric lambda, $\lambda(C|R)$, is interpreted as the probable improvement in predicting the column variable Y given knowledge of the row variable X. The range of asymmetric lambda is $0 \leq \lambda(C|R) \leq 1$. Asymmetric lambda $(C|R)$ is computed as

$$\lambda(C|R) = \frac{\sum_i r_i - r}{n - r}$$

and its asymptotic variance is

$$\text{Var}(\lambda(C|R)) = \frac{n - \sum_i r_i}{(n-r)^3} \left(\sum_i r_i + r - 2 \sum_i (r_i \mid l_i = l) \right)$$

where

$$r_i = \max_j(n_{ij})$$

$$r = \max_j(n_{.j})$$

$$c_j = \max_i(n_{ij})$$

$$c = \max_i(n_{i.})$$

The values of l_i and l are determined as follows. Denote by l_i the unique value of j such that $r_i = n_{ij}$, and let l be the unique value of j such that $r = n_{.j}$. Because of the uniqueness assumptions, ties in the frequencies or in the marginal totals must be broken in an arbitrary but consistent manner. In case of ties, l is defined as the smallest value of j such that $r = n_{.j}$.

For those columns containing a cell (i, j) for which $n_{ij} = r_i = c_j$, cs_j records the row in which c_j is assumed to occur. Initially cs_j is set equal to –1 for all j. Beginning with $i=1$, if there is at least one value j such that $n_{ij} = r_i = c_j$, and if $cs_j = -1$, l_i is defined to be the smallest such value of j, and cs_j is set equal to i. Otherwise, if $n_{il} = r_i$, l_i is defined to be equal to l. If neither condition is true, l_i is taken to be the smallest value of j such that $n_{ij} = r_i$.

The formulas for lambda asymmetric $(R|C)$ can be obtained by interchanging the indices.

See Goodman and Kruskal (1979) for more information.

Lambda (Symmetric)

The nondirectional lambda is the average of the two asymmetric lambdas, $\lambda(C|R)$ and $\lambda(R|C)$. Its range is $0 \leq \lambda \leq 1$. Lambda symmetric is computed as

$$\lambda = \frac{\sum_i r_i + \sum_j c_j - r - c}{2n - r - c} = \frac{w - v}{w}$$

and its asymptotic variance is computed as

$$\text{Var}(\lambda) = \frac{1}{w^4}\left(wvy - 2w^2\left(n - \sum_i \sum_j (n_{ij} \mid j = l_i, i = k_j)\right) - 2v^2(n - n_{kl})\right)$$

where

$$r_i = \max_j(n_{ij})$$

$$r = \max_j(n_{.j})$$

$$c_j = \max_i(n_{ij})$$

$$c = \max_i(n_{i.})$$

$$w = 2n - r - c$$

$$v = 2n - \sum_i r_i - \sum_j c_j$$

$$x = \sum_i (r_i \mid l_i = l) + \sum_j (c_j \mid k_j = k) + r_k + c_l$$

$$y = 8n - w - v - 2x$$

The definitions of l_i and l are given in the previous section. The values k_j and k are defined in a similar way for lambda asymmetric $(R|C)$.

See Goodman and Kruskal (1979) for more information.

Uncertainty Coefficients (Asymmetric)

The uncertainty coefficient $U(C|R)$ measures the proportion of uncertainty (entropy) in the column variable Y that is explained by the row variable X. Its range is $0 \leq U(C|R) \leq 1$. The uncertainty coefficient is computed as

$$U(C|R) = (H(X) + H(Y) - H(XY)) / H(Y) = v/w$$

and its asymptotic variance is

$$\text{Var}(U(C|R)) = \frac{1}{n^2 w^4} \sum_i \sum_j n_{ij} \left(H(Y) \ln\left(\frac{n_{ij}}{n_{i.}}\right) + (H(X) - H(XY)) \ln\left(\frac{n_{.j}}{n}\right) \right)^2$$

where

$$v = H(X) + H(Y) - H(XY)$$

$$w = H(Y)$$

$$H(X) = -\sum_i \left(\frac{n_{i.}}{n}\right) \ln\left(\frac{n_{i.}}{n}\right)$$

$$H(Y) = -\sum_j \left(\frac{n_{.j}}{n}\right) \ln\left(\frac{n_{.j}}{n}\right)$$

$$H(XY) = -\sum_i \sum_j \left(\frac{n_{ij}}{n}\right) \ln\left(\frac{n_{ij}}{n}\right)$$

The formulas for the uncertainty coefficient $U(R|C)$ can be obtained by interchanging the indices.

See Theil (1972, pp. 115–120) and Goodman and Kruskal (1979) for more information.

Uncertainty Coefficient (Symmetric)

The uncertainty coefficient U is the symmetric version of the two asymmetric uncertainty coefficients. Its range is $0 \leq U \leq 1$. The uncertainty coefficient is computed as

$$U = 2(H(X) + H(Y) - H(XY)) / (H(X) + H(Y))$$

and its asymptotic variance is

$$\text{Var}(U) = 4 \sum_i \sum_j \frac{n_{ij} \left(H(XY) \ln\left(\frac{n_{i.} n_{.j}}{n^2}\right) - (H(X) + H(Y)) \ln\left(\frac{n_{ij}}{n}\right) \right)^2}{n^2 (H(X) + H(Y))^4}$$

where $H(X)$, $H(Y)$, and $H(XY)$ are defined in the previous section. See Goodman and Kruskal (1979) for more information.

Binomial Proportion

If you specify the BINOMIAL option in the TABLES statement, PROC FREQ computes the binomial proportion for one-way tables. By default, this is the proportion of observations in the first variable level that appears in the output. (You can use the LEVEL= option to specify a different level for the proportion.) The binomial proportion is computed as

$$\hat{p} = n_1 / n$$

where n_1 is the frequency of the first (or designated) level and n is the total frequency of the one-way table. The standard error of the binomial proportion is computed as

$$\text{se}(\hat{p}) = \sqrt{\hat{p}(1 - \hat{p}) / n}$$

Binomial Confidence Limits

PROC FREQ provides Wald and exact (Clopper-Pearson) confidence limits for the binomial proportion. You can also request the following binomial confidence limit types by specifying the BINOMIAL(CL=) option: Agresti-Coull, Blaker, Jeffreys, exact mid-p, likelihood ratio, logit, and Wilson (score). For more information, see Brown, Cai, and DasGupta (2001), Agresti and Coull (1998), and Newcombe (1998b), in addition to the references cited for each confidence limit type.

Wald Confidence Limits Wald asymptotic confidence limits are based on the normal approximation to the binomial distribution. PROC FREQ computes the Wald confidence limits for the binomial proportion as

$$\hat{p} \pm (z_{\alpha/2} \times \text{se}(\hat{p}))$$

where $z_{\alpha/2}$ is the $100(1 - \alpha/2)$th percentile of the standard normal distribution. The confidence level α is determined by the ALPHA= option; by default, ALPHA=0.05, which produces 95% confidence limits.

If you specify CL=WALD(CORRECT) or the CORRECT *binomial-option*, PROC FREQ includes a continuity correction of $1/2n$ in the Wald asymptotic confidence limits. The purpose of this correction is to adjust for the difference between the normal approximation and the discrete binomial distribution. See Fleiss, Levin, and Paik (2003) for more information. The continuity-corrected Wald confidence limits for the binomial proportion are computed as

$$\hat{p} \pm (z_{\alpha/2} \times \text{se}(\hat{p}) + (1/2n))$$

Exact (Clopper-Pearson) Confidence Limits Exact (Clopper-Pearson) confidence limits for the binomial proportion are constructed by inverting the equal-tailed test based on the binomial distribution. This method is attributed to Clopper and Pearson (1934). The exact confidence limits P_L and P_U satisfy the following equations, for $n_1 = 1, 2, \ldots n - 1$:

$$\sum_{x=n_1}^{n} \binom{n}{x} P_L^x (1 - P_L)^{n-x} = \alpha/2$$

$$\sum_{x=0}^{n_1} \binom{n}{x} P_U^x (1 - P_U)^{n-x} = \alpha/2$$

The lower confidence limit is 0 when $n_1 = 0$, and the upper confidence limit is 1 when $n_1 = n$.

PROC FREQ computes the exact (Clopper-Pearson) confidence limits by using the F distribution as

$$P_L = \left(1 + \frac{n - n_1 + 1}{n_1 \; F(\alpha/2, \; 2n_1, \; 2(n - n_1 + 1))}\right)^{-1}$$

$$P_U = \left(1 + \frac{n - n_1}{(n_1 + 1) \; F(1 - \alpha/2, \; 2(n_1 + 1), \; 2(n - n_1))}\right)^{-1}$$

where $F(\alpha/2, b, c)$ is the $(\alpha/2)$th percentile of the F distribution with b and c degrees of freedom. See Leemis and Trivedi (1996) for a derivation of this expression. Also see Collett (1991) for more information about exact binomial confidence limits.

Because this is a discrete problem, the confidence coefficient (coverage probability) of the exact (Clopper-Pearson) interval is not exactly $(1 - \alpha)$ but is at least $(1 - \alpha)$. Thus, this confidence interval is conservative. Unless the sample size is large, the actual coverage probability can be much larger than the target value. For more information about the performance of these confidence limits, see Agresti and Coull (1998), Brown, Cai, and DasGupta (2001), and Leemis and Trivedi (1996).

Agresti-Coull Confidence Limits If you specify the CL=AGRESTICOULL *binomial-option*, PROC FREQ computes Agresti-Coull confidence limits for the binomial proportion as

$$\tilde{p} \pm \left(z_{\alpha/2} \times \sqrt{\tilde{p}(1 - \tilde{p})/\tilde{n}}\right)$$

where

$$\tilde{n}_1 = n_1 + z_{\alpha/2}^2/2$$
$$\tilde{n} = n + z_{\alpha/2}^2$$
$$\tilde{p} = \tilde{n}_1 / \tilde{n}$$

The Agresti-Coull confidence interval has the same general form as the standard Wald interval but uses $\tilde{p}$ in place of $\hat{p}$. For $\alpha = 0.05$, the value of $z_{\alpha/2}$ is close to 2, and this interval is the "add 2 successes and 2 failures" adjusted Wald interval of Agresti and Coull (1998).

Blaker Confidence Limits If you specify the CL=BLAKER *binomial-option*, PROC FREQ computes Blaker confidence limits for the binomial proportion, which are constructed by inverting the two-sided exact Blaker test (Blaker 2000). The $100(1 - \alpha)\%$ Blaker confidence interval consists of all values of the proportion p_0 for which the test statistic $B(p_0, n_1)$ falls in the acceptance region,

$$\{p_0 : B(p_0, n_1) > \alpha\}$$

where

$$B(p_0, n_1) = \text{Prob}(\gamma(p_0, X) \leq \gamma(p_0, n_1) \mid p_0)$$

$$\gamma(p_0, n_1) = \min(\text{Prob}(X \geq n_1 \mid p_0), \text{Prob}(X \leq n_1 \mid p_0))$$

and X is a binomial random variable. For more information, see Blaker (2000).

Jeffreys Confidence Limits If you specify the CL=JEFFREYS *binomial-option*, PROC FREQ computes Jeffreys confidence limits for the binomial proportion as

$$(\beta(\alpha/2, n_1 + 1/2, n - n_1 + 1/2), \beta(1 - \alpha/2, n_1 + 1/2, n - n_1 + 1/2))$$

where $\beta(\alpha, b, c)$ is the αth percentile of the beta distribution with shape parameters b and c. The lower confidence limit is set to 0 when $n_1 = 0$, and the upper confidence limit is set to 1 when $n_1 = n$. This is an equal-tailed interval based on the noninformative Jeffreys prior for a binomial proportion. For more information, see Brown, Cai, and DasGupta (2001). For information about using beta priors for inference on the binomial proportion, see Berger (1985).

Likelihood Ratio Confidence Limits If you specify the CL=LIKELIHOODRATIO *binomial-option*, PROC FREQ computes likelihood ratio confidence limits for the binomial proportion by inverting the likelihood ratio test. The likelihood ratio test statistic for the null hypothesis that the proportion equals p_0 can be expressed as

$$L(p_0) = -2(n_1 \log(\hat{p}/p_0) + (n - n_1) \log((1 - \hat{p})/(1 - p_0)))$$

The $100(1 - \alpha)\%$ likelihood ratio confidence interval consists of all values of p_0 for which the test statistic $L(p_0)$ falls in the acceptance region,

$$\{p_0 : L(p_0) < \chi^2_{1,\alpha}\}$$

where $\chi^2_{1,\alpha}$ is the $100(1 - \alpha)$th percentile of the chi-square distribution with 1 degree of freedom. PROC FREQ finds the confidence limits by iterative computation. For more information, see Fleiss, Levin, and Paik (2003), Brown, Cai, and DasGupta (2001), Agresti (2013), and Newcombe (1998b).

Logit Confidence Limits If you specify the CL=LOGIT *binomial-option*, PROC FREQ computes logit confidence limits for the binomial proportion, which are based on the logit transformation $Y = \log(\hat{p}/(1-\hat{p}))$. Approximate confidence limits for Y are computed as

$$Y_L = \log(\hat{p}/(1 - \hat{p})) - z_{\alpha/2}\sqrt{n/(n_1(n - n_1))}$$

$$Y_U = \log(\hat{p}/(1 - \hat{p})) + z_{\alpha/2}\sqrt{n/(n_1(n - n_1))}$$

The confidence limits for Y are inverted to produce $100(1 - \alpha)\%$ logit confidence limits P_L and P_U for the binomial proportion p as

$$P_L = \exp(Y_L/(1 + \exp(Y_L))$$

$$P_U = \exp(Y_U/(1 + \exp(Y_U))$$

For more information, see Brown, Cai, and DasGupta (2001) and Korn and Graubard (1998).

Mid-*p* Confidence Limits If you specify the CL=MIDP *binomial-option*, PROC FREQ computes exact mid-*p* confidence limits for the binomial proportion by inverting two one-sided binomial tests that include mid-*p* tail areas. The mid-*p* approach replaces the probability of the observed frequency by half of that probability in the Clopper-Pearson sum, which is described in the section "Exact (Clopper-Pearson) Confidence Limits" on page 170. The exact mid-p confidence limits P_L and P_U are the solutions to the equations

$$\sum_{x=n_1+1}^{n} \binom{n}{x} P_L^x (1-P_L)^{n-x} + \frac{1}{2}\binom{n}{n_1} P_L^{n_1}(1-P_L)^{n-n_1} = \alpha/2$$

$$\sum_{x=0}^{n_1-1} \binom{n}{x} P_U^x (1-P_U)^{n-x} + \frac{1}{2}\binom{n}{n_1} P_U^{n_1}(1-P_U)^{n-n_1} = \alpha/2$$

For more information, see Agresti and Gottard (2007), Agresti (2013), Newcombe (1998b), and Brown, Cai, and DasGupta (2001).

Wilson (Score) Confidence Limits If you specify the CL=WILSON *binomial-option*, PROC FREQ computes Wilson confidence limits for the binomial proportion. These are also known as score confidence limits (Wilson 1927). The confidence limits are based on inverting the normal test that uses the null proportion in the variance (the score test). Wilson confidence limits are the roots of

$$|p - \hat{p}| = z_{\alpha/2}\sqrt{p(1-p)/n}$$

and are computed as

$$\left(\hat{p} + z_{\alpha/2}^2/2n \pm z_{\alpha/2}\sqrt{\left(\hat{p}(1-\hat{p}) + z_{\alpha/2}^2/4n\right)/n}\right) / \left(1 + z_{\alpha/2}^2/n\right)$$

If you specify CL=WILSON(CORRECT) or the CORRECT *binomial-option*, PROC FREQ provides continuity-corrected Wilson confidence limits, which are computed as the roots of

$$|p - \hat{p}| - 1/2n = z_{\alpha/2}\sqrt{p(1-p)/n}$$

The Wilson interval has been shown to have better performance than the Wald interval and the exact (Clopper-Pearson) interval. For more information, see Agresti and Coull (1998), Brown, Cai, and DasGupta (2001), and Newcombe (1998b).

Binomial Tests

The BINOMIAL option provides an asymptotic equality test for the binomial proportion by default. You can also specify *binomial-options* to request tests of noninferiority, superiority, and equivalence for the binomial proportion. If you specify the BINOMIAL option in the EXACT statement, PROC FREQ also computes exact *p*-values for the tests that you request with the *binomial-options*.

Equality Test PROC FREQ computes an asymptotic test of the hypothesis that the binomial proportion equals p_0, where you can specify the value of p_0 with the P= *binomial-option*. If you do not specify a null value with P=, PROC FREQ uses $p_0 = 0.5$ by default. The binomial test statistic is computed as

$$z = (\hat{p} - p_0)/\text{se}$$

By default, the standard error is based on the null hypothesis proportion as

$$\text{se} = \sqrt{p_0(1-p_0)/n}$$

If you specify the VAR=SAMPLE *binomial-option*, the standard error is computed from the sample proportion as

$$se = \sqrt{\hat{p}(1-\hat{p})/n}$$

If you specify the CORRECT *binomial-option*, PROC FREQ includes a continuity correction in the asymptotic test statistic, towards adjusting for the difference between the normal approximation and the discrete binomial distribution. For more information, see Fleiss, Levin, and Paik (2003). The continuity correction of $(1/2n)$ is subtracted from the numerator of the test statistic if $(\hat{p} - p_0)$ is positive; otherwise, the continuity correction is added to the numerator.

PROC FREQ computes one-sided and two-sided *p*-values for this test. When the test statistic z is greater than 0 (its expected value under the null hypothesis), PROC FREQ computes the right-sided *p*-value, which is the probability of a larger value of the statistic occurring under the null hypothesis. A small right-sided *p*-value supports the alternative hypothesis that the true value of the proportion is greater than p_0. When the test statistic is less than or equal to 0, PROC FREQ computes the left-sided *p*-value, which is the probability of a smaller value of the statistic occurring under the null hypothesis. A small left-sided *p*-value supports the alternative hypothesis that the true value of the proportion is less than p_0. The one-sided *p*-value P_1 can be expressed as

$$P_1 = \begin{cases} \text{Prob}(Z > z) & \text{if } z > 0 \\ \text{Prob}(Z < z) & \text{if } z \leq 0 \end{cases}$$

where Z has a standard normal distribution. The two-sided *p*-value is computed as $P_2 = 2 \times P_1$.

If you specify the BINOMIAL option in the EXACT statement, PROC FREQ also computes an exact test of the null hypothesis $H_0\colon p = p_0$. To compute the exact test, PROC FREQ uses the binomial probability function,

$$\text{Prob}(X = x \mid p_0) = \binom{n}{x} p_0^x (1-p_0)^{(n-x)} \quad \text{for } x = 0, 1, 2, \ldots, n$$

where the variable X has a binomial distribution with parameters n and p_0. To compute the left-sided *p*-value, $\text{Prob}(X \leq n_1)$, PROC FREQ sums the binomial probabilities over x from 0 to n_1. To compute the right-sided *p*-value, $\text{Prob}(X \geq n_1)$, PROC FREQ sums the binomial probabilities over x from n_1 to n. The exact one-sided *p*-value is the minimum of the left-sided and right-sided *p*-values,

$$P_1 = \min\left(\,\text{Prob}(X \leq n_1 \mid p_0),\ \text{Prob}(X \geq n_1 \mid p_0)\,\right)$$

and the exact two-sided *p*-value is computed as $P_2 = 2 \times P_1$.

Noninferiority Test If you specify the NONINF *binomial-option*, PROC FREQ provides a noninferiority test for the binomial proportion. The null hypothesis for the noninferiority test is

$$H_0\colon p - p_0 \leq -\delta$$

versus the alternative

$$H_a\colon p - p_0 > -\delta$$

where δ is the noninferiority margin and p_0 is the null proportion. Rejection of the null hypothesis indicates that the binomial proportion is not inferior to the null value. See Chow, Shao, and Wang (2003) for more information.

You can specify the value of δ with the MARGIN= *binomial-option*, and you can specify p_0 with the P= *binomial-option*. By default, $\delta = 0.2$ and $p_0 = 0.5$.

PROC FREQ provides an asymptotic Wald test for noninferiority. The test statistic is computed as

$$z = (\hat{p} - p_0^*) / \text{se}$$

where p_0^* is the noninferiority limit,

$$p_0^* = p_0 - \delta$$

By default, the standard error is computed from the sample proportion as

$$\text{se} = \sqrt{\hat{p}(1-\hat{p})/n}$$

If you specify the VAR=NULL *binomial-option*, the standard error is based on the noninferiority limit (determined by the null proportion and the margin) as

$$\text{se} = \sqrt{p_0^*(1-p_0^*)/n}$$

If you specify the CORRECT *binomial-option*, PROC FREQ includes a continuity correction in the asymptotic test statistic z. The continuity correction of $(1/2n)$ is subtracted from the numerator of the test statistic if $(\hat{p} - p_0^*)$ is positive; otherwise, the continuity correction is added to the numerator.

The *p*-value for the noninferiority test is

$$P_z = \text{Prob}(Z > z)$$

where Z has a standard normal distribution.

As part of the noninferiority analysis, PROC FREQ provides asymptotic Wald confidence limits for the binomial proportion. These confidence limits are computed as described in the section "Wald Confidence Limits" on page 170 but use the same standard error (VAR=NULL or VAR=SAMPLE) as the noninferiority test statistic z. The confidence coefficient is $100(1 - 2\alpha)\%$ (Schuirmann 1999). By default, if you do not specify the ALPHA= option, the noninferiority confidence limits are 90% confidence limits. You can compare the confidence limits to the noninferiority limit, $p_0^* = p_0 - \delta$.

If you specify the BINOMIAL option in the EXACT statement, PROC FREQ provides an exact noninferiority test for the binomial proportion. The exact *p*-value is computed by using the binomial probability function with parameters p_0^* and n,

$$P_x = \sum_{k=n_1}^{k=n} \binom{n}{k} (p_0^*)^k (1-p_0^*)^{(n-k)}$$

For more information, see Chow, Shao, and Wang (2003, p. 116). If you request exact binomial statistics, PROC FREQ also includes exact (Clopper-Pearson) confidence limits for the binomial proportion in the equivalence analysis display. For more information, see the section "Exact (Clopper-Pearson) Confidence Limits" on page 170.

Superiority Test If you specify the SUP *binomial-option*, PROC FREQ provides a superiority test for the binomial proportion. The null hypothesis for the superiority test is

$$H_0: p - p_0 \leq \delta$$

versus the alternative

$$H_a: p - p_0 > \delta$$

where δ is the superiority margin and p_0 is the null proportion. Rejection of the null hypothesis indicates that the binomial proportion is superior to the null value. You can specify the value of δ with the MARGIN= *binomial-option*, and you can specify the value of p_0 with the P= *binomial-option*. By default, $\delta = 0.2$ and $p_0 = 0.5$.

The superiority analysis is identical to the noninferiority analysis but uses a positive value of the margin δ in the null hypothesis. The superiority limit equals $p_0 + \delta$. The superiority computations follow those in the section "Noninferiority Test" on page 174 but replace $-\delta$ with δ. See Chow, Shao, and Wang (2003) for more information.

Equivalence Test If you specify the EQUIV *binomial-option*, PROC FREQ provides an equivalence test for the binomial proportion. The null hypothesis for the equivalence test is

$$H_0: p - p_0 \leq \delta_L \quad \text{or} \quad p - p_0 \geq \delta_U$$

versus the alternative

$$H_a: \delta_L < p - p_0 < \delta_U$$

where δ_L is the lower margin, δ_U is the upper margin, and p_0 is the null proportion. Rejection of the null hypothesis indicates that the binomial proportion is equivalent to the null value. See Chow, Shao, and Wang (2003) for more information.

You can specify the value of the margins δ_L and δ_U with the MARGIN= *binomial-option*. If you do not specify MARGIN=, PROC FREQ uses lower and upper margins of –0.2 and 0.2 by default. If you specify a single margin value δ, PROC FREQ uses lower and upper margins of $-\delta$ and δ. You can specify the null proportion p_0 with the P= *binomial-option*. By default, $p_0 = 0.5$.

PROC FREQ computes two one-sided tests (TOST) for equivalence analysis (Schuirmann 1987). The TOST approach includes a right-sided test for the lower margin and a left-sided test for the upper margin. The overall *p*-value is taken to be the larger of the two *p*-values from the lower and upper tests.

For the lower margin, the asymptotic Wald test statistic is computed as

$$z_L = (\hat{p} - p_L^*) / \text{se}$$

where the lower equivalence limit is

$$p_L^* = p_0 + \delta_L$$

By default, the standard error is computed from the sample proportion as

$$\text{se} = \sqrt{\hat{p}(1-\hat{p})/n}$$

If you specify the VAR=NULL *binomial-option*, the standard error is based on the lower equivalence limit (determined by the null proportion and the lower margin) as

$$\text{se} = \sqrt{p_L^*(1 - p_L^*)/n}$$

If you specify the CORRECT *binomial-option*, PROC FREQ includes a continuity correction in the asymptotic test statistic z_L. The continuity correction of $(1/2n)$ is subtracted from the numerator of the test statistic $(\hat{p} - p_L^*)$ if the numerator is positive; otherwise, the continuity correction is added to the numerator.

The *p*-value for the lower margin test is

$$P_{z,L} = \text{Prob}(Z > z_L)$$

The asymptotic test for the upper margin is computed similarly. The Wald test statistic is

$$z_U = (\hat{p} - p_U^*) / \text{se}$$

where the upper equivalence limit is

$$p_U^* = p_0 + \delta_U$$

By default, the standard error is computed from the sample proportion. If you specify the VAR=NULL *binomial-option*, the standard error is based on the upper equivalence limit as

$$\text{se} = \sqrt{p_U^*(1 - p_U^*)/n}$$

If you specify the CORRECT *binomial-option*, PROC FREQ includes a continuity correction of $(1/2n)$ in the asymptotic test statistic z_U.

The *p*-value for the upper margin test is

$$P_{z,U} = \text{Prob}(Z < z_U)$$

Based on the two one-sided tests (TOST), the overall *p*-value for the test of equivalence equals the larger *p*-value from the lower and upper margin tests, which can be expressed as

$$P_z = \max(P_{z,L}, P_{z,U})$$

As part of the equivalence analysis, PROC FREQ provides asymptotic Wald confidence limits for the binomial proportion. These confidence limits are computed as described in the section "Wald Confidence Limits" on page 170, but use the same standard error (VAR=NULL or VAR=SAMPLE) as the equivalence test statistics and have a confidence coefficient of $100(1 - 2\alpha)\%$ (Schuirmann 1999). By default, if you do not specify the ALPHA= option, the equivalence confidence limits are 90% limits. If you specify VAR=NULL, separate standard errors are computed for the lower and upper margin tests, each based on the null proportion and the corresponding (lower or upper) margin. The confidence limits are computed by using the maximum of these two standard errors. You can compare the confidence limits to the equivalence limits, $(p_0 + \delta_L, p_0 + \delta_U)$.

If you specify the BINOMIAL option in the EXACT statement, PROC FREQ also provides an exact equivalence test by using two one-sided exact tests (TOST). The procedure computes lower and upper margin exact tests by using the binomial probability function as described in the section "Noninferiority Test" on page 174. The overall exact *p*-value for the equivalence test is taken to be the larger *p*-value from the lower and upper margin exact tests. If you request exact statistics, PROC FREQ also includes exact (Clopper-Pearson) confidence limits in the equivalence analysis display. The confidence coefficient is $100(1 - 2\alpha)\%$ (Schuirmann 1999). For more information, see the section "Exact (Clopper-Pearson) Confidence Limits" on page 170.

Risks and Risk Differences

The RISKDIFF option in the TABLES statement provides estimates of risks (binomial proportions) and risk differences for 2 × 2 tables. This analysis might be appropriate when comparing the proportion of some characteristic for two groups, where row 1 and row 2 correspond to the two groups, and the columns correspond to two possible characteristics or outcomes. For example, the row variable might be a treatment or dose, and the column variable might be the response. For more information, see Collett (1991); Fleiss, Levin, and Paik (2003); Stokes, Davis, and Koch (2012).

Let the frequencies of the 2 × 2 table be represented as follows.

	Column 1	Column 2	Total
Row 1	n_{11}	n_{12}	$n_{1\cdot}$
Row 2	n_{21}	n_{22}	$n_{2\cdot}$
Total	$n_{\cdot 1}$	$n_{\cdot 2}$	n

By default when you specify the RISKDIFF option, PROC FREQ provides estimates of the row 1 risk (proportion), the row 2 risk, the overall risk, and the risk difference for column 1 and for column 2 of the 2 × 2 table. The risk difference is defined as the row 1 risk minus the row 2 risk. The risks are binomial proportions of their rows (row 1, row 2, or overall), and the computation of their standard errors and Wald confidence limits follow the binomial proportion computations, which are described in the section "Binomial Proportion" on page 170.

The column 1 risk for row 1 is the proportion of row 1 observations classified in column 1,

$$\hat{p}_1 = n_{11} / n_{1\cdot}$$

which estimates the conditional probability of the column 1 response, given the first level of the row variable. The column 1 risk for row 2 is the proportion of row 2 observations classified in column 1,

$$\hat{p}_2 = n_{21} / n_{2\cdot}$$

The overall column 1 risk is the proportion of all observations classified in column 1,

$$\hat{p} = n_{\cdot 1} / n$$

The column 1 risk difference compares the risks for the two rows, and it is computed as the column 1 risk for row 1 minus the column 1 risk for row 2,

$$\hat{d} = \hat{p}_1 - \hat{p}_2$$

The standard error of the column 1 risk for row i is computed as

$$\text{se}(\hat{p}_i) = \sqrt{\hat{p}_i (1 - \hat{p}_i) / n_{i\cdot}}$$

The standard error of the overall column 1 risk is computed as

$$\text{se}(\hat{p}) = \sqrt{\hat{p} (1 - \hat{p}) / n}$$

Where the two rows represent independent binomial samples, the standard error of the column 1 risk difference is computed as

$$\text{se}(\hat{d}) = \sqrt{\hat{p}_1(1-\hat{p}_1)/n_{1\cdot} + \hat{p}_2(1-\hat{p}_2)/n_{2\cdot}}$$

The computations are similar for the column 2 risks and risk difference.

Confidence Limits

By default, the RISKDIFF option provides Wald asymptotic confidence limits for the risks (row 1, row 2, and overall) and the risk difference. By default, the RISKDIFF option also provides exact (Clopper-Pearson) confidence limits for the risks. You can suppress the display of this information by specifying the NORISKS *riskdiff-option*. You can specify *riskdiff-options* to request tests and other types of confidence limits for the risk difference. For more information, see the sections "Confidence Limits for the Risk Difference" on page 179 and "Risk Difference Tests" on page 183.

The risks are equivalent to the binomial proportions of their corresponding rows. This section describes the Wald confidence limits that are provided by default when you specify the RISKDIFF option. The BINOMIAL option provides additional confidence limit types and tests for risks (binomial proportions). For more information, see the sections "Binomial Confidence Limits" on page 170 and "Binomial Tests" on page 173.

The Wald confidence limits are based on the normal approximation to the binomial distribution. PROC FREQ computes the Wald confidence limits for the risks and risk differences as

$$\text{Est} \pm (z_{\alpha/2} \times \text{se(Est)})$$

where Est is the estimate, $z_{\alpha/2}$ is the $100(1 - \alpha/2)$th percentile of the standard normal distribution, and se(Est) is the standard error of the estimate. The confidence level α is determined by the value of the ALPHA= option; by default, ALPHA=0.05, which produces 95% confidence limits.

If you specify the CORRECT *riskdiff-option*, PROC FREQ includes continuity corrections in the Wald confidence limits for the risks and risk differences. The purpose of a continuity correction is to adjust for the difference between the normal approximation and the binomial distribution, which is discrete. See Fleiss, Levin, and Paik (2003) for more information. The continuity-corrected Wald confidence limits are computed as

$$\text{Est} \pm (z_{\alpha/2} \times \text{se(Est)} + cc)$$

where cc is the continuity correction. For the row 1 risk, $cc = (1/2n_1.)$; for the row 2 risk, $cc = (1/2n_2.)$; for the overall risk, $cc = (1/2n)$; and for the risk difference, $cc = ((1/n_1. + 1/n_2.)/2)$. The column 1 and column 2 risks use the same continuity corrections.

By default when you specify the RISKDIFF option, PROC FREQ also provides exact (Clopper-Pearson) confidence limits for the column 1, column 2, and overall risks. These confidence limits are constructed by inverting the equal-tailed test that is based on the binomial distribution. For more information, see the section "Exact (Clopper-Pearson) Confidence Limits" on page 170.

Confidence Limits for the Risk Difference PROC FREQ provides the following confidence limit types for the risk difference: Agresti-Caffo, exact unconditional, Hauck-Anderson, Miettinen-Nurminen (score), Newcombe (hybrid-score), and Wald confidence limits. Continuity-corrected forms of Newcombe and Wald confidence limits are also available.

The confidence coefficient for the confidence limits produced by the CL= *riskdiff-option* is $100(1 - \alpha)\%$, where the value of α is determined by the ALPHA= option. By default, ALPHA=0.05, which produces 95% confidence limits. This differs from the test-based confidence limits that are provided with the equivalence, noninferiority, and superiority tests, which have a confidence coefficient of $100(1 - 2\alpha)\%$ (Schuirmann 1999). For more information, see the section "Risk Difference Tests" on page 183.

Agresti-Caffo Confidence Limits

Agresti-Caffo confidence limits for the risk difference are computed as

$$\tilde{d} \pm (z_{\alpha/2} \times \text{se}(\tilde{d}))$$

where $\tilde{d} = \tilde{p}_1 - \tilde{p}_2$, $\tilde{p}_i = (n_{i1} + 1)/(n_{i\cdot} + 2)$,

$$\text{se}(\tilde{d}) = \sqrt{\tilde{p}_1(1 - \tilde{p}_2)/(n_{1\cdot} + 2) + \tilde{p}_2(1 - \tilde{p}_2)/(n_{2\cdot} + 2)}$$

and $z_{\alpha/2}$ is the $100(1 - \alpha/2)$th percentile of the standard normal distribution.

The Agresti-Caffo interval adjusts the Wald interval for the risk difference by adding a pseudo-observation of each type (success and failure) to each sample. See Agresti and Caffo (2000) and Agresti and Coull (1998) for more information.

Hauck-Anderson Confidence Limits

Hauck-Anderson confidence limits for the risk difference are computed as

$$\hat{d} \pm (cc + z_{\alpha/2} \times \text{se}(\hat{d}))$$

where $\hat{d} = \hat{p}_1 - \hat{p}_2$ and $z_{\alpha/2}$ is the $100(1 - \alpha/2)$th percentile of the standard normal distribution. The standard error is computed from the sample proportions as

$$\text{se}(\hat{d}) = \sqrt{\hat{p}_1(1 - \hat{p}_1)/(n_{1\cdot} - 1) + \hat{p}_2(1 - \hat{p}_2)/(n_{2\cdot} - 1)}$$

The Hauck-Anderson continuity correction cc is computed as

$$cc = 1 / \big(2 \min(n_{1\cdot}, n_{2\cdot})\big)$$

See Hauck and Anderson (1986) for more information. The subsection "Hauck-Anderson Test" in the section "Noninferiority Tests" on page 184 describes the corresponding noninferiority test.

Miettinen-Nurminen (Score) Confidence Limits

Miettinen-Nurminen (score) confidence limits for the risk difference (Miettinen and Nurminen 1985) are computed by inverting score tests for the risk difference. A score-based test statistic for the null hypothesis that the risk difference equals δ can be expressed as

$$T(\delta) = (\hat{d} - \delta)/\sqrt{\widetilde{\text{Var}}(\delta)}$$

where $\hat{d}$ is the observed value of the risk difference ($\hat{p}_1 - \hat{p}_2$),

$$\widetilde{\text{Var}}(\delta) = (n/(n-1))\ (\ \tilde{p}_1(\delta)(1 - \tilde{p}_1(\delta))/n_1 + \tilde{p}_2(\delta)(1 - \tilde{p}_2(\delta))/n_2\ )$$

and $\tilde{p}_1(\delta)$ and $\tilde{p}_2(\delta)$ are the maximum likelihood estimates of the row 1 and row 2 risks (proportions) under the restriction that the risk difference is δ. For more information, see Miettinen and Nurminen (1985, pp. 215–216) and Miettinen (1985, chapter 12).

The $100(1 - \alpha)\%$ confidence interval for the risk difference consists of all values of δ for which the score test statistic $T(\delta)$ falls in the acceptance region,

$$\{\delta : T(\delta) < z_{\alpha/2}\}$$

where $z_{\alpha/2}$ is the $100(1 - \alpha/2)$th percentile of the standard normal distribution. PROC FREQ finds the confidence limits by iterative computation, which stops when the iteration increment falls below the convergence criterion or when the maximum number of iterations is reached, whichever occurs first. By default, the convergence criterion is 0.00000001 and the maximum number of iterations is 100.

By default, the Miettinen-Nurminen confidence limits include the bias correction factor $n/(n-1)$ in the computation of $\widetilde{\text{Var}}(\delta)$ (Miettinen and Nurminen 1985, p. 216). For more information, see Newcombe and Nurminen (2011). If you specify the CL=MN(CORRECT=NO) *riskdiff-option*, PROC FREQ does not include the bias correction factor in this computation (Mee 1984). See also Agresti (2002, p. 77). The uncorrected confidence limits are labeled as "Miettinen-Nurminen-Mee" confidence limits in the displayed output.

The maximum likelihood estimates of p_1 and p_2, subject to the constraint that the risk difference is δ, are computed as

$$\tilde{p}_1 = 2u\cos(w) - b/3a \quad \text{and} \quad \tilde{p}_2 = \tilde{p}_1 - \delta$$

where

$$\begin{aligned}
w &= (\pi + \cos^{-1}(v/u^3))/3 \\
v &= b^3/(3a)^3 - bc/6a^2 + d/2a \\
u &= \text{sign}(v)\sqrt{b^2/(3a)^2 - c/3a} \\
a &= 1 + \theta \\
b &= -(1 + \theta + \hat{p}_1 + \theta\hat{p}_2 + \delta(\theta + 2)) \\
c &= \delta^2 + \delta(2\hat{p}_1 + \theta + 1) + \hat{p}_1 + \theta\hat{p}_2 \\
d &= -\hat{p}_1\delta(1 + \delta) \\
\theta &= n_{2.}/n_{1.}
\end{aligned}$$

For more information, see Farrington and Manning (1990, p. 1453).

Newcombe Confidence Limits

Newcombe (hybrid-score) confidence limits for the risk difference are constructed from the Wilson score confidence limits for each of the two individual proportions. The confidence limits for the individual proportions are used in the standard error terms of the Wald confidence limits for the proportion difference. See Newcombe (1998a) and Barker et al. (2001) for more information.

Wilson score confidence limits for p_1 and p_2 are the roots of

$$|p_i - \hat{p}_i| = z_{\alpha/2}\sqrt{p_i(1-p_i)/n_{i.}}$$

for $i = 1, 2$. The confidence limits are computed as

$$\left(\hat{p}_i + z_{\alpha/2}^2/2n_{i.} \pm z_{\alpha/2}\sqrt{\left(\hat{p}_i(1-\hat{p}_i) + z_\alpha^2/4n_{i.}\right)/n_{i.}}\right) / \left(1 + z_{\alpha/2}^2/n_{i.}\right)$$

For more information, see the section "Wilson (Score) Confidence Limits" on page 173.

Denote the lower and upper Wilson score confidence limits for p_1 as L_1 and U_1, and denote the lower and upper confidence limits for p_2 as L_2 and U_2. The Newcombe confidence limits for the proportion difference

$(d = p_1 - p_2)$ are computed as

$$d_L = (\hat{p}_1 - \hat{p}_2) - \sqrt{(\hat{p}_1 - L_1)^2 + (U_2 - \hat{p}_2)^2}$$

$$d_U = (\hat{p}_1 - \hat{p}_2) + \sqrt{(U_1 - \hat{p}_1)^2 + (\hat{p}_2 - L_2)^2}$$

If you specify the CORRECT *riskdiff-option*, PROC FREQ provides continuity-corrected Newcombe confidence limits. By including a continuity correction of $1/2n_i.$, the Wilson score confidence limits for the individual proportions are computed as the roots of

$$|p_i - \hat{p}_i| - 1/2n_i. = z_{\alpha/2}\sqrt{p_i(1 - p_i)/n_i.}$$

The continuity-corrected confidence limits for the individual proportions are then used to compute the proportion difference confidence limits d_L and d_U.

Wald Confidence Limits

Wald confidence limits for the risk difference are computed as

$$\hat{d} \pm (z_{\alpha/2} \times \text{se}(\hat{d}))$$

where $\hat{d} = \hat{p}_1 - \hat{p}_2$, $z_{\alpha/2}$ is the $100(1 - \alpha/2)$th percentile of the standard normal distribution. and the standard error is computed from the sample proportions as

$$\text{se}(\hat{d}) = \sqrt{\hat{p}_1(1 - \hat{p}_1)/n_1. + \hat{p}_2(1 - \hat{p}_2)/n_2.}$$

If you specify the CORRECT *riskdiff-option*, the Wald confidence limits include a continuity correction cc,

$$\hat{d} \pm (cc + z_{\alpha/2} \times \text{se}(\hat{d}))$$

where $cc = (1/n_1. + 1/n_2.)/2$.

The subsection "Wald Test" in the section "Noninferiority Tests" on page 184 describes the corresponding noninferiority test.

Exact Unconditional Confidence Limits

If you specify the RISKDIFF option in the EXACT statement, PROC FREQ provides exact unconditional confidence limits for the risk difference $(d = p_1 - p_2)$. The exact unconditional approach fixes the row margins of the 2 × 2 table and eliminates the nuisance parameter p_2 by using the maximum *p*-value (worst-case scenario) over all possible values of p_2 (Santner and Snell 1980). The conditional approach, which is described in the section "Exact Statistics" on page 221, does not apply to the risk difference because of the nuisance parameter (Agresti 1992).

By default, PROC FREQ computes the confidence limits by the tail method, which inverts two separate one-sided exact tests of the risk difference, where the tests are based on the score statistic (Chan and Zhang 1999). The size of each one-sided exact test is at most $\alpha/2$, and the confidence coefficient is at least $(1-\alpha)$. If you specify the RISKDIFF(METHOD=NOSCORE) option in the EXACT statement, PROC FREQ computes the confidence limits by inverting two separate one-sided exact tests that are based on the unstandardized risk difference. If you specify the RISKDIFF(METHOD=SCORE2) option in the EXACT statement, PROC

FREQ computes the confidence limits by inverting a single two-sided exact test that is based on the score statistic (Agresti and Min 2001).

The score statistic is a less discrete statistic than the unstandardized risk difference and produces less conservative confidence limits (Agresti and Min 2001). For more information, see Santner et al. (2007). The section "Miettinen-Nurminen (Score) Confidence Limits" describe computation of the risk difference score statistic. For more information, see Miettinen and Nurminen (1985) and Farrington and Manning (1990).

PROC FREQ computes the exact unconditional confidence limits as follows. The risk difference is defined as the difference between the row 1 and row 2 risks (proportions), $d = p_1 - p_2$, and n_1 and n_2 denote the row totals of the 2×2 table. The joint probability function for the table can be expressed in terms of the table cell frequencies, the risk difference, and the nuisance parameter p_2 as

$$f(n_{11}, n_{21}; n_1, n_2, d, p_2) = \binom{n_1}{n_{11}}(d + p_2)^{n_{11}}(1 - d - p_2)^{n_1 - n_{11}} \times \binom{n_2}{n_{21}} p_2^{n_{21}}(1 - p_2)^{n_2 - n_{21}}$$

For the tail method (which inverts two separate one-sided exact tests), the $100(1 - \alpha/2)\%$ confidence limits for the risk difference are computed as

$$d_L = \sup\,(d_* : P_U(d_*) > \alpha/2)$$
$$d_U = \inf\,(d_* : P_L(d_*) > \alpha/2)$$

where

$$P_U(d_*) = \sup_{p_2}\Big(\sum_{A, T(a) \geq t_0} f(n_{11}, n_{21}; n_1, n_2, d_*, p_2)\Big)$$

$$P_L(d_*) = \sup_{p_2}\Big(\sum_{A, T(a) \leq t_0} f(n_{11}, n_{21}; n_1, n_2, d_*, p_2)\Big)$$

The set A includes all 2×2 tables in which the row sums are n_1 and n_2, $T(a)$ denotes the value of the test statistic for table a in A, and t_0 is the value of the test statistic for the observed table. The test statistic is either the score statistic (by default) or the unstandardized risk difference. To compute $P_U(d_*)$, the sum includes probabilities of those tables for which $(T(a) \geq t_0)$. For a fixed value of d_*, $P_U(d_*)$ is defined as the maximum sum over all possible values of p_2.

The two-sided score method evaluates the p-values $P_U(d_*)$ and $P_L(d_*)$ by comparing $|T(a)|$ to $|t_0|$. To compute the confidence limits d_L and d_u, the two-sided method compares the p-values to α. For more information, see Agresti and Min (2001) and Santner et al. (2007).

Risk Difference Tests

PROC FREQ provides tests of equality, noninferiority, superiority, and equivalence for the risk (proportion) difference. The following analysis methods are available: Wald (with and without continuity correction), Hauck-Anderson, Farrington-Manning (score), and Newcombe (with and without continuity correction). You can specify the method by using the METHOD= *riskdiff-option*; by default, PROC FREQ provides Wald tests.

Equality Tests The equality test for the risk difference tests the null hypothesis that the risk difference equals the null value. You can specify a null value by using the EQUAL(NULL=) *riskdiff-option*; by default, the null value is 0. This test can be expressed as $H_0: d = d_0$ versus the alternative $H_a: d \neq d_0$, where $d = p_1 - p_2$ denotes the risk difference (for column 1 or column 2) and d_0 denotes the null value.

The test statistic is computed as

$$z = (\hat{d} - d_0)/\text{se}(\hat{d})$$

where the standard error $\text{se}(\hat{d})$ is computed by using the method that you specify. Available methods for the equality test include Wald (with and without continuity correction), Hauck-Anderson, and Farrington-Manning (score). For a description of the standard error computation, see the subsections "Wald Test," "Hauck-Anderson Test," and "Farrington-Manning (Score) Test," respectively, in the section "Noninferiority Tests" on page 184.

PROC FREQ computes one-sided and two-sided *p*-values for equality tests. When the test statistic z is greater than 0, PROC FREQ displays the right-sided *p*-value, which is the probability of a larger value occurring under the null hypothesis. The one-sided *p*-value can be expressed as

$$P_1 = \begin{cases} \text{Prob}(Z > z) & \text{if } z > 0 \\ \text{Prob}(Z < z) & \text{if } z \leq 0 \end{cases}$$

where Z has a standard normal distribution. The two-sided *p*-value is computed as $P_2 = 2 \times P_1$.

Noninferiority Tests If you specify the NONINF *riskdiff-option*, PROC FREQ provides a noninferiority test for the risk difference, or the difference between two proportions. The null hypothesis for the noninferiority test is

$$H_0: p_1 - p_2 \leq -\delta$$

versus the alternative

$$H_a: p_1 - p_2 > -\delta$$

where δ is the noninferiority margin. Rejection of the null hypothesis indicates that the row 1 risk is not inferior to the row 2 risk. See Chow, Shao, and Wang (2003) for more information.

You can specify the value of δ with the MARGIN= *riskdiff-option*. By default, $\delta = 0.2$. You can specify the test method with the METHOD= *riskdiff-option*. The following methods are available for the risk difference noninferiority analysis: Wald (with and without continuity correction), Hauck-Anderson, Farrington-Manning (score), and Newcombe (with and without continuity correction). The Wald, Hauck-Anderson, and Farrington-Manning methods provide tests and corresponding test-based confidence limits; the Newcombe method provides only confidence limits. If you do not specify METHOD=, PROC FREQ uses the Wald test by default.

The confidence coefficient for the test-based confidence limits is $100(1 - 2\alpha)\%$ (Schuirmann 1999). By default, if you do not specify the ALPHA= option, these are 90% confidence limits. You can compare the confidence limits to the noninferiority limit, $-\delta$.

The following sections describe the noninferiority analysis methods for the risk difference.

Wald Test

If you specify the METHOD=WALD *riskdiff-option*, PROC FREQ provides an asymptotic Wald test of noninferiority for the risk difference. This is also the default method. The Wald test statistic is computed as

$$z = (\hat{d} + \delta) / \text{se}(\hat{d})$$

where ($\hat{d} = \hat{p}_1 - \hat{p}_2$) estimates the risk difference and δ is the noninferiority margin.

By default, the standard error for the Wald test is computed from the sample proportions as

$$\text{se}(\hat{d}) = \sqrt{\hat{p}_1(1-\hat{p}_1)/n_1. + \hat{p}_2(1-\hat{p}_2)/n_2.}$$

If you specify the VAR=NULL *riskdiff-option*, the standard error is based on the null hypothesis that the risk difference equals $-\delta$ (Dunnett and Gent 1977). The standard error is computed as

$$\text{se}(\hat{d}) = \sqrt{\tilde{p}(1-\tilde{p})/n_2. + (\tilde{p}-\delta)(1-\tilde{p}+\delta)/n_1.}$$

where

$$\tilde{p} = (n_{11} + n_{21} + \delta n_1.)/n$$

If you specify the CORRECT *riskdiff-option*, the test statistic includes a continuity correction. The continuity correction is subtracted from the numerator of the test statistic if the numerator is greater than 0; otherwise, the continuity correction is added to the numerator. The value of the continuity correction is $(1/n_1. + 1/n_2.)/2$.

The *p*-value for the Wald noninferiority test is $P_z = \text{Prob}(Z > z)$, where Z has a standard normal distribution.

Hauck-Anderson Test

If you specify the METHOD=HA *riskdiff-option*, PROC FREQ provides the Hauck-Anderson test for noninferiority. The Hauck-Anderson test statistic is computed as

$$z = (\hat{d} + \delta \pm cc) / \text{se}(\hat{d})$$

where $\hat{d} = \hat{p}_1 - \hat{p}_2$ and the standard error is computed from the sample proportions as

$$\text{se}(\hat{d}) = \sqrt{\hat{p}_1(1-\hat{p}_1)/(n_1. - 1) + \hat{p}_2(1-\hat{p}_2)/(n_2. - 1)}$$

The Hauck-Anderson continuity correction *cc* is computed as

$$cc = 1 / \big(2 \min(n_1., n_2.)\big)$$

The *p*-value for the Hauck-Anderson noninferiority test is $P_z = \text{Prob}(Z > z)$, where Z has a standard normal distribution. See Hauck and Anderson (1986) and Schuirmann (1999) for more information.

Farrington-Manning (Score) Test

If you specify the METHOD=FM *riskdiff-option*, PROC FREQ provides the Farrington-Manning (score) test of noninferiority for the risk difference. A score test statistic for the null hypothesis that the risk difference equals $-\delta$ can be expressed as

$$z = (\hat{d} + \delta) / \text{se}(\hat{d})$$

where $\hat{d}$ is the observed value of the risk difference ($\hat{p}_1 - \hat{p}_2$),

$$\text{se}(\hat{d}) = \sqrt{\tilde{p}_1(1-\tilde{p}_1)/n_1 + \tilde{p}_2(1-\tilde{p}_2)/n_2}.$$

and $\tilde{p}_1$ and $\tilde{p}_2$ are the maximum likelihood estimates of the row 1 and row 2 risks (proportions) under the restriction that the risk difference is $-\delta$. The *p*-value for the noninferiority test is $P_z = \text{Prob}(Z > z)$, where Z has a standard normal distribution. For more information, see Miettinen and Nurminen (1985); Miettinen (1985); Farrington and Manning (1990); Dann and Koch (2005).

The maximum likelihood estimates of p_1 and p_1, subject to the constraint that the risk difference is $-\delta$, are computed as

$$\tilde{p}_1 = 2u\cos(w) - b/3a \quad \text{and} \quad \tilde{p}_2 = \tilde{p}_1 + \delta$$

where

$$\begin{aligned}
w &= (\pi + \cos^{-1}(v/u^3))/3 \\
v &= b^3/(3a)^3 - bc/6a^2 + d/2a \\
u &= \text{sign}(v)\sqrt{b^2/(3a)^2 - c/3a} \\
a &= 1 + \theta \\
b &= -(1 + \theta + \hat{p}_1 + \theta\hat{p}_2 - \delta(\theta + 2)) \\
c &= \delta^2 - \delta(2\hat{p}_1 + \theta + 1) + \hat{p}_1 + \theta\hat{p}_2 \\
d &= \hat{p}_1\delta(1 - \delta) \\
\theta &= n_2./n_1.
\end{aligned}$$

For more information, see Farrington and Manning (1990, p. 1453).

Newcombe Noninferiority Analysis

If you specify the METHOD=NEWCOMBE *riskdiff-option*, PROC FREQ provides a noninferiority analysis that is based on Newcombe hybrid-score confidence limits for the risk difference. The confidence coefficient for the confidence limits is $100(1-2\alpha)\%$ (Schuirmann 1999). By default, if you do not specify the ALPHA= option, these are 90% confidence limits. You can compare the confidence limits with the noninferiority limit, $-\delta$. If you specify the CORRECT *riskdiff-option*, the confidence limits includes a continuity correction. See the subsection "Newcombe Confidence Limits" in the section "Confidence Limits for the Risk Difference" on page 179 for more information.

Superiority Test If you specify the SUP *riskdiff-option*, PROC FREQ provides a superiority test for the risk difference. The null hypothesis is

$$H_0: : p_1 - p_2 \leq \delta$$

versus the alternative

$$H_a: p_1 - p_2 > \delta$$

where δ is the superiority margin. Rejection of the null hypothesis indicates that the row 1 proportion is superior to the row 2 proportion. You can specify the value of δ with the MARGIN= *riskdiff-option*. By default, $\delta = 0.2$.

The superiority analysis is identical to the noninferiority analysis but uses a positive value of the margin δ in the null hypothesis. The superiority computations follow those in the section "Noninferiority Tests" on page 184 by replacing $-\delta$ by δ. See Chow, Shao, and Wang (2003) for more information.

Equivalence Test If you specify the EQUIV *riskdiff-option*, PROC FREQ provides an equivalence test for the risk difference, or the difference between two proportions. The null hypothesis for the equivalence test is

$$H_0: p_1 - p_2 \leq -\delta_L \quad \text{or} \quad p_1 - p_2 \geq \delta_U$$

versus the alternative

$$H_a: \delta_L < p_1 - p_2 < \delta_U$$

where δ_L is the lower margin and δ_U is the upper margin. Rejection of the null hypothesis indicates that the two binomial proportions are equivalent. See Chow, Shao, and Wang (2003) for more information.

You can specify the value of the margins δ_L and δ_U with the MARGIN= *riskdiff-option*. If you do not specify MARGIN=, PROC FREQ uses lower and upper margins of –0.2 and 0.2 by default. If you specify a single margin value δ, PROC FREQ uses lower and upper margins of $-\delta$ and δ. You can specify the test method with the METHOD= *riskdiff-option*. The following methods are available for the risk difference equivalence analysis: Wald (with and without continuity correction), Hauck-Anderson, Farrington-Manning (score), and Newcombe (with and without continuity correction). The Wald, Hauck-Anderson, and Farrington-Manning methods provide tests and corresponding test-based confidence limits; the Newcombe method provides only confidence limits. If you do not specify METHOD=, PROC FREQ uses the Wald test by default.

PROC FREQ computes two one-sided tests (TOST) for equivalence analysis (Schuirmann 1987). The TOST approach includes a right-sided test for the lower margin δ_L and a left-sided test for the upper margin δ_U. The overall *p*-value is taken to be the larger of the two *p*-values from the lower and upper tests.

The section "Noninferiority Tests" on page 184 gives details about the Wald, Hauck-Anderson, Farrington-Manning (score), and Newcombe methods for the risk difference. The lower margin equivalence test statistic takes the same form as the noninferiority test statistic but uses the lower margin value δ_L in place of $-\delta$. The upper margin equivalence test statistic take the same form as the noninferiority test statistic but uses the upper margin value δ_U in place of $-\delta$.

The test-based confidence limits for the risk difference are computed according to the equivalence test method that you select. If you specify METHOD=WALD with VAR=NULL, or METHOD=FM, separate standard errors are computed for the lower and upper margin tests. In this case, the test-based confidence limits are computed by using the maximum of these two standard errors. These confidence limits have a confidence coefficient of $100(1 - 2\alpha)\%$ (Schuirmann 1999). By default, if you do not specify the ALPHA= option, these are 90% confidence limits. You can compare the test-based confidence limits to the equivalence limits, (δ_L, δ_U).

Barnard's Unconditional Exact Test

The BARNARD option in the EXACT statement provides an unconditional exact test for the risk (proportion) difference for 2×2 tables. The reference set for the unconditional exact test consists of all 2×2 tables that have the same row sums as the observed table (Barnard 1945, 1947, 1949). This differs from the reference set for exact conditional inference, which is restricted to the set of tables that have the same row sums and the same column sums as the observed table. See the sections "Fisher's Exact Test" on page 158 and "Exact Statistics" on page 221 for more information.

The test statistic is the standardized risk difference, which is computed as

$$T = d / \sqrt{p_{\cdot 1}(1 - p_{\cdot 1})(1/n_1 + 1/n_2)}$$

where the risk difference d is defined as the difference between the row 1 and row 2 risks (proportions), $d = (n_{11}/n_1 - n_{21}/n_2)$; n_1 and n_2 are the row 1 and row 2 totals, respectively; and $p_{\cdot 1}$ is the overall proportion in column 1, $(n_{11} + n_{21})/n$.

Under the null hypothesis that the risk difference is 0, the joint probability function for a table can be expressed in terms of the table cell frequencies, the row totals, and the unknown parameter π as

$$f(n_{11}, n_{21}; n_1, n_2, \pi) = \binom{n_1}{n_{11}} \binom{n_2}{n_{21}} \pi^{n_{11}+n_{21}} (1-\pi)^{n-n_{11}-n_{21}}$$

where π is the common value of the risk (proportion).

PROC FREQ sums the table probabilities over the reference set for those tables where the test statistic is greater than or equal to the observed value of the test statistic. This sum can be expressed as

$$\text{Prob}(\pi) = \sum_{A, T(a) \geq t_0} f(n_{11}, n_{21}; n_1, n_2, \pi)$$

where the set A contains all 2×2 tables with row sums equal to n_1 and n_2, and $T(a)$ denotes the value of the test statistic for table a in A. The sum includes probabilities of those tables for which $(T(a) \geq t_0)$, where t_0 is the value of the test statistic for the observed table.

The sum $\text{Prob}(\pi)$ depends on the unknown value of π. To compute the exact p-value, PROC FREQ eliminates the nuisance parameter π by taking the maximum value of $\text{Prob}(\pi)$ over all possible values of π,

$$\text{Prob} = \sup_{(0 \leq \pi \leq 1)} (\text{Prob}(\pi))$$

See Suissa and Shuster (1985) and Mehta and Senchaudhuri (2003).

Common Risk Difference

If you specify the COMMONRISKDIFF option in the TABLES statement, PROC FREQ provides estimates, confidence limits, and tests for the common (overall) risk difference for multiway 2×2 tables.

Mantel-Haenszel Confidence Limits and Test

PROC FREQ computes the Mantel-Haenszel estimate, confidence limits, and test for the common risk difference by using Mantel-Haenszel stratum weights (Mantel and Haenszel 1959) and the Sato variance estimator (Sato 1989). The Mantel-Haenszel estimate of the common risk difference is

$$\widehat{d}_{MH} = \sum_h \widehat{d}_h \, w_h$$

where $\widehat{d}_h$ is the risk difference in stratum h and

$$w_h = \frac{n_{h1\cdot} n_{h2\cdot}}{n_h} \bigg/ \sum_i \frac{n_{i1\cdot} n_{i2\cdot}}{n_i}$$

is the Mantel-Haenszel weight of stratum h. The column 1 risk difference in stratum (2×2 table) h is computed as

$$\widehat{d}_h = \hat{p}_{h1} - \hat{p}_{h2} = (n_{h11}/n_{h1\cdot}) - (n_{h21}/n_{h2\cdot})$$

where $\hat{p}_{h1}$ is the proportion of row 1 observations that are classified in column 1 and $\hat{p}_{h2}$ is the proportion or row 2 observations that are classified in column 1. The column 2 risk is computed in the same way. For more information, see Agresti (2013, p. 231).

PROC FREQ computes the variance of $\widehat{d}_{MH}$ (Sato 1989) as

$$\hat{\sigma}^2(\widehat{d}_{MH}) = \left(\widehat{d}_{MH} \sum_h P_h + \sum_h Q_h \right) / \left(\sum_h n_{h1\cdot} n_{h2\cdot} / n_h \right)^2$$

where

$$P_h = \left(n_{h1\cdot}^2 n_{h21} - n_{h2\cdot}^2 n_{h11} + n_{h1\cdot} n_{h2\cdot} (n_{h2\cdot} - n_{h1\cdot})/2 \right) / n_h^2$$

$$Q_h = (n_{h11}(n_{h2\cdot} - n_{h21}) + n_{h21}(n_{h1\cdot} - n_{h11}))/2n_h$$

The $100(1-\alpha)\%$ confidence limits for the common risk difference are

$$\widehat{d}_{MH} \pm \left(z_{\alpha/2} \times \hat{\sigma}(\widehat{d}_{MH}) \right)$$

If you specify the COMMONRISKDIFF(TEST=MH) option, PROC FREQ provides a Mantel-Haenszel test of the null hypothesis that the common risk difference is 0, which is computed as $z_{MH} = \widehat{d}_{MH} / \hat{\sigma}(\widehat{d}_{MH})$. The two-sided p-value is $\text{Prob}(|Z| > |z_{MH}|)$, where Z has a standard normal distribution.

Minimum Risk Confidence Limits and Test

PROC FREQ computes the minimum risk estimate, confidence limits, and test for the common risk difference by using the method of Mehrotra and Railkar (2000). The stratum estimates are weighted by minimum risk weights, which minimize the mean square error of the estimate of the common risk difference. Minimum risk weights are designed to improve precision and reduce bias (compared to other weighting strategies) and can minimize the power loss that can occur when underlying assumptions are not met. For more information, see Mehrotra (2001) and Dmitrienko et al. (2005, section 1.3.3).

The minimum risk estimate of the common risk difference is

$$\widehat{d}_{MR} = \sum_h \widehat{d}_h w_h^*$$

where $\widehat{d}_h$ is the risk difference in stratum h and w_h^* is the minimum risk weight of stratum h (which is described in the section "Minimum Risk Weights" on page 190). The variance of $\widehat{d}_{MR}$ is estimated by

$$\widehat{V}(\widehat{d}_{MR}) = \sum_h w_h^{*2} \widehat{V}_h$$

where $\widehat{V}_h$ (the variance estimate of the stratum h risk difference) is computed as

$$\widehat{V}_h = \hat{p}_{h1}(1 - \hat{p}_{h1})/n_{h1\cdot} + \hat{p}_{h2}(1 - \hat{p}_{h2})/n_{h2\cdot}$$

The $100(1-\alpha)\%$ minimum risk confidence limits for the common risk difference are

$$\widehat{d}_{MR} \pm \left(cc + z_{\alpha/2} \sqrt{\widehat{V}(\widehat{d}_{MR})} \right)$$

where the continuity correction is

$$cc = 0.1875 / \sum_h (n_{h1\cdot} n_{h2\cdot} / n_h)$$

The continuity correction is applied only when $cc < |\widehat{d}_{MR}|$ (Fleiss, Levin, and Paik 2003). You can remove the continuity correction by specifying the COMMONRISKDIFF(CORRECT=NO) option.

By default, the minimum risk test is computed as

$$z_{MR} = \left(\widehat{d}_{MR} \pm cc\right) / \sqrt{\widehat{V}_0(\widehat{d}_{MR})}$$

The continuity correction cc is subtracted from $\widehat{d}_{MR}$ if $\widehat{d}_{MR} > 0$ and added to $\widehat{d}_{MR}$ if $\widehat{d}_{MR} < 0$. The null variance of the common risk difference is estimated by

$$\widehat{V}_0(\widehat{d}_{MR}) = \sum_h w_h^{*2} \, \widehat{V}_0(\widehat{d}_h)$$

where $\widehat{V}_0(\widehat{d}_h)$ (an estimate of the variance of the stratum h risk difference under the null hypothesis) is

$$\widehat{V}_0(\widehat{d}_h) = \bar{p}_h(1 - \bar{p}_h) \left(1/n_{h1\cdot} + 1/n_{h2\cdot}\right)$$

and

$$\bar{p}_h = (n_{h1\cdot} \hat{p}_{h1\cdot} + n_{h2\cdot} \hat{p}_{h2\cdot}) / (n_{h1\cdot} + n_{h2\cdot})$$

The two-sided p-value is $\text{Prob}(|Z| > |z_{MR}|)$, where Z has a standard normal distribution.

If you specify the VAR=SAMPLE option for COMMONRISKDIFF(TEST=MR), PROC FREQ uses the sample variance estimate $\widehat{V}(\widehat{d}_{MR})$ instead of the null variance estimate $\widehat{V}_0(\widehat{d}_{MR})$ in the denominator of the test statistic z_{MR}. If you specify the COMMONRISKDIFF(CORRECT=NO) option, the continuity correction is not included in the test statistic.

Minimum Risk Weights The estimate of the minimum risk weight for stratum h is defined by Mehrotra and Railkar (2000). as

$$w_h^* = \frac{\beta_h}{\sum_i \widehat{V}_i^{-1}} - \left(\frac{\alpha_h \widehat{V}_h^{-1}}{\sum_i \widehat{V}_i^{-1} + \sum_k \alpha_i \widehat{d}_i \widehat{V}_i^{-1}}\right)\left(\frac{\sum_i \widehat{d}_i \beta_i}{\sum_i \widehat{V}_i^{-1}}\right)$$

where

$$\alpha_h = \sum_i \widehat{V}_i^{-1} - \sum_i \widehat{d}_i \widehat{V}_i^{-1}$$

$$\beta_h = \widehat{V}_h \left(1 + \alpha_h \sum_i f_i \widehat{d}_i\right)$$

and f_h is the fraction in stratum h

$$f_h = n_h / \sum_i n_h$$

All sums are over the s strata (2×2 tables) in the multiway table request, $\widehat{d}_i$ denotes the risk difference estimate in stratum i, and $\widehat{V}_i$ denotes the sample variance estimate of the risk difference in stratum i.

Summary Score Confidence Limits

PROC FREQ computes the summary score estimate of the common risk difference (Agresti 2013, p. 231) by using inverse-variance stratum weights and Miettinen-Nurminen (score) confidence limits for the stratum risk differences. For more information, see the section "Miettinen-Nurminen (Score) Confidence Limits."

The score confidence interval for the risk difference in stratum h can be expressed as $\widehat{d}'_h \pm z_{\alpha/2} \, s'_h$, where $\widehat{d}'_h$ is the midpoint of the score confidence interval and s'_h is the width of the confidence interval divided by $2z_{\alpha/2}$. The summary score estimate of the common risk difference is computed as

$$\widehat{d}_S = \sum_h \widehat{d}'_h \, w'_h$$

where

$$w'_h = (1/s'^2_h) / \sum_i (1/s'^2_i)$$

The variance of $\widehat{d}_S$ is computed as

$$\widehat{\sigma}^2(\widehat{d}_S) = 1 / \sum_h (1/s'^2_h)$$

The $100(1-\alpha)\%$ summary score confidence limits for the common risk difference are

$$\widehat{d}_S \pm \left(z_{\alpha/2} \times \widehat{\sigma}(\widehat{d}_S) \right)$$

If you specify the COMMONRISKDIFF(TEST=SCORE) option, PROC FREQ provides a summary score test of the null hypothesis that the common risk difference is 0. The test statistic is $z_S = \widehat{d}_S / \widehat{\sigma}(\widehat{d}_S)$ The two-sided p-value is $\text{Prob}(|Z| > |z_S|)$ where Z has a standard normal distribution.

Stratified Newcombe Confidence Limits

PROC FREQ computes stratified Newcombe confidence limits for the common risk (proportion) difference by using the method of Yan and Su (2010). The stratified Newcombe confidence limits are constructed from stratified Wilson confidence limits for the common (overall) row proportions. By default, the strata are weighted by Mantel-Haenszel weights; if you specify the COMMONRISKDIFF(CL=NEWCOMBEMR) option, the strata are weighted by minimum risk weights.

PROC FREQ first computes individual Wilson confidence limits for the row proportions in each 2×2 table (stratum), as described in the section "Wilson (Score) Confidence Limits" on page 173. These stratum Wilson confidence limits are then combined to form stratified Wilson confidence limits for the overall row proportions by using stratum weights (either Mantel-Haenszel or minimum risk). The confidence levels of the stratum Wilson confidence limits are chosen so that the overall confidence coefficient (for the stratified Wilson confidence limits) is $100(1-\alpha)\%$ (Yan and Su 2010).

Denote the lower and upper stratified Wilson score confidence limits for the common row 1 proportion as L_1 and U_1, respectively, and denote the lower and upper stratified Wilson confidence limits for the common row 2 proportion as L_2 and U_2, respectively. The $100(1-\alpha)\%$ stratified Newcombe confidence limits for the common risk (proportion) difference are

$$L = \widehat{d} - z_{\alpha/2} \sqrt{\lambda_1 L_1(1-L_1) + \lambda_2 U_2(1-U_2)}$$

$$U = \widehat{d} + z_{\alpha/2} \sqrt{\lambda_2 L_2(1-L_2) + \lambda_1 U_1(1-U_1)}$$

where $\widehat{d}$ is the weighted estimate of the common risk difference and

$$\lambda_1 = \sum_h w_h^2 / n_{h1\cdot}$$
$$\lambda_2 = \sum_h w_h^2 / n_{h2\cdot}$$

By default, the strata are weighted by Mantel-Haenszel weights, which are defined as

$$w_h = \frac{n_{h1\cdot} n_{h2\cdot}}{n_h} \bigg/ \sum_i \frac{n_{i1\cdot} n_{i2\cdot}}{n_i}$$

and the weighted estimate of the common risk difference is $\widehat{d}_{MH}$. For more information, see the section "Mantel-Haenszel Confidence Limits and Test" on page 188. Optionally, the strata are weighted by minimum risk weights, and the weighted estimate of the common risk difference is $\widehat{d}_{MR}$. For more information, see the section "Minimum Risk Confidence Limits and Test" on page 189.

When there is a single stratum, the stratified Newcombe confidence interval is equivalent to the (unstratified) Newcombe confidence interval. For more information, see the subsection "Newcombe Confidence Limits" in the section "Confidence Limits for the Risk Difference" on page 179. See also Kim and Won (2013).

Odds Ratio and Relative Risks for 2 × 2 Tables

Odds Ratio

The odds ratio is a useful measure of association for a variety of study designs. For a retrospective design called a *case-control study*, the odds ratio can be used to estimate the relative risk when the probability of positive response is small (Agresti 2002). In a case-control study, two independent samples are identified based on a binary (yes-no) response variable, and the conditional distribution of a binary explanatory variable is examined within fixed levels of the response variable. For more information, see Stokes, Davis, and Koch (2012), Agresti (2013), and Agresti (2007).

The odds of a positive response (column 1) in row 1 is n_{11}/n_{12}. Similarly, the odds of a positive response in row 2 is n_{21}/n_{22}. The odds ratio is formed as the ratio of the row 1 odds to the row 2 odds. The odds ratio for a 2 × 2 table is defined as

$$OR = \frac{n_{11}/n_{12}}{n_{21}/n_{22}} = \frac{n_{11}\, n_{22}}{n_{12}\, n_{21}}$$

The odds ratio can be any nonnegative number. When the row and column variables are independent, the true value of the odds ratio is 1. An odds ratio greater than 1 indicates that the odds of a positive response are higher in row 1 than in row 2. An odds ratio less than 1 indicates that the odds of a positive response are higher in row 2. The strength of association increases as the deviation from 1 increases.

The transformation $G = (OR - 1)/(OR + 1)$ transforms the odds ratio to the range (–1,1), where $G = 0$ when $OR = 1$; $G = -1$ when $OR = 0$; and G approaches 1 as OR approaches infinity. G is the gamma statistic, which PROC FREQ computes when you specify the MEASURES option.

Confidence Limits for the Odds Ratio The following types of confidence limits are available for the odds ratio: exact, exact mid-p, likelihood ratio, score, Wald, and Wald modified.

Wald Confidence Limits

The asymptotic Wald confidence limits are based on a log transformation of the odds ratio (Woolf 1955; Haldane 1955). PROC FREQ computes the Wald confidence limits as

$$(OR \times \exp(-z\sqrt{v}), \ OR \times \exp(z\sqrt{v}))$$

where

$$v = \text{Var}(\ln OR) = 1/n_{11} + 1/n_{12} + 1/n_{21} + 1/n_{22}$$

and z is the $100(1 - \alpha/2)$th percentile of the standard normal distribution. The confidence level α is determined by the ALPHA= option in the TABLES statement; by default, ALPHA=0.05, which produces 95% confidence limits for the odds ratio. If any of the four cell frequencies are 0, v is undefined and the Wald confidence limits cannot be computed. For more information, see Agresti (2013, p. 70).

Wald Modified Confidence Limits

PROC FREQ computes Wald modified confidence limits (Haldane 1955) for the odds ratio by replacing the n_{ij} by $(n_{ij} + 0.5)$ in the estimator OR and the variance v as follows:

$$OR = \frac{(n_{11} + 0.5)(n_{22} + 0.5)}{(n_{12} + 0.5)(n_{21} + 0.5)}$$

$$v = \text{Var}(\ln OR) = 1/(n_{11} + 0.5) + 1/(n_{12} + 0.5) + 1/(n_{21} + 0.5) + 1/(n_{22} + 0.5)$$

The modified confidence limits are then computed as

$$(OR \times \exp(-z\sqrt{v}), \ OR \times \exp(z\sqrt{v}))$$

where z is the $100(1 - \alpha/2)$th percentile of the standard normal distribution. For more information, see Fleiss, Levin, and Paik (2003) and Agresti (2013).

Score Confidence Limits

Score confidence limits for the odds ratio (Miettinen and Nurminen 1985) are computed by inverting score tests for the odds ratio. A score-based chi-square test statistic for the null hypothesis that the odds ratio equals θ can be expressed as

$$Q(\theta) = \{n_1.(\hat{p}_1 - \tilde{p}_1)\}^2 / \{n/(n-1)\} \{1/(n_1.\tilde{p}_1(1 - \tilde{p}_1)) + 1/(n_2.\tilde{p}_2(1 - \tilde{p}_2))\}^{-1}$$

where $\hat{p}_1$ is the observed row 1 risk ($n_{11}/n_1.$), and $\tilde{p}_1$ and $\tilde{p}_2$ are the maximum likelihood estimates of the row 1 and row 2 risks under the restriction that the odds ratio ($n_{11}n_{22}/n_{12}n_{21}$) is θ. For more information, see Miettinen and Nurminen (1985) and Miettinen (1985, chapter 14).

The $100(1 - \alpha)\%$ score confidence interval for the odds ratio consists of all values of θ for which the test statistic $Q(\theta)$ falls in the acceptance region,

$$\{\theta : Q(\theta) < \chi^2_{1,\alpha}\}$$

where $\chi^2_{1,\alpha}$ is the $100(1-\alpha)$th percentile of the chi-square distribution with 1 degree of freedom. For more information about score confidence limits, see Agresti (2013).

By default, the score confidence limits include the bias correction factor $n/(n-1)$ in the denominator of $Q(\theta)$ (Miettinen and Nurminen 1985, p. 217). If you specify the CL=SCORE(CORRECT=NO) option, PROC FREQ does not include this factor in the computation.

The maximum likelihood estimates of p_1 and p_2, subject to the constraint that the odds ratio is θ, are computed as

$$\tilde{p}_2 = \left(-b + \sqrt{b^2 - 4ac}\right)/2a \quad \text{and} \quad \tilde{p}_1 = \tilde{p}_2\theta/(1 + \tilde{p}_2(\theta - 1))$$

where

$$\begin{aligned} a &= n_{2\cdot}(\theta - 1) \\ b &= n_{1\cdot}\theta + n_{2\cdot} - \hat{p}_{\cdot 1}(\theta - 1) \\ c &= -\hat{p}_{\cdot 1} \end{aligned}$$

For more information, see Miettinen and Nurminen (1985, pp. 217–218) and Miettinen (1985, chapter 14).

Likelihood Ratio Confidence Limits

Likelihood ratio (profile likelihood) confidence limits for the odds ratio are computed by inverting likelihood ratio tests. The likelihood ratio test statistic for the null hypothesis that the odds ratio equals θ can be expressed as

$$G^2(\theta) = 2\left(n_{11}\ln(\hat{p}_1/\tilde{p}_1) + n_{12}\ln((1-\hat{p}_1)/(1-\tilde{p}_1)) + n_{21}\ln(\hat{p}_2/\tilde{p}_2) + n_{22}\ln((1-\hat{p}_2/(1-\tilde{p}_2))\right)$$

where $\hat{p}_i$ is the observed row i risk $(n_{11}/n_{1\cdot})$ and $\tilde{p}_i$ is the maximum likelihood estimate of the row i risk under the restriction that the odds ratio is θ. The computation of the maximum likelihood estimates is described in the subsection "Score Confidence Limits" in this section. For more information, see Agresti (2013), Miettinen and Nurminen (1985), and Miettinen (1985, chapter 14).

The $100(1-\alpha)\%$ likelihood ratio confidence interval for the odds ratio consists of all values of θ for which the test statistic $G^2(\theta)$ falls in the acceptance region,

$$\{\theta : G^2(\theta) < \chi^2_{1,\alpha}\}$$

where $\chi^2_{1,\alpha}$ is the $100(1-\alpha)$th percentile of the chi-square distribution with 1 degree of freedom.

Exact Confidence Limits

PROC FREQ computes exact confidence limits for the odds ratio by inverting two one-sided (equal-tail) exact tests that are based on the noncentral hypergeometric distribution, where the distribution is conditional on the observed marginal totals of the 2 × 2 table. The exact confidence limits ϕ_1 and ϕ_2 are the solutions to the equations

$$\sum_{i=n_{11}}^{n_{\cdot 1}} f(i : n_{\cdot 1}, n_{1\cdot}, n_{2\cdot}, \phi_1) = \alpha/2$$

$$\sum_{i=0}^{n_{11}} f(i : n_{\cdot 1}, n_{1\cdot}, n_{2\cdot}, \phi_2) = \alpha/2$$

where

$$f(i : n_{.1}, n_{1.}, n_{2.}, \phi) = \binom{n_{1.}}{i}\binom{n_{2.}}{n_{.1} - i} \phi^i \bigg/ \sum_{i=0}^{n_{.1}} \binom{n_{1.}}{i}\binom{n_{2.}}{n_{.1} - i} \phi^i$$

For more information, see Fleiss, Levin, and Paik (2003), Thomas (1971), and Gart (1971).

Because this is a discrete problem, the confidence coefficient for the exact confidence interval is not exactly $(1 - \alpha)$ but is at least $(1 - \alpha)$; thus, these confidence limits are conservative. For more information, see Agresti (1992).

When the odds ratio is 0, which occurs when either $n_{11} = 0$ or $n_{22} = 0$, PROC FREQ sets the lower exact confidence limit to 0 and determines the upper limit by using the level α (instead of $\alpha/2$). Similarly, when the odds ratio is infinity, which occurs when either $n_{12} = 0$ or $n_{21} = 0$, PROC FREQ sets the upper exact confidence limit to infinity and determines the lower limit by using level α.

Exact Mid-p Confidence Limits

PROC FREQ computes exact mid-*p* confidence limits for the odds ratio by inverting two one-sided hypergeometric tests that include mid-*p* tail areas. The mid-*p* approach replaces the probability of the observed table by half of that probability in the hypergeometric probability sums, which are described in the subsection "Exact Confidence Limits" in this section. The exact mid-*p* confidence limits ϕ_1 and ϕ_2 are the solutions to the equations

$$\sum_{i=n_{11}+1}^{n_{.1}} \left(f(i : n_{.1}, n_{1.}, n_{2.}, \phi_1) \right) + (1/2) f(n_{11} : n_{.1}, n_{1.}, n_{2.}, \phi_1) = \alpha/2$$

$$\sum_{i=0}^{n_{11}-1} \left(f(i : n_{.1}, n_{1.}, n_{2.}, \phi_2) \right) + (1/2) f(n_{11} : n_{.1}, n_{1.}, n_{2.}, \phi_2) = \alpha/2$$

where

$$f(i : n_{.1}, n_{1.}, n_{2.}, \phi) = \binom{n_{1.}}{i}\binom{n_{2.}}{n_{.1} - i} \phi^i \bigg/ \sum_{i=0}^{n_{.1}} \binom{n_{1.}}{i}\binom{n_{2.}}{n_{.1} - i} \phi^i$$

For more information, see Agresti (2013).

When the odds ratio is 0, which occurs when either $n_{11} = 0$ or $n_{22} = 0$, PROC FREQ sets the lower exact confidence limit to 0 and determines the upper limit by using the level α (instead of $\alpha/2$). Similarly, when the odds ratio is infinity, which occurs when either $n_{12} = 0$ or $n_{21} = 0$, PROC FREQ sets the upper exact confidence limit to infinity and determines the lower limit by using level α.

Relative Risks

Relative risks are useful measures in *cohort* (prospective) study designs, where two samples are identified based on the presence or absence of an explanatory factor. The two samples are observed in future time for the binary (yes-no) response variable under study. Relative risks are also useful in cross-sectional studies, where two variables are observed simultaneously. For more information, see Stokes, Davis, and Koch (2012) and Agresti (2007).

The relative risk is the ratio of the row 1 risk to the row 2 risk in a 2 × 2 table. The column 1 risk in row 1 is the proportion of row 1 observations that are classified in column 1, which can be expressed as

$$p_1 = n_{11} / n_{1\cdot}$$

Similarly, the column 1 risk in row 2 is

$$p_2 = n_{21} / n_{2\cdot}$$

The column 1 relative risk is computed as

$$R = p_1 / p_2$$

A relative risk greater than 1 indicates that the probability of positive response is greater in row 1 than in row 2. Similarly, a relative risk less than 1 indicates that the probability of positive response is less in row 1 than in row 2. The strength of association increases as the deviation from 1 increases.

Confidence Limits for the Relative Risk PROC FREQ provides the following types of confidence limits for the relative risk: exact unconditional, likelihood ratio, score, Wald, and Wald modified.

Wald Confidence Limits
The asymptotic Wald confidence limits are based on a log transformation of the relative risk. PROC FREQ computes the Wald confidence limits for the column 1 relative risk as

$$\left(\hat{r} \times \exp(-z\sqrt{v}),\ \hat{r} \times \exp(z\sqrt{v}) \right)$$

where $\hat{r}$ is the observed value of the relative risk, $\hat{p}_1/\hat{p}_2$, and

$$v = \mathrm{Var}(\ln(\hat{r})) = \left((1 - \hat{p}_1)/n_{11}\right) + \left((1 - \hat{p}_2)/n_{21}\right)$$

and z is the $100(1 - \alpha/2)$th percentile of the standard normal distribution. The confidence level α is determined by the ALPHA= option in the TABLES statement; by default, ALPHA=0.05, which produces 95% confidence limits. If either cell frequency n_{11} or n_{21} is 0, then v is undefined and the Wald confidence limits cannot be computed.

PROC FREQ computes the confidence limits for the column 2 relative risk in the same way.

Wald Modified Confidence Limits
PROC FREQ computes Wald modified confidence limits (Haldane 1955) for the relative risk by replacing the n_{ij} with $(n_{ij} + 0.5)$ and the $n_{i\cdot}$ with $(n_{i\cdot} + 0.5)$ in the estimator R and the variance v as follows:

$$\hat{r}_m = \hat{p}_1/\hat{p}_2 = \frac{(n_{11} + 0.5)/(n_{1\cdot} + 0.5)}{(n_{21} + 0.5)/(n_{2\cdot} + 0.5)}$$

$$v = \mathrm{Var}(\ln(\hat{r}_m)) = 1/(n_{11} + 0.5) + 1/(n_{21} + 0.5) - 1/(n_{1\cdot} + 0.5) - 1/(n_{2\cdot} + 0.5)$$

The confidence limits are computed as

$$\left(\hat{r}_m \times \exp(-z\sqrt{v}),\ \hat{r}_m im \times \exp(z\sqrt{v}) \right)$$

where z is the $100(1 - \alpha/2)$th percentile of the standard normal distribution. For more information, see Fleiss, Levin, and Paik (2003) and Agresti (2013).

Score Confidence Limits

Score confidence limits (Miettinen and Nurminen 1985; Farrington and Manning 1990) are computed by inverting score tests for the relative risk. A score-based chi-square test statistic for the null hypothesis that the relative risk is r_0 can be expressed as

$$Q(r_0) = (\hat{p}_1 - r_0 \hat{p}_2)^2 / \widetilde{\text{Var}}(r_0)$$

where $\hat{p}_1$ and $\hat{p}_2$ are the observed row 1 and row 2 risks (proportions), respectively,

$$\widetilde{\text{Var}}(r_0) = (n/(n-1)) \left(\tilde{p}_1(1-\tilde{p}_1)/n_1. + r_0^2 \, \tilde{p}_2(1-\tilde{p}_2)/n_2. \right)$$

where $\tilde{p}_1$ and $\tilde{p}_2$ are the maximum likelihood estimates of p_1 and p_2, respectively, under the null hypothesis that the relative risk is r_0. For more information, see Miettinen and Nurminen (1985) and Miettinen (1985, chapter 13).

The $100(1-\alpha)\%$ score confidence interval for the relative risk consists of all values of r_0 for which the test statistic $Q(r_0)$ falls in the acceptance region,

$$\{r_0 : Q(r_0) < \chi^2_{1,\alpha}\}$$

where $\chi^2_{1,\alpha}$ is the $100(1-\alpha)$th percentile of the chi-square distribution with 1 degree of freedom. For more information, see Agresti (2013).

By default, the score confidence limits include the bias correction factor $n/(n-1)$ in the denominator of $Q(r_0)$ (Miettinen and Nurminen 1985, p. 217). If you specify the CL=SCORE(CORRECT=NO) option, PROC FREQ does not include this factor in the computation.

The maximum likelihood estimates of p_1 and p_2, subject to the constraint that the relative risk is r_0, are computed as

$$\tilde{p}_1 = \left(-b - \sqrt{b^2 - 4ac}\right)/2a \quad \text{and} \quad \tilde{p}_2 = \tilde{p}_1/r_0$$

where

$$
\begin{aligned}
a &= 1 + \theta \\
b &= -(r_0(1 + \theta \hat{p}_2) + \theta + \hat{p}_1) \\
c &= r_0(\hat{p}_1 + \theta \hat{p}_2) \\
\theta &= n_2./n_1.
\end{aligned}
$$

For more information, see Farrington and Manning (1990, p. 1454) and Miettinen and Nurminen (1985, p. 217).

Likelihood Ratio Confidence Limits

Likelihood ratio (profile likelihood) confidence limits for the relative risk are computed by inverting likelihood ratio tests. The likelihood ratio test statistic for the null hypothesis that the relative risk ratio is r_0 can be expressed as

$$G^2(r_0) = 2 \left(n_{11} \ln(\hat{p}_1/\tilde{p}_1) + n_{12} \ln((1-\hat{p}_1)/(1-\tilde{p}_1)) + n_{21} \ln(\hat{p}_2/\tilde{p}_2) + n_{22} \ln((1-\hat{p}_2)/(1-\tilde{p}_2)) \right)$$

where $\hat{p}_i$ is the observed row i risk ($n_{i1}/n_{i.}$) and $\tilde{p}_i$ is the maximum likelihood estimate of the row i risk under the restriction that the relative risk is r_0. Expressions for the maximum likelihood estimates $\tilde{p}_1$ and $\tilde{p}_2$ are given in the subsection "Score Confidence Limits" in this section. For more information, see Miettinen and Nurminen (1985) and Miettinen (1985, chapter 13).

The $100(1-\alpha)\%$ likelihood ratio confidence interval for the relative risk consists of all values of r_0 for which the test statistic $G^2(r_0)$ falls in the acceptance region,

$$\{\theta : G^2(r_0) < \chi^2_{1,\alpha}\}$$

where $\chi^2_{1,\alpha}$ is the $100(1-\alpha)$th percentile of the chi-square distribution with 1 degree of freedom.

Exact Unconditional Confidence Limits

If you specify the RELRISK option in the EXACT statement, PROC FREQ provides exact unconditional confidence limits for the relative risk. The exact unconditional approach fixes the row margins of the 2×2 table and eliminates the nuisance parameter p_2 by using the maximum *p*-value (worst-case scenario) over all possible values of p_2 (Santner and Snell 1980). The conditional approach, which is described in the section "Exact Statistics" on page 221, does not apply to the relative risk because of the nuisance parameter (Agresti 1992).

By default, PROC FREQ computes the confidence limits by the tail method, which inverts two separate one-sided exact tests of the relative risk, where the tests are based on the score statistic (Chan and Zhang 1999). The size of each one-sided exact test is at most $\alpha/2$, and the confidence coefficient is at least $(1-\alpha)$. If you specify the RELRISK(METHOD=NOSCORE) option in the EXACT statement, PROC FREQ computes the confidence limits by inverting two separate one-sided exact tests that are based on the unstandardized relative risk. If you specify the RELRISK(METHOD=SCORE2) option in the EXACT statement, PROC FREQ computes the confidence limits by inverting a single two-sided exact test that is based on the score statistic (Agresti and Min 2001).

PROC FREQ uses the relative risk score statistic (or the modified form of the unstandardized relative risk) to compute the exact confidence limits as described in the subsection "Exact Unconditional Confidence Limits" in the section "Confidence Limits for the Risk Difference" on page 179.

The score statistic is a less discrete statistic than the unstandardized risk difference and produces less conservative confidence limits (Agresti and Min 2001). For more information, see Santner et al. (2007). The relative risk score statistic (Miettinen and Nurminen 1985; Farrington and Manning 1990) is computed as

$$z(r_0) = (\hat{p}_1 - r_0 \hat{p}_2) / \text{se}(r_0)$$

where

$$\text{se}(r_0) = \sqrt{\tilde{p}_1(1-\tilde{p}_1)/n_{1.} + r_0^2\, \tilde{p}_2(1-\tilde{p}_2)/n_{2.}}$$

where $\tilde{p}_1$ and $\tilde{p}_2$ are the maximum likelihood estimates of p_1 and p_2 under the restriction that the relative risk is r_0. Expressions for the maximum likelihood estimates $\tilde{p}_1$ and $\tilde{p}_2$ are given in the subsection "Score Confidence Limits" in this section. For more information, see Farrington and Manning (1990, p. 1454) and Miettinen and Nurminen (1985, p. 217).

When the confidence limits are computed by using the unstandardized relative risk as the test statistic (METHOD=NOSCORE), PROC FREQ uses a modified form of the relative risk to ensure that the statistic is defined when there are zero-frequency table cells. The modified form adds 0.5 to the table cell and row frequencies (Gart and Nam 1988) and is computed as:

$$\hat{r} = \frac{(n_{11} + 0.5) / (n_{1\cdot} + 0.5)}{(n_{21} + 0.5) / (n_{2\cdot} + 0.5)}$$

For more information, see the subsection "Wald Modified Confidence Limits" in this section.

Relative Risk Tests PROC FREQ provides tests of equality, noninferiority, superiority, and equivalence for the relative risk. The following analysis methods are available: Wald (which is based on a log transformation), Wald modified, score, and likelihood ratio. You can specify the method by using the METHOD= *relrisk-option*; by default, PROC FREQ provides Wald tests.

Equality Test

An equality test for the relative risk can be expressed as

$$H_0: R = r_0$$

versus the alternative

$$H_a: R \neq r_0$$

where $R = p_1/p_2$ denotes the relative risk (for column 1 or column 2) and r_0 denotes the null value. You can specify a null value by using the EQUAL(NULL=) *relrisk-option*; by default, the null value is 1.

The test statistic is computed by the method that you specify; by default, PROC FREQ uses the Wald test. For information about test statistic computation, see the subsections "Wald Test," "Wald Modified Test," "Farrington-Manning (Score) Test," and "Likelihood Ratio Test" in this section.

For the Wald and score methods, the test statistics z have standard normal distributions under the null hypothesis. For the likelihood ratio test, the test statistic G^2 has a chi-square distribution with 1 degree of freedom under the null hypothesis.

When the test statistic z is greater than 0, PROC FREQ displays the right-sided *p*-value, which is the probability of a larger value occurring under the null hypothesis. The one-sided *p*-value can be expressed as

$$P_1 = \begin{cases} \text{Prob}(Z > z) & \text{if } z > 0 \\ \text{Prob}(Z < z) & \text{if } z \leq 0 \end{cases}$$

where Z has a standard normal distribution. The two-sided *p*-value is computed as $P_2 = 2 \times P_1$.

Noninferiority Test

A noninferiority test for the relative risk can be expressed as

$$H_0: R \leq \delta$$

versus the alternative

$$H_a: R > \delta$$

where $R = p_1/p_2$ denotes the relative risk (for column 1 or column 2) and δ denotes the noninferiority margin (limit). You can specify the margin by using the MARGIN= *relrisk-option*; by default, the noninferiority margin is 0.8. The noninferiority margin for a relative risk test should be less than 1. Rejection of the null hypothesis indicates that the row 1 risk is not inferior to the row 2 risk. For more information, see Chow, Shao, and Wang (2008).

The test statistic z is computed by the method that you specify. For information about test statistic computation, see the subsections "Wald Test," "Wald Modified Test," "Farrington-Manning (Score) Test," and "Likelihood Ratio Test" in this section. The test statistic z is computed by using the noninferiority margin (limit) as the null value of the relative risk. Under the null hypothesis, the test statistic has a standard normal distribution. The p-value for the noninferiority test is the right-sided p-value (the probability that $Z > z$).

As part of the noninferiority analysis, PROC FREQ also provides confidence limits for the relative risk. The confidence coefficient is $100(1 - 2\alpha)\%$ (Schuirmann 1999). The confidence level α is determined by the ALPHA= option in the TABLES statement; by default, ALPHA=0.05, which produces 90% confidence limits for the noninferiority analysis. You can compare the confidence limits to the value of the noninferiority limit δ.

Superiority Test

A superiority test for the relative risk can be expressed as

$$H_0: R \leq \delta$$

versus the alternative

$$H_a: R > \delta$$

where $R = p_1/p_2$ denotes the relative risk (for column 1 or column 2) and δ denotes the superiority margin (limit). You can specify the margin by using the MARGIN= *relrisk-option*; by default, the superiority margin is 1.25. The superiority margin for a relative risk test should be greater than 1. Rejection of the null hypothesis indicates that the row 1 risk is superior to the row 2 risk. For more information, see Chow, Shao, and Wang (2008).

The test statistic z is computed by using the superiority margin (limit) as the null value of the relative risk. Under the null hypothesis, the test statistic has a standard normal distribution. The p-value for the superiority test is the right-sided p-value (the probability that $Z > z$).

The computations for the superiority analysis are the same as the computations for the noninferiority analysis, which are described in the subsection "Noninferiority Test" in this section.

Equivalence Test

An equivalence test for the relative risk can be expressed as

$$H_0: R \leq \delta_L \quad \text{or} \quad R \geq \delta_U$$

versus the alternative

$$H_a: \delta_L < R < \delta_U$$

where δ_L is the lower margin and δ_U is the upper margin. Rejection of the null hypothesis indicates that the two risks are equivalent. For more information, see Chow, Shao, and Wang (2008).

You can specify the margins by using the MARGIN= *relrisk-option*; by default, the lower margin is 0.8 and the upper margin is 1.25. If you specify a single margin value, PROC FREQ uses this value as the lower margin for the equivalence test and computes the upper margin as the inverse of the lower margin.

PROC FREQ computes two one-sided tests (TOST) for equivalence analysis (Schuirmann 1987), which include a right-sided test for the lower margin δ_L and a left-sided test for the upper margin δ_U. The lower test statistic uses the lower margin as the null relative risk value, and the p-value is the right-sided probability ($Z > z_L$). The upper test statistic uses the upper margin as the null value, and the p-value is the left-sided probability ($Z < z_U$). The overall p-value is taken to be the larger of the two p-values for the lower and upper tests.

The test statistics are computed by the method that you specify. For more information about the test statistic computation, see the subsections "Wald Test," "Wald Modified Test," "Farrington-Manning (Score) Test," and "Likelihood Ratio Test" in this section.

As part of the equivalence analysis, PROC FREQ also provides confidence limits for the relative risk. The confidence coefficient is $100(1 - 2\alpha)\%$ (Schuirmann 1999). The confidence level α is determined by the ALPHA= option in the TABLES statement; by default, ALPHA=0.05, which produces 90% confidence limits for the equivalence analysis. You can compare the confidence limits to the equivalence limits, which are δ_L and δ_U.

Wald Test

The Wald test statistic (which is based on a log transformation of the relative risk) is computed as $z(r_0) = (\ln(\hat{r}) - \ln(r_0))/\sqrt{(v)}$, where $\hat{r}$ is the relative risk estimate ($\hat{p}_1/\hat{p}_2$), r_0 is the null value of the relative risk, and

$$v = \mathrm{Var}(\ln(\hat{r})) = 1/n_{11} + 1/(n_{21} - 1/n_1. - 1/n_2.$$

The null value is determined by the type of test (equality, noninferiority, superiority, or equivalence) and the null or margin values that you specify. The side of the p-value and the interpretation of the test are also determined by the type of test; for more information, see the subsections "Equality Test," "Noninferiority Test," "Superiority Test," and "Equivalence Test" in this section.

Wald Modified Test

The Wald modified test statistic is computed by replacing the n_{ij} with ($n_{ij} + 0.5$) and the $n_i.$ with ($n_i. + 0.5$) in the relative risk estimator R and the variance v. The test statistic is computed as $z(r_0) = (\ln(\hat{r}) - \ln(r_0))/\sqrt{(v)}$, where r_0 is the null value of the relative risk,

$$\hat{r} = \hat{p}_1/\hat{p}_2 = \frac{(n_{11} + 0.5)/(n_1. + 0.5)}{(n_{21} + 0.5)/(n_2. + 0.5)}$$

$$v = \mathrm{Var}(\ln(\hat{r})) = 1/(n_{11} + 0.5) + 1/(n_{21} + 0.5) - 1/(n_1. + 0.5) - 1/(n_2. + 0.5)$$

The null value is determined by the type of test (equality, noninferiority, superiority, or equivalence) and the null or margin values that you specify. The side of the p-value and the interpretation of the test are also determined by the type of test; for more information, see the subsections "Equality Test," "Noninferiority Test," "Superiority Test," and "Equivalence Test" in this section.

Farrington-Manning (Score) Test

The relative risk score test statistic (Miettinen and Nurminen 1985; Farrington and Manning 1990) for the null value r_0 is computed as

$$z(r_0) = (\hat{p}_1 - r_0 \hat{p}_2) / \mathrm{se}(r_0)$$

where

$$\text{se}(r_0) = \sqrt{\tilde{p}_1(1-\tilde{p}_1)/n_1. + r_0^2\,\tilde{p}_2(1-\tilde{p}_2)/n_2.}$$

where $\tilde{p}_1$ and $\tilde{p}_2$ are the maximum likelihood estimates of p_1 and p_2 under the null value r_0. Expressions for the maximum likelihood estimates $\tilde{p}_1$ and $\tilde{p}_2$ are given in the subsection "Score Confidence Limits" in this section.

The null value is determined by the type of test (equality, noninferiority, superiority, or equivalence) and the null or margin values that you specify. The side of the p-value and the interpretation of the test are also determined by the type of test; for more information, see the subsections "Equality Test," "Noninferiority Test," "Superiority Test," and "Equivalence Test" in this section.

Likelihood Ratio Test

The likelihood ratio statistic for the null relative risk value r_0 is computed as

$$G^2(r_0) = 2\left(n_{11}\ln(\hat{p}_1/\tilde{p}_1) + n_{12}\ln((1-\hat{p}_1)/(1-\tilde{p}_1)) + n_{21}\ln(\hat{p}_2/\tilde{p}_2) + n_{22}\ln((1-\hat{p}_2/(1-\tilde{p}_2))\right)$$

where $\tilde{p}_1$ and $\tilde{p}_2$ are the maximum likelihood estimates of p_1 and p_2 under the null value r_0. Expressions for the maximum likelihood estimates $\tilde{p}_1$ and $\tilde{p}_2$ are given in the subsection "Score Confidence Limits" in this section. For more information, see Miettinen and Nurminen (1985) and Miettinen (1985, chapter 13).

PROC FREQ computes the likelihood ratio test statistic $z(r_0)$ for the noninferiority, superiority, and equivalence tests as $\sqrt{G^2(r_0)}$, where the sign is positive if the estimate is greater than the null value ($\hat{r} \geq r_0$) and negative otherwise ($\hat{r} < r_0$).

The null value is determined by the type of test (equality, noninferiority, superiority, or equivalence) and the null or margin values that you specify. The side of the p-value and the interpretation of the test are also determined by the type of test; for more information, see the subsections "Equality Test," "Noninferiority Test," "Superiority Test," and "Equivalence Test" in this section.

Cochran-Armitage Test for Trend

The TREND option in the TABLES statement provides the Cochran-Armitage test for trend, which tests for trend in binomial proportions across levels of a single factor or covariate. This test is appropriate for a two-way table where one variable has two levels and the other variable is ordinal. The two-level variable represents the response, and the other variable represents an explanatory variable with ordered levels. When the two-way has two columns and R rows, PROC FREQ tests for trend across the R levels of the row variable, and the binomial proportion is computed as the proportion of observations in the first column. When the table has two rows and C columns, PROC FREQ tests for trend across the C levels of the column variable, and the binomial proportion is computed as the proportion of observations in the first row.

The trend test is based on the regression coefficient for the weighted linear regression of the binomial proportions on the scores of the explanatory variable levels. For more information, see Margolin (1988) and Agresti (2002). If the table has two columns and R rows, the trend test statistic is computed as

$$T = \sum_{i=1}^{R} n_{i1}(R_i - \bar{R}) \Big/ \sqrt{p_{\cdot 1}(1 - p_{\cdot 1})s^2}$$

where R_i is the score of row i, $\bar{R}$ is the average row score, and

$$s^2 = \sum_{i=1}^{R} n_{i\cdot}(R_i - \bar{R})^2$$

The SCORES= option in the TABLES statement determines the type of row scores used in computing the trend test (and other score-based statistics). The default is SCORES=TABLE. For more information, see the section "Scores" on page 154. For character variables, the table scores for the row variable are the row numbers (for example, 1 for the first row, 2 for the second row, and so on). For numeric variables, the table score for each row is the numeric value of the row level. When you perform the trend test, the explanatory variable might be numeric (for example, dose of a test substance), and the variable values might be appropriate scores. If the explanatory variable has ordinal levels that are not numeric, you can assign meaningful scores to the variable levels. Sometimes equidistant scores, such as the table scores for a character variable, might be appropriate. For more information on choosing scores for the trend test, see Margolin (1988).

The null hypothesis for the Cochran-Armitage test is no trend, which means that the binomial proportion $p_{i1} = n_{i1}/n_i$. is the same for all levels of the explanatory variable. Under the null hypothesis, the trend statistic has an asymptotic standard normal distribution.

PROC FREQ computes one-sided and two-sided p-values for the trend test. When the test statistic is greater than its null hypothesis expected value of 0, PROC FREQ displays the right-sided p-value, which is the probability of a larger value of the statistic occurring under the null hypothesis. A small right-sided p-value supports the alternative hypothesis of increasing trend in proportions from row 1 to row R. When the test statistic is less than or equal to 0, PROC FREQ displays the left-sided p-value. A small left-sided p-value supports the alternative of decreasing trend.

The one-sided p-value for the trend test is computed as

$$P_1 = \begin{cases} \text{Prob}(Z > T) & \text{if } T > 0 \\ \text{Prob}(Z < T) & \text{if } T \leq 0 \end{cases}$$

where Z has a standard normal distribution. The two-sided p-value is computed as

$$P_2 = \text{Prob}(|Z| > |T|)$$

PROC FREQ also provides exact p-values for the Cochran-Armitage trend test. You can request the exact test by specifying the TREND option in the EXACT statement. See the section "Exact Statistics" on page 221 for more information.

Jonckheere-Terpstra Test

The JT option in the TABLES statement provides the Jonckheere-Terpstra test, which is a nonparametric test for ordered differences among classes. It tests the null hypothesis that the distribution of the response variable does not differ among classes. It is designed to detect alternatives of ordered class differences, which can be expressed as $\tau_1 \leq \tau_2 \leq \cdots \leq \tau_R$ (or $\tau_1 \geq \tau_2 \geq \cdots \geq \tau_R$), with at least one of the inequalities being strict, where τ_i denotes the effect of class i. For such ordered alternatives, the Jonckheere-Terpstra test can be preferable to tests of more general class difference alternatives, such as the Kruskal–Wallis test (produced by the WILCOXON option in the NPAR1WAY procedure). See Pirie (1983) and Hollander and Wolfe (1999) for more information about the Jonckheere-Terpstra test.

The Jonckheere-Terpstra test is appropriate for a two-way table in which an ordinal column variable represents the response. The row variable, which can be nominal or ordinal, represents the classification variable. The levels of the row variable should be ordered according to the ordering you want the test to detect. The order of variable levels is determined by the ORDER= option in the PROC FREQ statement. The default is

ORDER=INTERNAL, which orders by unformatted values. If you specify ORDER=DATA, PROC FREQ orders values according to their order in the input data set. For more information about how to order variable levels, see the ORDER= option.

The Jonckheere-Terpstra test statistic is computed by first forming $R(R-1)/2$ Mann-Whitney counts $M_{i,i'}$, where $i < i'$, for pairs of rows in the contingency table,

$$M_{i,i'} = \{ \text{ number of times } X_{i,j} < X_{i',j'}, \ j = 1, \ldots, n_{i.}; \ j' = 1, \ldots, n_{i'.} \}$$
$$+ \tfrac{1}{2} \{ \text{ number of times } X_{i,j} = X_{i',j'}, \ j = 1, \ldots, n_{i.}; \ j' = 1, \ldots, n_{i'.} \}$$

where $X_{i,j}$ is response j in row i. The Jonckheere-Terpstra test statistic is computed as

$$J = \sum_{1 \leq i < i' \leq R} \sum M_{i,i'}$$

This test rejects the null hypothesis of no difference among classes for large values of J. Asymptotic p-values for the Jonckheere-Terpstra test are obtained by using the normal approximation for the distribution of the standardized test statistic. The standardized test statistic is computed as

$$J^* = (J - E_0(J)) / \sqrt{\text{Var}_0(J)}$$

where $E_0(J)$ and $\text{Var}_0(J)$ are the expected value and variance of the test statistic under the null hypothesis,

$$E_0(J) = \left(n^2 - \sum_i n_{i.}^2 \right) / 4$$

$$\text{Var}_0(J) = A/72 + B/(36n(n-1)(n-2)) + C/(8n(n-1))$$

where

$$A = n(n-1)(2n+5) - \sum_i n_{i.}(n_{i.} - 1)(2n_{i.} + 5) - \sum_j n_{.j}(n_{.j} - 1)(2n_{.j} + 5)$$

$$B = \left(\sum_i n_{i.}(n_{i.} - 1)(n_{i.} - 2) \right) \left(\sum_j n_{.j}(n_{.j} - 1)(n_{.j} - 2) \right)$$

$$C = \left(\sum_i n_{i.}(n_{i.} - 1) \right) \left(\sum_j n_{.j}(n_{.j} - 1) \right)$$

PROC FREQ computes one-sided and two-sided p-values for the Jonckheere-Terpstra test. When the standardized test statistic is greater than its null hypothesis expected value of 0, PROC FREQ displays the right-sided p-value, which is the probability of a larger value of the statistic occurring under the null hypothesis. A small right-sided p-value supports the alternative hypothesis of increasing order from row 1 to row R. When the standardized test statistic is less than or equal to 0, PROC FREQ displays the left-sided p-value. A small left-sided p-value supports the alternative of decreasing order from row 1 to row R.

The one-sided p-value for the Jonckheere-Terpstra test, P_1, is computed as

$$P_1 = \begin{cases} \text{Prob}(Z > J^*) & \text{if } J^* > 0 \\ \text{Prob}(Z < J^*) & \text{if } J^* \leq 0 \end{cases}$$

where Z has a standard normal distribution. The two-sided p-value, P_2, is computed as

$$P_2 = \text{Prob}(|Z| > |J^*|)$$

PROC FREQ also provides exact p-values for the Jonckheere-Terpstra test. You can request the exact test by specifying the JT option in the EXACT statement. See the section "Exact Statistics" on page 221 for more information.

Tests and Measures of Agreement

When you specify the AGREE option in the TABLES statement, PROC FREQ computes tests and measures of agreement for square tables (for which the number of rows equals the number of columns). By default, these statistics include McNemar's test for 2×2 tables, Bowker's symmetry test, the simple kappa coefficient, and the weighted kappa coefficient. For multiple strata (n-way tables, where $n > 2$), the AGREE option provides the overall simple and weighted kappa coefficients, in addition to tests for equal kappas (simple and weighted) among strata. For multiple strata of 2×2 tables, the AGREE option provides Cochran's Q test.

Optionally, PROC FREQ provides kappa tests and other agreement statistics. In addition to the asymptotic tests described in this section, PROC FREQ provides exact p-values for McNemar's test, the simple kappa coefficient test, and the weighted kappa coefficient test. You can request these exact tests by specifying the corresponding options in the EXACT statement. For more information, see the section "Exact Statistics" on page 221.

The following sections provide the formulas that PROC FREQ uses to compute agreement statistics. For information about the use and interpretation of these statistics, see Agresti (2002, 2007); Fleiss, Levin, and Paik (2003), and the other references cited for each statistic.

McNemar's Test

PROC FREQ computes McNemar's test (McNemar 1947) for 2×2 tables when you specify the AGREE option. This test is appropriate when you are analyzing data from matched pairs of subjects with a dichotomous (yes-no) response. By default, the null hypothesis for McNemar's test is marginal homogeneity, which can be expressed as $p_{1.} = p_{.1}$; this is equivalent to a discordant proportion ratio (p_{12}/p_{21}) of 1. The corresponding test statistic is computed as

$$Q_M = (n_{12} - n_{21})^2 / (n_{12} + n_{21})$$

Under the null hypothesis, Q_M has an asymptotic chi-square distribution with 1 degree of freedom.

Optionally, you can specify the null ratio of discordant proportions (p_{12}/p_{21}) by using the AGREE(MNULLRATIO=) option. When the null ratio is r, McNemar's test is computed as

$$Q_M(r) = (n_{12} - e_{12})^2)/e_{12} + (n_{21} - e_{21})^2/e_{21}$$

where $e_{12} = D/(1 + 1/r)$, $e_{21} = D/(1 + r)$, and D is the number of discordant pairs, $(n_{12} + n_{21})$. Under the null hypothesis, $Q_M(r)$ has an asymptotic chi-square distribution with 1 degree of freedom.

PROC FREQ also computes an exact p-value for McNemar's test when you specify the MCNEM option in the EXACT statement.

Bowker's Symmetry Test

The null hypothesis for Bowker's symmetry test (Bowker 1948) is symmetric table-cell proportions, which can be expressed as $p_{ij} = p_{ji}$ for all off-diagonal pairs of table cells. For 2×2 tables, Bowker's test is identical to McNemar's test; therefore, PROC FREQ provides Bowker's test only for square tables that are larger than 2×2.

Bowker's symmetry test is computed as

$$Q_B = \sum_{i<j} \sum (n_{ij} - n_{ji})^2 / (n_{ij} + n_{ji})$$

For large samples, Q_B has an asymptotic chi-square distribution with $R(R-1)/2$ degrees of freedom under the null hypothesis of symmetry, where R is the dimension of the square, two-way table.

By default, the number of degrees of freedom for this test ($R(R-1)/2$) is the number of off-diagonal table-cell comparisons. You can specify the number of degrees of freedom in the AGREE(DFSYM=) option. Alternatively, you can specify the AGREE(DFSYM=ADJUST) option, which reduces the degrees of freedom by the number of off-diagonal table-cell pairs that have a total frequency of 0. For more information, see Hoenig, Morgan, and Brown (1995).

Exact Symmetry Test When you specify the SYMMETRY option in the EXACT statement, PROC FREQ provides an exact symmetry test by using the method of Krauth (1973). This exact test is computed by conditioning on the observed frequency sums of the complementary off-diagonal table-cell pairs ($n_{ij} + n_{ji}$). PROC FREQ evaluates the symmetry test statistic for all tables in the reference set, which includes all possible tables in which the frequency sums of the off-diagonal table-cell pairs match the corresponding frequency sums in the observed table. The exact *p*-value is then computed as the sum of the table probabilities for those tables for which the symmetry test statistic is greater than or equal to the observed test statistic. The table probabilities are computed as products of $R(R-1)/2$ binomial probabilities (which correspond to the off-diagonal table-cell pairs in tables of dimension R) by using the binomial proportion 0.5 under the null hypothesis of symmetry. For more information, see the section "Exact Statistics" on page 221.

Alternatively, you can request a Monte Carlo estimate of the exact *p*-value by specifying the SYMMETRY option together with the MC *computation-option* in the EXACT statement. The Monte Carlo computation for the exact symmetry test is conditional on the same reference set that the exact test uses (tables in which the frequency sums of the off-diagonal table-cell pairs match the corresponding sums in the observed table). For more information, see the section "Monte Carlo Estimation" on page 225.

Simple Kappa Coefficient

The simple kappa coefficient (Cohen 1960) is a measure of interrater agreement. PROC FREQ computes the simple kappa coefficient as

$$\hat{\kappa} = (P_o - P_e) / (1 - P_e)$$

where $P_o = \sum_i p_{ii}$ and $P_e = \sum_i p_{i.} p_{.i}$. The component P_o is the proportion of observed agreement, and the component P_e represents the proportion of chance-expected agreement.

If the two response variables are viewed as two independent ratings of the *n* subjects, the kappa coefficient is +1 when there is complete agreement of the raters. When the observed agreement exceeds the chance-expected agreement, the kappa coefficient is positive, and its magnitude reflects the strength of agreement. When the observed agreement is less than the chance-expected agreement, the kappa coefficient is negative. The minimum value of kappa is between −1 and 0, depending on the marginal proportions of the table.

PROC FREQ computes the asymptotic variance of the simple kappa coefficient as

$$\text{Var}(\hat{\kappa}) = (A + B - C) / (1 - P_e)^2 n$$

where

$$A = \sum_i p_{ii} \left(1 - (p_{i\cdot} + p_{\cdot i})(1 - \hat{\kappa})\right)^2$$

$$B = (1 - \hat{\kappa})^2 \sum\sum_{i \neq j} p_{ij}(p_{\cdot i} + p_{j\cdot})^2$$

$$C = (\hat{\kappa} - P_e(1 - \hat{\kappa}))^2$$

For more information, see Fleiss, Cohen, and Everitt (1969).

Confidence limits for the simple kappa coefficient are computed as

$$\hat{\kappa} \pm \left(z_{\alpha/2} \times \sqrt{\text{Var}(\hat{\kappa})}\right)$$

where $z_{\alpha/2}$ is the $100(1 - \alpha/2)$th percentile of the standard normal distribution. The value of α is determined by the ALPHA= option; by default ALPHA=0.05, which produces 95% confidence limits.

PROC FREQ provides an asymptotic test for the simple kappa coefficient. By default, the null hypothesis value of kappa is 0; alternatively, you can specify a nonzero null value of kappa (by using the AGREE(NULLKAPPA=) option in the TABLES statement). When the null value of kappa is nonzero, PROC FREQ computes the test statistic as

$$z = (\hat{\kappa} - \kappa_0) / \sqrt{\text{Var}(\hat{\kappa})}$$

where κ_0 is the null value that you specify and $\text{Var}(\hat{\kappa})$ is the variance of the kappa coefficient.

When the null value of kappa is 0, PROC FREQ computes the test statistic as

$$z = \hat{\kappa} / \sqrt{\text{Var}_0(\hat{\kappa})}$$

where $\text{Var}_0(\hat{\kappa})$ is the variance of the kappa coefficient under the null hypothesis (that kappa is 0) and is computed as

$$\text{Var}_0(\hat{\kappa}) = \left(P_e + P_e^2 - \sum_i p_{i\cdot} p_{\cdot i}(p_{i\cdot} + p_{\cdot i})\right) / (1 - P_e)^2 n$$

This test statistic has an asymptotic standard normal distribution under the null hypothesis. For more information, see Fleiss, Levin, and Paik (2003).

PROC FREQ also provides an exact test for the simple kappa coefficient. You can request the exact test by specifying the KAPPA or AGREE option in the EXACT statement. For more information, see the section "Exact Statistics" on page 221.

Kappa Details When you specify the AGREE(KAPPADETAILS) option, PROC FREQ displays the "Kappa Details" table, which includes the observed agreement P_o, chance-expected agreement P_e, maximum kappa, and B_n measure.

The maximum kappa, which is the maximum possible value of the kappa coefficient given the marginal proportions of the two-way table, is computed as

$$\max(\kappa) = (\max(P_o) - P_e) \,/\, (1 - P_e)$$

where

$$\max(P_o) = \left(\sum_i \min(n_{i\cdot}, n_{\cdot i}) \right) / n$$

The B_n measure (Bangdiwala 1988; Bangdiwala et al. 2008) is computed as

$$B_n = \left(\sum_i n_{ii}^2 \right) / \left(\sum_i \sum_j n_{i\cdot} n_{\cdot j} \right)$$

For 2 × 2 tables, the "Kappa Details" table also includes the prevalence index and the bias index. The prevalence index is the absolute difference between the agreement proportions, $|p_{11} - p_{22}|$. The bias index is the absolute difference between the disagreement proportions, $|p_{12} - p_{21}|$. For more information, see Sim and Wright (2005) and Byrt, Bishop, and Carlin (1993).

Weighted Kappa Coefficient

The weighted kappa coefficient is a generalization of the simple kappa coefficient that uses weights to quantify the relative differences between categories. For 2 × 2 tables, the weighted kappa coefficient is equivalent to the simple kappa coefficient; therefore, PROC FREQ displays the weighted kappa coefficient only for tables larger than 2 × 2. PROC FREQ computes the kappa weights from the column scores, by using either Cicchetti-Allison weights or Fleiss-Cohen weights, both of which are described in the section "Kappa Weights" on page 210. The kappa weights w_{ij} are constructed so that $0 \leq w_{ij} < 1$ for all $i \neq j$, $w_{ii} = 1$ for all i, and $w_{ij} = w_{ji}$. The weighted kappa coefficient is computed as

$$\hat{\kappa}_w = \left(P_{o(w)} - P_{e(w)} \right) / \left(1 - P_{e(w)} \right)$$

where

$$P_{o(w)} = \sum_i \sum_j w_{ij} p_{ij}$$

$$P_{e(w)} = \sum_i \sum_j w_{ij} p_{i\cdot} p_{\cdot j}$$

The component $P_{o(w)}$ is the proportion of observed (weighted) agreement, and the component $P_{e(w)}$ represents the proportion of chance-expected (weighted) agreement. When you specify the AGREE(WTKAPDETAILS) option, PROC FREQ displays these components in the "Weighted Kappa Details" table.

PROC FREQ computes the asymptotic variance of the weighted kappa coefficient as

$$\text{Var}(\hat{\kappa}_w) = \left(\sum_i \sum_j p_{ij} \left(w_{ij} - (\overline{w}_{i\cdot} + \overline{w}_{\cdot j})(1 - \hat{\kappa}_w) \right)^2 - \left(\hat{\kappa}_w - P_{e(w)}(1 - \hat{\kappa}_w) \right)^2 \right) / (1 - P_{e(w)})^2 \, n$$

where

$$\overline{w}_{i\cdot} = \sum_j p_{\cdot j} w_{ij}$$

$$\overline{w}_{\cdot j} = \sum_i p_{i\cdot} w_{ij}$$

For more information, see Fleiss, Cohen, and Everitt (1969).

Confidence limits for the weighted kappa coefficient are computed as

$$\hat{\kappa}_w \pm \left(z_{\alpha/2} \times \sqrt{\text{Var}(\hat{\kappa}_w)} \right)$$

where $z_{\alpha/2}$ is the $100(1 - \alpha/2)$th percentile of the standard normal distribution. The value of α is determined by the ALPHA= option; by default ALPHA=0.05, which produces 95% confidence limits.

PROC FREQ provides an asymptotic test for the weighted kappa coefficient. By default, the null hypothesis value of weighted kappa is 0; alternatively, you can specify a nonzero null value of weighted kappa (by using the AGREE(NULLWTKAPPA=) option in the TABLES statement). When the null value of weighted kappa is nonzero, PROC FREQ computes the test statistic as

$$z = (\hat{\kappa}_w - \kappa_{w(0)}) / \sqrt{\text{Var}(\hat{\kappa}_w)}$$

where $\kappa_{w(0)}$ is the null value that you specify and $\text{Var}(\hat{\kappa}_w)$ is the variance of the weighted kappa coefficient.

When the null value of weighted kappa is 0, PROC FREQ computes the test statistic as

$$z = \hat{\kappa}_w / \sqrt{\text{Var}_0(\hat{\kappa}_w)}$$

where $\text{Var}_0(\hat{\kappa}_w)$ is the variance of the weighted kappa coefficient under the null hypothesis (that weighted kappa is 0) and is computed as

$$\text{Var}_0(\hat{\kappa}_w) = \left(\sum_i \sum_j p_{i\cdot} p_{\cdot j} \left(w_{ij} - (\overline{w}_{i\cdot} + \overline{w}_{\cdot j}) \right)^2 - P_{e(w)}^2 \right) / (1 - P_{e(w)})^2 \, n$$

This test statistic has an asymptotic standard normal distribution under the null hypothesis. For more information, see Fleiss, Levin, and Paik (2003).

PROC FREQ also provides an exact test for the weighted kappa coefficient. You can request the exact test by specifying the KAPPA or AGREE option in the EXACT statement. For more information, see the section "Exact Statistics" on page 221.

Kappa Weights PROC FREQ computes kappa coefficient weights by using the column scores and one of the two available weight types. The column scores are determined by the SCORES= option in the TABLES statement. The two available types of kappa weights are Cicchetti-Allison and Fleiss-Cohen weights. By default, PROC FREQ uses Cicchetti-Allison weights. If you specify the AGREE(WT=FC) option, PROC FREQ uses Fleiss-Cohen weights to compute the weighted kappa coefficient.

PROC FREQ computes Cicchetti-Allison kappa coefficient weights as

$$w_{ij} = 1 - \frac{|C_i - C_j|}{C_C - C_1}$$

where C_i is the score for column i and C is the number of categories or columns. For more information, see Cicchetti and Allison (1971).

The SCORES= option in the TABLES statement determines the type of column scores used to compute the kappa weights (and other score-based statistics). The default is SCORES=TABLE. For more information, see the section "Scores" on page 154. For numeric variables, table scores are the values of the variable levels. You can assign numeric values to the levels in a way that reflects their level of similarity. For example, suppose you have four levels and order them according to similarity. If you assign them values of 0, 2, 4, and 10, the Cicchetti-Allison kappa weights take the following values: $w_{12} = 0.8$, $w_{13} = 0.6$, $w_{14} = 0$, $w_{23} = 0.8$, $w_{24} = 0.2$, and $w_{34} = 0.4$. Note that when there are only two categories (that is, $C = 2$), the weighted kappa coefficient is identical to the simple kappa coefficient.

If you specify the AGREE(WT=FC) option in the TABLES statement, PROC FREQ computes Fleiss-Cohen kappa coefficient weights as

$$w_{ij} = 1 - \frac{(C_i - C_j)^2}{(C_C - C_1)^2}$$

For more information, see Fleiss and Cohen (1973).

For the preceding example, the Fleiss-Cohen kappa weights are $w_{12} = 0.96$, $w_{13} = 0.84$, $w_{14} = 0$, $w_{23} = 0.96$, $w_{24} = 0.36$, and $w_{34} = 0.64$.

Prevalence-Adjusted Bias-Adjusted Kappa

When you specify the AGREE(PABAK) option, PROC FREQ provides the prevalence-adjusted bias-adjusted kappa coefficient (PABAK) (Byrt, Bishop, and Carlin 1993). This coefficient is computed as

$$\hat{\kappa}_a = (P_o - 1/R) \,/\, (1 - 1/R)$$

where $P_o = \sum_i p_{ii}$ and R is the dimension of the square, two-way table. The component P_o is the proportion of observed agreement, and the component $1/R$ represents the chance-expected agreement. When the table is 2×2, $\hat{\kappa}_a = 2P_o - 1$. For more information, see Sim and Wright (2005), Xie (2013), and Holley and Guilford (1964).

PROC FREQ computes the variance of the prevalence-adjusted bias-adjusted kappa as

$$\mathrm{Var}(\hat{\kappa}_a) = (R/(R-1))^2 \, (P_o(1 - P_o)/n)$$

Confidence limits are computed as

$$\hat{\kappa}_a \pm \left(z_{\alpha/2} \times \sqrt{\mathrm{Var}(\hat{\kappa}_a)} \right)$$

where $z_{\alpha/2}$ is the $100(1 - \alpha/2)$th percentile of the standard normal distribution. The value of α is determined by the ALPHA= option; by default ALPHA=0.05, which produces 95% confidence limits.

AC1 Agreement Coefficient

When you specify the AGREE(AC1) option, PROC FREQ provides Gwet's first-order agreement coefficient, AC1 (Gwet 2008). This coefficient is computed as

$$\hat{\gamma} = \left(P_o - P_{e(\gamma)}\right) / \left(1 - P_{e(\gamma)}\right)$$

where $P_o = \sum_i p_{ii}$, $P_e = \sum_i e_i(1-e_i)/(R-1)$, and $e_i = (p_{i\cdot} + p_{\cdot i})/2$ The component P_o is the proportion of observed agreement, and the component $P_{e(\gamma)}$ represents the proportion of chance-expected agreement. For more information, see Xie (2013) and Blood and Spratt (2007).

PROC FREQ computes the variance of AC1 as

$$\text{Var}(\hat{\gamma}) = \left(P_o(1-P_o) - 4(1-\hat{\gamma})A + 4(1-\hat{\gamma}^2)B\right) / n(1-P_{e(\gamma)})^2$$

where

$$A = \sum_i p_{ii}(1-e_i)/(R-1) - P_o P_{e(\gamma)}$$

$$B = \sum_i \sum_j p_{ij}\left(1 - (e_i + e_j)/2\right)^2 /(R-1)^2 - P_{e(\gamma)}^2$$

Confidence limits for AC1 are computed as

$$\hat{\gamma} \pm \left(z_{\alpha/2} \times \sqrt{\text{Var}(\hat{\gamma})}\right)$$

where $z_{\alpha/2}$ is the $100(1-\alpha/2)$th percentile of the standard normal distribution. The value of α is determined by the ALPHA= option; by default ALPHA=0.05, which produces 95% confidence limits.

Overall Kappa Coefficient

When there are multiple strata, PROC FREQ combines the stratum-level estimates of kappa into an overall estimate of the supposed common value of kappa. Assume there are q strata, indexed by $h = 1, 2, \ldots, q$, and let $\text{Var}(\hat{\kappa}_h)$ denote the variance of $\hat{\kappa}_h$. The estimate of the overall kappa coefficient is computed as

$$\hat{\kappa}_T = \sum_{h=1}^{q} \frac{\hat{\kappa}_h}{\text{Var}(\hat{\kappa}_h)} / \sum_{h=1}^{q} \frac{1}{\text{Var}(\hat{\kappa}_h)}$$

For more information, see Fleiss, Levin, and Paik (2003).

PROC FREQ computes an estimate of the overall weighted kappa in the same way.

Tests for Equal Kappa Coefficients

When there are multiple strata, the following chi-square statistic tests whether the stratum-level values of kappa are equal:

$$Q_K = \sum_{h=1}^{q} (\hat{\kappa}_h - \hat{\kappa}_T)^2 / \text{Var}(\hat{\kappa}_h)$$

Under the null hypothesis of equal kappas for the q strata, Q_K has an asymptotic chi-square distribution with $q-1$ degrees of freedom. See Fleiss, Levin, and Paik (2003) for more information. PROC FREQ computes a test for equal weighted kappa coefficients in the same way.

Cochran's Q Test

Cochran's Q is computed for multiway tables when each variable has two levels, that is, for $2 \times 2 \cdots \times 2$ tables. Cochran's Q statistic is used to test the homogeneity of the one-dimensional margins. Let m denote the number of variables and N denote the total number of subjects. Cochran's Q statistic is computed as

$$Q_C = m(m-1) \left(\sum_{j=1}^{m} T_j^2 - T^2 \right) \Big/ \left(mT - \sum_{k=1}^{N} S_k^2 \right)$$

where T_j is the number of positive responses for variable j, T is the total number of positive responses over all variables, and S_k is the number of positive responses for subject k. Under the null hypothesis, Cochran's Q has an asymptotic chi-square distribution with $m-1$ degrees of freedom. For more information, see Cochran (1950). When there are only two binary response variables ($m=2$), Cochran's Q simplifies to McNemar's test. When there are more than two response categories, you can test for marginal homogeneity by using the repeated measures capabilities of the CATMOD procedure.

Tables with Zero-Weight Rows or Columns

The AGREE statistics are defined only for square tables, where the number of rows equals the number of columns; if a table is not square, PROC FREQ does not compute AGREE statistics for the table. In the kappa statistic framework, where two independent raters assign ratings to each of n subjects, suppose one of the raters does not use all possible r rating levels. If the corresponding table contains r rows but only $r-1$ columns, the table is not square and PROC FREQ does not compute AGREE statistics. To create a square table in this situation, you can use the ZEROS option in the WEIGHT statement, which includes zero-weight observations in the analysis. You can include zero-weight observations in the input data set to represent any rating levels that are not used by a rater, so that the input data set has at least one observation for each possible rater and rating combination. When you use this input data set and specify the ZEROS option, the analysis includes all rating levels (even when all levels are not actually assigned by both raters). The resulting table (of rater 1 by rater 2) is a square table, and AGREE statistics can be computed.

For more information, see the description of the ZEROS option in the WEIGHT statement. By default, PROC FREQ does not process observations that have weights of 0 because these observations do not contribute to the total frequency count, and because many of the tests and measures of association are undefined for tables that contain zero-weight rows or columns. However, kappa statistics are defined for tables that contain zero-weight rows or columns, and the ZEROS option enables you to input zero-weight observations and construct the tables needed to compute kappa statistics.

Cochran-Mantel-Haenszel Statistics

The CMH option in the TABLES statement gives a stratified statistical analysis of the relationship between the row and column variables after controlling for the strata variables in a multiway table. For example, for the table request A*B*C*D, the CMH option provides an analysis of the relationship between C and D, after controlling for A and B. The stratified analysis provides a way to adjust for the possible confounding effects of A and B without being forced to estimate parameters for them.

The CMH analysis produces Cochran-Mantel-Haenszel statistics, which include the correlation statistic, the ANOVA (row mean scores) statistic, and the general association statistic. For 2×2 tables, the CMH option also provides Mantel-Haenszel and logit estimates of the common odds ratio and the common relative risks, in addition to the Breslow-Day test for homogeneity of the odds ratios.

Exact statistics are also available for stratified 2×2 tables. If you specify the EQOR option in the EXACT statement, PROC FREQ provides Zelen's exact test for equal odds ratios. If you specify the COMOR option in the EXACT statement, PROC FREQ provides exact confidence limits for the common odds ratio and an exact test that the common odds ratio equals one.

Let the number of strata be denoted by q, indexing the strata by $h = 1, 2, \ldots, q$. Each stratum contains a contingency table with X representing the row variable and Y representing the column variable. For table h, denote the cell frequency in row i and column j by n_{hij}, with corresponding row and column marginal totals denoted by $n_{hi\cdot}$ and $n_{h\cdot j}$, and the overall stratum total by n_h.

Because the formulas for the Cochran-Mantel-Haenszel statistics are more easily defined in terms of matrices, the following notation is used. Vectors are presumed to be column vectors unless they are transposed ($'$).

$$\begin{aligned}
\mathbf{n}'_{hi} &= (n_{hi1}, n_{hi2}, \ldots, n_{hiC}) & (1 \times C) \\
\mathbf{n}'_h &= (\mathbf{n}'_{h1}, \mathbf{n}'_{h2}, \ldots, \mathbf{n}'_{hR}) & (1 \times RC) \\
p_{hi\cdot} &= n_{hi\cdot} / n_h & (1 \times 1) \\
p_{h\cdot j} &= n_{h\cdot j} / n_h & (1 \times 1) \\
\mathbf{P}'_{h*\cdot} &= (p_{h1\cdot}, p_{h2\cdot}, \ldots, p_{hR\cdot}) & (1 \times R) \\
\mathbf{P}'_{h\cdot *} &= (p_{h\cdot 1}, p_{h\cdot 2}, \ldots, p_{h\cdot C}) & (1 \times C)
\end{aligned}$$

Assume that the strata are independent and that the marginal totals of each stratum are fixed. The null hypothesis, H_0, is that there is no association between X and Y in any of the strata. The corresponding model is the multiple hypergeometric; this implies that, under H_0, the expected value and covariance matrix of the frequencies are, respectively,

$$\mathbf{m}_h = \mathbf{E}[\mathbf{n}_h \mid H_0] = n_h (\mathbf{P}_{h\cdot *} \otimes \mathbf{P}_{h*\cdot})$$

$$\mathrm{Var}[\mathbf{n}_h \mid H_0] = c \left((\mathbf{D}_{\mathbf{P}h\cdot *} - \mathbf{P}_{h\cdot *}\mathbf{P}'_{h\cdot *}) \otimes (\mathbf{D}_{\mathbf{P}h*} - \mathbf{P}_{h*\cdot}\mathbf{P}'_{h*\cdot}) \right)$$

where

$$c = n_h^2 / (n_h - 1)$$

and where $\otimes$ denotes Kronecker product multiplication and $\mathbf{D}_\mathbf{a}$ is a diagonal matrix with the elements of $\mathbf{a}$ on the main diagonal.

The generalized CMH statistic (Landis, Heyman, and Koch 1978) is defined as

$$Q_{CMH} = \mathbf{G}' \mathbf{V}_\mathbf{G}^{-1} \mathbf{G}$$

where

$$\mathbf{G} = \sum_h \mathbf{B}_h (\mathbf{n}_h - \mathbf{m}_h)$$

$$\mathbf{V}_\mathbf{G} = \sum_h \mathbf{B}_h (\mathrm{Var}[\mathbf{n}_h \mid H_0]) \mathbf{B}'_h$$

and where

$$\mathbf{B}_h = \mathbf{C}_h \otimes \mathbf{R}_h$$

is a matrix of fixed constants based on column scores C_h and row scores R_h. When the null hypothesis is true, the CMH statistic has an asymptotic chi-square distribution with degrees of freedom equal to the rank of B_h. If V_G is found to be singular, PROC FREQ prints a message and sets the value of the CMH statistic to missing.

PROC FREQ computes three CMH statistics by using this formula for the generalized CMH statistic, with different row and column score definitions for each statistic. The CMH statistics that PROC FREQ computes are the correlation statistic, the ANOVA (row mean scores) statistic, and the general association statistic. These statistics test the null hypothesis of no association against different alternative hypotheses. The following sections describe the computation of these CMH statistics.

CAUTION: The CMH statistics have low power for detecting an association in which the patterns of association for some of the strata are in the opposite direction of the patterns displayed by other strata. Thus, a nonsignificant CMH statistic suggests either that there is no association or that no pattern of association has enough strength or consistency to dominate any other pattern.

Correlation Statistic

The correlation statistic, popularized by Mantel and Haenszel, has 1 degree of freedom and is known as the Mantel-Haenszel statistic (Mantel and Haenszel 1959; Mantel 1963).

The alternative hypothesis for the correlation statistic is that there is a linear association between X and Y in at least one stratum. If either X or Y does not lie on an ordinal (or interval) scale, this statistic is not meaningful.

To compute the correlation statistic, PROC FREQ uses the formula for the generalized CMH statistic with the row and column scores determined by the SCORES= option in the TABLES statement. See the section "Scores" on page 154 for more information about the available score types. The matrix of row scores R_h has dimension $1 \times R$, and the matrix of column scores C_h has dimension $1 \times C$.

When there is only one stratum, this CMH statistic reduces to $(n-1)r^2$, where r is the Pearson correlation coefficient between X and Y. When nonparametric (RANK or RIDIT) scores are specified, the statistic reduces to $(n-1)r_s^2$, where r_s is the Spearman rank correlation coefficient between X and Y. When there is more than one stratum, this CMH statistic becomes a stratum-adjusted correlation statistic.

ANOVA (Row Mean Scores) Statistic

The ANOVA statistic can be used only when the column variable Y lies on an ordinal (or interval) scale so that the mean score of Y is meaningful. For the ANOVA statistic, the mean score is computed for each row of the table, and the alternative hypothesis is that, for at least one stratum, the mean scores of the R rows are unequal. In other words, the statistic is sensitive to location differences among the R distributions of Y.

The matrix of column scores C_h has dimension $1 \times C$, and the column scores are determined by the SCORES= option.

The matrix of row scores R_h has dimension $(R-1) \times R$ and is created internally by PROC FREQ as

$$R_h = [I_{R-1}, -J_{R-1}]$$

where I_{R-1} is an identity matrix of rank $R-1$ and J_{R-1} is an $(R-1) \times 1$ vector of ones. This matrix has the effect of forming $R-1$ independent contrasts of the R mean scores.

When there is only one stratum, this CMH statistic is essentially an analysis of variance (ANOVA) statistic in the sense that it is a function of the variance ratio F statistic that would be obtained from a one-way ANOVA on the dependent variable Y. If nonparametric scores are specified in this case, the ANOVA statistic is a Kruskal-Wallis test.

When there is more than one stratum, this CMH statistic corresponds to a stratum-adjusted ANOVA or Kruskal-Wallis test. In the special case where there is one subject per row and one subject per column in the contingency table of each stratum, this CMH statistic is identical to Friedman's chi-square. See Example 3.9 for an illustration.

General Association Statistic

The alternative hypothesis for the general association statistic is that, for at least one stratum, there is some kind of association between X and Y. This statistic is always interpretable because it does not require an ordinal scale for either X or Y.

For the general association statistic, the matrix $\mathbf{R}_h$ is the same as the one used for the ANOVA statistic. The matrix $\mathbf{C}_h$ is defined similarly as

$$\mathbf{C}_h = [\mathbf{I}_{C-1}, -\mathbf{J}_{C-1}]$$

PROC FREQ generates both score matrices internally. When there is only one stratum, the general association CMH statistic reduces to $Q_P(n-1)/n$, where Q_P is the Pearson chi-square statistic. When there is more than one stratum, the CMH statistic becomes a stratum-adjusted Pearson chi-square statistic. Note that a similar adjustment can be made by summing the Pearson chi-squares across the strata. However, the latter statistic requires a large sample size in each stratum to support the resulting chi-square distribution with $q(R-1)(C-1)$ degrees of freedom. The CMH statistic requires only a large overall sample size because it has only $(R-1)(C-1)$ degrees of freedom.

See Cochran (1954); Mantel and Haenszel (1959); Mantel (1963); Birch (1965); Landis, Heyman, and Koch (1978).

Mantel-Fleiss Criterion

If you specify the CMH(MANTELFLEISS) option in the TABLES statement, PROC FREQ computes the Mantel-Fleiss criterion for stratified 2×2 tables. The Mantel-Fleiss criterion can be used to assess the validity of the chi-square approximation for the distribution of the Mantel-Haenszel statistic for 2×2 tables. For more information, see Mantel and Fleiss (1980); Mantel and Haenszel (1959); Stokes, Davis, and Koch (2012); Dmitrienko et al. (2005).

The Mantel-Fleiss criterion is computed as

$$MF = \min\left(\left[\sum_h m_{h11} - \sum_h (n_{h11})_L\right], \left[\sum_h (n_{h11})_U - \sum_h m_{h11}\right]\right)$$

where m_{h11} is the expected value of n_{h11} under the hypothesis of no association between the row and column variables in table h, $(n_{h11})_L$ is the minimum possible value of the table cell frequency, and $(n_{h11})_U$ is the maximum possible value,

$$m_{h11} = n_{h1\cdot} \, n_{h\cdot 1} \, / \, n_h$$

$$(n_{h11})_L = \max(\,0,\ n_{h1\cdot} - n_{h\cdot 2}\,)$$

$$(n_{h11})_U = \min(\,n_{h\cdot 1},\ n_{h1\cdot}\,)$$

The Mantel-Fleiss guideline accepts the validity of the Mantel-Haenszel approximation when the value of the criterion is at least 5. When the criterion is less than 5, PROC FREQ displays a warning.

Adjusted Odds Ratio and Relative Risk Estimates

The CMH option provides adjusted odds ratio and relative risk estimates for stratified 2 × 2 tables. For each of these measures, PROC FREQ computes a Mantel-Haenszel estimate and a logit estimate. These estimates apply to *n*-way table requests in the TABLES statement, when the row and column variables both have two levels.

For example, for the table request A*B*C*D, if the row and column variables C and D both have two levels, PROC FREQ provides odds ratio and relative risk estimates, adjusting for the confounding variables A and B.

The choice of an appropriate measure depends on the study design. For case-control (retrospective) studies, the odds ratio is appropriate. For cohort (prospective) or cross-sectional studies, the relative risk is appropriate. See the section "Odds Ratio and Relative Risks for 2 × 2 Tables" on page 192 for more information on these measures.

Throughout this section, z denotes the $100(1 - \alpha/2)$th percentile of the standard normal distribution.

Odds Ratio, Case-Control Studies PROC FREQ provides Mantel-Haenszel and logit estimates for the common odds ratio for stratified 2 × 2 tables.

Mantel-Haenszel Estimator

The Mantel-Haenszel estimate of the common odds ratio is computed as

$$OR_{MH} = \left(\sum_h n_{h11} \, n_{h22}/n_h \right) \bigg/ \left(\sum_h n_{h12} \, n_{h21}/n_h \right)$$

It is always computed unless the denominator is 0. For more information, see Mantel and Haenszel (1959) and Agresti (2002).

To compute confidence limits for the common odds ratio, PROC FREQ uses the Robins, Breslow, and Greenland (1986) variance estimate for $\ln(OR_{MH})$. The $100(1 - \alpha/2)\%$ confidence limits for the common odds ratio are

$$\left(OR_{MH} \times \exp(-z\hat{\sigma}), \ OR_{MH} \times \exp(z\hat{\sigma}) \right)$$

where

$$\begin{aligned}
\hat{\sigma}^2 &= \widehat{\text{Var}}(\ln(OR_{MH})) \\
&= \frac{\sum_h (n_{h11} + n_{h22})(n_{h11} \, n_{h22})/n_h^2}{2 \left(\sum_h n_{h11} \, n_{h22}/n_h \right)^2} \\
&\quad + \frac{\sum_h [(n_{h11} + n_{h22})(n_{h12} \, n_{h21}) + (n_{h12} + n_{h21})(n_{h11} \, n_{h22})]/n_h^2}{2 \left(\sum_h n_{h11} \, n_{h22}/n_h \right) \left(\sum_h n_{h12} \, n_{h21}/n_h \right)} \\
&\quad + \frac{\sum_h (n_{h12} + n_{h21})(n_{h12} \, n_{h21})/n_h^2}{2 \left(\sum_h n_{h12} \, n_{h21}/n_h \right)^2}
\end{aligned}$$

Note that the Mantel-Haenszel odds ratio estimator is less sensitive to small n_h than the logit estimator.

Logit Estimator

The adjusted logit estimate of the common odds ratio (Woolf 1955) is computed as

$$OR_L = \exp\left(\sum_h w_h \ln(OR_h) / \sum_h w_h\right)$$

and the corresponding $100(1-\alpha)\%$ confidence limits are

$$\left(OR_L \times \exp\left(-z/\sqrt{\sum_h w_h}\right), \quad OR_L \times \exp\left(z/\sqrt{\sum_h w_h}\right)\right)$$

where OR_h is the odds ratio for stratum h, and

$$w_h = 1/\text{Var}(\ln(OR_h))$$

If any table cell frequency in a stratum h is 0, PROC FREQ adds 0.5 to each cell of the stratum before computing OR_h and w_h (Haldane 1955) for the logit estimate. The procedure provides a warning when this occurs.

Relative Risks, Cohort Studies PROC FREQ provides Mantel-Haenszel and logit estimates of the common relative risks for stratified 2×2 tables.

Mantel-Haenszel Estimator

The Mantel-Haenszel estimate of the common relative risk for column 1 is computed as

$$RR_{MH} = \left(\sum_h n_{h11}\, n_{h2\cdot} / n_h\right) / \left(\sum_h n_{h21}\, n_{h1\cdot} / n_h\right)$$

It is always computed unless the denominator is 0. See Mantel and Haenszel (1959) and Agresti (2002) for more information.

To compute confidence limits for the common relative risk, PROC FREQ uses the Greenland and Robins (1985) variance estimate for $\log(RR_{MH})$. The $100(1 - \alpha/2)\%$ confidence limits for the common relative risk are

$$\left(RR_{MH} \times \exp(-z\hat{\sigma}), \quad RR_{MH} \times \exp(z\hat{\sigma})\right)$$

where

$$\hat{\sigma}^2 = \widehat{\text{Var}}(\ln(RR_{MH})) = \frac{\sum_h (n_{h1\cdot}\, n_{h2\cdot}\, n_{h\cdot 1} - n_{h11}\, n_{h21}\, n_h)/n_h^2}{\left(\sum_h n_{h11}\, n_{h2\cdot}/n_h\right)\left(\sum_h n_{h21}\, n_{h1\cdot}/n_h\right)}$$

Logit Estimator

The adjusted logit estimate of the common relative risk for column 1 is computed as

$$RR_L = \exp\left(\sum_h w_h \ln(RR_h) / \sum_h w_h\right)$$

and the corresponding $100(1-\alpha)\%$ confidence limits are

$$\left(RR_L \times \exp\left(-z \Big/ \sqrt{\sum_h w_h}\right), \ RR_L \times \exp\left(z \Big/ \sqrt{\sum_h w_h}\right) \right)$$

where RR_h is the column 1 relative risk estimate for stratum h and

$$w_h = 1 \ / \ \text{Var}(\ln(RR_h))$$

If n_{h11} or n_{h21} is 0, PROC FREQ adds 0.5 to each cell of the stratum before computing RR_h and w_h for the logit estimate. The procedure prints a warning when this occurs. For more information, see Kleinbaum, Kupper, and Morgenstern (1982, Sections 17.4 and 17.5).

Breslow-Day Test for Homogeneity of the Odds Ratios

When you specify the CMH option, PROC FREQ computes the Breslow-Day test for stratified 2×2 tables. It tests the null hypothesis that the odds ratios for the q strata are equal. When the null hypothesis is true, the statistic has approximately a chi-square distribution with $q-1$ degrees of freedom. See Breslow and Day (1980) and Agresti (2007) for more information.

The Breslow-Day statistic is computed as

$$Q_{BD} = \sum_h (n_{h11} - \text{E}(n_{h11} \mid OR_{MH}))^2 \ / \ \text{Var}(n_{h11} \mid OR_{MH})$$

where E and Var denote expected value and variance, respectively. The summation does not include any table that contains a row or column that has a total frequency of 0. If OR_{MH} equals 0 or if it is undefined, PROC FREQ does not compute the statistic and prints a warning message.

For the Breslow-Day test to be valid, the sample size should be relatively large in each stratum, and at least 80% of the expected cell counts should be greater than 5. Note that this is a stricter sample size requirement than the requirement for the Cochran-Mantel-Haenszel test for $q \times 2 \times 2$ tables, in that each stratum sample size (not just the overall sample size) must be relatively large. Even when the Breslow-Day test is valid, it might not be very powerful against certain alternatives, as discussed in Breslow and Day (1980).

If you specify the BDT option, PROC FREQ computes the Breslow-Day test with Tarone's adjustment, which subtracts an adjustment factor from Q_{BD} to make the resulting statistic asymptotically chi-square. The Breslow-Day-Tarone statistic is computed as

$$Q_{BDT} = Q_{BD} - \left(\sum_h (n_{h11} - \text{E}(n_{h11} \mid OR_{MH})) \right)^2 \ / \ \sum_h \text{Var}(n_{h11} \mid OR_{MH})$$

See Tarone (1985); Jones et al. (1989); Breslow (1996) for more information.

Zelen's Exact Test for Equal Odds Ratios

If you specify the EQOR option in the EXACT statement, PROC FREQ computes Zelen's exact test for equal odds ratios for stratified 2×2 tables. Zelen's test is an exact counterpart to the Breslow-Day asymptotic test for equal odds ratios. The reference set for Zelen's test includes all possible $q \times 2 \times 2$ tables with the same row, column, and stratum totals as the observed multiway table and with the same sum of cell (1,1) frequencies as the observed table. The test statistic is the probability of the observed $q \times 2 \times 2$ table conditional on the fixed margins, which is a product of hypergeometric probabilities.

The p-value for Zelen's test is the sum of all table probabilities that are less than or equal to the observed table probability, where the sum is computed over all tables in the reference set determined by the fixed margins and the observed sum of cell (1,1) frequencies. This test is similar to Fisher's exact test for two-way tables. For more information, see Zelen (1971); Hirji (2006); Agresti (1992). PROC FREQ computes Zelen's exact test by using the polynomial multiplication algorithm of Hirji et al. (1996).

Exact Confidence Limits for the Common Odds Ratio

If you specify the COMOR option in the EXACT statement, PROC FREQ computes exact confidence limits for the common odds ratio for stratified 2×2 tables. This computation assumes that the odds ratio is constant over all the 2×2 tables. Exact confidence limits are constructed from the distribution of $S = \sum_h n_{h11}$, conditional on the marginal totals of the 2×2 tables.

Because this is a discrete problem, the confidence coefficient for these exact confidence limits is not exactly $(1 - \alpha)$ but is at least $(1 - \alpha)$. Thus, these confidence limits are conservative. See Agresti (1992) for more information.

PROC FREQ computes exact confidence limits for the common odds ratio by using an algorithm based on Vollset, Hirji, and Elashoff (1991). See also Mehta, Patel, and Gray (1985).

Conditional on the marginal totals of 2×2 table h, let the random variable S_h denote the frequency of table cell (1,1). Given the row totals $n_{h1\cdot}$ and $n_{h2\cdot}$ and column totals $n_{h\cdot 1}$ and $n_{h\cdot 2}$, the lower and upper bounds for S_h are l_h and u_h,

$$l_h = \max(0, n_{h1\cdot} - n_{h\cdot 2})$$
$$u_h = \min(n_{h1\cdot}, n_{h\cdot 1})$$

Let C_{s_h} denote the hypergeometric coefficient,

$$C_{s_h} = \binom{n_{h\cdot 1}}{s_h} \binom{n_{h\cdot 2}}{n_{h1\cdot} - s_h}$$

and let ϕ denote the common odds ratio. Then the conditional distribution of S_h is

$$P(S_h = s_h \mid n_{1\cdot}, n_{\cdot 1}, n_{\cdot 2}) = C_{s_h} \phi^{s_h} / \sum_{x=l_h}^{x=u_h} C_x \phi^x$$

Summing over all the 2×2 tables, $S = \sum_h S_h$, and the lower and upper bounds of S are l and u,

$$l = \sum_h l_h \quad \text{and} \quad u = \sum_h u_h$$

The conditional distribution of the sum S is

$$P(S = s \mid n_{h1\cdot}, n_{h\cdot 1}, n_{h\cdot 2}; \; h = 1, \ldots, q) = C_s \phi^s / \sum_{x=l}^{x=u} C_x \phi^x$$

where

$$C_s = \sum_{s_1 + \cdots + s_q = s} \left(\prod_h C_{s_h} \right)$$

Let s_0 denote the observed sum of cell (1,1) frequencies over the q tables. The following two equations are solved iteratively for lower and upper confidence limits for the common odds ratio, ϕ_1 and ϕ_2:

$$\sum_{x=s_0}^{x=u} C_x \phi_1^x \Big/ \sum_{x=l}^{x=u} C_x \phi_1^x = \alpha/2$$

$$\sum_{x=l}^{x=s_0} C_x \phi_2^x \Big/ \sum_{x=l}^{x=u} C_x \phi_2^x = \alpha/2$$

When the observed sum s_0 equals the lower bound l, PROC FREQ sets the lower confidence limit to 0 and determines the upper limit with level α. Similarly, when the observed sum s_0 equals the upper bound u, PROC FREQ sets the upper confidence limit to infinity and determines the lower limit with level α.

When you specify the COMOR option in the EXACT statement, PROC FREQ also computes the exact test that the common odds ratio equals one. Setting $\phi = 1$, the conditional distribution of the sum S under the null hypothesis becomes

$$P_0(S = s \mid n_{h1\cdot}, n_{h\cdot 1}, n_{h\cdot 2}; \ h = 1,\ldots,q) = C_s \Big/ \sum_{x=l}^{x=u} C_x$$

The point probability for this exact test is the probability of the observed sum s_0 under the null hypothesis, conditional on the marginals of the stratified 2×2 tables, and is denoted by $P_0(s_0)$. The expected value of S under the null hypothesis is

$$E_0(S) = \sum_{x=l}^{x=u} x\, C_x \Big/ \sum_{x=l}^{x=u} C_x$$

The one-sided exact p-value is computed from the conditional distribution as $P_0(S >= s_0)$ or $P_0(S \leq s_0)$, depending on whether the observed sum s_0 is greater or less than $E_0(S)$,

$$P_1 = P_0(S >= s_0) = \sum_{x=s_0}^{x=u} C_x \Big/ \sum_{x=l}^{x=u} C_x \quad \text{if } s_0 > E_0(S)$$

$$P_1 = P_0(S <= s_0) = \sum_{x=l}^{x=s_0} C_x \Big/ \sum_{x=l}^{x=u} C_x \quad \text{if } s_0 \leq E_0(S)$$

PROC FREQ computes two-sided p-values for this test according to three different definitions. A two-sided p-value is computed as twice the one-sided p-value, setting the result equal to one if it exceeds one,

$$P_2^a = 2 \times P_1$$

In addition, a two-sided p-value is computed as the sum of all probabilities less than or equal to the point probability of the observed sum s_0, summing over all possible values of s, $l \leq s \leq u$,

$$P_2^b = \sum_{l \leq s \leq u:\, P_0(s) \leq P_0(s_0)} P_0(s)$$

Also, a two-sided p-value is computed as the sum of the one-sided p-value and the corresponding area in the opposite tail of the distribution, equidistant from the expected value,

$$P_2^c = P_0\left(|S - E_0(S)| \geq |s_0 - E_0(S)|\right)$$

Gail-Simon Test for Qualitative Interactions

The GAILSIMON option in the TABLES statement provides the Gail-Simon test for qualitative interaction for stratified 2 × 2 tables. For more information, see Gail and Simon (1985); Silvapulle (2001); Dmitrienko et al. (2005).

The Gail-Simon test is based on the risk differences in stratified 2 × 2 tables, where the risk difference is defined as the row 1 risk (proportion in column 1) minus the row 2 risk. For more information, see the section "Risks and Risk Differences" on page 178. By default, PROC FREQ uses column 1 risks to compute the Gail-Simon test. If you specify the GAILSIMON(COLUMN=2) option, PROC FREQ uses column 2 risks.

PROC FREQ computes the Gail-Simon test statistics as described in Gail and Simon (1985),

$$Q- = \sum_h (d_h/s_h)^2 \, I(d_h > 0)$$

$$Q+ = \sum_h (d_h/s_h)^2 \, I(d_h < 0)$$

$$Q = \min(Q-, Q+)$$

where d_h is the risk difference in table h, s_h is the standard error of the risk difference, and $I(d_h > 0)$ equals 1 if $d_h > 0$ and 0 otherwise. Similarly, $I(d_h < 0)$ equals 1 if $d_h < 0$ and 0 otherwise. The q 2 × 2 tables (strata) are indexed by $h = 1, 2, \ldots, q$.

The p-values for the Gail-Simon statistics are computed as

$$p(Q-) = \sum_h (1 - F_h(Q-)) \, B(h; n = q, p = 0.5)$$

$$p(Q+) = \sum_h (1 - F_h(Q+)) \, B(h; n = q, p = 0.5)$$

$$p(Q) = \sum_{h=1}^{q-1} (1 - F_h(Q)) \, B(h; n = (q-1), p = 0.5)$$

where $F_h(\cdot)$ is the cumulative chi-square distribution function with h degrees of freedom and $B(h; n, p)$ is the binomial probability function with parameters n and p. The statistic Q tests the null hypothesis of no qualitative interaction. The statistic $Q-$ tests the null hypothesis of positive risk differences. A small p-value for $Q-$ indicates negative differences; similarly, a small p-value for $Q+$ indicates positive risk differences.

Exact Statistics

Exact statistics can be useful in situations where the asymptotic assumptions are not met, and so the asymptotic p-values are not close approximations for the true p-values. Standard asymptotic methods involve the assumption that the test statistic follows a particular distribution when the sample size is sufficiently large. When the sample size is not large, asymptotic results might not be valid, with the asymptotic p-values differing perhaps substantially from the exact p-values. Asymptotic results might also be unreliable when the distribution of the data is sparse, skewed, or heavily tied. See Agresti (2007) and Bishop, Fienberg, and Holland (1975) for more information. Exact computations are based on the statistical theory of exact conditional inference for contingency tables, reviewed by Agresti (1992).

In addition to computation of exact *p*-values, PROC FREQ provides the option of estimating exact *p*-values by Monte Carlo simulation. This can be useful for problems that are so large that exact computations require a great amount of time and memory, but for which asymptotic approximations might not be sufficient.

Exact statistics are available for many PROC FREQ tests. For one-way tables, PROC FREQ provides exact *p*-values for the binomial proportion tests and the chi-square goodness-of-fit test. Exact (Clopper-Pearson) confidence limits are available for the binomial proportion. For two-way tables, PROC FREQ provides exact *p*-values for the following tests: Pearson chi-square test, likelihood ratio chi-square test, Mantel-Haenszel chi-square test, Fisher's exact test, Jonckheere-Terpstra test, Cochran-Armitage test for trend, and the symmetry test. PROC FREQ also computes exact *p*-values for tests of the following statistics: Kendall's tau-*b*, Stuart's tau-*c*, Somers' $D(C|R)$, Somers' $D(R|C)$, Pearson correlation coefficient, Spearman correlation coefficient, simple kappa coefficient, and weighted kappa coefficient. For 2×2 tables, PROC FREQ provides McNemar's exact test and exact confidence limits for the odds ratio. PROC FREQ also provides exact unconditional confidence limits for the proportion (risk) difference and for the relative risk. For stratified 2×2 tables, PROC FREQ provides Zelen's exact test for equal odds ratios, exact confidence limits for the common odds ratio, and an exact test for the common odds ratio.

The following sections summarize the exact computational algorithms, define the exact *p*-values that PROC FREQ computes, discuss the computational resource requirements, and describe the Monte Carlo estimation option.

Computational Algorithms

PROC FREQ computes exact *p*-values for general $R \times C$ tables by using the network algorithm developed by Mehta and Patel (1983). This algorithm provides a substantial advantage over direct enumeration, which can be very time-consuming and feasible only for small problems. See Agresti (1992) for a review of algorithms for computation of exact *p*-values, and see Mehta, Patel, and Tsiatis (1984) and Mehta, Patel, and Senchaudhuri (1991) for information about the performance of the network algorithm.

The reference set for a given contingency table is the set of all contingency tables with the observed marginal row and column sums. Corresponding to this reference set, the network algorithm forms a directed acyclic network consisting of nodes in a number of stages. A path through the network corresponds to a distinct table in the reference set. The distances between nodes are defined so that the total distance of a path through the network is the corresponding value of the test statistic. At each node, the algorithm computes the shortest and longest path distances for all the paths that pass through that node. For statistics that can be expressed as a linear combination of cell frequencies multiplied by increasing row and column scores, PROC FREQ computes shortest and longest path distances by using the algorithm of Agresti, Mehta, and Patel (1990). For statistics of other forms, PROC FREQ computes an upper bound for the longest path and a lower bound for the shortest path by following the approach of Valz and Thompson (1994).

The longest and shortest path distances or bounds for a node are compared to the value of the test statistic to determine whether all paths through the node contribute to the *p*-value, none of the paths through the node contribute to the *p*-value, or neither of these situations occurs. If all paths through the node contribute, the *p*-value is incremented accordingly, and these paths are eliminated from further analysis. If no paths contribute, these paths are eliminated from the analysis. Otherwise, the algorithm continues, still processing this node and the associated paths. The algorithm finishes when all nodes have been accounted for.

In applying the network algorithm, PROC FREQ uses full numerical precision to represent all statistics, row and column scores, and other quantities involved in the computations. Although it is possible to use rounding to improve the speed and memory requirements of the algorithm, PROC FREQ does not do this because it can result in reduced accuracy of the p-values.

For one-way tables, PROC FREQ computes the exact chi-square goodness-of-fit test by the method of Radlow and Alf (1975). PROC FREQ generates all possible one-way tables with the observed total sample size and number of categories. For each possible table, PROC FREQ compares its chi-square value with the value for the observed table. If the table's chi-square value is greater than or equal to the observed chi-square, PROC FREQ increments the exact p-value by the probability of that table, which is calculated under the null hypothesis by using the multinomial frequency distribution. By default, the null hypothesis states that all categories have equal proportions. If you specify null hypothesis proportions or frequencies by using the TESTP= or TESTF= option in the TABLES statement, PROC FREQ calculates the exact chi-square test based on that null hypothesis.

Other exact computations are described in sections about the individual statistics. For information about the computation of exact confidence limits and tests for the binomial proportion, see the section "Binomial Proportion" on page 170. For information about computation of exact confidence limits for the odds ratio, see the subsection "Exact Confidence Limits" in the section "Confidence Limits for the Odds Ratio" on page 193. For information about other exact computations, see the subsection "Exact Unconditional Confidence Limits" in the section "Confidence Limits for the Risk Difference" on page 179, the subsection "Exact Unconditional Confidence Limits" in the section "Confidence Limits for the Relative Risk" on page 196, and the sections "Exact Symmetry Test" on page 206, "Exact Confidence Limits for the Common Odds Ratio" on page 219 and "Zelen's Exact Test for Equal Odds Ratios" on page 218.

Definition of p-Values

For several tests in PROC FREQ, the test statistic is nonnegative, and large values of the test statistic indicate a departure from the null hypothesis. Such nondirectional tests include the Pearson chi-square, the likelihood ratio chi-square, the Mantel-Haenszel chi-square, Fisher's exact test for tables larger than 2×2, McNemar's test, the symmetry test, and the one-way chi-square goodness-of-fit test. The exact p-value for a nondirectional test is the sum of probabilities for those tables having a test statistic greater than or equal to the value of the observed test statistic.

There are other tests where it might be appropriate to test against either a one-sided or a two-sided alternative hypothesis. For example, when you test the null hypothesis that the true parameter value equals 0 ($T = 0$), the alternative of interest might be one-sided ($T \leq 0$, or $T \geq 0$) or two-sided ($T \neq 0$). Such tests include the Pearson correlation coefficient, Spearman correlation coefficient, Jonckheere-Terpstra test, Cochran-Armitage test for trend, simple kappa coefficient, and weighted kappa coefficient. For these tests, PROC FREQ displays the right-sided p-value when the observed value of the test statistic is greater than its expected value. The right-sided p-value is the sum of probabilities for those tables for which the test statistic is greater than or equal to the observed test statistic. Otherwise, when the observed test statistic is less than or equal to the expected value, PROC FREQ displays the left-sided p-value. The left-sided p-value is the sum of probabilities for those tables for which the test statistic is less than or equal to the one observed. The one-sided p-value P_1 can be expressed as

$$P_1 = \begin{cases} \text{Prob}(\text{ Test Statistic} \geq t\ ) & \text{if } t > E_0(T) \\ \text{Prob}(\text{ Test Statistic} \leq t\ ) & \text{if } t \leq E_0(T) \end{cases}$$

where t is the observed value of the test statistic and $E_0(T)$ is the expected value of the test statistic under the null hypothesis. PROC FREQ computes the two-sided p-value as the sum of the one-sided p-value and the corresponding area in the opposite tail of the distribution of the statistic, equidistant from the expected value. The two-sided p-value P_2 can be expressed as

$$P_2 = \text{Prob}\left(\,|\text{Test Statistic} - E_0(T)| \geq |t - E_0(T)|\,\right)$$

If you specify the POINT option in the EXACT statement, PROC FREQ provides exact point probabilities for the exact tests. The exact point probability is the exact probability that the test statistic equals the observed value.

If you specify the MIDP option in the EXACT statement, PROC FREQ provides exact mid-p-values. The exact mid p-value is defined as the exact p-value minus half the exact point probability, which equals the average of $\text{Prob}(\text{Test Statistic} \geq t)$ and $\text{Prob}(\text{Test Statistic} > t)$ for a right-sided test. The exact mid p-value is smaller and less conservative than the non-adjusted exact p-value. For more information, see Agresti (2013, section 1.1.4) and Hirji (2006, sections 2.5 and 2.11.1).

Computational Resources

PROC FREQ uses relatively fast and efficient algorithms for exact computations. These recently developed algorithms, together with improvements in computer power, now make it feasible to perform exact computations for data sets where previously only asymptotic methods could be applied. Nevertheless, there are still large problems that might require a prohibitive amount of time and memory for exact computations, depending on the speed and memory available on your computer. For large problems, consider whether exact methods are really needed or whether asymptotic methods might give results quite close to the exact results and require much less computer time and memory. When asymptotic methods might not be sufficient for such large problems, consider using Monte Carlo estimation of exact p-values, as described in the section "Monte Carlo Estimation" on page 225.

A formula does not exist that can predict in advance how much time and memory are needed to compute an exact p-value for a certain problem. The time and memory required depend on several factors, including which test is being performed, the total sample size, the number of rows and columns, and the specific arrangement of the observations into table cells. Generally, larger problems (in terms of total sample size, number of rows, and number of columns) tend to require more time and memory. For a fixed total sample size, time and memory requirements tend to increase as the number of rows and number of columns increase because the number of tables in the reference set increases. Also for a fixed sample size, time and memory requirements tend to increase as the marginal row and column totals become more homogeneous. For more information, see Agresti, Mehta, and Patel (1990) and Gail and Mantel (1977).

When PROC FREQ is computing exact p-values, you can terminate the computations by pressing the system interrupt key sequence (see the *SAS Companion* for your system) and choosing to stop computations. After you terminate exact computations, PROC FREQ completes all other remaining tasks. The procedure produces the requested output and reports missing values for any exact p-values that were not computed by the time of termination.

You can also use the MAXTIME= option in the EXACT statement to limit the amount of time PROC FREQ uses for exact computations. You specify a MAXTIME= value that is the maximum amount of clock time (in seconds) that PROC FREQ can use to compute an exact p-value. If PROC FREQ does not finish computing an exact p-value within that time, it terminates the computation and completes all other remaining tasks.

Monte Carlo Estimation

If you specify the option MC in the EXACT statement, PROC FREQ computes Monte Carlo estimates of the exact *p*-values instead of directly computing the exact *p*-values. Monte Carlo estimation can be useful for large problems that require a great amount of time and memory for exact computations but for which asymptotic approximations might not be sufficient. To describe the precision of each Monte Carlo estimate, PROC FREQ provides the asymptotic standard error and $100(1-\alpha)\%$ confidence limits. The ALPHA= option in the EXACT statement determines the confidence level α; by default, ALPHA=0.01, which produces 99% confidence limits. The N=*n* option in the EXACT statement specifies the number of samples that PROC FREQ uses for Monte Carlo estimation; the default is 10000 samples. You can specify a larger value for *n* to improve the precision of the Monte Carlo estimates. Because larger values of *n* generate more samples, the computation time increases. Alternatively, you can specify a smaller value of *n* to reduce the computation time.

To compute a Monte Carlo estimate of an exact *p*-value, PROC FREQ generates a random sample of tables with the same total sample size, row totals, and column totals as the observed table. PROC FREQ uses the algorithm of Agresti, Wackerly, and Boyett (1979), which generates tables in proportion to their hypergeometric probabilities conditional on the marginal frequencies. For each sample table, PROC FREQ computes the value of the test statistic and compares it to the value for the observed table. When estimating a right-sided *p*-value, PROC FREQ counts all sample tables for which the test statistic is greater than or equal to the observed test statistic. Then the *p*-value estimate equals the number of these tables divided by the total number of tables sampled.

$$\hat{P}_{MC} = M / N$$
$$M = \text{number of samples with (Test Statistic} \geq t)$$
$$N = \text{total number of samples}$$
$$t = \text{observed Test Statistic}$$

PROC FREQ computes left-sided and two-sided *p*-value estimates in a similar manner. For left-sided *p*-values, PROC FREQ evaluates whether the test statistic for each sampled table is less than or equal to the observed test statistic. For two-sided *p*-values, PROC FREQ examines the sample test statistics according to the expression for P_2 given in the section "Definition of *p*-Values" on page 223.

The variable *M* is a binomially distributed variable with *N* trials and success probability *p*. It follows that the asymptotic standard error of the Monte Carlo estimate is

$$\text{se}(\hat{P}_{MC}) = \sqrt{\hat{P}_{MC}(1-\hat{P}_{MC})/(N-1)}$$

PROC FREQ constructs asymptotic confidence limits for the *p*-values according to

$$\hat{P}_{MC} \pm \left(z_{\alpha/2} \times \text{se}(\hat{P}_{MC})\right)$$

where $z_{\alpha/2}$ is the $100(1-\alpha/2)$th percentile of the standard normal distribution and the confidence level α is determined by the ALPHA= option in the EXACT statement.

When the Monte Carlo estimate $\hat{P}_{MC}$ is 0, PROC FREQ computes the confidence limits for the *p*-value as

$$(0, \ 1-\alpha^{(1/N)})$$

When the Monte Carlo estimate $\hat{P}_{MC}$ is 1, PROC FREQ computes the confidence limits as

$(\alpha^{(1/N)}, 1)$

Computational Resources

For each variable in a table request, PROC FREQ stores all of the levels in memory. If all variables are numeric and not formatted, this requires about 84 bytes for each variable level. When there are character variables or formatted numeric variables, the memory that is required depends on the formatted variable lengths, with longer formatted lengths requiring more memory. The number of levels for each variable is limited only by the largest integer that your operating environment can store.

For any single crosstabulation table requested, PROC FREQ builds the entire table in memory, regardless of whether the table has cell frequencies of 0. Thus, if the numeric variables A, B, and C each have 10 levels, PROC FREQ requires 2520 bytes to store the variable levels for the table request A*B*C, as follows:

```
3 variables * 10 levels/variable * 84 bytes/level
```

In addition, PROC FREQ requires 8000 bytes to store the table cell frequencies

```
1000 cells * 8 bytes/cell
```

even though there might be only 10 observations.

When the variables have many levels or when there are many multiway tables, your computer might not have enough memory to construct the tables. If PROC FREQ runs out of memory while constructing tables, it stops collecting levels for the variable with the most levels and returns the memory that is used by that variable. The procedure then builds the tables that do not contain the disabled variables.

If there is not enough memory for your table request and if increasing the available memory is impractical, you can reduce the number of multiway tables or variable levels. If you are not using the CMH or AGREE option in the TABLES statement to compute statistics across strata, reduce the number of multiway tables by using PROC SORT to sort the data set by one or more of the variables or by using the DATA step to create an index for the variables. Then remove the sorted or indexed variables from the TABLES statement and include a BY statement that uses these variables. You can also reduce memory requirements by using a FORMAT statement in the PROC FREQ step to reduce the number of levels. In addition, reducing the formatted variable lengths reduces the amount of memory that is needed to store the variable levels. For more information about using formats, see the section "Grouping with Formats" on page 149.

Output Data Sets

PROC FREQ produces two types of output data sets that you can use with other statistical and reporting procedures. You can request these data sets as follows:

- Specify the OUT= option in a TABLES statement. This creates an output data set that contains frequency or crosstabulation table counts and percentages

- Specify an OUTPUT statement. This creates an output data set that contains statistics.

PROC FREQ does not display the output data sets. Use PROC PRINT, PROC REPORT, or any other SAS reporting tool to display an output data set.

In addition to these two output data sets, you can create a SAS data set from any piece of PROC FREQ output by using the Output Delivery System. See the section "ODS Table Names" on page 239 for more information.

Contents of the TABLES Statement Output Data Set

The OUT= option in the TABLES statement creates an output data set that contains one observation for each combination of variable values (or table cell) in the last table request. By default, each observation contains the frequency and percentage for the table cell. When the input data set contains missing values, the output data set also contains an observation with the frequency of missing values. The output data set includes the following variables:

- BY variables
- table request variables, such as A, B, C, and D in the table request A*B*C*D
- COUNT, which contains the table cell frequency
- PERCENT, which contains the table cell percentage

If you specify the OUTEXPECT option in the TABLES statement for a two-way or multiway table, the output data set also includes expected frequencies. If you specify the OUTPCT option for a two-way or multiway table, the output data set also includes row, column, and table percentages. The additional variables are as follows:

- EXPECTED, which contains the expected frequency
- PCT_TABL, which contains the percentage of two-way table frequency, for n-way tables where $n > 2$
- PCT_ROW, which contains the percentage of row frequency
- PCT_COL, which contains the percentage of column frequency

If you specify the OUTCUM option in the TABLES statement for a one-way table, the output data set also includes cumulative frequencies and cumulative percentages. The additional variables are as follows:

- CUM_FREQ, which contains the cumulative frequency
- CUM_PCT, which contains the cumulative percentage

The OUTCUM option has no effect for two-way or multiway tables.

The following PROC FREQ statements create an output data set of frequencies and percentages:

```
proc freq;
   tables A A*B / out=D;
run;
```

The output data set D contains frequencies and percentages for the table of A by B, which is the last table request listed in the TABLES statement. If A has two levels (1 and 2), B has three levels (1,2, and 3), and no table cell count is 0 or missing, the output data set D includes six observations, one for each combination of A and B levels. The first observation corresponds to A=1 and B=1; the second observation corresponds to A=1 and B=2; and so on. The data set includes the variables COUNT and PERCENT. The value of COUNT is the number of observations with the given combination of A and B levels. The value of PERCENT is the percentage of the total number of observations with that A and B combination.

When PROC FREQ combines different variable values into the same formatted level, the output data set contains the smallest internal value for the formatted level. For example, suppose a variable X has the values 1.1., 1.4, 1.7, 2.1, and 2.3. When you submit the statement

```
format X 1.;
```

in a PROC FREQ step, the formatted levels listed in the frequency table for X are 1 and 2. If you create an output data set with the frequency counts, the internal values of the levels of X are 1.1 and 1.7. To report the internal values of X when you display the output data set, use a format of 3.1 for X.

Contents of the OUTPUT Statement Output Data Set

The OUTPUT statement creates a SAS data set that contains statistics computed by PROC FREQ. Table 3.7 lists the statistics that can be stored in the output data set. You identify which statistics to include by specifying *output-options*. For more information, see the description of the OUTPUT statement.

If you specify multiple TABLES statements or multiple table requests in a single TABLES statement, the contents of the output data set correspond to the last table request.

For a one-way table or a two-way table, the output data set contains one observation that stores the requested statistics for the table. For a multiway table, the output data set contains an observation for each two-way table (stratum) of the multiway crosstabulation. If you request summary statistics for the multiway table, the output data set also contains an observation that stores the across-strata summary statistics. If you use a BY statement, the output data set contains an observation (for one-way or two-way tables) or set of observations (for multiway tables) for each BY group.

The OUTPUT data set can include the following variables:

- BY variables

- Variables that identify the stratum for multiway tables, such as A and B in the table request A*B*C*D

- Variables that contain the specified statistics

In addition to the specified estimate or test statistic, the output data set includes associated values such as standard errors, confidence limits, *p*-values, and degrees of freedom.

PROC FREQ constructs variable names for the statistics in the output data set by enclosing the *output-option* names in underscores. Variable names for the corresponding standard errors, confidence limits, *p*-values, and degrees of freedom are formed by combining the *output-option* names with prefixes that identify the associated values. Table 3.21 lists the prefixes and their descriptions.

Table 3.21 Output Data Set Variable Name Prefixes

Prefix	Description
E_	Asymptotic standard error (ASE)
L_	Lower confidence limit
U_	Upper confidence limit
E0_	Null hypothesis ASE
Z_	Standardized value
DF_	Degrees of freedom
P_	*p*-value
P2_	Two-sided *p*-value
PL_	Left-sided *p*-value
PR_	Right-sided *p*-value
XP_	Exact *p*-value
XP2_	Exact two-sided *p*-value
XPL_	Exact left-sided *p*-value
XPR_	Exact right-sided *p*-value
XPT_	Exact point probability
XMP_	Exact mid *p*-value
XL_	Exact lower confidence limit
XU_	Exact upper confidence limit

For example, the PCHI *output-option* in the OUTPUT statement includes the Pearson chi-square test in the output data set. The variable names for the Pearson chi-square statistic, its degrees of freedom, and the corresponding *p*-value are _PCHI_, DF_PCHI, and P_PCHI, respectively. For variables that were added to the output data set before SAS/STAT 8.2, PROC FREQ truncates the variable name to eight characters when the length of the prefix plus the *output-option* name exceeds eight characters.

Displayed Output

Number of Variable Levels Table

If you specify the NLEVELS option in the PROC FREQ statement, PROC FREQ displays the "Number of Variable Levels" table. This table provides the number of levels for all variables named in the TABLES statements. PROC FREQ determines the variable levels from the formatted variable values. For more information, see the section "Grouping with Formats" on page 149. The "Number of Variable Levels" table contains the following information:

- Variable name
- Levels, which is the total number of levels of the variable

- Number of Nonmissing Levels, if there are missing levels for any of the variables
- Number of Missing Levels, if there are missing levels for any of the variables

One-Way Frequency Tables

PROC FREQ displays one-way frequency tables for all one-way table requests in the TABLES statements, unless you specify the NOPRINT option in the PROC FREQ statement or the NOPRINT option in the TABLES statement. For a one-way table showing the frequency distribution of a single variable, PROC FREQ displays the name of the variable and its values. For each variable value or level, PROC FREQ displays the following information:

- Frequency count, which is the number of observations in the level
- Test Frequency count, if you specify the CHISQ and TESTF= options to request a chi-square goodness-of-fit test for specified frequencies
- Percent, which is the percentage of the total number of observations. (The NOPERCENT option suppresses this information.)
- Test Percent, if you specify the CHISQ and TESTP= options to request a chi-square goodness-of-fit test for specified percents. (The NOPERCENT option suppresses this information.)
- Cumulative Frequency count, which is the sum of the frequency counts for that level and all other levels listed above it in the table. The last cumulative frequency is the total number of nonmissing observations. (The NOCUM option suppresses this information.)
- Cumulative Percent, which is the percentage of the total number of observations in that level and in all other levels listed above it in the table. (The NOCUM or the NOPERCENT option suppresses this information.)

The one-way table also displays the Frequency Missing, which is the number of observations with missing values.

Statistics for One-Way Frequency Tables

For one-way tables, two statistical options are available in the TABLES statement. The CHISQ option provides a chi-square goodness-of-fit test, and the BINOMIAL option provides binomial proportion statistics and tests. PROC FREQ displays the following information, unless you specify the NOPRINT option in the PROC FREQ statement:

- If you specify the CHISQ option for a one-way table, PROC FREQ provides a chi-square goodness-of-fit test, displaying the Chi-Square statistic, the degrees of freedom (DF), and the probability value (Pr > ChiSq). If you specify the CHISQ option in the EXACT statement, PROC FREQ also displays the exact probability value for this test. If you specify the POINT option with the CHISQ option in the EXACT statement, PROC FREQ displays the exact point probability for the test statistic. If you specify the MIDP option in the EXACT statement, PROC FREQ displays the exact mid p-value for the chi-square test.

- If you specify the BINOMIAL option for a one-way table, PROC FREQ displays the estimate of the binomial Proportion, which is the proportion of observations in the first class listed in the one-way table. PROC FREQ also displays the asymptotic standard error (ASE) and the asymptotic (Wald) and exact (Clopper-Pearson) confidence limits by default. For the binomial proportion test, PROC FREQ displays the asymptotic standard error under the null hypothesis (ASE Under H0), the standardized test statistic (Z), and the one-sided and two-sided probability values.

  If you specify the BINOMIAL option in the EXACT statement, PROC FREQ also displays the exact one-sided and two-sided probability values for this test. If you specify the POINT option with the BINOMIAL option in the EXACT statement, PROC FREQ displays the exact point probability for the test. If you specify the MIDP option in the EXACT statement, PROC FREQ displays the exact mid p-value for the binomial proportion test.

- If you request binomial confidence limits by specifying the BINOMIAL(CL=) option, PROC FREQ displays the "Binomial Confidence Limits" table, which includes the Lower and Upper Confidence Limits for each confidence limit Type that you request. In addition to Wald and Clopper-Pearson (Exact) confidence limits, you can request the following confidence limit types for the binomial proportion: Agresti-Coull, Blaker, Jeffreys, Likelihood Ratio, Logit, Mid-p, and Wilson (score).

- If you request a binomial noninferiority or superiority test by specifying the NONINF or SUP *binomial-option*, PROC FREQ displays a Noninferiority Analysis or Superiority Analysis table that contains the following information: the binomial Proportion, the test ASE (under H0 or Sample), the test statistic Z, the probability value, the noninferiority or superiority limit, and the test confidence limits. If you specify the BINOMIAL option in the EXACT statement, PROC FREQ also provides the exact probability value for the test, and exact test confidence limits.

- If you request a binomial equivalence test by specifying the EQUIV *binomial-option*, PROC FREQ displays an Equivalence Analysis table that contains the following information: binomial Proportion and the test ASE (under H0 or Sample). PROC FREQ displays two one-sided tests (TOST) for equivalence, which include test statistics (Z) and probability values for the Lower and Upper tests, together with the Overall probability value. PROC FREQ also displays the equivalence limits and the test-based confidence limits. If you specify the BINOMIAL option in the EXACT statement, PROC FREQ provides exact probability values for the TOST and exact test-based confidence limits.

Two-Way and Multiway Tables

PROC FREQ displays all multiway table requests in the TABLES statements, unless you specify the NOPRINT option in the PROC FREQ statement or the NOPRINT option in the TABLES statement.

For two-way to multiway crosstabulation tables, the values of the last variable in the table request form the table columns. The values of the next-to-last variable form the rows. Each level (or combination of levels) of the other variables forms one stratum.

There are three ways to display multiway tables in PROC FREQ. By default, PROC FREQ displays multiway tables as separate two-way crosstabulation tables for each stratum of the multiway table. Also by default, PROC FREQ displays these two-way crosstabulation tables in table cell format. Alternatively, if you specify the CROSSLIST option, PROC FREQ displays the two-way crosstabulation tables in ODS column format. If you specify the LIST option, PROC FREQ displays multiway tables in list format, which presents the entire multiway crosstabulation in a single table.

Crosstabulation Tables

By default, PROC FREQ displays two-way crosstabulation tables in table cell format. The row variable values are listed down the side of the table, the column variable values are listed across the top of the table, and each row and column variable level combination forms a table cell.

Each cell of a crosstabulation table can contain the following information:

- Frequency, which is the number of observations in the table cell. (The NOFREQ option suppresses this information.)

- Expected frequency under the hypothesis of independence, if you specify the EXPECTED option

- Deviation of the cell frequency from the expected value, if you specify the DEVIATION option

- Cell Chi-Square, which is the cell's contribution to the total chi-square statistic, if you specify the CELLCHI2 option

- Tot Pct, which is the cell's percentage of the total multiway table frequency, for *n*-way tables when *n* > 2, if you specify the TOTPCT option

- Percent, which is the cell's percentage of the total (two-way table) frequency. (The NOPERCENT option suppresses this information.)

- Row Pct, or the row percentage, which is the cell's percentage of the total frequency for its row. (The NOROW option suppresses this information.)

- Col Pct, or column percentage, which is the cell's percentage of the total frequency for its column. (The NOCOL option suppresses this information.)

- Cumulative Col%, or cumulative column percentage, if you specify the CUMCOL option

The table also displays the Frequency Missing, which is the number of observations with missing values.

CROSSLIST Tables

If you specify the CROSSLIST option, PROC FREQ displays two-way crosstabulation tables in ODS column format. The CROSSLIST column format is different from the default crosstabulation table cell format, but the CROSSLIST table provides the same information (frequencies, percentages, and other statistics) as the default crosstabulation table.

In the CROSSLIST table format, the rows of the display correspond to the crosstabulation table cells, and the columns of the display correspond to descriptive statistics such as frequencies and percentages. Each table cell is identified by the values of its TABLES row and column variable levels, with all column variable levels listed within each row variable level. The CROSSLIST table also provides row totals, column totals, and overall table totals.

For a crosstabulation table in CROSSLIST format, PROC FREQ displays the following information:

- the row variable name and values

- the column variable name and values

- Frequency, which is the number of observations in the table cell. (The NOFREQ option suppresses this information.)

- Expected cell frequency under the hypothesis of independence, if you specify the EXPECTED option

- Deviation of the cell frequency from the expected value, if you specify the DEVIATION option

- Standardized Residual, if you specify the CROSSLIST(STDRES) option

- Pearson Residual, if you specify the CROSSLIST(PEARSONRES) option

- Cell Chi-Square, which is the cell's contribution to the total chi-square statistic, if you specify the CELLCHI2 option

- Total Percent, which is the cell's percentage of the total multiway table frequency, for n-way tables when $n > 2$, if you specify the TOTPCT option

- Percent, which is the cell's percentage of the total (two-way table) frequency. (The NOPERCENT option suppresses this information.)

- Row Percent, which is the cell's percentage of the total frequency for its row. (The NOROW option suppresses this information.)

- Column Percent, the cell's percentage of the total frequency for its column. (The NOCOL option suppresses this information.)

The table also displays the Frequency Missing, which is the number of observations with missing values.

LIST Tables

If you specify the LIST option in the TABLES statement, PROC FREQ displays multiway tables in a list format rather than as crosstabulation tables. The LIST option displays the entire multiway table in one table, instead of displaying a separate two-way table for each stratum. The LIST option is not available when you also request statistical options. Unlike the default crosstabulation output, the LIST output does not display row percentages, column percentages, and optional information such as expected frequencies and cell chi-squares.

For a multiway table in list format, PROC FREQ displays the following information:

- the variable names and values

- Frequency, which is the number of observations in the level (with the indicated variable values)

- Percent, which is the level's percentage of the total number of observations. (The NOPERCENT option suppresses this information.)

- Cumulative Frequency, which is the accumulated frequency of the level and all other levels listed above it in the table. The last cumulative frequency in the table is the total number of nonmissing observations. (The NOCUM option suppresses this information.)

- Cumulative Percent, which is the accumulated percentage of the level and all other levels listed above it in the table. (The NOCUM or the NOPERCENT option suppresses this information.)

The table also displays the Frequency Missing, which is the number of observations with missing values.

Statistics for Two-Way and Multiway Tables

PROC FREQ computes statistical tests and measures for crosstabulation tables, depending on which statements and options you specify. You can suppress the display of these results by specifying the NOPRINT option in the PROC FREQ statement. With any of the following information, PROC FREQ also displays the Sample Size and the Frequency Missing.

- If you specify the SCOROUT option in the TABLES statement, PROC FREQ displays the Row Scores and Column Scores that it uses for statistical computations. The Row Scores table displays the row variable values and the Score corresponding to each value. The Column Scores table displays the column variable values and the corresponding Scores. PROC FREQ also identifies the score type used to compute the row and column scores. You can specify the score type with the SCORES= option in the TABLES statement.

- If you specify the CHISQ option, PROC FREQ displays the following statistics for each two-way table: Pearson Chi-Square, Likelihood Ratio Chi-Square, Continuity-Adjusted Chi-Square (for 2 × 2 tables), Mantel-Haenszel Chi-Square, the Phi Coefficient, the Contingency Coefficient, and Cramér's V. For each test statistic, PROC FREQ also displays the degrees of freedom (DF) and the probability value (Prob).

- If you specify the CHISQ option for 2 × 2 tables, PROC FREQ also displays Fisher's exact test. The test output includes the cell (1,1) frequency (F), the exact left-sided and right-sided probability values, the table probability (P), and the exact two-sided probability value. If you specify the POINT option in the EXACT statement, PROC FREQ displays the exact point probability for Fisher's exact test. If you specify the MIDP option in the EXACT statement, PROC FREQ displays the Mid p-Value for the test.

- If you specify the FISHER option in the TABLES statement (or, equivalently, the FISHER option in the EXACT statement), PROC FREQ displays Fisher's exact test for tables larger than 2 × 2. The test output includes the table probability (P) and the probability value. If you specify the POINT option in the EXACT statement, PROC FREQ displays the exact point probability for Fisher's exact test. If you specify the MIDP option in the EXACT statement, PROC FREQ displays the Mid p-Value for the test.

- If you specify the PCHI, LRCHI, or MHCHI option in the EXACT statement, PROC FREQ displays the corresponding exact test: Pearson Chi-Square, Likelihood Ratio Chi-Square, or Mantel-Haenszel Chi-Square, respectively. The test output includes the test statistic, the degrees of freedom (DF), and the asymptotic and exact probability values. If you also specify the POINT option in the EXACT statement, PROC FREQ displays the point probability for each exact test requested. If you specify the MIDP option in the EXACT statement, PROC FREQ displays the exact mid p-value for each test. If you specify the CHISQ option in the EXACT statement, PROC FREQ displays exact probability values for all three of these chi-square tests.

- If you specify the MC option in the EXACT statement, PROC FREQ displays Monte Carlo estimates for all exact p-values that you request in the EXACT statement. The Monte Carlo output includes the p-value Estimate, its Confidence Limits, the Number of Samples used to compute the Monte Carlo estimate, and the Initial Seed for random number generation.

- If you specify the MEASURES option, PROC FREQ displays the following statistics and their asymptotic standard errors (ASE) for each two-way table: Gamma, Kendall's Tau-b, Stuart's Tau-c, Somers' $D(C|R)$, Somers' $D(R|C)$, Pearson Correlation, Spearman Correlation, Lambda Asymmetric $(C|R)$, Lambda Asymmetric $(R|C)$, Lambda Symmetric, Uncertainty Coefficient $(C|R)$, Uncertainty

Coefficient ($R|C$), and Uncertainty Coefficient Symmetric. If you specify the CL option, PROC FREQ also displays confidence limits for these measures.

- If you specify the PLCORR option, PROC FREQ displays the polychoric correlation and its asymptotic standard error (ASE). For 2×2 tables, this statistic is known as the tetrachoric correlation (and is labeled as such in the displayed output). If you specify the CL option, PROC FREQ also displays confidence limits for the polychoric correlation. If you specify the PLCORR option in the TEST statement, PROC FREQ displays the polychoric correlation, asymptotic standard error (ASE), confidence limits, and the following: the standardized test statistic (Z), the corresponding one-sided and two-sided probability values, the likelihood ratio (LR) chi-square, and the probability value (Pr > ChiSq).

- If you specify the GAMMA, KENTB, STUTC, SMDCR, SMDRC, PCORR, or SCORR option in the TEST statement, PROC FREQ displays asymptotic tests for Gamma, Kendall's Tau-b, Stuart's Tau-c, Somers' $D(C|R)$, Somers' $D(R|C)$, the Pearson Correlation, or the Spearman Correlation, respectively. If you specify the MEASURES option in the TEST statement, PROC FREQ displays all these asymptotic tests. The test output includes the statistic, its asymptotic standard error (ASE), Confidence Limits, the ASE under the null hypothesis H0, the standardized test statistic (Z), and the one-sided and two-sided probability values.

- If you specify the KENTB, STUTC, SMDCR, SMDRC, PCORR, or SCORR option in the EXACT statement, PROC FREQ displays asymptotic and exact tests for the corresponding measure of association: Kendall's Tau-b, Stuart's Tau-c, Somers' $D(C|R)$, Somers' $D(R|C)$, the Pearson Correlation, or the Spearman Correlation, respectively. The test output includes the correlation, its asymptotic standard error (ASE), Confidence Limits, the ASE under the null hypothesis H0, the standardized test statistic (Z), and the asymptotic and exact one-sided and two-sided probability values. If you also specify the POINT option in the EXACT statement, PROC FREQ displays the point probability for each exact test requested. If you specify the MIDP option in the EXACT statement, PROC FREQ displays the exact Mid p-Value for each test.

- If you specify the RISKDIFF option for 2 × 2 tables, PROC FREQ displays the Column 1 and Column 2 Risk Estimates. For each column, PROC FREQ displays the Row 1 Risk, Row 2 Risk, Total Risk, and Risk Difference, together with their asymptotic standard errors (ASE) and Asymptotic Confidence Limits. PROC FREQ also displays Exact Confidence Limits for the Row 1 Risk, Row 2 Risk, and Total Risk. If you specify the RISKDIFF option in the EXACT statement, PROC FREQ provides unconditional Exact Confidence Limits for the Risk Difference. You can suppress this table by specifying the RISKDIFF(NORISKS) option.

- If you specify the RISKDIFF(CL=) option for 2 × 2 tables, PROC FREQ displays the "Confidence Limits for the Proportion (Risk) Difference" table, which includes the Lower and Upper Confidence Limits for each confidence limit Type that you request (Agresti-Caffo, Exact, Hauck-Anderson, Miettinen-Nurminen, Newcombe, or Wald).

- If you specify the RISKDIFF(NONINF) option for 2×2 tables, PROC FREQ displays the "Noninferiority Analysis for the Risk Difference" table, which includes the Risk Difference, test ASE, standardized test statistic Z, probability value (Pr > Z), Noninferiority Limit, and (test-based) Confidence Limits.

- If you specify the RISKDIFF(SUP) option for 2 × 2 tables, PROC FREQ displays the "Superiority Analysis for the Risk Difference" table, which includes the Risk Difference, test ASE, standardized test statistic Z, probability value (Pr > Z), Superiority Limit, and (test-based) Confidence Limits.

- If you specify the RISKDIFF(EQUIV) option for 2 × 2 tables, PROC FREQ displays the "Equivalence Analysis for the Risk Difference" table, which includes the Risk Difference, test ASE, Equivalence Limits, and (test-based) Confidence Limits. PROC FREQ also displays the "Two One-Sided Tests (TOST)" table, which includes test statistics (Z) and P-Values for the Lower Margin and Upper Margin tests, along with the Overall P-Value.

- If you specify the RISKDIFF(EQUAL) option for 2 × 2 tables, PROC FREQ displays the "Risk Difference Test" table, which includes the Risk Difference, test ASE, standardized test statistic Z, One-sided probability value (Pr > Z or Pr < Z), and Two-sided probability value (Pr > |Z|).

- If you specify the MEASURES option or the RELRISK option for 2 × 2 tables, PROC FREQ displays the "Odds Ratio and Relative Risks" table, which includes the following statistics with their confidence limits: Odds Ratio, Relative Risk (Column 1), and Relative Risk (Column 2). If you specify the OR option in the EXACT statement, PROC FREQ also displays the "Exact Confidence Limits for the Odds Ratio" table. If you specify the RELRISK option in the EXACT statement, PROC FREQ displays the "Exact Confidence Limits for the Relative Risk" table.

- If you specify the OR(CL=) option for 2 × 2 tables, PROC FREQ displays the "Confidence Limits for the Odds Ratio" table, which includes the Lower and Upper Confidence Limits for each confidence limit Type that you request (Exact, Mid-p, Likelihood Ratio, Score, Wald, or Wald Modified).

- If you specify the RELRISK(CL=) option for 2 × 2 tables, PROC FREQ displays the "Confidence Limits for the Relative Risk" table, which includes the Lower and Upper Confidence Limits for each confidence limit Type that you request (Exact, Likelihood Ratio, Score, Wald, or Wald Modified).

- If you specify the RELRISK(NONINF) option, PROC FREQ displays the "Noninferiority Analysis for the Relative Risk" table, which includes the Relative Risk, standardized test statistic Z, probability value (Pr > Z), Noninferiority Limit, and Confidence Limits.

- If you specify the RELRISK(SUP) option, PROC FREQ displays the "Superiority Analysis for the Relative Risk" table, which includes the Relative Risk, standardized test statistic Z, probability value (Pr > Z), Superiority Limit, and Confidence Limits.

- If you specify the RELRISK(EQUIV) option, PROC FREQ displays the "Equivalence Analysis for the Relative Risk" table, which includes the Relative Risk, Equivalence Limits, and Confidence Limits. PROC FREQ also displays the "Two One-Sided Tests(TOST)" table, which includes test statistics (Z) and P-Values for the Lower Margin and Upper Margin tests, along with the Overall P-Value.

- If you specify the RELRISK(EQUAL) option, PROC FREQ displays the "Relative Risk Test" table, which includes the Relative Risk, standardized test statistic Z, One-sided probability value (Pr > Z or Pr < Z), and Two-sided probability value (Pr > |Z|).

- If you specify the TREND option, PROC FREQ displays the Cochran-Armitage Trend Test for tables that are $2 \times C$ or $R \times 2$. For this test, PROC FREQ gives the Statistic (Z) and the one-sided and two-sided probability values. If you specify the TREND option in the EXACT statement, PROC FREQ also displays the exact one-sided and two-sided probability values for this test. If you specify the POINT option with the TREND option in the EXACT statement, PROC FREQ displays the exact point probability for the test statistic. If you specify the MIDP option in the EXACT statement, PROC FREQ displays the exact Mid p-Value for the trend test.

- If you specify the JT option, PROC FREQ displays the Jonckheere-Terpstra Test, showing the Statistic (JT), the standardized test statistic (Z), and the one-sided and two-sided probability values. If you specify the JT option in the EXACT statement, PROC FREQ also displays the exact one-sided and two-sided probability values for this test. If you specify the POINT option with the JT option in the EXACT statement, PROC FREQ displays the exact point probability for the test statistic. If you specify the MIDP option in the EXACT statement, PROC FREQ displays the exact Mid p-Value for the Jonckheere-Terpstra test.

- If you specify the AGREE option for a 2 × 2 table, PROC FREQ displays the "McNemar's Test" table. This table includes the McNemar test statistic (chi-square), the degrees of freedom, and the p-value. If you specify the MCNEM option in the EXACT statement, this table also includes the exact p-value. If you specify the POINT option or the MIDP option in the EXACT statement, the "McNemar's Test" table includes the exact point probability or the exact mid p-value, respectively.

- If you specify the AGREE option for a square table of dimension greater than 2, PROC FREQ displays the "Symmetry Test" table. This table displays Bowker's symmetry test statistic (chi-square), the degrees of freedom, and the p-value. If you specify the SYMMETRY option in the EXACT statement, this table also includes the exact p-value. If you specify the POINT option or the MIDP option in the EXACT statement, the "Symmetry Test" table includes the exact point probability or the exact mid p-value, respectively.

- The AGREE option also produces the "Kappa Statistics" table, which displays the simple kappa coefficient. If the dimension of the two-way table is greater than 2, the "Kappa Statistics" table includes the weighted kappa coefficient. If you specify the AGREE(AC1) option or the AGREE(PABAK) option, this table includes the AC1 agreement coefficient or the prevalence-adjusted bias-adjusted kappa (PABAK), respectively. The "Kappa Statistics" table displays the standard error and confidence limits for each agreement statistic.

- If you specify the AGREE(KAPPADETAILS) option, PROC FREQ displays the "Kappa Details" table, which includes the observed agreement, the chance-expected agreement, the maximum kappa, and the B_N measure. For 2 × 2 tables, the "Kappa Details" table also includes the prevalence index and the bias index.

- If you specify the AGREE(WTKAPPADETAILS) or AGREE(KAPPADETAILS) option for a square table of dimension greater than 2, PROC FREQ produces the "Weighted Kappa Details" table, which displays the observed agreement and the chance-expected agreement components of the weighted kappa coefficient.

- If you specify the AGREE(PRINTKWTS) option for a square table of dimension greater than 2, PROC FREQ displays the matrix of agreement weights in the "Kappa Coefficient Weights" table.

- If you request a simple kappa coefficient test, PROC FREQ produces the "Kappa Test" table. You can request this test by specifying the KAPPA option in the TEST statement, the KAPPA option in the EXACT statement, or the AGREE(NULLKAPPA=) option in the TABLES statement. The "Kappa Test" table displays the kappa coefficient, null test value, standard error (when the null value is 0), standardized test statistic (Z), and one-sided and two-sided p-values.

  If you request an exact test (by specifying the KAPPA option in the EXACT statement), the "Kappa Test" table also includes the exact one-sided and two-sided p-values. If you specify the POINT option or the MIDP option in the EXACT statement, the "Kappa Test" table includes the point probability or the exact mid p-value, respectively.

- If you request a weighted kappa coefficient test for a square table of dimension greater than 2, PROC FREQ produces the "Weighted Kappa Test" table. You can request this test by specifying the WTKAPPA option in the TEST statement, the WTKAPPA option in the EXACT statement, or the AGREE(NULLWTKAPPA=) option in the TABLES statement. The "Weighted Kappa Test" table displays the weighted kappa coefficient, null test value, standard error (when the null value is 0), standardized test statistic (Z), and one-sided and two-sided *p*-values.

  If you request an exact test (by specifying the WTKAPPA option in the EXACT statement), the "Weighted Kappa Test" table also includes the exact one-sided and two-sided *p*-values. If you specify the POINT option or the MIDP option in the EXACT statement, the "Weighted Kappa Test" table includes the point probability or the exact mid *p*-value, respectively.

- If you specify the AGREE option for a multiway square table, PROC FREQ displays the "Overall Kappa Coefficients" table, which includes the overall simple kappa coefficient together with its standard error and confidence limits. This table also includes the overall weighted kappa coefficient if the two-way table dimension is greater than 2.

- For multiway square tables, the AGREE option also produces the "Tests for Equal Kappa Coefficients" table. This table includes the chi-square statistic, degrees of freedom, and *p*-value for the test of equal simple kappa coefficients (over all strata). If the two-way table dimension is greater than 2, this table also includes the test for equal weighted kappa coefficients.

- For multiway 2×2 tables, the AGREE option displays the "Cochran's Q" table, which includes Cochran's Q statistic (to test for marginal homogeneity), the degrees of freedom, and the *p*-value.

- If you specify the COMMONRISKDIFF option for a multiway 2×2 table, PROC FREQ displays the "Confidence Limits for the Common Risk Difference" table, which includes the Method, Value of the common risk difference, Standard Error, and Confidence Limits for each confidence limit type that you request (Mantel-Haenszel, Minimum Risk, Newcombe, Newcombe MR, or Summary Score).

- If you specify the COMMONRISKDIFF(TEST) option for a multiway 2×2 table, PROC FREQ displays the "Common Risk Difference Tests" table, which includes Method, Risk Difference, Z, and Pr > |Z| for each test that you request (Mantel-Haenszel, Minimum Risk, or Summary Score).

- If you specify the COMMONRISKDIFF(PRINTWTS) option for a multiway 2×2 table, PROC FREQ displays the "Stratum Weights" table, which includes the following information for each stratum (2×2 table): Stratum index, variable levels, Risk Difference, Frequency, Fraction, Mantel-Haenszel Weight, and Minimum Risk Weight.

- If you specify the CMH option, PROC FREQ displays Cochran-Mantel-Haenszel Statistics for the following three alternative hypotheses: Nonzero Correlation, Row Mean Scores Differ (ANOVA Statistic), and General Association. For each of these statistics, PROC FREQ gives the degrees of freedom (DF) and the probability value (Prob). If you specify the MANTELFLEISS option, PROC FREQ displays the Mantel-Fleiss Criterion for 2×2 tables. For 2×2 tables, PROC FREQ also displays Estimates of the Common Relative Risk for Case-Control and Cohort studies, together with their confidence limits. These include both Mantel-Haenszel and Logit stratum-adjusted estimates of the common Odds Ratio, Column 1 Relative Risk, and Column 2 Relative Risk. Also for 2×2 tables, PROC FREQ displays the Breslow-Day Test for Homogeneity of the Odds Ratios. For this test, PROC FREQ gives the Chi-Square, the degrees of freedom (DF), and the probability value (Pr > ChiSq).

- If you specify the CMH option in the TABLES statement and also specify the COMOR option in the EXACT statement for a multiway 2×2 table, PROC FREQ displays exact confidence limits for the Common Odds Ratio. PROC FREQ also displays the Exact Test of H0: Common Odds Ratio = 1. The test output includes the Cell (1,1) Sum (S), Mean of S Under H0, One-sided Pr <= S, and Point Pr = S. PROC FREQ also provides exact two-sided probability values for the test, computed according to the following three methods: 2 * One-sided, Sum of probabilities <= Point probability, and Pr >= |S - Mean|. If you specify the MIDP option in the EXACT statement, PROC FREQ provides the exact Mid p-Value for the common odds ratio test.

- If you specify the CMH option in the TABLES statement and also specify the EQOR option in the EXACT statement for a multiway 2×2 table, PROC FREQ computes Zelen's exact test for equal odds ratios. PROC FREQ displays Zelen's test along with the asymptotic Breslow-Day test produced by the CMH option. PROC FREQ displays the test statistic, Zelen's Exact Test (P), and the probability value, Exact Pr <= P.

- If you specify the GAILSIMON option in the TABLES statement for a multiway 2×2 tables, PROC FREQ displays the Gail-Simon test for qualitative interactions. The display include the following statistics and their p-values: Q+ (Positive Risk Differences), Q- (Negative Risk Differences), and Q (Two-Sided).

ODS Table Names

PROC FREQ assigns a name to each table that it creates. You can use these names to refer to tables when you use the Output Delivery System (ODS) to select tables and create output data sets. For more information about ODS, see Chapter 20, "Using the Output Delivery System" (*SAS/STAT User's Guide*).

Table 3.22 lists the ODS table names together with their descriptions and the options required to produce the tables. Note that the ALL option in the TABLES statement invokes the CHISQ, MEASURES, and CMH options.

Table 3.22 ODS Tables Produced by PROC FREQ

ODS Table Name	Description	Statement	Option
BarnardsTest	Barnard's exact test	EXACT	BARNARD
BinomialCLs	Binomial confidence limits	TABLES	BINOMIAL(CL=)
BinomialEquiv	Binomial equivalence analysis	TABLES	BINOMIAL(EQUIV)
BinomialEquivLimits	Binomial equivalence limits	TABLES	BINOMIAL(EQUIV)
BinomialEquivTest	Binomial equivalence test	TABLES	BINOMIAL(EQUIV)
BinomialNoninf	Binomial noninferiority test	TABLES	BINOMIAL(NONINF)
Binomial	Binomial proportion	TABLES	BINOMIAL
BinomialTest	Binomial proportion test	TABLES	BINOMIAL
BinomialSup	Binomial superiority test	TABLES	BINOMIAL(SUP)
BnMeasure	Agreement measures	TABLES	PLOTS=AGREEPLOT(STATS)
BreslowDayTest	Breslow-Day test	TABLES	CMH ($h \times 2 \times 2$ table)
CMH	Cochran-Mantel-Haenszel statistics	TABLES	CMH
ChiSq	Chi-square tests	TABLES	CHISQ

Table 3.22 *continued*

ODS Table Name	Description	Statement	Option
CochransQ	Cochran's Q	TABLES	AGREE ($h \times 2 \times 2$ table)
ColScores	Column scores	TABLES	SCOROUT
CommonOdds-RatioCl	Exact confidence limits for the common odds ratio	EXACT	COMOR ($h \times 2 \times 2$ table)
CommonOdds-RatioTest	Common odds ratio exact test	EXACT	COMOR ($h \times 2 \times 2$ table)
CommonPdiff	Common risk difference confidence limits	TABLES	COMMONRISKDIFF ($h \times 2 \times 2$ table)
CommonPdiffTests	Common risk difference tests	TABLES	COMMONRISKDIFF(TESTS) ($h \times 2 \times 2$ table)
CommonRelRisks	Common relative risks	TABLES	CMH ($h \times 2 \times 2$ table)
CrossList	Crosstabulation table in column format	TABLES	CROSSLIST (n-way table, $n > 1$)
CrossTabFreqs	Crosstabulation table	TABLES	(n-way table, $n > 1$)
EqualKappaTest	Test for equal simple kappas	TABLES	AGREE ($h \times 2 \times 2$ table)
EqualKappaTests	Tests for equal kappas	TABLES	AGREE ($h \times r \times r$ table, $r > 2$)
EqualOddsRatios	Tests for equal odds ratios	EXACT	EQOR ($h \times 2 \times 2$ table)
GailSimon	Gail-Simon test	TABLES	GAILSIMON ($h \times 2 \times 2$ table)
FishersExact	Fisher's exact test	EXACT	FISHER
		or TABLES	FISHER or EXACT
		or TABLES	CHISQ (2×2 table)
FishersExactMC	Monte Carlo estimates for Fisher's exact test	EXACT	FISHER / MC
Gamma	Gamma	TEST	GAMMA
GammaTest	Gamma test	TEST	GAMMA
JTTest	Jonckheere-Terpstra test	TABLES	JT
JTTestMC	Monte Carlo estimates for Jonckheere-Terpstra exact test	EXACT	JT / MC
KappaDetails	Kappa details	TABLES	AGREE(KAPPADETAILS)
KappaMC	Monte Carlo exact test for simple kappa coefficient	EXACT	KAPPA / MC
KappaStatistics	Kappa statistics	TABLES	AGREE
KappaTest	Simple kappa test	TEST	KAPPA
		or EXACT	KAPPA
		or TABLES	AGREE(NULLKAPPA=)
KappaWeights	Kappa weights	TABLES	AGREE(PRINTKWTS)
List	List format multiway table	TABLES	LIST
LRChiSq	Likelihood ratio chi-square exact test	EXACT	LRCHI

Table 3.22 *continued*

ODS Table Name	Description	Statement	Option
LRChiSqMC	Monte Carlo exact test for likelihood ratio chi-square	EXACT	LRCHI / MC
MantelFleiss	Mantel-Fleiss criterion	TABLES	CMH(MANTELFLEISS) ($h \times 2 \times 2$ table)
McNemarsTest	McNemar's test	TABLES	AGREE (2×2 table)
Measures	Measures of association	TABLES	MEASURES
MHChiSq	Mantel-Haenszel chi-square exact test	EXACT	MHCHI
MHChiSqMC	Monte Carlo exact test for Mantel-Haenszel chi-square	EXACT	MHCHI / MC
NLevels	Number of variable levels	PROC	NLEVELS
OddsRatioCLs	Odds ratio confidence limits	TABLES	OR(CL=) (2×2 table)
OddsRatioExactCL	Exact confidence limits for the odds ratio	EXACT	OR (2×2 table)
OneWayChiSq	One-way chi-square test	TABLES	CHISQ (one-way table)
OneWayChiSqMC	Monte Carlo exact test for one-way chi-square	EXACT	CHISQ / MC (one-way table)
OneWayFreqs	One-way frequencies	PROC or TABLES	(no TABLES stmt) (one-way table)
OneWayLRChiSq	One-way likelihood ratio chi-square test	TABLES	CHISQ(LRCHI) (one-way table)
OverallKappa	Overall simple kappa	TABLES	AGREE ($h \times 2 \times 2$ table)
OverallKappas	Overall kappa coefficients	TABLES	AGREE ($h \times r \times r$ table, $r > 2$)
PdiffCLs	Risk difference confidence limits	TABLES	RISKDIFF(CL=) (2×2 table)
PdiffEquiv	Equivalence analysis for the risk difference	TABLES	RISKDIFF(EQUIV) (2×2 table)
PdiffEquivTest	Equivalence test for the risk difference	TABLES	RISKDIFF(EQUIV) (2×2 table)
PdiffNoninf	Noninferiority test for the risk difference	TABLES	RISKDIFF(NONINF) (2×2 table)
PdiffSup	Superiority test for the risk difference	TABLES	RISKDIFF(SUP) (2×2 table)
PdiffTest	Risk difference test	TABLES	RISKDIFF(EQUAL) (2×2 table)
PearsonChiSq	Pearson chi-square exact test	EXACT	PCHI
PearsonChiSqMC	Monte Carlo exact test for Pearson chi-square	EXACT	PCHI / MC
PearsonCorr	Pearson correlation	TEST or EXACT	PCORR PCORR

Table 3.22 *continued*

ODS Table Name	Description	Statement	Option
PearsonCorrMC	Monte Carlo exact test for Pearson correlation	EXACT	PCORR / MC
PearsonCorrTest	Pearson correlation test	TEST or EXACT	PCORR PCORR
PlCorr	Polychoric correlation	TEST	PLCORR
PlCorrTest	Polychoric correlation test	TEST	PLCORR
RelativeRiskCLs	Relative risk confidence limits	TABLES	RELRISK(CL=) (2×2 table)
RelativeRisks	Relative risk estimates	TABLES	RELRISK or MEASURES (2×2 table)
RelRisk1ExactCL	Exact confidence limits for column 1 relative risk	EXACT	RELRISK (2×2 table)
RelRisk2ExactCL	Exact confidence limits for column 2 relative risk	EXACT	RELRISK (2×2 table)
RelriskEquiv	Equivalence analysis for the relative risk	TABLES	RELRISK(EQUIV) (2×2 table)
RelriskEquivTest	Equivalence test for the relative risk	TABLES	RELRISK(EQUIV) (2×2 table)
RelriskNoninf	Noninferiority test for the relative risk	TABLES	RELRISK(NONINF) (2×2 table)
RelriskSup	Superiority test for the relative risk	TABLES	RELRISK(SUP) (2×2 table)
RelriskTest	Relative risk test	TABLES	RELRISK(EQUAL) (2×2 table)
RiskDiffCol1	Column 1 risk estimates	TABLES	RISKDIFF (2×2 table)
RiskDiffCol2	Column 2 risk estimates	TABLES	RISKDIFF (2×2 table)
RowScores	Row scores	TABLES	SCOROUT
SomersDCR	Somers' $D(C\|R)$	TEST or EXACT	SMDCR SMDCR
SomersDCRMC	Monte Carlo exact test for Somers' $D(C\|R)$	EXACT	SMDCR / MC
SomersDCRTest	Somers' $D(C\|R)$ test	TEST or EXACT	SMDCR SMDCR
SomersDRC	Somers' $D(R\|C)$	TEST or EXACT	SMDRC SMDRC
SomersDRCMC	Monte Carlo exact test for Somers' $D(R\|C)$	EXACT	SMDRC / MC
SomersDRCTest	Somers' $D(R\|C)$ test	TEST or EXACT	SMDRC SMDRC
SpearmanCorr	Spearman correlation	TEST or EXACT	SCORR SCORR
SpearmanCorrMC	Monte Carlo exact test for Spearman correlation	EXACT	SCORR / MC

Table 3.22 *continued*

ODS Table Name	Description	Statement	Option
SpearmanCorrTest	Spearman correlation test	TEST	SCORR
		or EXACT	SCORR
StratumWeights	Stratum weights and risk differences	TABLES	COMMONRISKDIFF ($h \times 2 \times 2$ table)
SymmetryTest	Symmetry test	TABLES	AGREE
SymmetryMC	Monte Carlo exact symmetry test	EXACT	SYMMETRY / MC
TauB	Kendall's tau-b	TEST	KENTB
		or EXACT	KENTB
TauBMC	Monte Carlo exact test for Kendall's tau-b	EXACT	KENTB / MC
TauBTest	Kendall's tau-b test	TEST	KENTB
		or EXACT	KENTB
TauC	Stuart's tau-c	TEST	STUTC
		or EXACT	STUTC
TauCMC	Monte Carlo exact test for Stuart's tau-c	EXACT	STUTC / MC
TauCTest	Stuart's tau-c test	TEST	STUTC
		or EXACT	STUTC
TrendTest	Cochran-Armitage trend test	TABLES	TREND
TrendTestMC	Monte Carlo exact test for trend	EXACT	TREND / MC
WtKappaDetails	Weighted kappa details	TABLES	AGREE(WTKAPPADETAILS)
WtKappaMC	Monte Carlo exact test for weighted kappa coefficient	EXACT	WTKAPPA / MC
WtKappaTest	Weighted kappa test	TEST	WTKAPPA
		or EXACT	WTKAPPA
		or TABLES	AGREE(NULLWTKAPPA=)

ODS Graphics

Statistical procedures use ODS Graphics to create graphs as part of their output. ODS Graphics is described in detail in Chapter 21, "Statistical Graphics Using ODS" (*SAS/STAT User's Guide*).

Before you create graphs, ODS Graphics must be enabled (for example, with the ODS GRAPHICS ON statement). For more information about enabling and disabling ODS Graphics, see the section "Enabling and Disabling ODS Graphics" in that chapter.

The overall appearance of graphs is controlled by ODS styles. Styles and other aspects of using ODS Graphics are discussed in the section "A Primer on ODS Statistical Graphics" in that chapter.

When ODS Graphics is enabled, you can request specific plots with the PLOTS= option in the TABLES statement. To produce a frequency plot or cumulative frequency plot, you must specify the FREQPLOT or

CUMFREQPLOT *plot-request*, respectively, in the PLOTS= option. To produce a mosaic plot, you must specify the MOSAICPLOT *plot-request* in the PLOTS= option. You can also produce frequency, cumulative frequency, and mosaic plots by specifying the PLOTS=ALL option. By default, PROC FREQ produces all other plots that are associated with the analyses that you request in the TABLES statement. You can suppress the default plots and request specific plots by using the PLOTS(ONLY)= option. See the description of the PLOTS= option for details.

PROC FREQ assigns a name to each graph that it creates with ODS Graphics. You can use these names to refer to the graphs. Table 3.23 lists the names of the graphs that PROC FREQ generates together with their descriptions, their PLOTS= options (*plot-requests*), and the TABLES statement options that are required to produce the graphs.

Table 3.23 Graphs Produced by PROC FREQ

ODS Graph Name	Description	PLOTS= Option	TABLES Statement Option
AgreePlot	Agreement plot	AGREEPLOT	AGREE ($r \times r$ table)
CumFreqPlot	Cumulative frequency plot	CUMFREQPLOT	One-way table request
DeviationPlot	Deviation plot	DEVIATIONPLOT	CHISQ (one-way table)
FreqPlot	Frequency plot	FREQPLOT	Any table request
KappaPlot	Kappa plot	KAPPAPLOT	AGREE ($h \times r \times r$ table)
MosaicPlot	Mosaic plot	MOSAICPLOT	Two-way or multiway table request
ORPlot	Odds ratio plot	ODDSRATIOPLOT	MEASURES, OR, or RELRISK ($h \times 2 \times 2$ table)
RelRiskPlot	Relative risk plot	RELRISKPLOT	MEASURES or RELRISK ($h \times 2 \times 2$ table)
RiskDiffPlot	Risk difference plot	RISKDIFFPLOT	RISKDIFF ($h \times 2 \times 2$ table)
WtKappaPlot	Weighted kappa plot	WTKAPPAPLOT	AGREE ($h \times r \times r$ table, $r > 2$)

Examples: FREQ Procedure

Example 3.1: Output Data Set of Frequencies

The eye and hair color of children from two different regions of Europe are recorded in the data set Color. Instead of recording one observation per child, the data are recorded as cell counts, where the variable Count contains the number of children exhibiting each of the 15 eye and hair color combinations. The data set does not include missing combinations.

The following DATA step statements create the SAS data set Color:

```
data Color;
   input Region Eyes $ Hair $ Count @@;
   label Eyes ='Eye Color'
         Hair ='Hair Color'
         Region='Geographic Region';
   datalines;
1 blue   fair    23  1 blue   red     7  1 blue   medium  24
```

```
1 blue    dark    11   1 green fair    19   1 green red      7
1 green   medium  18   1 green dark    14   1 brown fair    34
1 brown   red      5   1 brown medium  41   1 brown dark    40
1 brown   black    3   2 blue  fair    46   2 blue  red     21
2 blue    medium  44   2 blue  dark    40   2 blue  black    6
2 green   fair    50   2 green red     31   2 green medium  37
2 green   dark    23   2 brown fair    56   2 brown red     42
2 brown   medium  53   2 brown dark    54   2 brown black   13
;
```

The following PROC FREQ statements read the Color data set and create an output data set that contains the frequencies, percentages, and expected cell frequencies of the two-way table of Eyes by Hair. The TABLES statement requests three tables: a frequency table for Eyes, a frequency table for Hair, and a crosstabulation table for Eyes by Hair. The OUT= option creates the FreqCount data set, which contains the crosstabulation table frequencies. The OUTEXPECT option outputs the expected table cell frequencies to FreqCount, and the SPARSE option includes cell frequencies of 0 in the output data set. The WEIGHT statement specifies that the variable Count contains the observation weights. These statements create Output 3.1.1 through Output 3.1.3.

```
proc freq data=Color;
   tables Eyes Hair Eyes*Hair / out=FreqCount outexpect sparse;
   weight Count;
   title 'Eye and Hair Color of European Children';
run;

proc print data=FreqCount noobs;
   title2 'Output Data Set from PROC FREQ';
run;
```

Output 3.1.1 displays the two frequency tables produced by PROC FREQ: one showing the distribution of eye color, and one showing the distribution of hair color. By default, PROC FREQ lists the variables values in alphabetical order. The 'Eyes*Hair' specification produces a crosstabulation table, shown in Output 3.1.2, with eye color defining the table rows and hair color defining the table columns. A cell frequency of 0 for green eyes and black hair indicates that this eye and hair color combination does not occur in the data.

The output data set FreqCount (Output 3.1.3) contains frequency counts and percentages for the last table requested in the TABLES statement, Eyes by Hair. Because the SPARSE option is specified, the data set includes the observation that has a frequency of 0. The variable Expected contains the expected frequencies, as requested by the OUTEXPECT option.

Output 3.1.1 Frequency Tables

Eye and Hair Color of European Children

The FREQ Procedure

		Eye Color		
Eyes	Frequency	Percent	Cumulative Frequency	Cumulative Percent
blue	222	29.13	222	29.13
brown	341	44.75	563	73.88
green	199	26.12	762	100.00

Output 3.1.1 *continued*

Hair Color

Hair	Frequency	Percent	Cumulative Frequency	Cumulative Percent
black	22	2.89	22	2.89
dark	182	23.88	204	26.77
fair	228	29.92	432	56.69
medium	217	28.48	649	85.17
red	113	14.83	762	100.00

Output 3.1.2 Crosstabulation Table

Frequency
Percent
Row Pct
Col Pct

Table of Eyes by Hair

Eyes(Eye Color)	Hair(Hair Color)					Total
	black	dark	fair	medium	red	
blue	6	51	69	68	28	222
	0.79	6.69	9.06	8.92	3.67	29.13
	2.70	22.97	31.08	30.63	12.61	
	27.27	28.02	30.26	31.34	24.78	
brown	16	94	90	94	47	341
	2.10	12.34	11.81	12.34	6.17	44.75
	4.69	27.57	26.39	27.57	13.78	
	72.73	51.65	39.47	43.32	41.59	
green	0	37	69	55	38	199
	0.00	4.86	9.06	7.22	4.99	26.12
	0.00	18.59	34.67	27.64	19.10	
	0.00	20.33	30.26	25.35	33.63	
Total	22	182	228	217	113	762
	2.89	23.88	29.92	28.48	14.83	100.00

Output 3.1.3 Output Data Set of Frequencies

Eye and Hair Color of European Children
Output Data Set from PROC FREQ

Eyes	Hair	COUNT	EXPECTED	PERCENT
blue	black	6	6.409	0.7874
blue	dark	51	53.024	6.6929
blue	fair	69	66.425	9.0551
blue	medium	68	63.220	8.9239
blue	red	28	32.921	3.6745
brown	black	16	9.845	2.0997
brown	dark	94	81.446	12.3360
brown	fair	90	102.031	11.8110
brown	medium	94	97.109	12.3360
brown	red	47	50.568	6.1680
green	black	0	5.745	0.0000
green	dark	37	47.530	4.8556
green	fair	69	59.543	9.0551
green	medium	55	56.671	7.2178
green	red	38	29.510	4.9869

Example 3.2: Frequency Dot Plots

This example produces frequency dot plots for the children's eye and hair color data from Example 3.1.

PROC FREQ produces plots by using ODS Graphics to create graphs as part of the procedure output. Frequency plots are available for any frequency or crosstabulation table request. You can display frequency plots as bar charts or dot plots. You can use *plot-options* to specify the orientation (vertical or horizontal), scale, and layout of the plots.

The following PROC FREQ statements request frequency tables and dot plots. The first TABLES statement requests a one-way frequency table of Hair and a crosstabulation table of Eyes by Hair. The PLOTS= option requests frequency plots for the tables, and the TYPE=DOTPLOT *plot-option* specifies dot plots. By default, frequency plots are produced as bar charts. ODS Graphics must be enabled before producing plots.

The second TABLES statement requests a crosstabulation table of Region by Hair and a frequency dot plot for this table. The SCALE=PERCENT *plot-option* plots percentages instead of frequency counts. SCALE=LOG and SCALE=SQRT *plot-options* are also available to plot log frequencies and square roots of frequencies, respectively.

The ORDER=FREQ option in the PROC FREQ statement orders the variable levels by frequency. This order applies to the frequency and crosstabulation table displays and also to the corresponding frequency plots.

```
ods graphics on;
proc freq data=Color order=freq;
   tables Hair Hair*Eyes / plots=freqplot(type=dotplot);
   tables Hair*Region / plots=freqplot(type=dotplot scale=percent);
   weight Count;
   title 'Eye and Hair Color of European Children';
run;
ods graphics off;
```

248 ✦ *Chapter 3: The FREQ Procedure*

Output 3.2.1, Output 3.2.2, and Output 3.2.3 display the dot plots produced by PROC FREQ. By default, the orientation of dot plots is horizontal, which places the variable levels on the Y axis. You can specify the ORIENT=VERTICAL *plot-option* to request a vertical orientation. For two-way plots, you can use the TWOWAY= *plot-option* to specify the plot layout. The default layout (shown in Output 3.2.2 and Output 3.2.3) is GROUPVERTICAL. Two-way layouts STACKED and GROUPHORIZONTAL are also available.

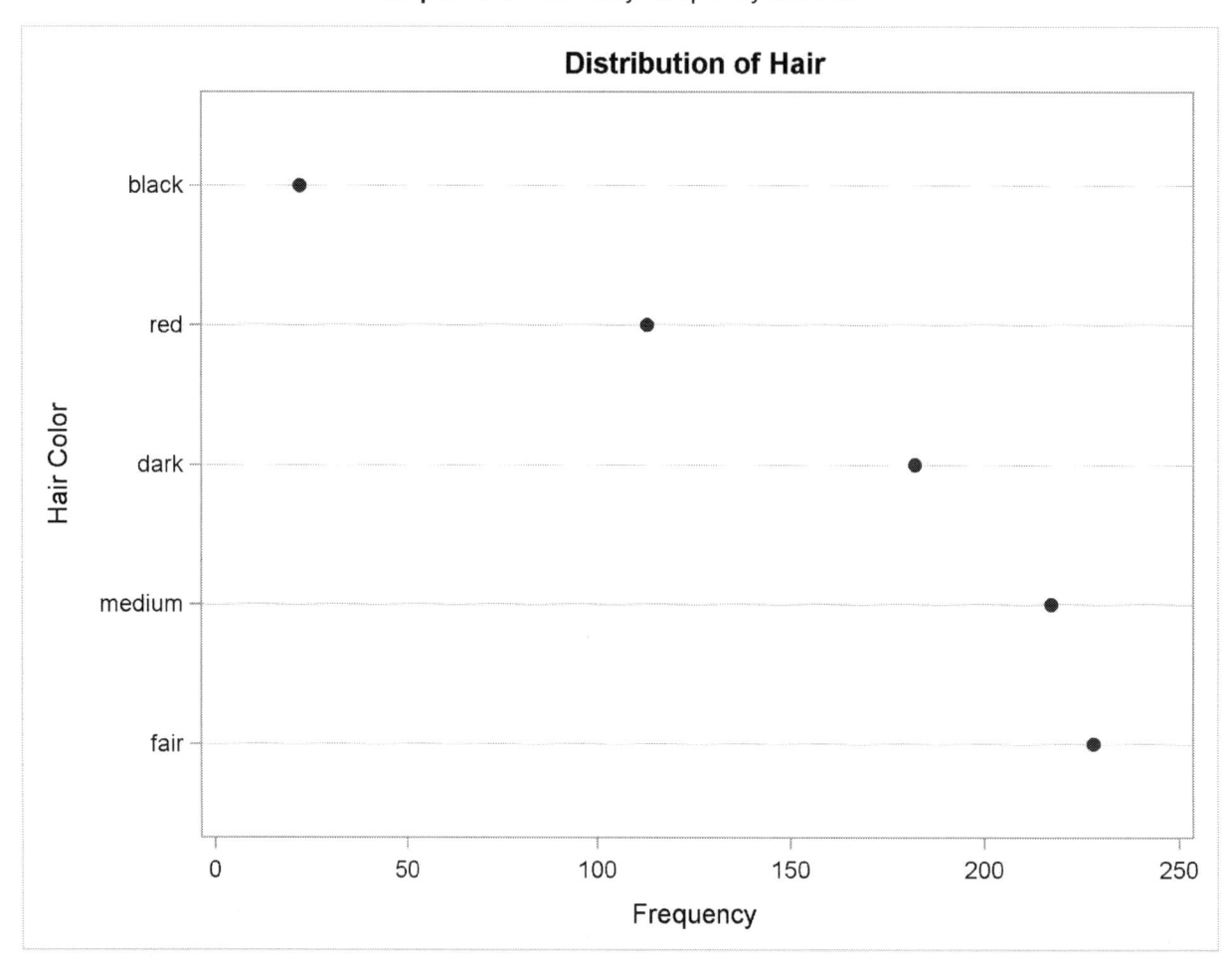

Output 3.2.1 One-Way Frequency Dot Plot

Output 3.2.2 Two-Way Frequency Dot Plot

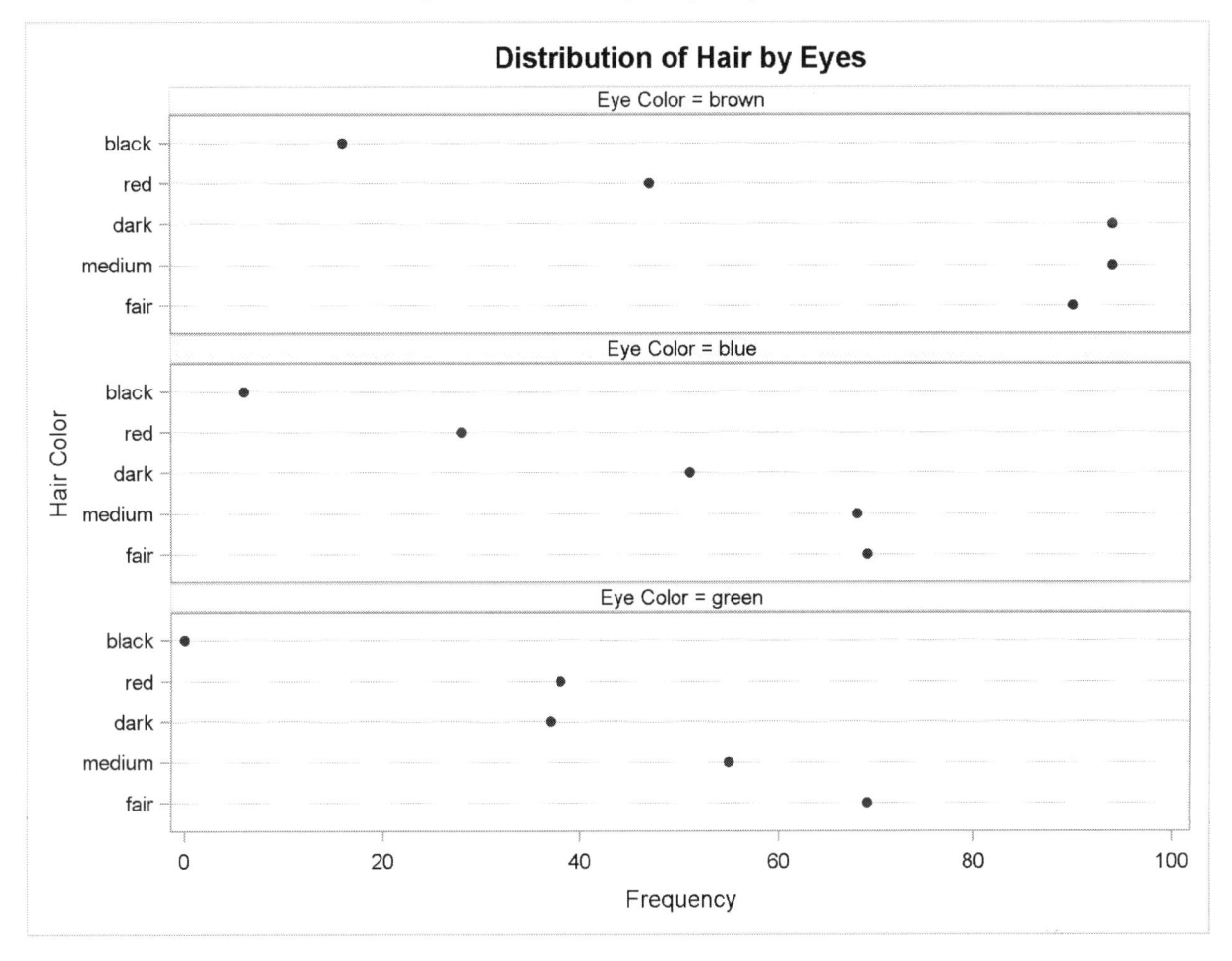

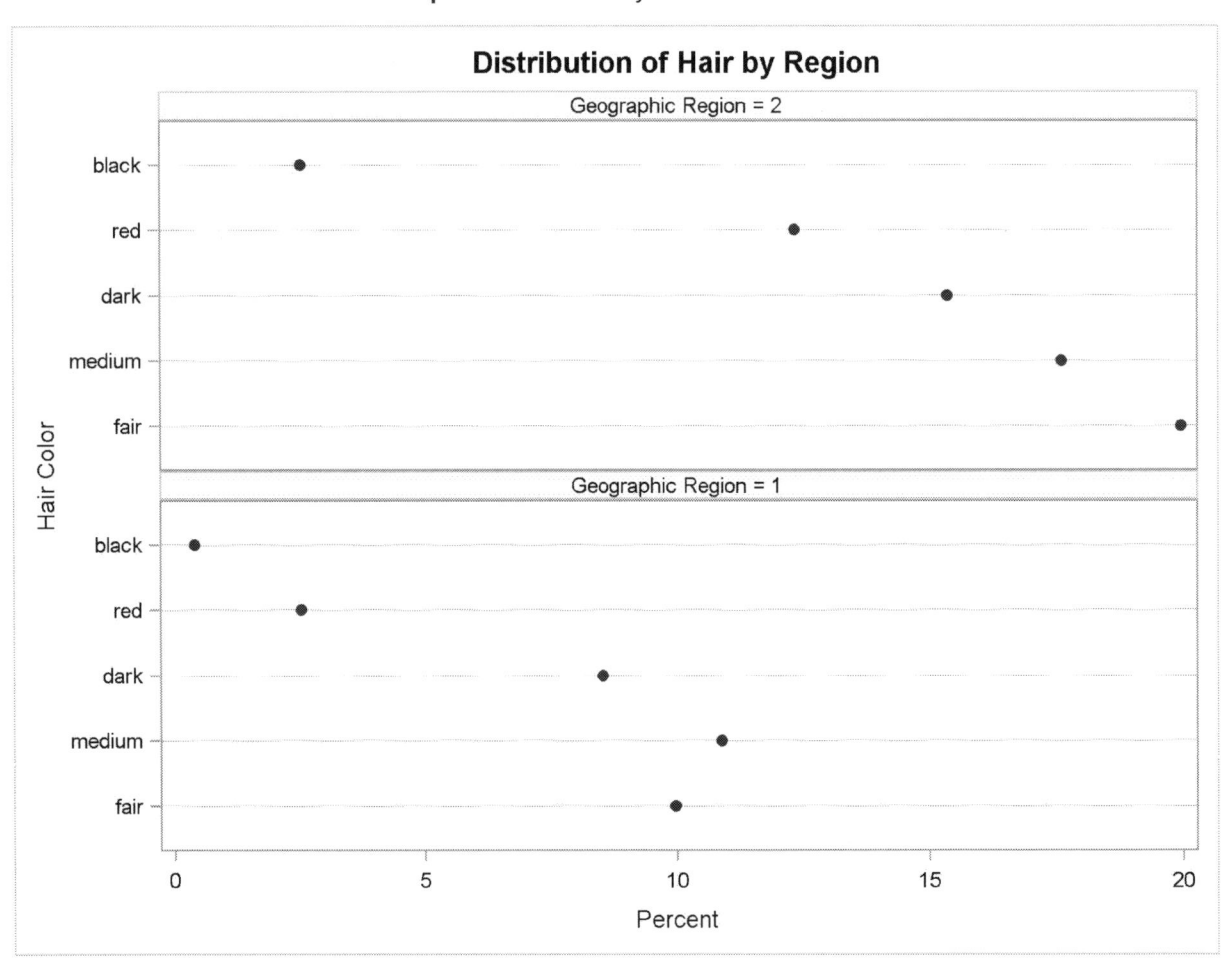

Output 3.2.3 Two-Way Percent Dot Plot

Distribution of Hair by Region

Example 3.3: Chi-Square Goodness-of-Fit Tests

This example examines whether the children's hair color (from Example 3.1) has a specified multinomial distribution for the two geographical regions. The hypothesized distribution of hair color is 30% fair, 12% red, 30% medium, 25% dark, and 3% black.

In order to test the hypothesis for each region, the data are first sorted by Region. Then the FREQ procedure uses a BY statement to produce a separate table for each BY group (Region). The option ORDER=DATA orders the variable values (hair color) in the frequency table by their order in the input data set. The TABLES statement requests a frequency table for hair color, and the option NOCUM suppresses the display of the cumulative frequencies and percentages.

The CHISQ option requests a chi-square goodness-of-fit test for the frequency table of Hair. The TESTP= option specifies the hypothesized (or test) percentages for the chi-square test; the number of percentages listed equals the number of table levels, and the percentages sum to 100%. The TESTP= percentages are listed in the same order as the corresponding variable levels appear in frequency table.

The PLOTS= option requests a deviation plot, which is associated with the CHISQ option and displays the relative deviations from the test frequencies. The TYPE=DOTPLOT *plot-option* requests a dot plot instead of the default type, which is a bar chart. ODS Graphics must be enabled before producing plots. These statements produce Output 3.3.1 through Output 3.3.4.

```
proc sort data=Color;
   by Region;
run;

ods graphics on;
proc freq data=Color order=data;
   tables Hair / nocum chisq testp=(30 12 30 25 3)
               plots(only)=deviationplot(type=dotplot);
   weight Count;
   by Region;
   title 'Hair Color of European Children';
run;
ods graphics off;
```

Output 3.3.1 Frequency Table and Chi-Square Test for Region 1

Hair Color of European Children

The FREQ Procedure

Geographic Region=1

Hair Color

Hair	Frequency	Percent	Test Percent
fair	76	30.89	30.00
red	19	7.72	12.00
medium	83	33.74	30.00
dark	65	26.42	25.00
black	3	1.22	3.00

Geographic Region=1

Chi-Square Test for Specified Proportions	
Chi-Square	7.7602
DF	4
Pr > ChiSq	0.1008

Output 3.3.1 shows the frequency table and chi-square test for Region 1. The frequency table lists the variable values (hair color) in the order in which they appear in the data set. The "Test Percent" column lists the hypothesized percentages for the chi-square test. Always check that you have ordered the TESTP= percentages to correctly match the order of the variable levels.

Output 3.3.2 shows the deviation plot for Region 1, which displays the relative deviations from the hypothesized values. The relative deviation for a level is the difference between the observed and hypothesized (test) percentage divided by the test percentage. You can suppress the chi-square *p*-value that is displayed by default in the deviation plot by specifying the NOSTATS *plot-option*.

Output 3.3.2 Deviation Plot for Region 1

Deviations of Hair

Pr > ChiSq 0.1008

Output 3.3.3 and Output 3.3.4 show the results for Region 2. PROC FREQ computes a chi-square statistic for each region. The chi-square statistic is significant at the 0.05 level for Region 2 ($p=0.0003$) but not for Region 1. This indicates a significant departure from the hypothesized percentages in Region 2.

Output 3.3.3 Frequency Table and Chi-Square Test for Region 2

Hair Color of European Children

The FREQ Procedure

Geographic Region=2

Hair Color

Hair	Frequency	Percent	Test Percent
fair	152	29.46	30.00
red	94	18.22	12.00
medium	134	25.97	30.00
dark	117	22.67	25.00
black	19	3.68	3.00

Output 3.3.3 *continued*

Geographic Region=2

Chi-Square Test for Specified Proportions	
Chi-Square	21.3824
DF	4
Pr > ChiSq	0.0003

Output 3.3.4 Deviation Plot for Region 2

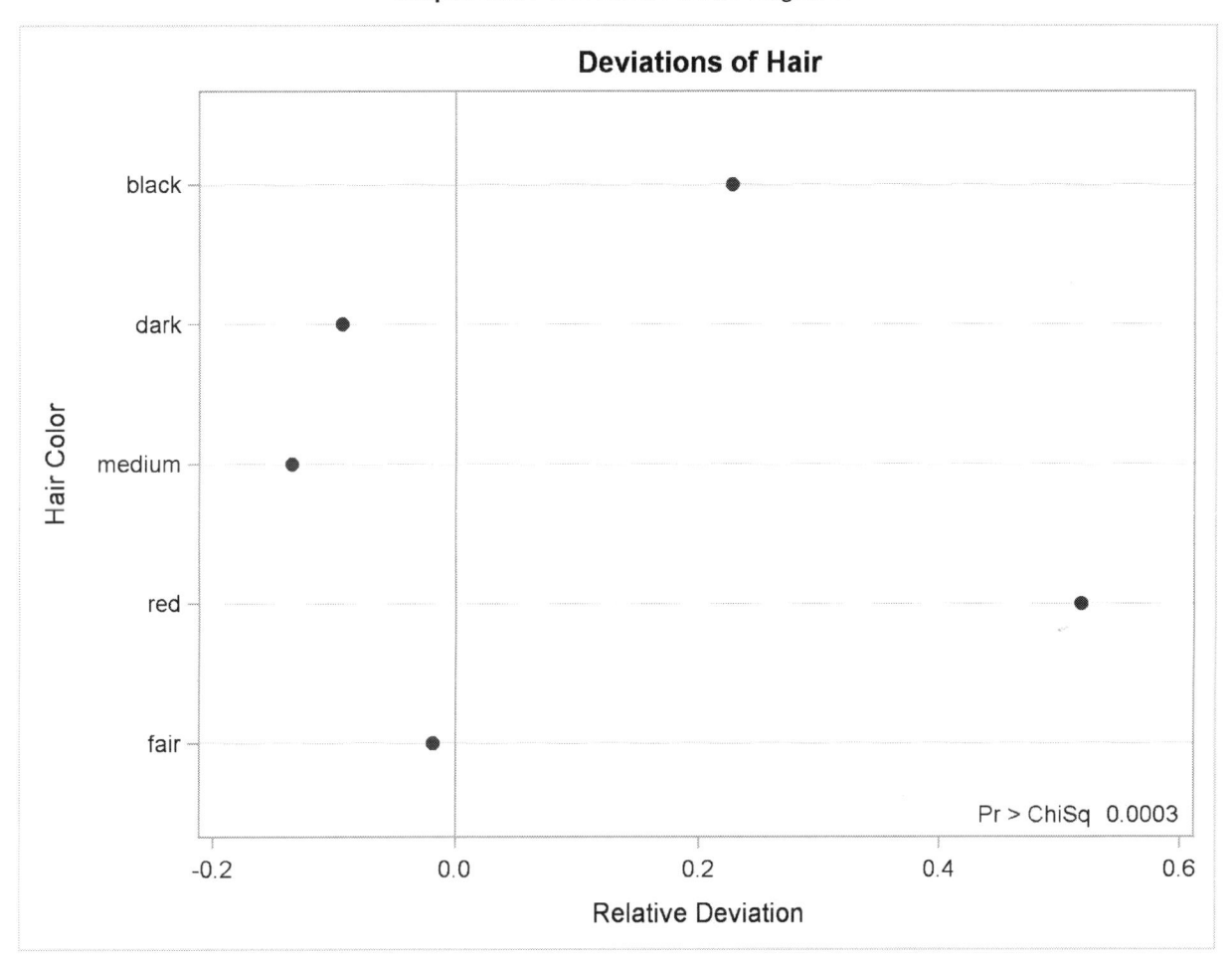

Example 3.4: Binomial Proportions

In this example, PROC FREQ computes binomial proportions, confidence limits, and tests. The example uses the eye and hair color data from Example 3.1. By default, PROC FREQ computes the binomial proportion as the proportion of observations in the first level of the one-way table. You can designate a different level by using the LEVEL= *binomial-option*.

The following PROC FREQ statements compute the proportion of children with brown eyes (from the data set in Example 3.1) and test the null hypothesis that the population proportion equals 50%. These statements also compute an equivalence test for the proportion of children with fair hair.

The first TABLES statement requests a one-way frequency table for the variable Eyes. The BINOMIAL option requests the binomial proportion, confidence limits, and test. PROC FREQ computes the proportion with Eyes = 'brown', which is the first level displayed in the table. The AC, WILSON, and EXACT *binomial-options* request the following confidence limits types: Agresti-Coull, Wilson (score), and exact (Clopper-Pearson). By default, PROC FREQ provides Wald and exact (Clopper-Pearson) confidence limits for the binomial proportion. The BINOMIAL option also produces an asymptotic Wald test that the proportion is 0.5. You can specify a different test proportion in the P= *binomial-option*. The ALPHA=0.1 option specifies that $\alpha = 10\%$, which produces 90% confidence limits.

The second TABLES statement requests a one-way frequency table for the variable Hair. The BINOMIAL option requests the proportion for the first level, Hair = 'fair'. The EQUIV *binomial-option* requests an equivalence test for the binomial proportion. The P=.28 option specifies 0.28 as the null hypothesis proportion, and the MARGIN=.1 option specifies 0.1 as the equivalence test margin.

```
proc freq data=Color order=freq;
   tables Eyes / binomial(ac wilson exact) alpha=.1;
   tables Hair / binomial(equiv p=.28 margin=.1);
   weight Count;
   title 'Hair and Eye Color of European Children';
run;
```

Output 3.4.1 displays the results for eye color, and Output 3.4.2 displays the results for hair color.

Output 3.4.1 Binomial Proportion for Eye Color

Hair and Eye Color of European Children

The FREQ Procedure

Eye Color

Eyes	Frequency	Percent	Cumulative Frequency	Cumulative Percent
brown	341	44.75	341	44.75
blue	222	29.13	563	73.88
green	199	26.12	762	100.00

Binomial Proportion
Eyes = brown

Proportion	0.4475
ASE	0.0180

Confidence Limits for the Binomial Proportion
Proportion = 0.4475

Type	90% Confidence Limits	
Agresti-Coull	0.4181	0.4773
Clopper-Pearson (Exact)	0.4174	0.4779
Wilson	0.4181	0.4773

Output 3.4.1 *continued*

Test of H0: Proportion = 0.5	
ASE under H0	0.0181
Z	-2.8981
One-sided Pr < Z	0.0019
Two-sided Pr > |Z|	0.0038

The frequency table in Output 3.4.1 displays the values of Eyes in order of descending frequency count. PROC FREQ computes the proportion of children in the first level displayed in the frequency table, Eyes = 'brown'. Output 3.4.1 displays the binomial proportion confidence limits and test. The confidence limits are 90% confidence limits. If you do not specify the ALPHA= option, PROC FREQ computes 95% confidence limits by default. Because the value of Z is less than 0, PROC FREQ displays the a left-sided *p*-value (0.0019). This small *p*-value supports the alternative hypothesis that the true value of the proportion of children with brown eyes is less than 50%.

Output 3.4.2 displays the equivalence test results produced by the second TABLES statement. The null hypothesis proportion is 0.28 and the equivalence margins are –0.1 and 0.1, which yield equivalence limits of 0.18 and 0.38. PROC FREQ provides two one-sided tests (TOST) for equivalence. The small *p*-value indicates rejection of the null hypothesis in favor of the alternative that the proportion is equivalent to the null value.

Output 3.4.2 Binomial Proportion for Hair Color

Hair Color

Hair	Frequency	Percent	Cumulative Frequency	Cumulative Percent
fair	228	29.92	228	29.92
medium	217	28.48	445	58.40
dark	182	23.88	627	82.28
red	113	14.83	740	97.11
black	22	2.89	762	100.00

Equivalence Analysis
H0: P - p0 <= Lower Margin or >= Upper Margin
Ha: Lower Margin < P - p0 < Upper Margin
p0 = 0.28   Lower Margin = -0.1   Upper Margin = 0.1

Proportion	ASE (Sample)
0.2992	0.0166

Two One-Sided Tests (TOST)

Test	Z	P-Value
Lower Margin	7.1865	Pr > Z <.0001
Upper Margin	-4.8701	Pr < Z <.0001
Overall		<.0001

Equivalence Limits		90% Confidence Limits	
0.1800	0.3800	0.2719	0.3265

Example 3.5: Analysis of a 2x2 Contingency Table

This example computes chi-square tests and Fisher's exact test to compare the probability of coronary heart disease for two types of diet. It also estimates the relative risks and computes exact confidence limits for the odds ratio.

The data set FatComp contains hypothetical data for a case-control study of high fat diet and the risk of coronary heart disease. The data are recorded as cell counts, where the variable Count contains the frequencies for each exposure and response combination. The data set is sorted in descending order by the variables Exposure and Response, so that the first cell of the 2×2 table contains the frequency of positive exposure and positive response. The FORMAT procedure creates formats to identify the type of exposure and response with character values.

```
proc format;
   value ExpFmt 1='High Cholesterol Diet'
                0='Low Cholesterol Diet';
   value RspFmt 1='Yes'
                0='No';
run;

data FatComp;
   input Exposure Response Count;
   label Response='Heart Disease';
   datalines;
0 0  6
0 1  2
1 0  4
1 1 11
;

proc sort data=FatComp;
   by descending Exposure descending Response;
run;
```

In the following PROC FREQ statements, ORDER=DATA option orders the contingency table values by their order in the input data set. The TABLES statement requests a two-way table of Exposure by Response. The CHISQ option produces several chi-square tests, and the RELRISK option produces relative risk measures. The EXACT statement requests the exact Pearson chi-square test and exact confidence limits for the odds ratio.

```
proc freq data=FatComp order=data;
   format Exposure ExpFmt. Response RspFmt.;
   tables Exposure*Response / chisq relrisk;
   exact pchi or;
   weight Count;
   title 'Case-Control Study of High Fat/Cholesterol Diet';
run;
```

The contingency table in Output 3.5.1 displays the variable values so that the first table cell contains the frequency for the first cell in the data set (the frequency of positive exposure and positive response).

Output 3.5.1 Contingency Table

Case-Control Study of High Fat/Cholesterol Diet

The FREQ Procedure

Frequency Percent Row Pct Col Pct	Table of Exposure by Response			
		Response(Heart Disease)		
	Exposure	Yes	No	Total
	High Cholesterol Diet	11 47.83 73.33 84.62	4 17.39 26.67 40.00	15 65.22
	Low Cholesterol Diet	2 8.70 25.00 15.38	6 26.09 75.00 60.00	8 34.78
	Total	13 56.52	10 43.48	23 100.00

Output 3.5.2 displays the chi-square statistics. Because the expected counts in some of the table cells are small, PROC FREQ gives a warning that the asymptotic chi-square tests might not be appropriate. In this case, the exact tests are appropriate. The alternative hypothesis for this analysis states that coronary heart disease is more likely to be associated with a high fat diet, and therefore a one-sided test is appropriate. Fisher's exact right-sided test analyzes whether the probability of heart disease in the high fat group exceeds the probability of heart disease in the low fat group; because this *p*-value is small, the alternative hypothesis is supported.

The odds ratio, displayed in Output 3.5.3, provides an estimate of the relative risk when an event is rare. This estimate indicates that the odds of heart disease is 8.25 times higher in the high fat diet group; however, the wide confidence limits indicate that this estimate has low precision.

Output 3.5.2 Chi-Square Statistics

Statistic	DF	Value	Prob
Chi-Square	1	4.9597	0.0259
Likelihood Ratio Chi-Square	1	5.0975	0.0240
Continuity Adj. Chi-Square	1	3.1879	0.0742
Mantel-Haenszel Chi-Square	1	4.7441	0.0294
Phi Coefficient		0.4644	
Contingency Coefficient		0.4212	
Cramer's V		0.4644	

WARNING: 50% of the cells have expected counts less than 5.
(Asymptotic) Chi-Square may not be a valid test.

Pearson Chi-Square Test	
Chi-Square	4.9597
DF	1
Asymptotic Pr > ChiSq	0.0259
Exact Pr >= ChiSq	0.0393

Output 3.5.2 *continued*

Fisher's Exact Test	
Cell (1,1) Frequency (F)	11
Left-sided Pr <= F	0.9967
Right-sided Pr >= F	0.0367
Table Probability (P)	0.0334
Two-sided Pr <= P	0.0393

Output 3.5.3 Relative Risk

Odds Ratio and Relative Risks			
Statistic	Value	95% Confidence Limits	
Odds Ratio	8.2500	1.1535	59.0029
Relative Risk (Column 1)	2.9333	0.8502	10.1204
Relative Risk (Column 2)	0.3556	0.1403	0.9009

Odds Ratio	
Odds Ratio	8.2500
Asymptotic Conf Limits	
95% Lower Conf Limit	1.1535
95% Upper Conf Limit	59.0029
Exact Conf Limits	
95% Lower Conf Limit	0.8677
95% Upper Conf Limit	105.5488

Example 3.6: Output Data Set of Chi-Square Statistics

This example uses the Color data from Example 3.1 to output the Pearson chi-square and the likelihood ratio chi-square statistics to a SAS data set. The following PROC FREQ statements create a two-way table of eye color versus hair color.

```
proc freq data=Color order=data;
   tables Eyes*Hair / expected cellchi2 norow nocol chisq;
   output out=ChiSqData n nmiss pchi lrchi;
   weight Count;
   title 'Chi-Square Tests for 3 by 5 Table of Eye and Hair Color';
run;

proc print data=ChiSqData noobs;
   title1 'Chi-Square Statistics for Eye and Hair Color';
   title2 'Output Data Set from the FREQ Procedure';
run;
```

The EXPECTED option displays expected cell frequencies in the crosstabulation table, and the CELLCHI2 option displays the cell contribution to the overall chi-square. The NOROW and NOCOL options suppress the display of row and column percents in the crosstabulation table. The CHISQ option produces chi-square tests.

The OUTPUT statement creates the ChiSqData output data set and specifies the statistics to include. The N option requests the number of nonmissing observations, the NMISS option stores the number of missing observations, and the PCHI and LRCHI options request Pearson and likelihood ratio chi-square statistics, respectively, together with their degrees of freedom and *p*-values.

The preceding statements produce Output 3.6.1 and Output 3.6.2. The contingency table in Output 3.6.1 displays eye and hair color in the order in which they appear in the Color data set. The Pearson chi-square statistic in Output 3.6.2 provides evidence of an association between eye and hair color (*p*=0.0073). The cell chi-square values show that most of the association is due to more green-eyed children with fair or red hair and fewer with dark or black hair. The opposite occurs with the brown-eyed children.

Output 3.6.3 displays the output data set created by the OUTPUT statement. It includes one observation that contains the sample size, the number of missing values, and the chi-square statistics and corresponding degrees of freedom and *p*-values as in Output 3.6.2.

Output 3.6.1 Contingency Table

Chi-Square Tests for 3 by 5 Table of Eye and Hair Color

The FREQ Procedure

Frequency Expected Cell Chi-Square Percent	Table of Eyes by Hair					
	Eyes(Eye Color)	Hair(Hair Color)				
		fair	red	medium	dark	black	Total
	blue	69	28	68	51	6	222
		66.425	32.921	63.22	53.024	6.4094	
		0.0998	0.7357	0.3613	0.0772	0.0262	
		9.06	3.67	8.92	6.69	0.79	29.13
	green	69	38	55	37	0	199
		59.543	29.51	56.671	47.53	5.7454	
		1.5019	2.4422	0.0492	2.3329	5.7454	
		9.06	4.99	7.22	4.86	0.00	26.12
	brown	90	47	94	94	16	341
		102.03	50.568	97.109	81.446	9.8451	
		1.4187	0.2518	0.0995	1.935	3.8478	
		11.81	6.17	12.34	12.34	2.10	44.75
	Total	228	113	217	182	22	762
		29.92	14.83	28.48	23.88	2.89	100.00

Output 3.6.2 Chi-Square Statistics

Statistic	DF	Value	Prob
Chi-Square	8	20.9248	0.0073
Likelihood Ratio Chi-Square	8	25.9733	0.0011
Mantel-Haenszel Chi-Square	1	3.7838	0.0518
Phi Coefficient		0.1657	
Contingency Coefficient		0.1635	
Cramer's V		0.1172	

Output 3.6.3 Output Data Set

**Chi-Square Statistics for Eye and Hair Color
Output Data Set from the FREQ Procedure**

N	NMISS	_PCHI_	DF_PCHI	P_PCHI	_LRCHI_	DF_LRCHI	P_LRCHI
762	0	20.9248	8	.007349898	25.9733	8	.001061424

Example 3.7: Cochran-Mantel-Haenszel Statistics

The data set Migraine contains hypothetical data for a clinical trial of migraine treatment. Subjects of both genders receive either a new drug therapy or a placebo. Their response to treatment is coded as 'Better' or 'Same'. The data are recorded as cell counts, and the number of subjects for each treatment and response combination is recorded in the variable Count.

```
data Migraine;
   input Gender $ Treatment $ Response $ Count @@;
   datalines;
female Active  Better 16   female Active  Same 11
female Placebo Better  5   female Placebo Same 20
male   Active  Better 12   male   Active  Same 16
male   Placebo Better  7   male   Placebo Same 19
;
```

The following PROC FREQ statements create a multiway table stratified by Gender, where Treatment forms the rows and Response forms the columns. The RELRISK option in the TABLES statement requests the odds ratio and relative risks for the two-way tables of Treatment by Response. The PLOTS= option requests a relative risk plot, which shows the relative risk and its confidence limits for each level of Gender and overall. The CMH option requests Cochran-Mantel-Haenszel statistics for the multiway table. For this stratified 2×2 table, the CMH option also produces estimates of the common relative risk and the Breslow-Day test for homogeneity of the odds ratios. The NOPRINT option suppresses the display of the crosstabulation tables.

```
ods graphics on;
proc freq data=Migraine;
   tables Gender*Treatment*Response /
         relrisk plots(only)=relriskplot(stats) cmh noprint;
   weight Count;
   title 'Clinical Trial for Treatment of Migraine Headaches';
run;
ods graphics off;
```

Output 3.7.1 through Output 3.7.4 show the results of the analysis. The relative risk plot (Output 3.7.1) displays the relative risks and confidence limits for the two levels of Gender and for the overall (common) relative risk. Output 3.7.2 displays the CMH statistics. For a stratified 2×2 table, the three CMH statistics test the same hypothesis. The significant *p*-value (0.004) indicates that the association between treatment and response remains strong after adjusting for gender.

The CMH option also produces a table of overall relative risks, as shown in Output 3.7.3. Because this is a prospective study, the relative risk estimate assesses the effectiveness of the new drug; the "Cohort (Col1 Risk)" values are the appropriate estimates for the first column (the risk of improvement). The probability of migraine improvement with the new drug is just over two times the probability of improvement with the placebo.

The large *p*-value for the Breslow-Day test (0.2218) in Output 3.7.4 indicates no significant gender difference in the odds ratios.

Output 3.7.1 Relative Risk Plot

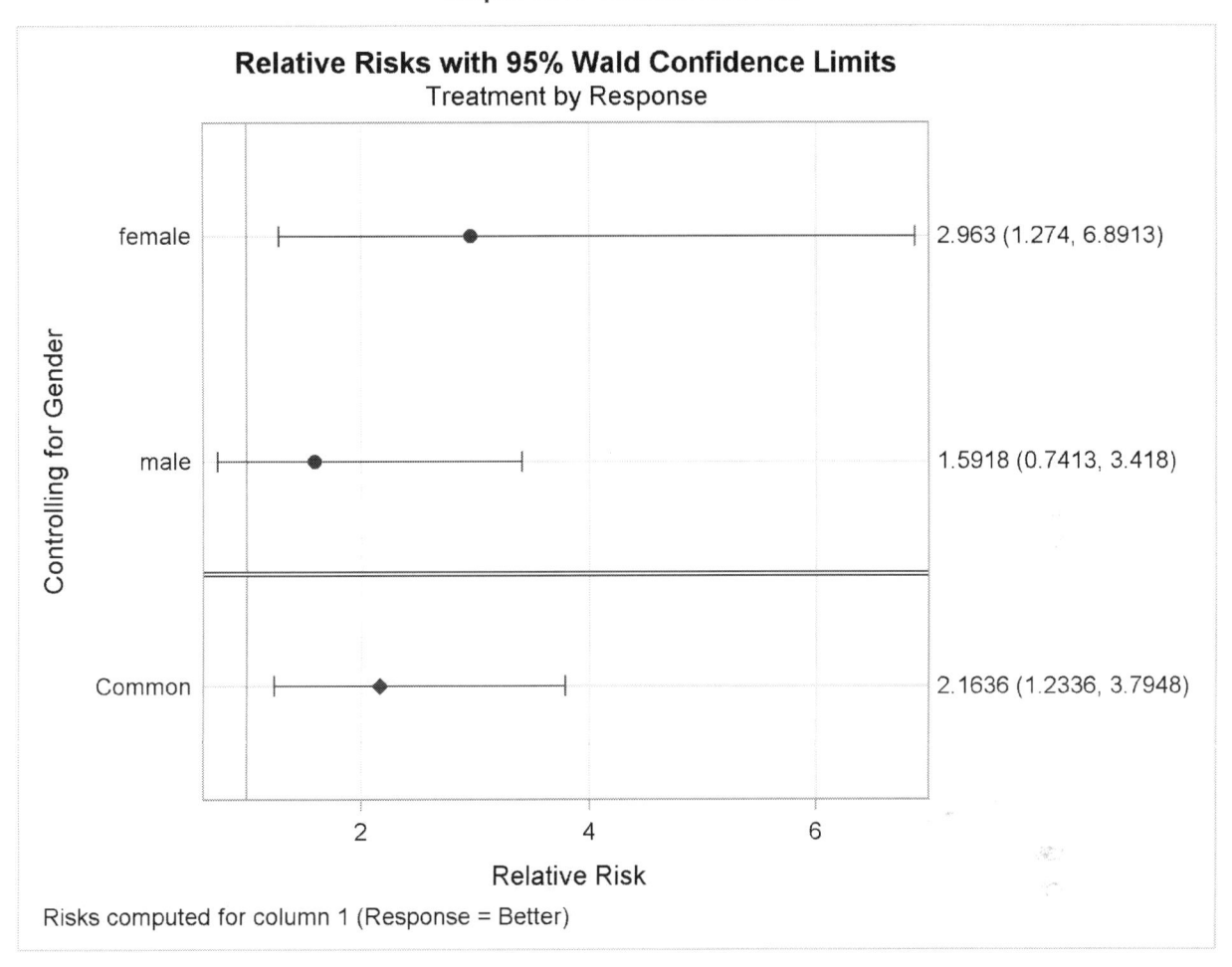

Relative Risks with 95% Wald Confidence Limits
Treatment by Response

Controlling for Gender:
- female: 2.963 (1.274, 6.8913)
- male: 1.5918 (0.7413, 3.418)
- Common: 2.1636 (1.2336, 3.7948)

Risks computed for column 1 (Response = Better)

Output 3.7.2 Cochran-Mantel-Haenszel Statistics

Cochran-Mantel-Haenszel Statistics (Based on Table Scores)				
Statistic	Alternative Hypothesis	DF	Value	Prob
1	Nonzero Correlation	1	8.3052	0.0040
2	Row Mean Scores Differ	1	8.3052	0.0040
3	General Association	1	8.3052	0.0040

Output 3.7.3 CMH Option: Common Relative Risks

	Common Odds Ratio and Relative Risks			
Statistic	Method	Value	95% Confidence Limits	
Odds Ratio	Mantel-Haenszel	3.3132	1.4456	7.5934
	Logit	3.2941	1.4182	7.6515
Relative Risk (Column 1)	Mantel-Haenszel	2.1636	1.2336	3.7948
	Logit	2.1059	1.1951	3.7108
Relative Risk (Column 2)	Mantel-Haenszel	0.6420	0.4705	0.8761
	Logit	0.6613	0.4852	0.9013

Output 3.7.4 CMH Option: Breslow-Day Test

Breslow-Day Test for Homogeneity of the Odds Ratios	
Chi-Square	1.4929
DF	1
Pr > ChiSq	0.2218

Example 3.8: Cochran-Armitage Trend Test

The data set Pain contains hypothetical data for a clinical trial of a drug therapy to control pain. The clinical trial investigates whether adverse responses increase with larger drug doses. Subjects receive either a placebo or one of four drug doses. An adverse response is recorded as Adverse='Yes'; otherwise, it is recorded as Adverse='No'. The number of subjects for each drug dose and response combination is contained in the variable Count.

```
data pain;
   input Dose Adverse $ Count @@;
   datalines;
0 No 26   0 Yes  6
1 No 26   1 Yes  7
2 No 23   2 Yes  9
3 No 18   3 Yes 14
4 No  9   4 Yes 23
;
```

The following PROC FREQ statements provide a trend analysis. The TABLES statement requests a table of Adverse by Dose. The MEASURES option produces measures of association, and the CL option produces confidence limits for these measures. The TREND option tests for a trend across the ordinal values of the variable Dose with the Cochran-Armitage test. The PLOTS= option requests a mosaic plot of Adverse by Dose.

The EXACT statement produces exact *p*-values for this test, and the MAXTIME= option terminates the exact computations if they do not complete within 60 seconds. The TEST statement computes an asymptotic test for Somers' $D(R|C)$.

```
ods graphics on;
proc freq data=Pain;
   tables Adverse*Dose / trend measures cl
          plots=mosaicplot;
   test smdrc;
   exact trend / maxtime=60;
   weight Count;
   title 'Clinical Trial for Treatment of Pain';
run;
ods graphics off;
```

Output 3.8.1 through Output 3.8.4 display the results of the analysis. The "Col Pct" values in Output 3.8.1 show the expected increasing trend in the proportion of adverse effects with the increasing dosage (from 18.75% to 71.88%). The corresponding mosaic plot (Output 3.8.2) also shows this increasing trend.

Output 3.8.1 Contingency Table

Clinical Trial for Treatment of Pain

The FREQ Procedure

Frequency Percent Row Pct Col Pct			Table of Adverse by Dose Dose				
	Adverse	0	1	2	3	4	Total
	No	26	26	23	18	9	102
		16.15	16.15	14.29	11.18	5.59	63.35
		25.49	25.49	22.55	17.65	8.82	
		81.25	78.79	71.88	56.25	28.13	
	Yes	6	7	9	14	23	59
		3.73	4.35	5.59	8.70	14.29	36.65
		10.17	11.86	15.25	23.73	38.98	
		18.75	21.21	28.13	43.75	71.88	
	Total	32	33	32	32	32	161
		19.88	20.50	19.88	19.88	19.88	100.00

Output 3.8.2 Mosaic Plot

Distribution of Adverse by Dose

Output 3.8.3 displays the measures of association produced by the MEASURES option. Somers' $D(R|C)$ measures the association treating the row variable (**Adverse**) as the response and the column variable (**Dose**) as a predictor. Because the asymptotic 95% confidence limits do not contain 0, this indicates a strong positive association. Similarly, the Pearson and Spearman correlation coefficients show evidence of a strong positive association, as hypothesized.

The Cochran-Armitage test (Output 3.8.4) supports the trend hypothesis. The small left-sided *p*-values for the Cochran-Armitage test indicate that the probability of the Row 1 level (**Adverse**='No') decreases as **Dose** increases or, equivalently, that the probability of the Row 2 level (**Adverse**='Yes') increases as **Dose** increases. The two-sided *p*-value tests against either an increasing or decreasing alternative. This is an appropriate hypothesis when you want to determine whether the drug has progressive effects on the probability of adverse effects but the direction is unknown.

Output 3.8.3 Measures of Association

Statistic	Value	ASE	95% Confidence Limits	
Gamma	0.5313	0.0935	0.3480	0.7146
Kendall's Tau-b	0.3373	0.0642	0.2114	0.4631
Stuart's Tau-c	0.4111	0.0798	0.2547	0.5675
Somers' D C|R	0.4427	0.0837	0.2786	0.6068
Somers' D R|C	0.2569	0.0499	0.1592	0.3547
Pearson Correlation	0.3776	0.0714	0.2378	0.5175
Spearman Correlation	0.3771	0.0718	0.2363	0.5178
Lambda Asymmetric C|R	0.1250	0.0662	0.0000	0.2547
Lambda Asymmetric R|C	0.2373	0.0837	0.0732	0.4014
Lambda Symmetric	0.1604	0.0621	0.0388	0.2821
Uncertainty Coefficient C|R	0.0515	0.0191	0.0140	0.0890
Uncertainty Coefficient R|C	0.1261	0.0467	0.0346	0.2175
Uncertainty Coefficient Symmetric	0.0731	0.0271	0.0199	0.1262

Somers' D R|C	
Somers' D R|C	0.2569
ASE	0.0499
95% Lower Conf Limit	0.1592
95% Upper Conf Limit	0.3547

Test of H0: Somers' D R|C = 0	
ASE under H0	0.0499
Z	5.1511
One-sided Pr > Z	<.0001
Two-sided Pr > |Z|	<.0001

Output 3.8.4 Trend Test

Cochran-Armitage Trend Test	
Statistic (Z)	-4.7918
Asymptotic Test	
One-sided Pr < Z	<.0001
Two-sided Pr > |Z|	<.0001
Exact Test	
One-sided Pr <= Z	<.0001
Two-sided Pr >= |Z|	<.0001

Example 3.9: Friedman's Chi-Square Test

Friedman's test is a nonparametric test for treatment differences in a randomized complete block design. Each block of the design might be a subject or a homogeneous group of subjects. If blocks are groups of subjects, the number of subjects in each block must equal the number of treatments. Treatments are randomly assigned to subjects within each block. If there is one subject per block, the subjects are repeatedly measured once under each treatment. The order of treatments is randomized for each subject.

In this setting, Friedman's test is identical to the ANOVA (row means scores) CMH statistic when the analysis uses rank scores (SCORES=RANK). The three-way table uses subject (or subject group) as the stratifying variable, treatment as the row variable, and response as the column variable. PROC FREQ handles ties by assigning midranks to tied response values. If there are multiple subjects per treatment in each block, the ANOVA CMH statistic is a generalization of Friedman's test.

The data set Hypnosis contains data from a study investigating whether hypnosis has the same effect on skin potential (measured in millivolts) for four emotions (Lehmann and D'Abrera 2006, p. 264). Eight subjects are asked to display fear, joy, sadness, and calmness under hypnosis. The data are recorded as one observation per subject for each emotion.

```
data Hypnosis;
   length Emotion $ 10;
   input Subject Emotion $ SkinResponse @@;
   datalines;
1 fear 23.1  1 joy 22.7  1 sadness 22.5  1 calmness 22.6
2 fear 57.6  2 joy 53.2  2 sadness 53.7  2 calmness 53.1
3 fear 10.5  3 joy  9.7  3 sadness 10.8  3 calmness  8.3
4 fear 23.6  4 joy 19.6  4 sadness 21.1  4 calmness 21.6
5 fear 11.9  5 joy 13.8  5 sadness 13.7  5 calmness 13.3
6 fear 54.6  6 joy 47.1  6 sadness 39.2  6 calmness 37.0
7 fear 21.0  7 joy 13.6  7 sadness 13.7  7 calmness 14.8
8 fear 20.3  8 joy 23.6  8 sadness 16.3  8 calmness 14.8
;
```

In the following PROC FREQ statements, the TABLES statement creates a three-way table stratified by Subject and a two-way table; the variables Emotion and SkinResponse form the rows and columns of each table. The CMH2 option produces the first two Cochran-Mantel-Haenszel statistics, the option SCORES=RANK specifies that rank scores are used to compute these statistics, and the NOPRINT option suppresses the contingency tables. These statements produce Output 3.9.1 and Output 3.9.2.

```
proc freq data=Hypnosis;
   tables Subject*Emotion*SkinResponse /
          cmh2 scores=rank noprint;
run;

proc freq data=Hypnosis;
   tables Emotion*SkinResponse /
          cmh2 scores=rank noprint;
run;
```

Because the CMH statistics in Output 3.9.1 are based on rank scores, the Row Mean Scores Differ statistic is identical to Friedman's chi-square ($Q = 6.45$). The p-value of 0.0917 indicates that differences in skin potential response for different emotions are significant at the 10% level but not at the 5% level.

When you do not stratify by subject, the Row Mean Scores Differ CMH statistic is identical to a Kruskal-Wallis test and is not significant ($p = 0.9038$ in Output 3.9.2). Thus, adjusting for subject is critical to reducing the background variation due to subject differences.

Output 3.9.1 CMH Statistics: Stratifying by Subject

The FREQ Procedure

Summary Statistics for Emotion by SkinResponse Controlling for Subject

	Cochran-Mantel-Haenszel Statistics (Based on Rank Scores)			
Statistic	Alternative Hypothesis	DF	Value	Prob
1	Nonzero Correlation	1	0.2400	0.6242
2	Row Mean Scores Differ	3	6.4500	0.0917

Output 3.9.2 CMH Statistics: No Stratification

The FREQ Procedure

Summary Statistics for Emotion by SkinResponse

	Cochran-Mantel-Haenszel Statistics (Based on Rank Scores)			
Statistic	Alternative Hypothesis	DF	Value	Prob
1	Nonzero Correlation	1	0.0001	0.9933
2	Row Mean Scores Differ	3	0.5678	0.9038

Example 3.10: Cochran's Q Test

When a binary response is measured several times or under different conditions, Cochran's Q tests that the marginal probability of a positive response is unchanged across the times or conditions. When there are more than two response categories, you can use the CATMOD procedure to fit a repeated-measures model.

The data set Drugs contains data for a study of three drugs to treat a chronic disease (Agresti 2002). Forty-six subjects receive drugs A, B, and C. The response to each drug is either favorable ('F') or unfavorable ('U').

```
proc format;
   value $ResponseFmt 'F'='Favorable'
                      'U'='Unfavorable';
run;

data drugs;
   input Drug_A $ Drug_B $ Drug_C $ Count @@;
   datalines;
F F F  6   U F F  2
F F U 16   U F U  4
F U F  2   U U F  6
F U U  4   U U U  6
;
```

The following statements create one-way frequency tables of the responses to each drug. The AGREE option produces Cochran's Q and other measures of agreement for the three-way table. These statements produce Output 3.10.1 through Output 3.10.5.

```
proc freq data=Drugs;
   tables Drug_A Drug_B Drug_C / nocum;
   tables Drug_A*Drug_B*Drug_C / agree noprint;
   format Drug_A Drug_B Drug_C $ResponseFmt.;
   weight Count;
   title 'Study of Three Drug Treatments for a Chronic Disease';
run;
```

The one-way frequency tables in Output 3.10.1 provide the marginal response for each drug. For drugs A and B, 61% of the subjects reported a favorable response; for drug C, 35% of the subjects reported a favorable response. Output 3.10.2 and Output 3.10.3 display measures of agreement for the 'Favorable' and 'Unfavorable' levels of drug A, respectively. McNemar's test shows a strong discordance between drugs B and C when the response to drug A is favorable.

Output 3.10.1 One-Way Frequency Tables

Study of Three Drug Treatments for a Chronic Disease

The FREQ Procedure

Drug_A	Frequency	Percent
Favorable	28	60.87
Unfavorable	18	39.13

Drug_B	Frequency	Percent
Favorable	28	60.87
Unfavorable	18	39.13

Drug_C	Frequency	Percent
Favorable	16	34.78
Unfavorable	30	65.22

Output 3.10.2 Measures of Agreement for Drug A Favorable

McNemar's Test		
Chi-Square	DF	Pr > ChiSq
10.8889	1	0.0010

Simple Kappa Coefficient			
Estimate	Standard Error	95% Confidence Limits	
-0.0328	0.1167	-0.2615	0.1960

Output 3.10.3 Measures of Agreement for Drug A Unfavorable

McNemar's Test		
Chi-Square	DF	Pr > ChiSq
0.4000	1	0.5271

Simple Kappa Coefficient			
Estimate	Standard Error	95% Confidence Limits	
-0.1538	0.2230	-0.5909	0.2832

Output 3.10.4 displays the overall kappa coefficient. The small negative value of kappa indicates no agreement between drug B response and drug C response.

Output 3.10.4 Overall Measures of Agreement

Overall Kappa Coefficient			
Estimate	Standard Error	95% Confidence Limits	
-0.0588	0.1034	-0.2615	0.1439

Test for Equal Kappas		
Chi-Square	DF	Pr > ChiSq
0.2314	1	0.6305

Cochran's Q is statistically significant (p=0.0145 in Output 3.10.5), which leads to rejection of the hypothesis that the probability of favorable response is the same for the three drugs.

Output 3.10.5 Cochran's Q Test

Cochran's Q, for Drug_A by Drug_B by Drug_C		
Chi-Square	DF	Pr > ChiSq
8.4706	2	0.0145

References

Agresti, A. (1992). "A Survey of Exact Inference for Contingency Tables." *Statistical Science* 7:131–177.

Agresti, A. (2002). *Categorical Data Analysis*. 2nd ed. New York: John Wiley & Sons.

Agresti, A. (2007). *An Introduction to Categorical Data Analysis*. 2nd ed. New York: John Wiley & Sons.

Agresti, A. (2013). *Categorical Data Analysis*. 3rd ed. Hoboken, NJ: John Wiley & Sons.

Agresti, A., and Caffo, B. (2000). "Simple and Effective Confidence Intervals for Proportions and Differences of Proportions Result from Adding Two Successes and Two Failures." *American Statistician* 54:280–288.

Agresti, A., and Coull, B. A. (1998). "Approximate Is Better Than 'Exact' for Interval Estimation of Binomial Proportions." *American Statistician* 52:119–126.

Agresti, A., and Gottard, A. (2007). "Nonconservative Exact Small-Sample Inference for Discrete Data." *Computational Statistics and Data Analysis* 51:6447–6458.

Agresti, A., Mehta, C. R., and Patel, N. R. (1990). "Exact Inference for Contingency Tables with Ordered Categories." *Journal of the American Statistical Association* 85:453–458.

Agresti, A., and Min, Y. (2001). "On Small-Sample Confidence Intervals for Parameters in Discrete Distributions." *Biometrics* 57:963–971.

Agresti, A., Wackerly, D., and Boyett, J. M. (1979). "Exact Conditional Tests for Cross-Classifications: Approximation of Attained Significance Levels." *Psychometrika* 44:75–83.

Bangdiwala, S. I. (1988). *The Agreement Chart*. Technical report, Department of Biostatistics, University of North Carolina at Chapel Hill.

Bangdiwala, S. I., and Bryan, H. E. (1987). "Using SAS Software Graphical Procedures for the Observer Agreement Chart." In *Proceedings of the Twelfth Annual SAS Users Group International Conference*, 1083–1088. Cary, NC: SAS Institute Inc.

Bangdiwala, S. I., Haedo, A. S., Natal, M. L., and Villaveces, A. (2008). "The Agreement Chart as an Alternative to the Receiver-Operating Characteristic Curve for Diagnostic Tests." *Journal of Clinical Epidemiology* 61:866–874.

Barker, L., Rolka, H., Rolka, D., and Brown, C. (2001). "Equivalence Testing for Binomial Random Variables: Which Test to Use?" *American Statistician* 55:279–287.

Barnard, G. A. (1945). "A New Test for 2×2 Tables." *Nature* 156:177.

Barnard, G. A. (1947). "Significance Tests for 2×2 Tables." *Biometrika* 34:123–138.

Barnard, G. A. (1949). "Statistical Inference." *Journal of the Royal Statistical Society, Series B* 11:115–139.

Berger, J. O. (1985). *Statistical Decision Theory and Bayesian Analysis*. 2nd ed. New York: Springer-Verlag.

Birch, M. W. (1965). "The Detection of Partial Association, Part 2: The General Case." *Journal of the Royal Statistical Society, Series B* 27:111–124.

Bishop, Y. M. M., Fienberg, S. E., and Holland, P. W. (1975). *Discrete Multivariate Analysis: Theory and Practice*. Cambridge, MA: MIT Press.

Blaker, H. (2000). "Confidence Curves and Improved Exact Confidence Intervals for Discrete Distributions." *Canadian Journal of Statistics* 28:783–798.

Blood, E., and Spratt, K. F. (2007). "Disagreement on Agreement: Two Alternative Agreement Coefficients." In *Proceedings of the SAS Global Forum 2007 Conference*. Cary, NC: SAS Institute Inc. http://www2.sas.com/proceedings/forum2007/186-2007.pdf.

Bowker, A. H. (1948). "Bowker's Test for Symmetry." *Journal of the American Statistical Association* 43:572–574.

Breslow, N. E. (1996). "Statistics in Epidemiology: The Case-Control Study." *Journal of the American Statistical Association* 91:14–26.

Breslow, N. E., and Day, N. E. (1980). *The Analysis of Case-Control Studies*. Statistical Methods in Cancer Research, IARC Scientific Publications, vol. 1, no. 32. Lyon: International Agency for Research on Cancer.

Breslow, N. E., and Day, N. E. (1987). *The Design and Analysis of Cohort Studies*. Statistical Methods in Cancer Research, IARC Scientific Publications, vol. 2, no. 82. Lyon: International Agency for Research on Cancer.

Bross, I. D. J. (1958). "How to Use Ridit Analysis." *Biometrics* 14:18–38.

Brown, L. D., Cai, T. T., and DasGupta, A. (2001). "Interval Estimation for a Binomial Proportion." *Statistical Science* 16:101–133.

Brown, M. B., and Benedetti, J. K. (1977a). "On the Mean and Variance of the Tetrachoric Correlation Coefficient." *Psychometrika* 42:347–355.

Brown, M. B., and Benedetti, J. K. (1977b). "Sampling Behavior of Tests for Correlation in Two-Way Contingency Tables." *Journal of the American Statistical Association* 72:309–315.

Byrt, T., Bishop, J., and Carlin, J. B. (1993). "Bias, Prevalence, and Kappa." *Journal of Clinical Epidemiology* 46:423–429.

Chan, I. S. F. (1998). "Exact Tests of Equivalence and Efficacy with a Non-zero Lower Bound for Comparative Studies." *Statistics in Medicine* 17:1403–1413.

Chan, I. S. F. (2003). "Proving Non-inferiority or Equivalence of Two Treatments with Dichotomous Endpoints Using Exact Methods." *Statistical Methods in Medical Research* 12:37–58.

Chan, I. S. F., and Zhang, Z. (1999). "Test-Based Exact Confidence Intervals for the Difference of Two Binomial Proportions." *Biometrics* 55:1202–1209.

Chow, S.-C., Shao, J., and Wang, H. (2003). *Sample Size Calculations in Clinical Research*. Boca Raton, FL: CRC Press.

Chow, S.-C., Shao, J., and Wang, H. (2008). *Sample Size Calculations in Clinical Research*. 2nd ed. Boca Raton, FL: Chapman & Hall/CRC.

Cicchetti, D. V., and Allison, T. (1971). "A New Procedure for Assessing Reliability of Scoring EEG Sleep Recordings." *American Journal of EEG Technology* 11:101–109.

Clopper, C. J., and Pearson, E. S. (1934). "The Use of Confidence or Fiducial Limits Illustrated in the Case of the Binomial." *Biometrika* 26:404–413.

Cochran, W. G. (1950). "The Comparison of Percentages in Matched Samples." *Biometrika* 37:256–266.

Cochran, W. G. (1954). "Some Methods for Strengthening the Common χ^2 Tests." *Biometrics* 10:417–451.

Cohen, J. (1960). "A Coefficient of Agreement for Nominal Scales." *Educational and Psychological Measurement* 20:37–46.

Collett, D. (1991). *Modelling Binary Data*. London: Chapman & Hall.

Dann, R. S., and Koch, G. G. (2005). "Review and Evaluation of Methods for Computing Confidence Intervals for the Ratio of Two Proportions and Considerations for Non-inferiority Clinical Trials." *Journal of Biopharmaceutical Statistics* 15:85–107.

Dmitrienko, A., Molenberghs, G., Chuang-Stein, C., and Offen, W. (2005). *Analysis of Clinical Trials Using SAS: A Practical Guide.* Cary, NC: SAS Institute Inc.

Drasgow, F. (1986). "Polychoric and Polyserial Correlations." In *Encyclopedia of Statistical Sciences*, vol. 7, edited by S. Kotz, N. L. Johnson, and C. B. Read. New York: John Wiley & Sons.

Dunnett, C. W., and Gent, M. (1977). "Significance Testing to Establish Equivalence between Treatments, with Special Reference to Data in the Form of 2 × 2 Tables." *Biometrics* 33:593–602.

Farrington, C. P., and Manning, G. (1990). "Test Statistics and Sample Size Formulae for Comparative Binomial Trials with Null Hypothesis of Non-zero Risk Difference or Non-unity Relative Risk." *Statistics in Medicine* 9:1447–1454.

Fienberg, S. E. (1980). *The Analysis of Cross-Classified Categorical Data.* 2nd ed. Cambridge, MA: MIT Press.

Fleiss, J. L., and Cohen, J. (1973). "The Equivalence of Weighted Kappa and the Intraclass Correlation Coefficient as Measures of Reliability." *Educational and Psychological Measurement* 33:613–619.

Fleiss, J. L., Cohen, J., and Everitt, B. S. (1969). "Large-Sample Standard Errors of Kappa and Weighted Kappa." *Psychological Bulletin* 72:323–327.

Fleiss, J. L., Levin, B., and Paik, M. C. (2003). *Statistical Methods for Rates and Proportions.* 3rd ed. Hoboken, NJ: John Wiley & Sons.

Freeman, G. H., and Halton, J. H. (1951). "Note on an Exact Treatment of Contingency, Goodness of Fit, and Other Problems of Significance." *Biometrika* 38:141–149.

Friendly, M. (2000). *Visualizing Categorical Data.* Cary, NC: SAS Institute Inc.

Gail, M. H., and Mantel, N. (1977). "Counting the Number of $r \times c$ Contingency Tables with Fixed Margins." *Journal of the American Statistical Association* 72:859–862.

Gail, M. H., and Simon, R. (1985). "Tests for Qualitative Interactions between Treatment Effects and Patient Subsets." *Biometrics* 41:361–372.

Gart, J. J. (1971). "The Comparison of Proportions: A Review of Significance Tests, Confidence Intervals, and Adjustments for Stratification." *Review of the International Statistical Institute* 39:148–169.

Gart, J. J., and Nam, J. (1988). "Approximate Interval Estimation of the Ratio of Binomial Parameters: A Review and Corrections for Skewness." *Biometrics* 44:323–338.

Goodman, L. A., and Kruskal, W. H. (1979). *Measures of Association for Cross Classification.* New York: Springer-Verlag.

Greenland, S., and Robins, J. M. (1985). "Estimation of a Common Effect Parameter from Sparse Follow-Up Data." *Biometrics* 41:55–68.

Gwet, K. L. (2008). "Computing Inter-rater Reliability and Its Variance in the Presence of High Agreement." *British Journal of Mathematical and Statistical Psychology* 61:29–48.

Haldane, J. B. S. (1955). "The Estimation and Significance of the Logarithm of a Ratio of Frequencies." *Annals of Human Genetics* 20:309–314.

Hauck, W. W., and Anderson, S. (1986). "A Comparison of Large-Sample Confidence Interval Methods for the Difference of Two Binomial Probabilities." *American Statistician* 40:318–322.

Hirji, K. F. (2006). *Exact Analysis of Discrete Data*. Boca Raton, FL: Chapman & Hall/CRC.

Hirji, K. F., Vollset, S. E., Reis, I. M., and Afifi, A. A. (1996). "Exact Tests for Interaction in Several 2 × 2 Tables." *Journal of Computational and Graphical Statistics* 5:209–224.

Hoenig, J. M., Morgan, M. J., and Brown, C. A. (1995). "Analysing Differences between Two Age Determination Methods by Tests of Symmetry." *Canadian Journal of Fisheries and Aquatic Sciences* 52:364–368.

Hollander, M., and Wolfe, D. A. (1999). *Nonparametric Statistical Methods*. 2nd ed. New York: John Wiley & Sons.

Holley, J. W., and Guilford, J. P. (1964). "A Note on the G Index of Agreement." *Educational and Psychological Measurement* 24:749–753.

Jones, M. P., O'Gorman, T. W., Lemka, J. H., and Woolson, R. F. (1989). "A Monte Carlo Investigation of Homogeneity Tests of the Odds Ratio under Various Sample Size Configurations." *Biometrics* 45:171–181.

Kendall, M. G. (1955). *Rank Correlation Methods*. 2nd ed. London: Charles Griffin.

Kendall, M. G., and Stuart, A. (1979). *The Advanced Theory of Statistics*. 4th ed. Vol. 2. New York: Macmillan.

Kim, Y., and Won, S. (2013). "Adjusted Proportion Difference and Confidence Interval in Stratified Randomized Trials." In *Proceedings of PharmaSUG 2013 (Pharmaceutical Industry SAS Users Group)*. Paper SP04. Cary, NC: SAS Institute Inc.

Kleinbaum, D. G., Kupper, L. L., and Morgenstern, H. (1982). *Epidemiologic Research: Principles and Quantitative Methods*. New York: Van Nostrand Reinhold.

Korn, E. L., and Graubard, B. I. (1998). "Confidence Intervals for Proportions with Small Expected Number of Positive Counts Estimated from Survey Data." *Survey Methodology* 24:193–201.

Krauth, J. (1973). "Nichtparametrische Ansätze zur Auswertung von Verlaufskurven [Nonparametric Approaches to Analyzing Time Effect Curves]." *Biometrische Zeitschrift* 15:557–566.

Landis, J. R., Heyman, E. R., and Koch, G. G. (1978). "Average Partial Association in Three-Way Contingency Tables: A Review and Discussion of Alternative Tests." *International Statistical Review* 46:237–254.

Leemis, L. M., and Trivedi, K. S. (1996). "A Comparison of Approximate Interval Estimators for the Bernoulli Parameter." *American Statistician* 50:63–68.

Lehmann, E. L., and D'Abrera, H. J. M. (2006). *Nonparametrics: Statistical Methods Based on Ranks*. Rev. ed. New York: Springer.

Liebetrau, A. M. (1983). *Measures of Association*. Vol. 32 of Quantitative Applications in the Social Sciences. Beverly Hills, CA: Sage Publications.

Mack, G. A., and Skillings, J. H. (1980). "A Friedman-Type Rank Test for Main Effects in a Two-Factor ANOVA." *Journal of the American Statistical Association* 75:947–951.

Mantel, N. (1963). "Chi-Square Tests with One Degree of Freedom: Extensions of the Mantel-Haenszel Procedure." *Journal of the American Statistical Association* 58:690–700.

Mantel, N., and Fleiss, J. L. (1980). "Minimum Expected Cell Size Requirements for the Mantel-Haenszel One-Degree-of-Freedom Chi-Square Test and a Related Rapid Procedure." *American Journal of Epidemiology* 112:129–134.

Mantel, N., and Haenszel, W. (1959). "Statistical Aspects of Analysis of Data from Retrospective Studies of Disease." *Journal of the National Cancer Institute* 22:719–748.

Margolin, B. H. (1988). "Test for Trend in Proportions." In *Encyclopedia of Statistical Sciences*, vol. 9, edited by S. Kotz, N. L. Johnson, and C. B. Read. New York: John Wiley & Sons.

McNemar, Q. (1947). "Note on the Sampling Error of the Difference between Correlated Proportions or Percentages." *Psychometrika* 12:153–157.

Mee, R. W. (1984). "Confidence Bounds for the Difference between Two Probabilities." *Biometrics* 40:1175–1176.

Mehrotra, D. V. (2001). "Stratification Issues with Binary Endpoints." *Drug Information Journal* 35:1343–1350.

Mehrotra, D. V., and Railkar, R. (2000). "Minimum Risk Weights for Comparing Treatments in Stratified Binomial Trials." *Statistics in Medicine* 19:811–825.

Mehta, C. R., and Patel, N. R. (1983). "A Network Algorithm for Performing Fisher's Exact Test in $r \times c$ Contingency Tables." *Journal of the American Statistical Association* 78:427–434.

Mehta, C. R., Patel, N. R., and Gray, R. J. (1985). "Computing an Exact Confidence Interval for the Common Odds Ratio in Several 2×2 Contingency Tables." *Journal of the American Statistical Association* 80:969–973.

Mehta, C. R., Patel, N. R., and Senchaudhuri, P. (1991). "Exact Stratified Linear Rank Tests for Binary Data." In *Computing Science and Statistics: Proceedings of the Twenty-Third Symposium on the Interface*, edited by E. M. Keramidas, 200–207. Fairfax Station, VA: Interface Foundation.

Mehta, C. R., Patel, N. R., and Tsiatis, A. A. (1984). "Exact Significance Testing to Establish Treatment Equivalence with Ordered Categorical Data." *Biometrics* 40:819–825.

Mehta, C. R., and Senchaudhuri, P. (2003). "Conditional versus Unconditional Exact Tests for Comparing Two Binomials." Cambridge, MA: Cytel Software Corporation.

Miettinen, O. S. (1985). *Theoretical Epidemiology: Principles of Occurrence in Research Medicine*. New York: John Wiley & Sons.

Miettinen, O. S., and Nurminen, M. M. (1985). "Comparative Analysis of Two Rates." *Statistics in Medicine* 4:213–226.

Newcombe, R. G. (1998a). "Interval Estimation for the Difference between Independent Proportions: Comparison of Eleven Methods." *Statistics in Medicine* 17:873–890.

Newcombe, R. G. (1998b). "Two-Sided Confidence Intervals for the Single Proportion: Comparison of Seven Methods." *Statistics in Medicine* 17:857–872.

Newcombe, R. G., and Nurminen, M. M. (2011). "In Defence of Score Intervals for Proportions and Their Differences." *Communications in Statistics—Theory and Methods* 40:1271–1282.

Olsson, U. (1979). "Maximum Likelihood Estimation of the Polychoric Correlation Coefficient." *Psychometrika* 12:443–460.

Pirie, W. (1983). "Jonckheere Tests for Ordered Alternatives." In *Encyclopedia of Statistical Sciences*, vol. 4, edited by S. Kotz, N. L. Johnson, and C. B. Read. New York: John Wiley & Sons.

Radlow, R., and Alf, E. F. (1975). "An Alternate Multinomial Assessment of the Accuracy of the Chi-Square Test of Goodness of Fit." *Journal of the American Statistical Association* 70:811–813.

Robins, J. M., Breslow, N., and Greenland, S. (1986). "Estimators of the Mantel-Haenszel Variance Consistent in Both Sparse Data and Large-Strata Limiting Models." *Biometrics* 42:311–323.

Santner, T. J., Pradhan, V., Senchaudhuri, P., Mehta, C. R., and Tamhane, A. (2007). "Small-Sample Comparisons of Confidence Intervals for the Difference of Two Independent Binomial Proportions." *Computational Statistics and Data Analysis* 51:5791–5799.

Santner, T. J., and Snell, M. K. (1980). "Small-Sample Confidence Intervals for $p_1 - p_2$ and p_1/p_2 in 2×2 Contingency Tables." *Journal of the American Statistical Association* 75:386–394.

Sato, T. (1989). "On the Variance Estimator of the Mantel-Haenszel Risk Difference." *Biometrics* 45:1323–1324. Letter to the editor.

Schuirmann, D. J. (1987). "A Comparison of the Two One-Sided Tests Procedure and the Power Approach for Assessing the Equivalence of Average Bioavailability." *Journal of Pharmacokinetics and Biopharmaceutics* 15:657–680.

Schuirmann, D. J. (1999). "Confidence Interval Methods for Bioequivalence Testing with Binomial Endpoints." In *Proceedings of the Biopharmaceutical Section*, 227–232. Alexandria, VA: American Statistical Association.

Silvapulle, M. J. (2001). "Tests against Qualitative Interaction: Exact Critical Values and Robust Tests." *Biometrics* 57:1157–1165.

Sim, J., and Wright, C. C. (2005). "The Kappa Statistic in Reliability Studies: Use, Interpretation, and Sample Size Requirements." *Physical Therapy* 85:257–268.

Snedecor, G. W., and Cochran, W. G. (1989). *Statistical Methods*. 8th ed. Ames: Iowa State University Press.

Somers, R. H. (1962). "A New Asymmetric Measure of Association for Ordinal Variables." *American Sociological Review* 27:799–811.

Stokes, M. E., Davis, C. S., and Koch, G. G. (2012). *Categorical Data Analysis Using SAS*. 3rd ed. Cary, NC: SAS Institute Inc.

Suissa, S., and Shuster, J. J. (1985). "Exact Unconditional Sample Sizes for the 2×2 Binomial Trial." *Journal of the Royal Statistical Society, Series A* 148:317–327.

Tarone, R. E. (1985). "On Heterogeneity Tests Based on Efficient Scores." *Biometrika* 72:91–95.

Theil, H. (1972). *Statistical Decomposition Analysis*. Amsterdam: North-Holland.

Thomas, D. G. (1971). "Algorithm AS-36: Exact Confidence Limits for the Odds Ratio in a 2 × 2 Table." *Journal of the Royal Statistical Society, Series C* 20:105–110.

Valz, P. D., and Thompson, M. E. (1994). "Exact Inference for Kendall's S and Spearman's ρ with Extensions to Fisher's Exact Test in $r \times c$ Contingency Tables." *Journal of Computational and Graphical Statistics* 3:459–472.

Van Elteren, P. H. (1960). "On the Combination of Independent Two-Sample Tests of Wilcoxon." *Bulletin of the International Statistical Institute* 37:351–361.

Vollset, S. E., Hirji, K. F., and Elashoff, R. M. (1991). "Fast Computation of Exact Confidence Limits for the Common Odds Ratio in a Series of 2 × 2 Tables." *Journal of the American Statistical Association* 86:404–409.

Wilson, E. B. (1927). "Probable Inference, the Law of Succession, and Statistical Inference." *Journal of the American Statistical Association* 22:209–212.

Woolf, B. (1955). "On Estimating the Relationship between Blood Group and Disease." *Annals of Human Genetics* 19:251–253.

Xie, Q. (2013). "Agree or Disagree? A Demonstration of an Alternative Statistic to Cohen's Kappa for Measuring the Extent and Reliability of Agreement between Observers." Paper presented at FCSM Research Conference, Nov. 7–9, Washington, DC.

Yan, X., and Su, X. G. (2010). "Stratified Wilson and Newcombe Confidence Intervals for Multiple Binomial Proportions." *Statistics in Biopharmaceutical Research* 2:329–335.

Zelen, M. (1971). "The Analysis of Several 2 × 2 Contingency Tables." *Biometrika* 58:129–137.

Chapter 4
The UNIVARIATE Procedure

Contents

Overview: UNIVARIATE Procedure	**279**
Getting Started: UNIVARIATE Procedure	**280**
Capabilities of PROC UNIVARIATE	280
Summarizing a Data Distribution	281
Exploring a Data Distribution	282
Modeling a Data Distribution	286
Syntax: UNIVARIATE Procedure	**288**
PROC UNIVARIATE Statement	288
BY Statement	296
CDFPLOT Statement	297
CLASS Statement	309
FREQ Statement	311
HISTOGRAM Statement	311
ID Statement	330
INSET Statement	330
OUTPUT Statement	340
PPPLOT Statement	346
PROBPLOT Statement	360
QQPLOT Statement	371
VAR Statement	383
WEIGHT Statement	383
Dictionary of Common Options	384
Details: UNIVARIATE Procedure	**392**
Missing Values	392
Rounding	393
Descriptive Statistics	394
Calculating the Mode	397
Calculating Percentiles	397
Tests for Location	399
Confidence Limits for Parameters of the Normal Distribution	401
Robust Estimators	402
Creating Line Printer Plots	405
Creating High-Resolution Graphics	407
Using the CLASS Statement to Create Comparative Plots	411
Positioning Insets	412
Formulas for Fitted Continuous Distributions	416

Goodness-of-Fit Tests	428
Kernel Density Estimates	432
Construction of Quantile-Quantile and Probability Plots	434
Interpretation of Quantile-Quantile and Probability Plots	435
Distributions for Probability and Q-Q Plots	436
Estimating Shape Parameters Using Q-Q Plots	441
Estimating Location and Scale Parameters Using Q-Q Plots	442
Estimating Percentiles Using Q-Q Plots	443
Input Data Sets	444
OUT= Output Data Set in the OUTPUT Statement	444
OUTHISTOGRAM= Output Data Set	446
OUTKERNEL= Output Data Set	447
OUTTABLE= Output Data Set	448
Tables for Summary Statistics	450
ODS Table Names	451
ODS Tables for Fitted Distributions	452
ODS Graphics	452
Computational Resources	453
Examples: UNIVARIATE Procedure	**454**
Example 4.1: Computing Descriptive Statistics for Multiple Variables	454
Example 4.2: Calculating Modes	456
Example 4.3: Identifying Extreme Observations and Extreme Values	457
Example 4.4: Creating a Frequency Table	460
Example 4.5: Creating Basic Summary Plots	461
Example 4.6: Analyzing a Data Set With a FREQ Variable	469
Example 4.7: Saving Summary Statistics in an OUT= Output Data Set	470
Example 4.8: Saving Percentiles in an Output Data Set	472
Example 4.9: Computing Confidence Limits for the Mean, Standard Deviation, and Variance	473
Example 4.10: Computing Confidence Limits for Quantiles and Percentiles	475
Example 4.11: Computing Robust Estimates	476
Example 4.12: Testing for Location	478
Example 4.13: Performing a Sign Test Using Paired Data	479
Example 4.14: Creating a Histogram	480
Example 4.15: Creating a One-Way Comparative Histogram	482
Example 4.16: Creating a Two-Way Comparative Histogram	485
Example 4.17: Adding Insets with Descriptive Statistics	487
Example 4.18: Binning a Histogram	489
Example 4.19: Adding a Normal Curve to a Histogram	492
Example 4.20: Adding Fitted Normal Curves to a Comparative Histogram	495
Example 4.21: Fitting a Beta Curve	497
Example 4.22: Fitting Lognormal, Weibull, and Gamma Curves	500
Example 4.23: Computing Kernel Density Estimates	504
Example 4.24: Fitting a Three-Parameter Lognormal Curve	506

Example 4.25: Annotating a Folded Normal Curve 507
Example 4.26: Creating Lognormal Probability Plots 513
Example 4.27: Creating a Histogram to Display Lognormal Fit 517
Example 4.28: Creating a Normal Quantile Plot 519
Example 4.29: Adding a Distribution Reference Line 521
Example 4.30: Interpreting a Normal Quantile Plot 523
Example 4.31: Estimating Three Parameters from Lognormal Quantile Plots 524
Example 4.32: Estimating Percentiles from Lognormal Quantile Plots 528
Example 4.33: Estimating Parameters from Lognormal Quantile Plots 530
Example 4.34: Comparing Weibull Quantile Plots 532
Example 4.35: Creating a Cumulative Distribution Plot 534
Example 4.36: Creating a P-P Plot 536
References **538**

Overview: UNIVARIATE Procedure

The UNIVARIATE procedure provides the following:

- descriptive statistics based on moments (including skewness and kurtosis), quantiles or percentiles (such as the median), frequency tables, and extreme values

- histograms that optionally can be fitted with probability density curves for various distributions and with kernel density estimates

- cumulative distribution function plots (cdf plots). Optionally, these can be superimposed with probability distribution curves for various distributions.

- quantile-quantile plots (Q-Q plots), probability plots, and probability-probability plots (P-P plots). These plots facilitate the comparison of a data distribution with various theoretical distributions.

- goodness-of-fit tests for a variety of distributions including the normal

- the ability to inset summary statistics on plots

- the ability to analyze data sets with a frequency variable

- the ability to create output data sets containing summary statistics, histogram intervals, and parameters of fitted curves

You can use the PROC UNIVARIATE statement, together with the VAR statement, to compute summary statistics. See the section "Getting Started: UNIVARIATE Procedure" on page 280 for introductory examples. In addition, you can use the following statements to request plots:

- the CDFPLOT statement for creating cdf plots

- the HISTOGRAM statement for creating histograms

- the PPPLOT statement for creating P-P plots
- the PROBPLOT statement for creating probability plots
- the QQPLOT statement for creating Q-Q plots
- the CLASS statement together with any of these plot statements for creating comparative plots
- the INSET statement with any of the plot statements for enhancing the plot with an inset table of summary statistics

The UNIVARIATE procedure produces two kinds of graphical output:

- ODS Statistical Graphics output, which is produced when ODS Graphics is enabled prior to your procedure statements
- traditional graphics, which are produced when ODS Graphics is not enabled

See the section "Creating High-Resolution Graphics" on page 407 for more information about producing traditional graphics and ODS Graphics output.

Getting Started: UNIVARIATE Procedure

The following examples demonstrate how you can use the UNIVARIATE procedure to analyze the distributions of variables through the use of descriptive statistical measures and graphical displays, such as histograms.

Capabilities of PROC UNIVARIATE

The UNIVARIATE procedure provides a variety of descriptive measures, graphical displays, and statistical methods, which you can use to summarize, visualize, analyze, and model the statistical distributions of numeric variables. These tools are appropriate for a broad range of tasks and applications:

- Exploring the distributions of the variables in a data set is an important preliminary step in data analysis, data warehousing, and data mining. With the UNIVARIATE procedure you can use tables and graphical displays, such as histograms and nonparametric density estimates, to find key features of distributions, identify outliers and extreme observations, determine the need for data transformations, and compare distributions.

- Modeling the distributions of data and validating distributional assumptions are basic steps in statistical analysis. You can use the UNIVARIATE procedure to fit parametric distributions (beta, exponential, gamma, Gumbel, inverse Gaussian, lognormal, normal, generalized Pareto, power function, Rayleigh, Johnson S_B, Johnson S_U, and Weibull) and to compute probabilities and percentiles from these models. You can assess goodness of fit with hypothesis tests and with graphical displays such as probability plots and quantile-quantile plots. You can also use the UNIVARIATE procedure to validate distributional assumptions for other types of statistical analysis. When standard assumptions are not met, you can use the UNIVARIATE procedure to perform nonparametric tests and compute robust estimates of location and scale.

- Summarizing the distribution of the data is often helpful for creating effective statistical reports and presentations. You can use the UNIVARIATE procedure to create tables of summary measures, such as means and percentiles, together with graphical displays, such as histograms and comparative histograms, which facilitate the interpretation of the report.

The following examples illustrate a few of the tasks that you can carry out with the UNIVARIATE procedure.

Summarizing a Data Distribution

Figure 4.1 shows a table of basic summary measures and a table of extreme observations for the loan-to-value ratios of 5,840 home mortgages. The ratios are saved as values of the variable LoanToValueRatio in a data set named HomeLoans. The following statements request a univariate analysis:

```
ods select BasicMeasures ExtremeObs;
proc univariate data=HomeLoans;
   var LoanToValueRatio;
run;
```

The ODS SELECT statement restricts the default output to the tables for basic statistical measures and extreme observations.

Figure 4.1 Basic Measures and Extreme Observations

The UNIVARIATE Procedure
Variable: LoanToValueRatio (Loan to Value Ratio)

Basic Statistical Measures			
Location		Variability	
Mean	0.292512	Std Deviation	0.16476
Median	0.248050	Variance	0.02715
Mode	0.250000	Range	1.24780
		Interquartile Range	0.16419

Extreme Observations			
Lowest		Highest	
Value	Obs	Value	Obs
0.0651786	1	1.13976	5776
0.0690157	3	1.14209	5791
0.0699755	59	1.14286	5801
0.0702412	84	1.17090	5799
0.0704787	4	1.31298	5811

The tables in Figure 4.1 show, in particular, that the average ratio is 0.2925 and the minimum and maximum ratios are 0.06518 and 1.313, respectively.

Exploring a Data Distribution

Figure 4.2 shows a histogram of the loan-to-value ratios. The histogram reveals features of the ratio distribution, such as its skewness and the peak at 0.175, which are not evident from the tables in the previous example. The following statements create the histogram:

```
title 'Home Loan Analysis';
ods graphics on;
proc univariate data=HomeLoans noprint;
   histogram LoanToValueRatio / odstitle = title;
   inset n = 'Number of Homes' / position=ne;
run;
```

The ODS GRAPHICS ON statement enables ODS Graphics, which causes PROC UNIVARIATE to produce ODS Graphics output. (See the section "Alternatives for Producing Graphics" on page 407 for information about traditional graphics and ODS Graphics.)

The NOPRINT option suppresses the display of summary statistics, and the ODSTITLE= option uses the title that is specified in the SAS TITLE statement as the graph title. The INSET statement inserts the total number of analyzed home loans in the upper right (northeast) corner of the plot.

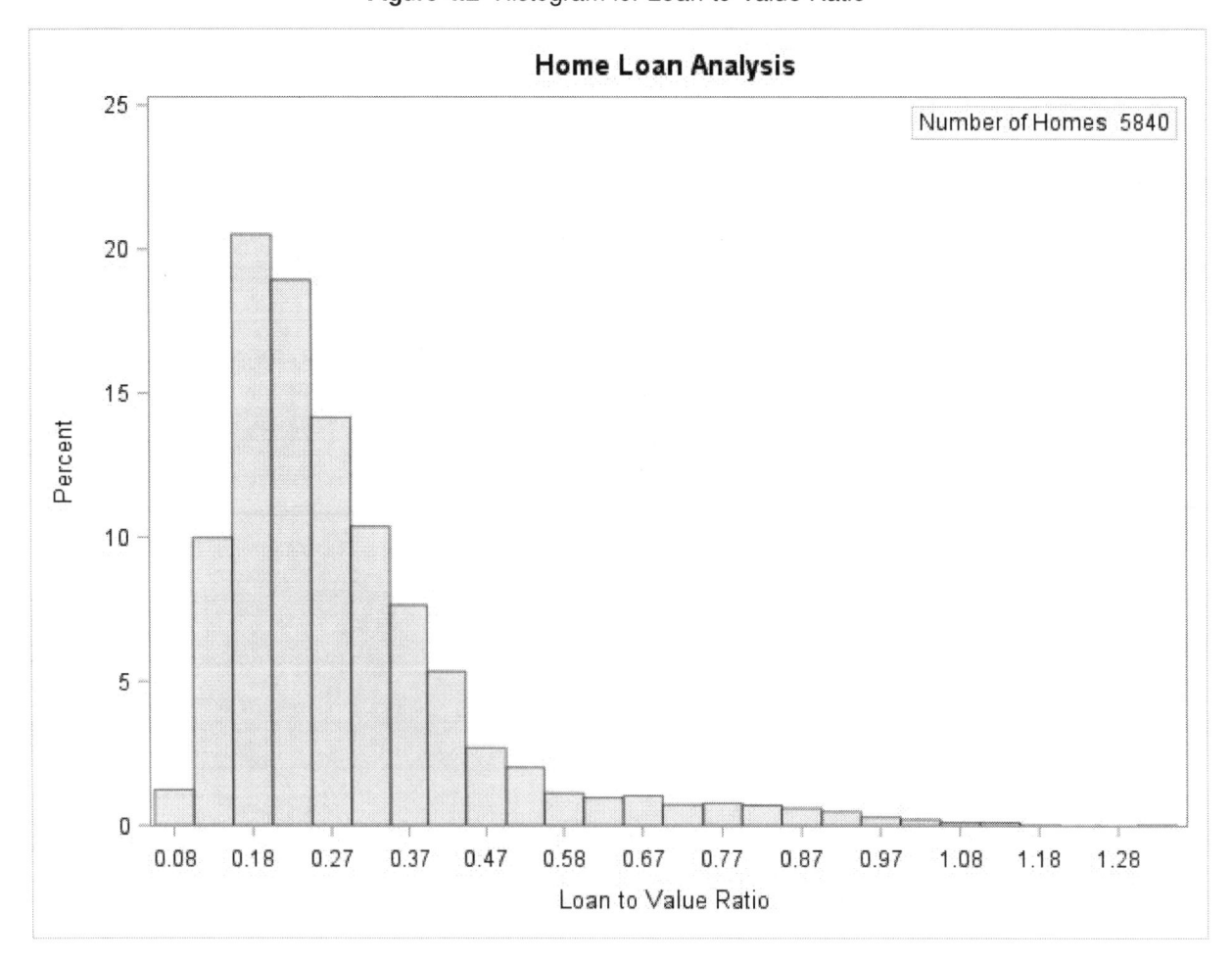

Figure 4.2 Histogram for Loan-to-Value Ratio

The data set HomeLoans contains a variable named LoanType that classifies the loans into two types: Gold and Platinum. It is useful to compare the distributions of LoanToValueRatio for the two types. The following statements request quantiles for each distribution and a comparative histogram, which are shown in Figure 4.3 and Figure 4.4.

```
title 'Comparison of Loan Types';
ods select Histogram Quantiles;
proc univariate data=HomeLoans;
   var LoanToValueRatio;
   class LoanType;
   histogram LoanToValueRatio / kernel
                                odstitle = title;
   inset n='Number of Homes' median='Median Ratio' (5.3) / position=ne;
   label LoanType = 'Type of Loan';
run;
options gstyle;
```

The ODS SELECT statement restricts the default output to the tables of quantiles and the graph produced by the HISTOGRAM statement. The CLASS statement specifies LoanType as a classification variable for the quantile computations and comparative histogram. The KERNEL option adds a smooth nonparametric estimate of the ratio density to each histogram. The INSET statement specifies summary statistics to be displayed directly in the graph.

Figure 4.3 Quantiles for Loan-to-Value Ratio

Comparison of Loan Types

The UNIVARIATE Procedure
Variable: LoanToValueRatio (Loan to Value Ratio)
LoanType = Gold

Quantiles (Definition 5)	
Level	Quantile
100% Max	1.0617647
99%	0.8974576
95%	0.6385908
90%	0.4471369
75% Q3	0.2985099
50% Median	0.2217033
25% Q1	0.1734568
10%	0.1411130
5%	0.1213079
1%	0.0942167
0% Min	0.0651786

Figure 4.3 *continued*

Comparison of Loan Types

The UNIVARIATE Procedure
Variable: LoanToValueRatio   (Loan to Value Ratio)
LoanType = Platinum

Quantiles (Definition 5)	
Level	Quantile
100% Max	1.312981
99%	1.050000
95%	0.691803
90%	0.549273
75% Q3	0.430160
50% Median	0.366168
25% Q1	0.314452
10%	0.273670
5%	0.253124
1%	0.231114
0% Min	0.215504

The output in Figure 4.3 shows that the median ratio for Platinum loans (0.366) is greater than the median ratio for Gold loans (0.222). The comparative histogram in Figure 4.4 enables you to compare the two distributions more easily. It shows that the ratio distributions are similar except for a shift of about 0.14.

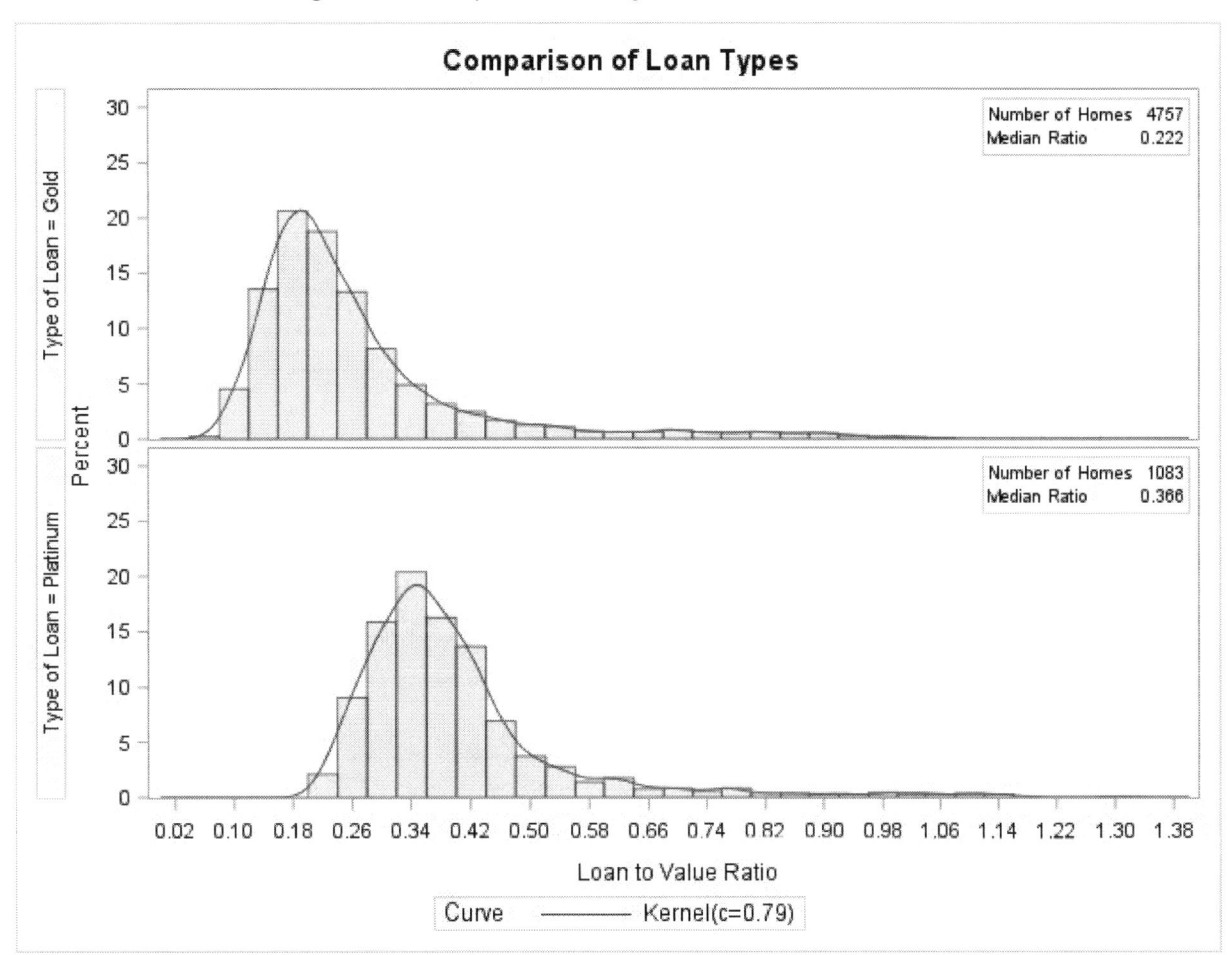

Figure 4.4 Comparative Histogram for Loan-to-Value Ratio

A sample program for this example, *univar1.sas*, is available in the SAS Sample Library for Base SAS software.

Modeling a Data Distribution

In addition to summarizing a data distribution as in the preceding example, you can use PROC UNIVARIATE to statistically model a distribution based on a random sample of data. The following statements create a data set named Aircraft that contains the measurements of a position deviation for a sample of 30 aircraft components.

```
data Aircraft;
   input Deviation @@;
   label Deviation = 'Position Deviation';
   datalines;
-.00653 0.00141 -.00702 -.00734 -.00649 -.00601
-.00631 -.00148 -.00731 -.00764 -.00275 -.00497
-.00741 -.00673 -.00573 -.00629 -.00671 -.00246
-.00222 -.00807 -.00621 -.00785 -.00544 -.00511
-.00138 -.00609 0.00038 -.00758 -.00731 -.00455
;
```

An initial question in the analysis is whether the measurement distribution is normal. The following statements request a table of moments, the tests for normality, and a normal probability plot, which are shown in Figure 4.5 and Figure 4.6:

```
title 'Position Deviation Analysis';
ods graphics on;
ods select Moments TestsForNormality ProbPlot;
proc univariate data=Aircraft normaltest;
   var Deviation;
   probplot Deviation / normal(mu=est sigma=est)
                    square
                    odstitle = title;
   label Deviation = 'Position Deviation';
   inset  mean std / format=6.4;
run;
```

PROC UNIVARIATE uses the label associated with the variable Deviation as the vertical axis label in the probability plot. The INSET statement displays the sample mean and standard deviation on the probability plot.

Figure 4.5 Moments and Tests for Normality

Position Deviation Analysis

The UNIVARIATE Procedure
Variable: Deviation (Position Deviation)

Moments			
N	30	Sum Weights	30
Mean	-0.0053067	Sum Observations	-0.1592
Std Deviation	0.00254362	Variance	6.47002E-6
Skewness	1.2562507	Kurtosis	0.69790426
Uncorrected SS	0.00103245	Corrected SS	0.00018763
Coeff Variation	-47.932613	Std Error Mean	0.0004644

Figure 4.5 *continued*

	Tests for Normality			
Test		Statistic		p Value
Shapiro-Wilk	W	0.845364	Pr < W	0.0005
Kolmogorov-Smirnov	D	0.208921	Pr > D	<0.0100
Cramer-von Mises	W-Sq	0.329274	Pr > W-Sq	<0.0050
Anderson-Darling	A-Sq	1.784881	Pr > A-Sq	<0.0050

All four goodness-of-fit tests in Figure 4.5 reject the hypothesis that the measurements are normally distributed.

Figure 4.6 shows a normal probability plot for the measurements. A linear pattern of points following the diagonal reference line would indicate that the measurements are normally distributed. Instead, the curved point pattern suggests that a skewed distribution, such as the lognormal, is more appropriate than the normal distribution.

A lognormal distribution for Deviation is fitted in Example 4.26.

A sample program for this example, *univar2.sas*, is available in the SAS Sample Library for Base SAS software.

Figure 4.6 Normal Probability Plot

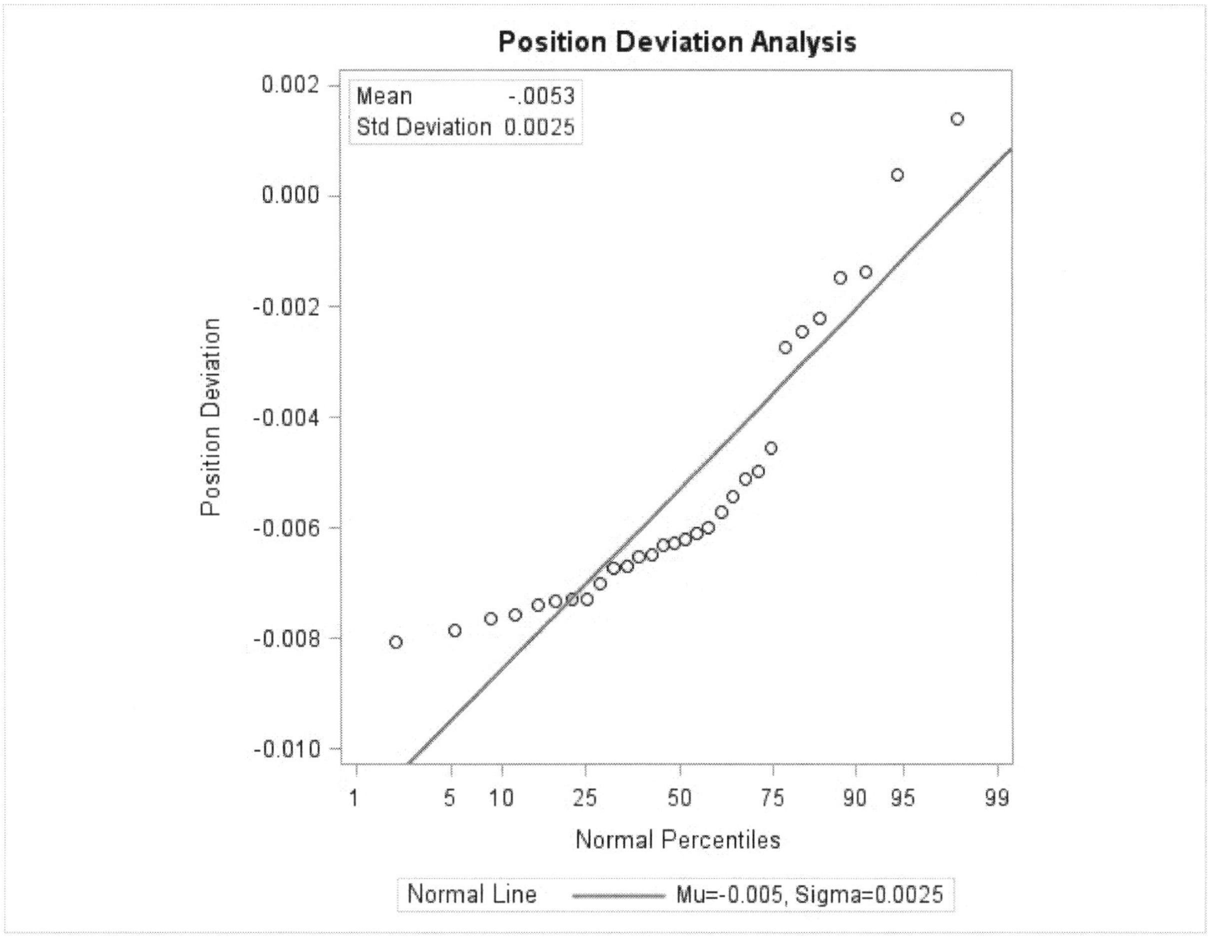

Syntax: UNIVARIATE Procedure

PROC UNIVARIATE < *options* > ;
    BY *variables* ;
    CDFPLOT < *variables* > < / *options* > ;
    CLASS *variable-1* < (*v-options*) > < *variable-2* < (*v-options*) > >
        < / **KEYLEVEL=** *value1* | (*value1 value2*) > ;
    FREQ *variable* ;
    HISTOGRAM < *variables* > < / *options* > ;
    ID *variables* ;
    INSET *keyword-list* < / *options* > ;
    OUTPUT < **OUT=**SAS-data-set > < *keyword1=names ... keywordk=names* > < *percentile-options* > ;
    PPPLOT < *variables* > < / *options* > ;
    PROBPLOT < *variables* > < / *options* > ;
    QQPLOT < *variables* > < / *options* > ;
    VAR *variables* ;
    WEIGHT *variable* ;

The PROC UNIVARIATE statement invokes the procedure. The VAR statement specifies the numeric variables to be analyzed, and it is required if the OUTPUT statement is used to save summary statistics in an output data set. If you do not use the VAR statement, all numeric variables in the data set are analyzed. The plot statements CDFPLOT, HISTOGRAM, PPPLOT, PROBPLOT, and QQPLOT create graphical displays, and the INSET statement enhances these displays by adding a table of summary statistics directly on the graph. You can specify one or more of each of the plot statements, the INSET statement, and the OUTPUT statement. If you use a VAR statement, the variables listed in a plot statement must be a subset of the variables listed in the VAR statement.

You can specify a BY statement to obtain separate analyses for each BY group. The FREQ statement specifies a variable whose values provide the frequency for each observation. The ID statement specifies one or more variables to identify the extreme observations. The WEIGHT statement specifies a variable whose values are used to weight certain statistics.

You can use a CLASS statement to specify one or two variables that group the data into classification levels. The analysis is carried out for each combination of levels in the input data set, or within each BY group if you also specify a BY statement. You can use the CLASS statement with plot statements to create comparative displays, in which each cell contains a plot for one combination of classification levels.

PROC UNIVARIATE Statement

PROC UNIVARIATE < *options* > ;

The PROC UNIVARIATE statement is required to invoke the UNIVARIATE procedure. You can use the PROC UNIVARIATE statement by itself to request a variety of statistics for summarizing the data distribution of each analysis variable:

- sample moments

- basic measures of location and variability
- confidence intervals for the mean, standard deviation, and variance
- tests for location
- tests for normality
- trimmed and Winsorized means
- robust estimates of scale
- quantiles and related confidence intervals
- extreme observations and extreme values
- frequency counts for observations
- missing values

In addition, you can use options in the PROC UNIVARIATE statement to do the following:

- specify the input data set to be analyzed
- specify a graphics catalog for saving traditional graphics output
- specify rounding units for variable values
- specify the definition used to calculate percentiles
- specify the divisor used to calculate variances and standard deviations
- request that plots be produced on line printers and define special printing characters used for features
- suppress tables
- save statistics in an output data set

The following are the options that can be used with the PROC UNIVARIATE statement:

ALL
> requests all statistics and tables that the FREQ, MODES, NEXTRVAL=5, PLOTS, and CIBASIC options generate. If the analysis variables are not weighted, this option also requests the statistics and tables generated by the CIPCTLDF, CIPCTLNORMAL, LOCCOUNT, NORMAL, ROBUSTSCALE, TRIMMED=.25, and WINSORIZED=.25 options. PROC UNIVARIATE also uses any values that you specify for ALPHA=, MU0=, NEXTRVAL=, CIBASIC, CIPCTLDF, CIPCTLNORMAL, TRIMMED=, or WINSORIZED= to produce the output.

ALPHA=α
> specifies the level of significance α for $100(1-\alpha)\%$ confidence intervals. The value α must be between 0 and 1; the default value is 0.05, which results in 95% confidence intervals.

> Note that specialized ALPHA= options are available for a number of confidence interval options. For example, you can specify CIBASIC(ALPHA=0.10) to request a table of basic confidence limits at the 90% level. The default value of these options is the value of the ALPHA= option in the PROC statement.

ANNOTATE=SAS-data-set

ANNO=SAS-data-set

> specifies an input data set that contains annotate variables as described in *SAS/GRAPH: Help*. You can use this data set to add features to your traditional graphics. PROC UNIVARIATE adds the features in this data set to every graph that is produced in the procedure. PROC UNIVARIATE does not use the ANNOTATE= data set unless you create a traditional graph with a plot statement. The option does not apply to ODS Graphics output. Use the ANNOTATE= option in the plot statement if you want to add a feature to a specific graph produced by that statement.

CIBASIC < (< TYPE=*keyword* > < ALPHA=α >) >

> requests confidence limits for the mean, standard deviation, and variance based on the assumption that the data are normally distributed. If you use the CIBASIC option, you must use the default value of VARDEF=, which is DF.
>
> **TYPE=***keyword*
>
> > specifies the type of confidence limit, where *keyword* is LOWER, UPPER, or TWOSIDED. The default value is TWOSIDED.
>
> **ALPHA=**α
>
> > specifies the level of significance α for $100(1-\alpha)\%$ confidence intervals. The value α must be between 0 and 1; the default value is 0.05, which results in 95% confidence intervals. The default value is the value of ALPHA= given in the PROC statement.

CIPCTLDF < (< TYPE=*keyword* > < ALPHA=α >) >

CIQUANTDF < (< TYPE=*keyword* > < ALPHA=α >) >

> requests confidence limits for quantiles based on a method that is distribution-free. In other words, no specific parametric distribution such as the normal is assumed for the data. PROC UNIVARIATE uses order statistics (ranks) to compute the confidence limits as described by Hahn and Meeker (1991). This option does not apply if you use a WEIGHT statement.
>
> **TYPE=***keyword*
>
> > specifies the type of confidence limit, where *keyword* is LOWER, UPPER, SYMMETRIC, or ASYMMETRIC. The default value is SYMMETRIC.
>
> **ALPHA=**α
>
> > specifies the level of significance α for $100(1-\alpha)\%$ confidence intervals. The value α must be between 0 and 1; the default value is 0.05, which results in 95% confidence intervals. The default value is the value of ALPHA= given in the PROC statement.

CIPCTLNORMAL < (< TYPE=*keyword* > < ALPHA=α >) >

CIQUANTNORMAL < (< TYPE=*keyword* > < ALPHA=α >) >

> requests confidence limits for quantiles based on the assumption that the data are normally distributed. The computational method is described in Section 4.4.1 of Hahn and Meeker (1991) and uses the noncentral t distribution as given by Odeh and Owen (1980). This option does not apply if you use a WEIGHT statement
>
> **TYPE=***keyword*
>
> > specifies the type of confidence limit, where *keyword* is LOWER, UPPER, or TWOSIDED. The default is TWOSIDED.

ALPHA=α
> specifies the level of significance α for $100(1-\alpha)\%$ confidence intervals. The value α must be between 0 and 1; the default value is 0.05, which results in 95% confidence intervals. The default value is the value of ALPHA= given in the PROC statement.

DATA=SAS-data-set
> specifies the input SAS data set to be analyzed. If the DATA= option is omitted, the procedure uses the most recently created SAS data set.

EXCLNPWGT
EXCLNPWGTS
> excludes observations with nonpositive weight values (zero or negative) from the analysis. By default, PROC UNIVARIATE counts observations with negative or zero weights in the total number of observations. This option applies only when you use a WEIGHT statement.

FORCEQN
> forces calculation of the robust estimate of scale Q_n. Because this calculation is very computationally intensive, by default Q_n is not computed for a variable that has more than 65,526 nonmissing observations. On some hosts, Q_n cannot be computed at all when there are more than 65,526 nonmissing observations.

FORCESN
> forces calculation of the robust estimate of scale S_n. Because this calculation is computationally intensive, by default S_n is not computed for a variable that has more than 1 million nonmissing observations.

FREQ
> requests a frequency table that consists of the variable values, frequencies, cell percentages, and cumulative percentages.
>
> If you specify the WEIGHT statement, PROC UNIVARIATE includes the weighted count in the table and uses this value to compute the percentages.

GOUT=graphics-catalog
> specifies the SAS catalog that PROC UNIVARIATE uses to save traditional graphics output. If you omit the libref in the name of the *graphics-catalog*, PROC UNIVARIATE looks for the catalog in the temporary library called WORK and creates the catalog if it does not exist. The option does not apply to ODS Graphics output.

IDOUT
> includes ID variables in the output data set created by an OUTPUT statement. The value of an ID variable in the output data set is its first value from the input data set or BY group. By default, ID variables are not included in OUTPUT statement data sets.

LOCCOUNT
> requests a table that shows the number of observations greater than, not equal to, and less than the value of MU0=. PROC UNIVARIATE uses these values to construct the sign test and the signed rank test. This option does not apply if you use a WEIGHT statement.

MODES|MODE

requests a table of all possible modes. By default, when the data contain multiple modes, PROC UNIVARIATE displays the lowest mode in the table of basic statistical measures. When all the values are unique, PROC UNIVARIATE does not produce a table of modes.

MU0=*values*
LOCATION=*values*

specifies the value of the mean or location parameter (μ_0) in the null hypothesis for tests of location summarized in the table labeled "Tests for Location: Mu0=value." If you specify one value, PROC UNIVARIATE tests the same null hypothesis for all analysis variables. If you specify multiple values, a VAR statement is required, and PROC UNIVARIATE tests a different null hypothesis for each analysis variable, matching variables and location values by their order in the two lists. The default *value* is 0.

The following statement tests the hypothesis $\mu_0 = 0$ for the first variable and the hypothesis $\mu_0 = 0.5$ for the second variable.

```
proc univariate mu0=0 0.5;
```

NEXTROBS=*n*

specifies the number of extreme observations that PROC UNIVARIATE lists in the table of extreme observations. The table lists the *n* lowest observations and the *n* highest observations. The default value is 5. You can specify NEXTROBS=0 to suppress the table of extreme observations.

NEXTRVAL=*n*

specifies the number of extreme values that PROC UNIVARIATE lists in the table of extreme values. The table lists the *n* lowest unique values and the *n* highest unique values. By default, *n* = 0 and no table is displayed.

NOBYPLOT

suppresses side-by-side line printer box plots that are created by default when you use the BY statement and either the ALL option or the PLOTS option in the PROC statement.

NOPRINT

suppresses all the tables of descriptive statistics that the PROC UNIVARIATE statement creates. NOPRINT does not suppress the tables that the HISTOGRAM statement creates. You can use the NOPRINT option in the HISTOGRAM statement to suppress the creation of its tables. Use NOPRINT when you want to create an OUT= or OUTTABLE= output data set only.

NORMAL
NORMALTEST

requests tests for normality that include a series of goodness-of-fit tests based on the empirical distribution function. The table provides test statistics and *p*-values for the Shapiro-Wilk test (provided the sample size is less than or equal to 2000), the Kolmogorov-Smirnov test, the Anderson-Darling test, and the Cramér–von Mises test. This option does not apply if you use a WEIGHT statement.

NOTABCONTENTS

suppresses the table of contents entries for tables of summary statistics produced by the PROC UNIVARIATE statement.

NOVARCONTENTS
   suppresses grouping entries associated with analysis variables in the table of contents. By default, the table of contents lists results associated with an analysis variable in a group with the variable name.

OUTTABLE=SAS-data-set
   creates an output data set that contains univariate statistics arranged in tabular form, with one observation per analysis variable. See the section "OUTTABLE= Output Data Set" on page 448 for details.

PCTLDEF=value

DEF=value
   specifies the definition that PROC UNIVARIATE uses to calculate quantiles. The default value is 5. Values can be 1, 2, 3, 4, or 5. You cannot use PCTLDEF= when you compute weighted quantiles. See the section "Calculating Percentiles" on page 397 for details on quantile definitions.

PLOTS | PLOT< (<plot-options> <**SSPLOT**(plot-options)>) >
   produces a panel of plots for each analysis variable. If ODS Graphics is enabled, the panel contains a horizontal histogram, a box plot, and a normal probability plot. Otherwise, the procedure produces a stem-and-leaf plot (or a horizontal bar chart), a box plot, and a normal probability plot by using line printer output. If you specify a BY statement, side-by-side box plots of the data from the BY groups are displayed following the univariate output for the last BY group.

   You can specify the following plot options to produce titles and footnotes for the plots when ODS Graphics is enabled. Plot options that are specified within the SSPLOT suboption apply to the side-by-side box plots of BY group data.

   ODSFOOTNOTE=FOOTNOTE | FOOTNOTE1 | 'string'
      adds a footnote to ODS Graphics output. If you specify the FOOTNOTE (or FOOTNOTE1) keyword, the value of the SAS FOOTNOTE statement is used as the graph footnote. If you specify a quoted string, that is used as the footnote. The quoted string can contain either of the following escaped characters, which are replaced with the appropriate values from the analysis:

      \n   is replaced by the analysis variable name.

      \l   is replaced by the analysis variable label (or name if the analysis variable has no label).

   ODSFOOTNOTE2=FOOTNOTE2 | 'string'
      adds a secondary footnote to ODS Graphics output. If you specify the FOOTNOTE2 keyword, the value of the SAS FOOTNOTE2 statement is used as the secondary graph footnote. If you specify a quoted string, that is used as the secondary footnote. The quoted string can contain any of the following escaped characters, which are replaced with the appropriate values from the analysis:

      \n   is replaced by the analysis variable name.

      \l   is replaced by the analysis variable label (or name if the analysis variable has no label).

ODSTITLE=TITLE | TITLE1 | NONE | DEFAULT | LABELFMT | '*string*'
specifies a title for ODS Graphics output.

TITLE (or TITLE1)   uses the value of SAS TITLE statement as the graph title.

NONE   suppresses all titles from the graph.

DEFAULT   uses the default ODS Graphics title (a descriptive title that consists of the plot type and the analysis variable name).

LABELFMT   uses the default ODS Graphics title with the variable label instead of the variable name.

If you specify a quoted string, that is used as the graph title. The quoted string can contain the following escaped characters, which are replaced with the appropriate values from the analysis:

\n   is replaced by the analysis variable name.

\l   is replaced by the analysis variable label (or name if the analysis variable has no label).

ODSTITLE2=TITLE2 | '*string*'
specifies a secondary title for ODS Graphics output. If you specify the TITLE2 keyword, the value of the SAS TITLE2 statement is used as the secondary graph title. If you specify a quoted string, that is used as the secondary title. The quoted string can contain the following escaped characters, which are replaced with the appropriate values from the analysis:

\n   is replaced by the analysis variable name.

\l   is replaced by the analysis variable label (or name if the analysis variable has no label).

NOTE: ODSTITLE=LABELFMT and the substitution of analysis variable names and labels are not supported for plot options specified within the SSPLOT suboption.

PLOTSIZE=*n*

specifies the approximate number of rows used in line-printer plots requested with the PLOTS option. If *n* is larger than the value of the SAS system option PAGESIZE=, PROC UNIVARIATE uses the value of PAGESIZE=. If *n* is less than 8, PROC UNIVARIATE uses eight rows to draw the plots.

ROBUSTSCALE

produces a table with robust estimates of scale. The statistics include the interquartile range, Gini's mean difference, the median absolute deviation about the median (*MAD*), and two statistics proposed by Rousseeuw and Croux (1993), Q_n, and S_n. See the section "Robust Estimates of Scale" on page 404 for details. This option does not apply if you use a WEIGHT statement.

ROUND=*units*

specifies the units to use to round the analysis variables prior to computing statistics. If you specify one unit, PROC UNIVARIATE uses this unit to round all analysis variables. If you specify multiple units, a VAR statement is required, and each unit rounds the values of the corresponding analysis variable. If ROUND=0, no rounding occurs. The ROUND= option reduces the number of unique variable values, thereby reducing memory requirements for the procedure. For example, to make the rounding unit 1 for the first analysis variable and 0.5 for the second analysis variable, submit the statement

```
proc univariate round=1 0.5;
   var Yieldstrength tenstren;
run;
```

When a variable value is midway between the two nearest rounded points, the value is rounded to the nearest even multiple of the roundoff value. For example, with a roundoff value of 1, the variable values of –2.5, –2.2, and –1.5 are rounded to –2; the values of –0.5, 0.2, and 0.5 are rounded to 0; and the values of 0.6, 1.2, and 1.4 are rounded to 1.

SUMMARYCONTENTS=*'string'*
    specifies the table of contents entry used for grouping the summary statistics produced by the PROC UNIVARIATE statement. You can specify SUMMARYCONTENTS='' to suppress the grouping entry.

TRIMMED=*values* < (< **TYPE=***keyword* > < **ALPHA=**α >) >

TRIM=*values* < (< **TYPE=***keyword* > < **ALPHA=**α >) >
    requests a table of trimmed means, where *value* specifies the number or the proportion of observations that PROC UNIVARIATE trims. If the *value* is the number n of trimmed observations, n must be between 0 and half the number of nonmissing observations. If *value* is a proportion p between 0 and ½, the number of observations that PROC UNIVARIATE trims is the smallest integer that is greater than or equal to np, where n is the number of observations. To include confidence limits for the mean and the Student's t test in the table, you must use the default value of VARDEF=, which is DF. For details concerning the computation of trimmed means, see the section "Trimmed Means" on page 403. The TRIMMED= option does not apply if you use a WEIGHT statement.

   TYPE=*keyword*
        specifies the type of confidence limit for the mean, where *keyword* is LOWER, UPPER, or TWOSIDED. The default value is TWOSIDED.

   ALPHA=α
        specifies the level of significance α for $100(1 - \alpha)\%$ confidence intervals. The value α must be between 0 and 1; the default value is 0.05, which results in 95% confidence intervals.

VARDEF=*divisor*
    specifies the divisor to use in the calculation of variances and standard deviation. By default, VARDEF=DF. Table 4.1 shows the possible values for *divisor* and associated divisors.

Table 4.1 Possible Values for VARDEF=

Value	Divisor	Formula for Divisor
DF	degrees of freedom	$n - 1$
N	number of observations	n
WDF	sum of weights minus one	$(\Sigma_i w_i) - 1$
WEIGHT | WGT	sum of weights	$\Sigma_i w_i$

The procedure computes the variance as $\frac{CSS}{\text{divisor}}$ where CSS is the corrected sums of squares and equals $\sum_{i=1}^{n}(x_i - \bar{x})^2$. When you weight the analysis variables, $CSS = \sum_{i=1}^{n} w_i(x_i - \bar{x}_w)^2$ where $\bar{x}_w$ is the weighted mean.

The default value is DF. To compute the standard error of the mean, confidence limits, and Student's t test, use the default value of VARDEF=.

When you use the WEIGHT statement and VARDEF=DF, the variance is an estimate of σ^2 where the variance of the ith observation is $var(x_i) = \frac{\sigma^2}{w_i}$ and w_i is the weight for the ith observation. This yields an estimate of the variance of an observation with unit weight.

When you use the WEIGHT statement and VARDEF=WGT, the computed variance is asymptotically (for large n) an estimate of $\frac{\sigma^2}{\bar{w}}$ where $\bar{w}$ is the average weight. This yields an asymptotic estimate of the variance of an observation with average weight.

WINSORIZED=values < (< **TYPE=**keyword > < **ALPHA=**α >) >

WINSOR=values < (< **TYPE=**keyword > < **ALPHA=**α >) >

requests of a table of Winsorized means, where value is the number or the proportion of observations that PROC UNIVARIATE uses to compute the Winsorized mean. If the value is the number n of Winsorized observations, n must be between 0 and half the number of nonmissing observations. If value is a proportion p between 0 and ½, the number of observations that PROC UNIVARIATE uses is equal to the smallest integer that is greater than or equal to np, where n is the number of observations. To include confidence limits for the mean and the Student t test in the table, you must use the default value of VARDEF=, which is DF. For details concerning the computation of Winsorized means, see the section "Winsorized Means" on page 402. The WINSORIZED= option does not apply if you use a WEIGHT statement.

TYPE=keyword

specifies the type of confidence limit for the mean, where keyword is LOWER, UPPER, or TWOSIDED. The default is TWOSIDED.

ALPHA=α

specifies the level of significance α for $100(1 - \alpha)\%$ confidence intervals. The value α must be between 0 and 1; the default value is 0.05, which results in 95% confidence intervals.

BY Statement

BY variables ;

You can specify a BY statement with PROC UNIVARIATE to obtain separate analyses of observations in groups that are defined by the BY variables. When a BY statement appears, the procedure expects the input data set to be sorted in order of the BY variables. If you specify more than one BY statement, only the last one specified is used.

If your input data set is not sorted in ascending order, use one of the following alternatives:

- Sort the data by using the SORT procedure with a similar BY statement.

- Specify the NOTSORTED or DESCENDING option in the BY statement for the UNIVARIATE procedure. The NOTSORTED option does not mean that the data are unsorted but rather that the data are arranged in groups (according to values of the BY variables) and that these groups are not necessarily in alphabetical or increasing numeric order.

- Create an index on the BY variables by using the DATASETS procedure (in Base SAS software).

For more information about BY-group processing, see the discussion in *SAS Language Reference: Concepts*. For more information about the DATASETS procedure, see the discussion in the *SAS Visual Data Management and Utility Procedures Guide*.

CDFPLOT Statement

CDFPLOT < *variables* > < / *options* > ;

The CDFPLOT statement plots the observed cumulative distribution function (cdf) of a variable, defined as

$$F_N(x) = \text{percent of nonmissing values} \leq x$$
$$= \frac{\text{number of values} \leq x}{N} \times 100\%$$

where N is the number of nonmissing observations. The cdf is an increasing step function that has a vertical jump of $\frac{1}{N}$ at each value of x equal to an observed value. The cdf is also referred to as the empirical cumulative distribution function (ECDF).

You can use any number of CDFPLOT statements in the UNIVARIATE procedure. The components of the CDFPLOT statement are as follows.

variables

specify variables for which to create cdf plots. If you specify a VAR statement, the *variables* must also be listed in the VAR statement. Otherwise, the *variables* can be any numeric variables in the input data set. If you do not specify a list of *variables*, then by default the procedure creates a cdf plot for each variable listed in the VAR statement, or for each numeric variable in the DATA= data set if you do not specify a VAR statement.

For example, suppose a data set named Steel contains exactly three numeric variables: Length, Width, and Height. The following statements create a cdf plot for each of the three variables:

```
proc univariate data=Steel;
   cdfplot;
run;
```

The following statements create a cdf plot for Length and a cdf plot for Width:

```
proc univariate data=Steel;
   var Length Width;
   cdfplot;
run;
```

The following statements create a cdf plot for Width:

```
proc univariate data=Steel;
   var Length Width;
   cdfplot Width;
run;
```

options

specify the theoretical distribution for the plot or add features to the plot. If you specify more than one variable, the *options* apply equally to each variable. Specify all *options* after the slash (/) in the CDFPLOT statement. You can specify only one *option* that names a distribution in each CDFPLOT statement, but you can specify any number of other *options*. The distributions available are listed in Table 4.2. By default, the procedure produces a plot for the normal distribution.

Table 4.2 through Table 4.4 list the CDFPLOT *options* by function. For complete descriptions, see the sections "Dictionary of Options" on page 302 and "Dictionary of Common Options" on page 384. *Options* can be any of the following:

- primary options
- secondary options
- general options

Distribution Options

Table 4.2 lists primary options for requesting a theoretical distribution.

Table 4.2 Primary Options for Theoretical Distribution

Option	Description
BETA(*beta-options*)	plots two-parameter beta distribution function, parameters θ and σ assumed known
EXPONENTIAL(*exponential-options*)	plots one-parameter exponential distribution function, parameter θ assumed known
GAMMA(*gamma-options*)	plots two-parameter gamma distribution function, parameter θ assumed known
GUMBEL(*Gumbel-options*)	plots Gumbel distribution with location parameter μ and scale parameter σ
IGAUSS(*iGauss-options*)	plots inverse Gaussian distribution with mean μ and shape parameter λ
LOGNORMAL(*lognormal-options*)	plots two-parameter lognormal distribution function, parameter θ assumed known
NORMAL(*normal-options*)	plots normal distribution function

Table 4.2 (continued)

Option	Description
PARETO(*Pareto-options*)	plots generalized Pareto distribution with threshold parameter θ, scale parameter σ, and shape parameter α
POWER(*power-options*)	plots power function distribution with threshold parameter θ, scale parameter σ, and shape parameter α
RAYLEIGH(*Rayleigh-options*)	plots Rayleigh distribution with threshold parameter θ and scale parameter σ
WEIBULL(*Weibull-options*)	plots two-parameter Weibull distribution function, parameter θ assumed known

Table 4.3 lists secondary options that specify distribution parameters and control the display of a theoretical distribution function. Specify these options in parentheses after the distribution keyword. For example, you can request a normal probability plot with a distribution reference line by specifying the NORMAL option as follows:

```
proc univariate;
   cdfplot / normal(mu=10 sigma=0.5 color=red);
run;
```

The COLOR= option specifies the color for the curve, and the *normal-options* MU= and SIGMA= specify the parameters $\mu = 10$ and $\sigma = 0.5$ for the distribution function. If you do not specify these parameters, maximum likelihood estimates are computed.

Table 4.3 Secondary Distribution Options

Option	Description
Options Used with All Distributions	
COLOR=	specifies color of theoretical distribution function
L=	specifies line type of theoretical distribution function
W=	specifies width of theoretical distribution function
Beta-Options	
ALPHA=	specifies first shape parameter α for beta distribution function
BETA=	specifies second shape parameter β for beta distribution function
SIGMA=	specifies scale parameter σ for beta distribution function
THETA=	specifies lower threshold parameter θ for beta distribution function
Exponential-Options	
SIGMA=	specifies scale parameter σ for exponential distribution function
THETA=	specifies threshold parameter θ for exponential distribution function
Gamma-Options	
ALPHA=	specifies shape parameter α for gamma distribution function
ALPHADELTA=	specifies change in successive estimates of α at which the Newton-Raphson approximation of $\hat{\alpha}$ terminates

Table 4.3 (continued)

Option	Description
ALPHAINITIAL=	specifies initial value for α in the Newton-Raphson approximation of $\hat{\alpha}$
MAXITER=	specifies maximum number of iterations in the Newton-Raphson approximation of $\hat{\alpha}$
SIGMA=	specifies scale parameter σ for gamma distribution function
THETA=	specifies threshold parameter θ for gamma distribution function
Gumbel-Options	
MU=	specifies location parameter μ for Gumbel distribution function
SIGMA=	specifies scale parameter σ for Gumbel distribution function
IGauss-Options	
LAMBDA=	specifies shape parameter λ for inverse Gaussian distribution function
MU=	specifies mean μ for inverse Gaussian distribution function
Lognormal-Options	
SIGMA=	specifies shape parameter σ for lognormal distribution function
THETA=	specifies threshold parameter θ for lognormal distribution function
ZETA=	specifies scale parameter ζ for lognormal distribution function
Normal-Options	
MU=	specifies mean μ for normal distribution function
SIGMA=	specifies standard deviation σ for normal distribution function
Pareto-Options	
ALPHA=	specifies shape parameter α for generalized Pareto distribution function
SIGMA=	specifies scale parameter σ for generalized Pareto distribution function
THETA=	specifies threshold parameter θ for generalized Pareto distribution function
Power-Options	
ALPHA=	specifies shape parameter α for power function distribution
SIGMA=	specifies scale parameter σ for power function distribution
THETA=	specifies threshold parameter θ for power function distribution
Rayleigh-Options	
SIGMA=	specifies scale parameter σ for Rayleigh distribution function
THETA=	specifies threshold parameter θ for Rayleigh distribution function
Secondary Weibull-Options	
C=	specifies shape parameter c for Weibull distribution function
ITPRINT	requests table of iteration history and optimizer details
MAXITER=	specifies maximum number of iterations in the Newton-Raphson approximation of $\hat{c}$
SIGMA=	specifies scale parameter σ for Weibull distribution function
THETA=	specifies threshold parameter θ for Weibull distribution function

General Options

Table 4.4 summarizes general options for enhancing cdf plots.

Table 4.4 General Graphics Options

Option	Description
General Graphics Options	
HREF=	specifies reference lines perpendicular to the horizontal axis
HREFLABELS=	specifies labels for HREF= lines
HREFLABPOS=	specifies position for HREF= line labels
NOECDF	suppresses plot of empirical (observed) distribution function
NOHLABEL	suppresses label for horizontal axis
NOVLABEL	suppresses label for vertical axis
NOVTICK	suppresses tick marks and tick mark labels for vertical axis
STATREF=	specifies reference lines at values of summary statistics
STATREFLABELS=	specifies labels for STATREF= lines
STATREFSUBCHAR=	specifies substitution character for displaying statistic values in STATREFLABELS= labels
VAXISLABEL=	specifies label for vertical axis
VREF=	specifies reference lines perpendicular to the vertical axis
VREFLABELS=	specifies labels for VREF= lines
VREFLABPOS=	specifies position for VREF= line labels
VSCALE=	specifies scale for vertical axis
Options for Traditional Graphics Output	
ANNOTATE=	specifies annotate data set
CAXIS=	specifies color for axis
CFRAME=	specifies color for frame
CHREF=	specifies colors for HREF= lines
CSTATREF=	specifies colors for STATREF= lines
CTEXT=	specifies color for text
CVREF=	specifies colors for VREF= lines
DESCRIPTION=	specifies description for graphics catalog member
FONT=	specifies text font
HAXIS=	specifies AXIS statement for horizontal axis
HEIGHT=	specifies height of text used outside framed areas
HMINOR=	specifies number of horizontal axis minor tick marks
INFONT=	specifies software font for text inside framed areas
INHEIGHT=	specifies height of text inside framed areas
LHREF=	specifies line types for HREF= lines
LSTATREF=	specifies line types for STATREF= lines
LVREF=	specifies line types for VREF= lines
NAME=	specifies name for plot in graphics catalog
NOFRAME	suppresses frame around plotting area
TURNVLABELS	turns and vertically strings out characters in labels for vertical axis
VAXIS=	specifies AXIS statement for vertical axis
VMINOR=	specifies number of vertical axis minor tick marks
WAXIS=	specifies line thickness for axes and frame
Options for ODS Graphics Output	
NOCDFLEGEND	suppresses legend for superimposed theoretical cdf
ODSFOOTNOTE=	specifies footnote displayed on plot

Table 4.4 *continued*

Option	Description
ODSFOOTNOTE2=	specifies secondary footnote displayed on plot
ODSTITLE=	specifies title displayed on plot
ODSTITLE2=	specifies secondary title displayed on plot
OVERLAY	overlays plots for different class levels
Options for Comparative Plots	
ANNOKEY	applies annotation requested in ANNOTATE= data set to key cell only
CFRAMESIDE=	specifies color for filling row label frames
CFRAMETOP=	specifies color for filling column label frames
CPROP=	specifies color for proportion of frequency bar
CTEXTSIDE=	specifies color for row labels
CTEXTTOP=	specifies color for column labels
INTERTILE=	specifies distance between tiles in comparative plot
NCOLS=	specifies number of columns in comparative plot
NROWS=	specifies number of rows in comparative plot
Miscellaneous Options	
CONTENTS=	specifies table of contents entry for cdf plot grouping

Dictionary of Options

The following entries provide detailed descriptions of the options specific to the CDFPLOT statement. See the section "Dictionary of Common Options" on page 384 for detailed descriptions of options common to all plot statements.

ALPHA=*value*

specifies the shape parameter α for distribution functions requested with the BETA, GAMMA, PARETO, and POWER options. Enclose the ALPHA= option in parentheses after the distribution keyword. If you do not specify a value for α, the procedure calculates a maximum likelihood estimate. For examples, see the entries for the BETA and GAMMA options.

BETA<(*beta-options***)>**

displays a fitted beta distribution function on the cdf plot. The equation of the fitted cdf is

$$F(x) = \begin{cases} 0 & \text{for } x \leq \theta \\ I_{\frac{x-\theta}{\sigma}}(\alpha, \beta) & \text{for } \theta < x < \theta + \sigma \\ 1 & \text{for } x \geq \sigma + \theta \end{cases}$$

where $I_y(\alpha, \beta)$ is the incomplete beta function and

θ = lower threshold parameter (lower endpoint)

σ = scale parameter ($\sigma > 0$)

α = shape parameter ($\alpha > 0$)

β = shape parameter ($\beta > 0$)

The beta distribution is bounded below by the parameter θ and above by the value $\theta + \sigma$. You can specify θ and σ by using the THETA= and SIGMA= *beta-options*, as illustrated in the following statements, which fit a beta distribution bounded between 50 and 75. The default values for θ and σ are 0 and 1, respectively.

```
proc univariate;
   cdfplot / beta(theta=50 sigma=25);
run;
```

The beta distribution has two shape parameters: α and β. If these parameters are known, you can specify their values with the ALPHA= and BETA= *beta-options*. If you do not specify values for α and β, the procedure calculates maximum likelihood estimates.

The BETA option can appear only once in a CDFPLOT statement. Table 4.3 lists options you can specify with the BETA distribution option.

BETA=*value*

B=*value*

    specifies the second shape parameter β for beta distribution functions requested by the BETA option. Enclose the BETA= option in parentheses after the BETA keyword. If you do not specify a value for β, the procedure calculates a maximum likelihood estimate. For examples, see the preceding entry for the BETA option.

C=*value*

    specifies the shape parameter c for Weibull distribution functions requested with the WEIBULL option. Enclose the C= option in parentheses after the WEIBULL keyword. If you do not specify a value for c, the procedure calculates a maximum likelihood estimate. You can specify the SHAPE= option as an alias for the C= option.

EXPONENTIAL<(*exponential-options***)>**

EXP<(*exponential-options***)>**

    displays a fitted exponential distribution function on the cdf plot. The equation of the fitted cdf is

$$F(x) = \begin{cases} 0 & \text{for } x \leq \theta \\ 1 - \exp\left(-\frac{x-\theta}{\sigma}\right) & \text{for } x > \theta \end{cases}$$

where

    θ = threshold parameter

    σ = scale parameter ($\sigma > 0$)

The parameter θ must be less than or equal to the minimum data value. You can specify θ with the THETA= *exponential-option*. The default value for θ is 0. You can specify σ with the SIGMA= *exponential-option*. By default, a maximum likelihood estimate is computed for σ. For example, the following statements fit an exponential distribution with $\theta = 10$ and a maximum likelihood estimate for σ:

```
proc univariate;
   cdfplot / exponential(theta=10 l=2 color=green);
run;
```

The exponential curve is green and has a line type of 2.

The EXPONENTIAL option can appear only once in a CDFPLOT statement. Table 4.3 lists the options you can specify with the EXPONENTIAL option.

GAMMA<(gamma-options)>

displays a fitted gamma distribution function on the cdf plot. The equation of the fitted cdf is

$$F(x) = \begin{cases} 0 & \text{for } x \leq \theta \\ \frac{1}{\Gamma(\alpha)\sigma} \int_\theta^x \left(\frac{t-\theta}{\sigma}\right)^{\alpha-1} \exp\left(-\frac{t-\theta}{\sigma}\right) dt & \text{for } x > \theta \end{cases}$$

where

θ = threshold parameter

σ = scale parameter ($\sigma > 0$)

α = shape parameter ($\alpha > 0$)

The parameter θ for the gamma distribution must be less than the minimum data value. You can specify θ with the THETA= *gamma-option*. The default value for θ is 0. In addition, the gamma distribution has a shape parameter α and a scale parameter σ. You can specify these parameters with the ALPHA= and SIGMA= *gamma-options*. By default, maximum likelihood estimates are computed for α and σ. For example, the following statements fit a gamma distribution function with $\theta = 4$ and maximum likelihood estimates for α and σ:

```
proc univariate;
   cdfplot / gamma(theta=4);
run;
```

Note that the maximum likelihood estimate of α is calculated iteratively using the Newton-Raphson approximation. The *gamma-options* ALPHADELTA=, ALPHAINITIAL=, and MAXITER= control the approximation.

The GAMMA option can appear only once in a CDFPLOT statement. Table 4.3 lists the options you can specify with the GAMMA option.

GUMBEL<(Gumbel-options)>

displays a fitted Gumbel distribution (also known as Type 1 extreme value distribution) function on the cdf plot. The equation of the fitted cdf is

$$F(x) = \exp\left(-e^{-(x-\mu)/\sigma}\right)$$

where

μ = location parameter

σ = scale parameter ($\sigma > 0$)

You can specify known values for μ and σ with the MU= and SIGMA= *Gumbel-options*. By default, maximum likelihood estimates are computed for μ and σ.

The GUMBEL option can appear only once in a CDFPLOT statement. Table 4.3 lists secondary options you can specify with the GUMBEL option.

IGAUSS< (*iGauss-options***) >**
   displays a fitted inverse Gaussian distribution function on the cdf plot. The equation of the fitted cdf is

$$F(x) = \Phi\left\{\sqrt{\frac{\lambda}{x}}\left(\frac{x}{\mu}-1\right)\right\} + e^{2\lambda/\mu}\Phi\left\{-\sqrt{\frac{\lambda}{x}}\left(\frac{x}{\mu}+1\right)\right\}$$

where $\Phi(\cdot)$ is the standard normal cumulative distribution function, and

μ = mean parameter ($\mu > 0$)
λ = shape parameter ($\lambda > 0$)

You can specify known values for μ and λ with the MU= and LAMBDA= *iGauss-options*. By default, maximum likelihood estimates are computed for μ and λ.

The IGAUSS option can appear only once in a CDFPLOT statement. Table 4.3 lists secondary options you can specify with the IGAUSS option.

LAMBDA=*value*
   specifies the shape parameter λ for distribution functions requested with the IGAUSS option. Enclose the LAMBDA= option in parentheses after the IGAUSS distribution keyword. If you do not specify a value for λ, the procedure calculates a maximum likelihood estimate.

LOGNORMAL< (*lognormal options***) >**
   displays a fitted lognormal distribution function on the cdf plot. The equation of the fitted cdf is

$$F(x) = \begin{cases} 0 & \text{for } x \leq \theta \\ \Phi\left(\frac{\log(x-\theta)-\zeta}{\sigma}\right) & \text{for } x > \theta \end{cases}$$

where $\Phi(\cdot)$ is the standard normal cumulative distribution function and

θ = threshold parameter
ζ = scale parameter
σ = shape parameter ($\sigma > 0$)

The parameter θ for the lognormal distribution must be less than the minimum data value. You can specify θ with the THETA= *lognormal-option*. The default value for θ is 0. In addition, the lognormal distribution has a shape parameter σ and a scale parameter ζ. You can specify these parameters with the SIGMA= and ZETA= *lognormal-options*. By default, maximum likelihood estimates are computed for σ and ζ. For example, the following statements fit a lognormal distribution function with $\theta = 10$ and maximum likelihood estimates for σ and ζ:

```
proc univariate;
   cdfplot / lognormal(theta = 10);
run;
```

The LOGNORMAL option can appear only once in a CDFPLOT statement. Table 4.3 lists options that you can specify with the LOGNORMAL option.

MU=value

specifies the parameter μ for theoretical cumulative distribution functions requested with the GUMBEL, IGAUSS, and NORMAL option. Enclose the MU= option in parentheses after the distribution keyword. For the inverse Gaussian and normal distributions, the default value is the sample mean. If you do not specify a value for μ for the Gumbel distribution, the procedure calculates a maximum likelihood estimate. For an example, see the entry for the NORMAL option.

NOCDFLEGEND

NOLEGEND

suppresses the legend for the superimposed theoretical cumulative distribution function. The NOCDFLEGEND option applies only to ODS Graphics output.

NOECDF

suppresses the observed distribution function (the empirical cumulative distribution function) of the variable, which is drawn by default. This option enables you to create theoretical cdf plots without displaying the data distribution. The NOECDF option can be used only with a theoretical distribution (such as the NORMAL option).

NORMAL< (*normal-options*) **>**

displays a fitted normal distribution function on the cdf plot. The equation of the fitted cdf is

$$F(x) = \Phi\left(\frac{x-\mu}{\sigma}\right) \quad \text{for } -\infty < x < \infty$$

where $\Phi(\cdot)$ is the standard normal cumulative distribution function and

μ = mean

σ = standard deviation ($\sigma > 0$)

You can specify known values for μ and σ with the MU= and SIGMA= *normal-options*, as shown in the following statements:

```
proc univariate;
   cdfplot / normal(mu=14 sigma=.05);
run;
```

By default, the sample mean and sample standard deviation are calculated for μ and σ. The NORMAL option can appear only once in a CDFPLOT statement. Table 4.3 lists options that you can specify with the NORMAL option.

PARETO< (*Pareto-options*) **>**

displays a fitted generalized Pareto distribution function on the cdf plot. The equation of the fitted cdf is

$$F(x) = 1 - \left(1 - \frac{\alpha(x-\theta)}{\sigma}\right)^{\frac{1}{\alpha}}$$

where

θ = threshold parameter

σ = scale parameter ($\sigma > 0$)
α = shape parameter

The parameter θ for the generalized Pareto distribution must be less than the minimum data value. You can specify θ with the THETA= *Pareto-option*. The default value for θ is 0. In addition, the generalized Pareto distribution has a shape parameter α and a scale parameter σ. You can specify these parameters with the ALPHA= and SIGMA= *Pareto-options*. By default, maximum likelihood estimates are computed for α and σ.

The PARETO option can appear only once in a CDFPLOT statement. Table 4.3 lists options that you can specify with the PARETO option.

POWER< (*power-options*) >

displays a fitted power function distribution on the cdf plot. The equation of the fitted cdf is

$$F(x) = \begin{cases} 0 & \text{for } x \leq \theta \\ \left(\frac{x-\theta}{\sigma}\right)^\alpha & \text{for } \theta < x < \theta + \sigma \\ 1 & \text{for } x \geq \theta + \sigma \end{cases}$$

where

θ = lower threshold parameter (lower endpoint)
σ = scale parameter ($\sigma > 0$)
α = shape parameter ($\alpha > 0$)

The power function distribution is bounded below by the parameter θ and above by the value $\theta + \sigma$. You can specify θ and σ by using the THETA= and SIGMA= *power-options*. The default values for θ and σ are 0 and 1, respectively.

You can specify a value for the shape parameter, α, with the ALPHA= *power-option*. If you do not specify a value for α, the procedure calculates a maximum likelihood estimate.

The power function distribution is a special case of the beta distribution with its second shape parameter, $\beta = 1$.

The POWER option can appear only once in a CDFPLOT statement. Table 4.3 lists options that you can specify with the POWER option.

RAYLEIGH< (*Rayleigh-options*) >

displays a fitted Rayleigh distribution function on the cdf plot. The equation of the fitted cdf is

$$F(x) = 1 - e^{-(x-\theta)^2/(2\sigma^2)}$$

where

θ = threshold parameter
σ = scale parameter ($\sigma > 0$)

The parameter θ for the Rayleigh distribution must be less than the minimum data value. You can specify θ with the THETA= *Rayleigh-option*. The default value for θ is 0. You can specify σ with the SIGMA= *Rayleigh-option*. By default, a maximum likelihood estimate is computed for σ.

The RAYLEIGH option can appear only once in a CDFPLOT statement. Table 4.3 lists options that you can specify with the RAYLEIGH option.

SIGMA=*value* **| EST**

specifies the parameter σ for distribution functions requested by the BETA, EXPONENTIAL, GAMMA, LOGNORMAL, NORMAL, and WEIBULL options. Enclose the SIGMA= option in parentheses after the distribution keyword. The following table summarizes the use of the SIGMA= option:

Distribution Option	SIGMA= Specifies	Default Value	Alias
BETA	scale parameter σ	1	SCALE=
EXPONENTIAL	scale parameter σ	maximum likelihood estimate	SCALE=
GAMMA	scale parameter σ	maximum likelihood estimate	SCALE=
GUMBEL	scale parameter σ	maximum likelihood estimate	
LOGNORMAL	shape parameter σ	maximum likelihood estimate	SHAPE=
NORMAL	scale parameter σ	standard deviation	
PARETO	scale parameter σ	maximum likelihood estimate	
POWER	scale parameter σ	1	
RAYLEIGH	scale parameter σ	maximum likelihood estimate	
WEIBULL	scale parameter σ	maximum likelihood estimate	SCALE=

THETA=*value* **| EST**

THRESHOLD=*value* **| EST**

specifies the lower threshold parameter θ for theoretical cumulative distribution functions requested with the BETA, EXPONENTIAL, GAMMA, LOGNORMAL, PARETO, POWER, RAYLEIGH, and WEIBULL options. Enclose the THETA= option in parentheses after the distribution keyword. The default value is 0.

VSCALE=PERCENT | PROPORTION

specifies the scale of the vertical axis. The value PERCENT scales the data in units of percent of observations per data unit. The value PROPORTION scales the data in units of proportion of observations per data unit. The default is PERCENT.

WEIBULL<(*Weibull-options***)>**

displays a fitted Weibull distribution function on the cdf plot. The equation of the fitted cdf is

$$F(x) = \begin{cases} 0 & \text{for } x \leq \theta \\ 1 - \exp\left(-\left(\frac{x-\theta}{\sigma}\right)^c\right) & \text{for } x > \theta \end{cases}$$

where

θ = threshold parameter

σ = scale parameter ($\sigma > 0$)

c = shape parameter ($c > 0$)

The parameter θ must be less than the minimum data value. You can specify θ with the THETA= *Weibull-option*. The default value for θ is 0. In addition, the Weibull distribution has a shape parameter c and a scale parameter σ. You can specify these parameters with the SIGMA= and C= *Weibull-options*. By default, maximum likelihood estimates are computed for c and σ. For example, the following statements fit a Weibull distribution function with $\theta = 15$ and maximum likelihood estimates for σ and c:

```
proc univariate;
   cdfplot / weibull(theta=15);
run;
```

The WEIBULL option can appear only once in a CDFPLOT statement. Table 4.3 lists options that you can specify with the WEIBULL option.

ZETA=*value*

specifies a value for the scale parameter ζ for a lognormal distribution function requested with the LOGNORMAL option. Enclose the ZETA= option in parentheses after the LOGNORMAL keyword. If you do not specify a value for ζ, a maximum likelihood estimate is computed. You can specify the SCALE= option as an alias for the ZETA= option.

CLASS Statement

CLASS *variable-1* < (*v-options*) > < *variable-2* < (*v-options*) > >
< / **KEYLEVEL=** *value1* | (*value1 value2*) > ;

The CLASS statement specifies one or two variables used to group the data into classification levels. Variables in a CLASS statement are referred to as *CLASS variables*. CLASS variables can be numeric or character. Class variables can have floating point values, but they typically have a few discrete values that define levels of the variable. You do not have to sort the data by CLASS variables. PROC UNIVARIATE uses the formatted values of the CLASS variables to determine the classification levels.

You can specify the following *v-options* enclosed in parentheses after the CLASS variable:

MISSING

specifies that missing values for the CLASS variable are to be treated as valid classification levels. Special missing values that represent numeric values ('.A' through '.Z' and '._') are each considered as a separate value. If you omit MISSING, PROC UNIVARIATE excludes the observations with a missing CLASS variable value from the analysis. Enclose this option in parentheses after the CLASS variable.

ORDER=DATA | FORMATTED | FREQ | INTERNAL

specifies the display order for the CLASS variable values. The default value is INTERNAL. You can specify the following values with the ORDER=*option*:

DATA orders values according to their order in the input data set. When you use a plot statement, PROC UNIVARIATE displays the rows (columns) of the comparative plot from top to bottom (left to right) in the order that the CLASS variable values first appear in the input data set.

FORMATTED orders values by their ascending formatted values. This order might depend on your operating environment. When you use a plot statement, PROC UNIVARIATE displays the rows (columns) of the comparative plot from top to bottom (left to right) in increasing order of the formatted CLASS variable values. For example, suppose a numeric CLASS variable Day (with values 1, 2, and 3) has a user-defined format that assigns Wednesday to the value 1, Thursday to the value 2, and Friday to the value 3. The rows of the comparative plot will appear in alphabetical order (Friday, Thursday, Wednesday) from top to bottom.

If there are two or more distinct internal values with the same formatted value, then PROC UNIVARIATE determines the order by the internal value that occurs first in the input data set. For numeric variables without an explicit format, the levels are ordered by their internal values.

FREQ
orders values by descending frequency count so that levels with the most observations are listed first. If two or more values have the same frequency count, PROC UNIVARIATE uses the formatted values to determine the order.

When you use a plot statement, PROC UNIVARIATE displays the rows (columns) of the comparative plot from top to bottom (left to right) in order of decreasing frequency count for the CLASS variable values.

INTERNAL
orders values by their unformatted values, which yields the same order as PROC SORT. This order may depend on your operating environment.

When you use a plot statement, PROC UNIVARIATE displays the rows (columns) of the comparative plot from top to bottom (left to right) in increasing order of the internal (unformatted) values of the CLASS variable. The first CLASS variable is used to label the rows of the comparative plots (top to bottom). The second CLASS variable is used to label the columns of the comparative plots (left to right). For example, suppose a numeric CLASS variable Day (with values 1, 2, and 3) has a user-defined format that assigns Wednesday to the value 1, Thursday to the value 2, and Friday to the value 3. The rows of the comparative plot will appear in day-of-the-week order (Wednesday, Thursday, Friday) from top to bottom.

You can specify the following *option* after the slash (/) in the CLASS statement.

KEYLEVEL=*value* | (*value1 value2*)

specifies the *key cells* in comparative plots. For each plot, PROC UNIVARIATE first determines the horizontal axis scaling for the key cell, and then extends the axis using the established tick interval to accommodate the data ranges for the remaining cells, if necessary. Thus, the choice of the key cell determines the uniform horizontal axis that PROC UNIVARIATE uses for all cells.

If you specify only one CLASS variable and use a plot statement, KEYLEVEL=*value* identifies the key cell as the level for which the CLASS variable is equal to *value*. By default, PROC UNIVARIATE sorts the levels in the order determined by the ORDER= option, and the key cell is the first occurrence of a level in this order. The cells display in order from top to bottom or left to right. Consequently, the key cell appears at the top (or left). When you specify a different key cell with the KEYLEVEL= option, this cell appears at the top (or left).

If you specify two CLASS variables, use KEYLEVEL= *(value1 value2)* to identify the key cell as the level for which CLASS variable *n* is equal to *valuen*. By default, PROC UNIVARIATE sorts the levels of the first CLASS variable in the order that is determined by its ORDER= option. Then, within each of these levels, it sorts the levels of the second CLASS variable in the order that is determined by its ORDER= option. The default key cell is the first occurrence of a combination of levels for the two variables in this order. The cells display in the order of the first CLASS variable from top to bottom and in the order of the second CLASS variable from left to right. Consequently, the default key cell appears at the upper left corner. When you specify a different key cell with the KEYLEVEL= option, this cell appears at the upper left corner.

The length of the KEYLEVEL= value cannot exceed 16 characters and you must specify a formatted value.

The KEYLEVEL= option has no effect unless you specify a plot statement.

NOKEYMOVE

specifies that the location of the key cell in a comparative plot be unchanged by the CLASS statement KEYLEVEL= option. By default, the key cell is positioned as the first cell in a comparative plot.

The NOKEYMOVE option has no effect unless you specify a plot statement.

FREQ Statement

FREQ *variable* ;

The FREQ statement specifies a numeric variable whose value represents the frequency of the observation. If you use the FREQ statement, the procedure assumes that each observation represents *n* observations, where *n* is the value of the variable. If the variable is not an integer, the SAS System truncates it. If the variable is less than 1 or is missing, the procedure excludes that observation from the analysis. See Example 4.6.

NOTE: The FREQ statement affects the degrees of freedom, but the WEIGHT statement does not.

HISTOGRAM Statement

HISTOGRAM < *variables* > < / *options* > ;

The HISTOGRAM statement creates histograms and optionally superimposes estimated parametric and nonparametric probability density curves. You cannot use the WEIGHT statement with the HISTOGRAM statement. You can use any number of HISTOGRAM statements after a PROC UNIVARIATE statement. The components of the HISTOGRAM statement are follows.

variables

are the variables for which histograms are to be created. If you specify a VAR statement, the *variables* must also be listed in the VAR statement. Otherwise, the *variables* can be any numeric variables in the input data set. If you do not specify *variables* in a VAR statement or in the HISTOGRAM statement, then by default, a histogram is created for each numeric variable in the DATA= data set. If you use a VAR statement and do not specify any *variables* in the HISTOGRAM statement, then by default, a histogram is created for each variable listed in the VAR statement.

For example, suppose a data set named Steel contains exactly two numeric variables named Length and Width. The following statements create two histograms, one for Length and one for Width:

```
proc univariate data=Steel;
   histogram;
run;
```

312 ✦ *Chapter 4: The UNIVARIATE Procedure*

Likewise, the following statements create histograms for Length and Width:

```
proc univariate data=Steel;
   var Length Width;
   histogram;
run;
```

The following statements create a histogram for Length only:

```
proc univariate data=Steel;
   var Length Width;
   histogram Length;
run;
```

options

add features to the histogram. Specify all *options* after the slash (/) in the HISTOGRAM statement. *Options* can be one of the following:

- primary options for fitted parametric distributions and kernel density estimates
- secondary options for fitted parametric distributions and kernel density estimates
- general options for graphics and output data sets

For example, in the following statements, the NORMAL option displays a fitted normal curve on the histogram, the MIDPOINTS= option specifies midpoints for the histogram, and the CTEXT= option specifies the color of the text:

```
proc univariate data=Steel;
   histogram Length / normal
                     midpoints = 5.6 5.8 6.0 6.2 6.4
                     ctext     = blue;
run;
```

Table 4.5 through Table 4.8 list the HISTOGRAM *options* by function. For complete descriptions, see the sections "Dictionary of Options" on page 319 and "Dictionary of Common Options" on page 384.

Parametric Density Estimation Options

Table 4.5 lists primary options that display parametric density estimates on the histogram. You can specify each primary option once in a given HISTOGRAM statement, and each primary option can display multiple curves from its family on the histogram.

Table 4.5 Primary Options for Parametric Fitted Distribution

Option	Description
BETA(*beta-options*)	fits beta distribution with threshold parameter θ, scale parameter σ, and shape parameters α and β

Table 4.5 (continued)

Option	Description
EXPONENTIAL(*exponential-options*)	fits exponential distribution with threshold parameter θ and scale parameter σ
GAMMA(*gamma-options*)	fits gamma distribution with threshold parameter θ, scale parameter σ, and shape parameter α
GUMBEL(*Gumbel-options*)	fits gumbel distribution with location parameter μ, and scale parameter σ
IGAUSS(*iGauss-options*)	fits inverse Gaussian distribution with location parameter μ, and shape parameter λ
LOGNORMAL(*lognormal-options*)	fits lognormal distribution with threshold parameter θ, scale parameter ζ, and shape parameter σ
NORMAL(*normal-options*)	fits normal distribution with mean μ and standard deviation σ
PARETO(*Pareto-options*)	fits generalized Pareto distribution with threshold parameter θ, scale parameter σ, and shape parameter α
POWER(*power-options*)	fits power function distribution with threshold parameter θ, scale parameter σ, and shape parameter α
RAYLEIGH(*Rayleigh-options*)	fits Rayleigh distribution with threshold parameter θ, and scale parameter σ
SB(S_B-options)	fits Johnson S_B distribution with threshold parameter θ, scale parameter σ, and shape parameters δ and γ
SU(S_U-options)	fits Johnson S_U distribution with threshold parameter θ, scale parameter σ, and shape parameters δ and γ
WEIBULL(*Weibull-options*)	fits Weibull distribution with threshold parameter θ, scale parameter σ, and shape parameter c

Table 4.6 lists secondary options that specify parameters for fitted parametric distributions and that control the display of fitted curves. Specify these secondary options in parentheses after the primary distribution option. For example, you can fit a normal curve by specifying the NORMAL option as follows:

```
proc univariate;
   histogram / normal(color=red mu=10 sigma=0.5);
run;
```

The COLOR= *normal-option* draws the curve in red, and the MU= and SIGMA= *normal-options* specify the parameters $\mu = 10$ and $\sigma = 0.5$ for the curve. Note that the sample mean and sample standard deviation are used to estimate μ and σ, respectively, when the MU= and SIGMA= *normal-options* are not specified.

You can specify lists of values for secondary options to display more than one fitted curve from the same distribution family on a histogram. Option values are matched by list position. You can specify the value *EST* in a list of distribution parameter values to use an estimate of the parameter.

For example, the following code displays two normal curves on a histogram:

```
proc univariate;
   histogram / normal(color=(red blue) mu=10 est sigma=0.5 est);
run;
```

The first curve is red, with $\mu = 10$ and $\sigma = 0.5$. The second curve is blue, with μ equal to the sample mean and σ equal to the sample standard deviation.

See the section "Formulas for Fitted Continuous Distributions" on page 416 for detailed information about the families of parametric distributions that you can fit with the HISTOGRAM statement.

Table 4.6   Secondary Options for Parametric Distributions

Option	Description
Options Used with All Distributions	
COLOR=	specifies colors of density curves
CONTENTS=	specifies table of contents entry for density curve grouping
FILL	fills area under density curve
L=	specifies line types of density curves
MIDPERCENTS	prints table of midpoints of histogram intervals
NOPRINT	suppresses tables summarizing curves
PERCENTS=	lists percents for which quantiles calculated from data and quantiles estimated from curves are tabulated
W=	specifies widths of density curves
Beta-Options	
ALPHA=	specifies first shape parameter α for beta curve
BETA=	specifies second shape parameter β for beta curve
SIGMA=	specifies scale parameter σ for beta curve
THETA=	specifies lower threshold parameter θ for beta curve
Exponential-Options	
SIGMA=	specifies scale parameter σ for exponential curve
THETA=	specifies threshold parameter θ for exponential curve
Gamma-Options	
ALPHA=	specifies shape parameter α for gamma curve
ALPHADELTA=	specifies change in successive estimates of α at which the Newton-Raphson approximation of $\hat{\alpha}$ terminates
ALPHAINITIAL=	specifies initial value for α in the Newton-Raphson approximation of $\hat{\alpha}$

Table 4.6 (continued)

Option	Description
MAXITER=	specifies maximum number of iterations in the Newton-Raphson approximation of $\hat{\alpha}$
SIGMA=	specifies scale parameter σ for gamma curve
THETA=	specifies threshold parameter θ for gamma curve
Gumbel-Options	
EDFNSAMPLES=	specifies number of samples for EDF goodness-of-fit simulation
EDFSEED=	specifies seed value for EDF goodness-of-fit simulation
MU=	specifies location parameter μ for gumbel curve
SIGMA=	specifies scale parameter σ for gumbel curve
IGauss-Options	
EDFNSAMPLES=	specifies number of samples for EDF goodness-of-fit simulation
EDFSEED=	specifies seed value for EDF goodness-of-fit simulation
LAMBDA=	specifies shape parameter λ for inverse Gaussian curve
MU=	specifies location parameter μ for inverse Gaussian curve
Lognormal-Options	
SIGMA=	specifies shape parameter σ for lognormal curve
THETA=	specifies threshold parameter θ for lognormal curve
ZETA=	specifies scale parameter ζ for lognormal curve
Normal-Options	
MU=	specifies mean μ for normal curve
SIGMA=	specifies standard deviation σ for normal curve
Pareto-Options	
EDFNSAMPLES=	specifies number of samples for EDF goodness-of-fit simulation
EDFSEED=	specifies seed value for EDF goodness-of-fit simulation
ALPHA=	specifies shape parameter α for generalized Pareto curve
SIGMA=	specifies scale parameter σ for generalized Pareto curve
THETA=	specifies threshold parameter θ for generalized Pareto curve
Power-Options	
ALPHA=	specifies shape parameter α for power function curve
SIGMA=	specifies scale parameter σ for power function curve
THETA=	specifies threshold parameter θ for power function curve
Rayleigh-Options	
EDFNSAMPLES=	specifies number of samples for EDF goodness-of-fit simulation
EDFSEED=	specifies seed value for EDF goodness-of-fit simulation
SIGMA=	specifies scale parameter σ for Rayleigh curve
THETA=	specifies threshold parameter θ for Rayleigh curve
Johnson S_B-Options	
DELTA=	specifies first shape parameter δ for Johnson S_B curve
FITINTERVAL=	specifies z-value for method of percentiles
FITMETHOD=	specifies method of parameter estimation
FITTOLERANCE=	specifies tolerance for method of percentiles
GAMMA=	specifies second shape parameter γ for Johnson S_B curve
SIGMA=	specifies scale parameter σ for Johnson S_B curve
THETA=	specifies lower threshold parameter θ for Johnson S_B curve

Table 4.6 (continued)

Option	Description
Johnson S_U-Options	
DELTA=	specifies first shape parameter δ for Johnson S_U curve
FITINTERVAL=	specifies z-value for method of percentiles
FITMETHOD=	specifies method of parameter estimation
FITTOLERANCE=	specifies tolerance for method of percentiles
GAMMA=	specifies second shape parameter γ for Johnson S_U curve
OPTBOUNDRANGE=	specifies the sampling range for parameter starting values in MLE optimization
OPTMAXITER=	specifies an iteration limit for MLE optimization
OPTMAXSTARTS=	specifies the maximum number of starting points to be used for MLE optimization
OPTPRINT	prints an iteration history for MLE optimization
OPTSEED=	specifies a seed value for MLE optimization
OPTTOLERANCE=	specifies the optimality tolerance for MLE optimization
SIGMA=	specifies scale parameter σ for Johnson S_U curve
THETA=	specifies lower threshold parameter θ for Johnson S_U curve
Weibull-Options	
C=	specifies shape parameter c for Weibull curve
ITPRINT	requests table of iteration history and optimizer details
MAXITER=	specifies maximum number of iterations in the Newton-Raphson approximation of $\hat{c}$
SIGMA=	specifies scale parameter σ for Weibull curve
THETA=	specifies threshold parameter θ for Weibull curve

Nonparametric Density Estimation Options

Use the *option* KERNEL(*kernel-options*) to compute kernel density estimates. Specify the following secondary options in parentheses after the KERNEL option to control features of density estimates requested with the KERNEL option.

Table 4.7 Kernel-Options

Option	Description
C=	specifies standardized bandwidth parameter c
COLOR=	specifies color of the kernel density curve
FILL	fills area under kernel density curve
K=	specifies type of kernel function
L=	specifies line type used for kernel density curve
LOWER=	specifies lower bound for kernel density curve
UPPER=	specifies upper bound for kernel density curve
W=	specifies line width for kernel density curve

General Options

Table 4.8 summarizes options for enhancing histograms.

Table 4.8 General Graphics Options

Option	Description
General Graphics Options	
BARLABEL=	produces labels above histogram bars
CLIPCURVES	scales vertical axis without considering fitted curves
ENDPOINTS=	lists endpoints for histogram intervals
GRID	creates a grid
HANGING	constructs hanging histogram
HREF=	specifies reference lines perpendicular to the horizontal axis
HREFLABELS=	specifies labels for HREF= lines
HREFLABPOS=	specifies vertical position of labels for HREF= lines
MIDPOINTS=	specifies midpoints for histogram intervals
NENDPOINTS=	specifies number of histogram interval endpoints
NMIDPOINTS=	specifies number of histogram interval midpoints
NOBARS	suppresses histogram bars
NOHLABEL	suppresses label for horizontal axis
NOPLOT	suppresses plot
NOVLABEL	suppresses label for vertical axis
NOVTICK	suppresses tick marks and tick mark labels for vertical axis
RTINCLUDE	includes right endpoint in interval
STATREF=	specifies reference lines at values of summary statistics
STATREFLABELS=	specifies labels for STATREF= lines
STATREFSUBCHAR=	specifies substitution character for displaying statistic values in STATREFLABELS= labels
VAXISLABEL=	specifies label for vertical axis
VREF=	specifies reference lines perpendicular to the vertical axis
VREFLABELS=	specifies labels for VREF= lines
VREFLABPOS=	specifies horizontal position of labels for VREF= lines
VSCALE=	specifies scale for vertical axis
Options for Traditional Graphics Output	
ANNOTATE=	specifies annotate data set
BARWIDTH=	specifies width for the bars
CAXIS=	specifies color for axis
CBARLINE=	specifies color for outlines of histogram bars
CFILL=	specifies color for filling under curve
CFRAME=	specifies color for frame
CGRID=	specifies color for grid lines
CHREF=	specifies colors for HREF= lines
CLIPREF	draws reference lines behind histogram bars
CSTATREF=	specifies colors for STATREF= lines
CTEXT=	specifies color for text
CVREF=	specifies colors for VREF= lines
DESCRIPTION=	specifies description for plot in graphics catalog

Table 4.8 *continued*

Option	Description
FONT=	specifies software font for text
FRONTREF	draws reference lines in front of histogram bars
HAXIS=	specifies AXIS statement for horizontal axis
HEIGHT=	specifies height of text used outside framed areas
HMINOR=	specifies number of horizontal minor tick marks
HOFFSET=	specifies offset for horizontal axis
INFONT=	specifies software font for text inside framed areas
INHEIGHT=	specifies height of text inside framed areas
INTERBAR=	specifies space between histogram bars
LGRID=	specifies a line type for grid lines
LHREF=	specifies line types for HREF= lines
LSTATREF=	specifies line types for STATREF= lines
LVREF=	specifies line types for VREF= lines
NAME=	specifies name for plot in graphics catalog
NOFRAME	suppresses frame around plotting area
PFILL=	specifies pattern for filling under curve
TURNVLABELS	turns and vertically strings out characters in labels for vertical axis
VAXIS=	specifies AXIS statement or values for vertical axis
VMINOR=	specifies number of vertical minor tick marks
VOFFSET=	specifies length of offset at upper end of vertical axis
WAXIS=	specifies line thickness for axes and frame
WBARLINE=	specifies line thickness for bar outlines
WGRID=	specifies line thickness for grid
Options for ODS Graphics Output	
NOCURVELEGEND	suppresses legend for curves
ODSFOOTNOTE=	specifies footnote displayed on histogram
ODSFOOTNOTE2=	specifies secondary footnote displayed on histogram
ODSTITLE=	specifies title displayed on histogram
ODSTITLE2=	specifies secondary title displayed on histogram
OVERLAY	overlays histograms for different class levels
Options for Comparative Plots	
ANNOKEY	applies annotation requested in ANNOTATE= data set to key cell only
CFRAMESIDE=	specifies color for filling frame for row labels
CFRAMETOP=	specifies color for filling frame for column labels
CPROP=	specifies color for proportion of frequency bar
CTEXTSIDE=	specifies color for row labels of comparative histograms
CTEXTTOP=	specifies color for column labels of comparative histograms
INTERTILE=	specifies distance between tiles
MAXNBIN=	specifies maximum number of bins to display
MAXSIGMAS=	limits the number of bins that display to within a specified number of standard deviations above and below mean of data in key cell
NCOLS=	specifies number of columns in comparative histogram
NROWS=	specifies number of rows in comparative histogram

HISTOGRAM Statement ✦ 319

Table 4.8 *continued*

Option	Description
Miscellaneous Options	
CONTENTS=	specifies table of contents entry for histogram grouping
MIDPERCENTS	creates table of histogram intervals
NOTABCONTENTS	suppresses table of contents entries for tables produced by HISTOGRAM statement
OUTHISTOGRAM=	creates a data set containing information about histogram intervals
OUTKERNEL=	creates a data set containing kernel density estimates

Dictionary of Options

The following entries provide detailed descriptions of options in the HISTOGRAM statement. Options marked with † are applicable only when traditional graphics are produced. See the section "Dictionary of Common Options" on page 384 for detailed descriptions of options common to all plot statements.

ALPHA=*value-list*

specifies the shape parameter α for fitted curves requested with the BETA, GAMMA, PARETO, and POWER options. Enclose the ALPHA= option in parentheses after the distribution keyword. By default, or if you specify the value *EST*, the procedure calculates a maximum likelihood estimate for α. You can specify A= as an alias for ALPHA= if you use it as a *beta-option*. You can specify SHAPE= as an alias for ALPHA= if you use it as a *gamma-option*.

BARLABEL=COUNT | PERCENT | PROPORTION

displays labels above the histogram bars. If you specify BARLABEL=COUNT, the label shows the number of observations associated with a given bar. If you specify BARLABEL=PERCENT, the label shows the percentage of observations represented by that bar. If you specify BARLABEL=PROPORTION, the label displays the proportion of observations associated with the bar.

† BARWIDTH=*value*

specifies the width of the histogram bars in percentage screen units. If both the BARWIDTH= and INTERBAR= options are specified, the INTERBAR= option takes precedence.

BETA < (*beta-options***) >**

displays fitted beta density curves on the histogram. The BETA option can occur only once in a HISTOGRAM statement, but it can request any number of beta curves. The beta distribution is bounded below by the parameter θ and above by the value $\theta + \sigma$. Use the THETA= and SIGMA= *beta-options* to specify these parameters. By default, THETA=0 and SIGMA=1. You can specify THETA=EST and SIGMA=EST to request maximum likelihood estimates for θ and σ.

The beta distribution has two shape parameters: α and β. If these parameters are known, you can specify their values with the ALPHA= and BETA= *beta-options*. By default, the procedure computes maximum likelihood estimates for α and β. **NOTE:** Three- and four-parameter maximum likelihood estimation may not always converge.

Table 4.6 lists secondary options you can specify with the BETA option. See the section "Beta Distribution" on page 416 for details and Example 4.21 for an example that uses the BETA option.

BETA=value-list

B=value-list

specifies the second shape parameter β for beta density curves requested with the BETA option. Enclose the BETA= option in parentheses after the BETA option. By default, or if you specify the value *EST*, the procedure calculates a maximum likelihood estimate for β.

C=value-list

specifies the shape parameter c for Weibull density curves requested with the WEIBULL option. Enclose the C= *Weibull-option* in parentheses after the WEIBULL option. By default, or if you specify the value *EST*, the procedure calculates a maximum likelihood estimate for c. You can specify the SHAPE= *Weibull-option* as an alias for the C= *Weibull-option*.

C=value-list

specifies the standardized bandwidth parameter c for kernel density estimates requested with the KERNEL option. Enclose the C= *kernel-option* in parentheses after the KERNEL option. You can specify a list of values to request multiple estimates. You can specify the value *MISE* to produce the estimate with a bandwidth that minimizes the approximate mean integrated square error (MISE), or *SJPI* to select the bandwidth by using the Sheather-Jones plug-in method.

You can also use the C= *kernel-option* with the K= *kernel-option* (which specifies the kernel function) to compute multiple estimates. If you specify more kernel functions than bandwidths, the last bandwidth in the list is repeated for the remaining estimates. Similarly, if you specify more bandwidths than kernel functions, the last kernel function is repeated for the remaining estimates. If you do not specify the C= *kernel-option*, the bandwidth that minimizes the approximate MISE is used for all the estimates.

See the section "Kernel Density Estimates" on page 432 for more information about kernel density estimates.

† CBARLINE=color

specifies the color for the outline of the histogram bars when producing traditional graphics. The option does not apply to ODS Graphics output.

† CFILL=color

specifies the color to fill the bars of the histogram (or the area under a fitted density curve if you also specify the FILL option) when producing traditional graphics. See the entries for the FILL and PFILL= options for additional details. See *SAS/GRAPH: Help* for a list of colors. The option does not apply to ODS Graphics output.

† CGRID=color

specifies the color for grid lines when a grid displays on the histogram in traditional graphics. This option also produces a grid if the GRID= option is not specified.

CLIPCURVES

scales the vertical axis without taking fitted curves into consideration. Curves that extend above the tallest histogram bar can be clipped. You can use this option to avoid compression of the histogram bars due to extremely high fitted curve peaks.

† CLIPREF

draws reference lines requested with the HREF= and VREF= options behind the histogram bars. When the GSTYLE system option is in effect for traditional graphics, reference lines are drawn in front of the bars by default.

CONTENTS=
> specifies the table of contents grouping entry for tables associated with a density curve. Enclose the CONTENTS= option in parentheses after the distribution option. You can specify CONTENTS='' to suppress the grouping entry.

DELTA=value-list
> specifies the first shape parameter δ for Johnson S_B and Johnson S_U distribution functions requested with the SB and SU options. Enclose the DELTA= option in parentheses after the SB or SU option. If you do not specify a value for δ, or if you specify the value *EST*, the procedure calculates an estimate.

EDFNSAMPLES=value
> specifies the number of simulation samples used to compute *p*-values for EDF goodness-of-fit statistics for density curves requested with the GUMBEL, IGAUSS, PARETO, and RAYLEIGH options. Enclose the EDFNSAMPLES= option in parentheses after the distribution option. The default value is 500.

EDFSEED=value
> specifies an integer value used to start the pseudo-random number generator when creating simulation samples for computing EDF goodness-of-fit statistic *p*-values for density curves requested with the GUMBEL, IGAUSS, PARETO, and RAYLEIGH options. Enclose the EDFSEED= option in parentheses after the distribution option. By default, the procedure uses a random number seed generated from reading the time of day from the computer's clock.

ENDPOINTS < =values | KEY | UNIFORM >
> uses histogram bin endpoints as the tick mark values for the horizontal axis and determines how to compute the bin width of the histogram bars. The *values* specify both the left and right endpoint of each histogram interval. The width of the histogram bars is the difference between consecutive endpoints. The procedure uses the same values for all variables.
>
> The range of endpoints must cover the range of the data. For example, if you specify
>
> ```
> endpoints=2 to 10 by 2
> ```
>
> then all of the observations must fall in the intervals [2,4) [4,6) [6,8) [8,10]. You also must use evenly spaced endpoints which you list in increasing order.
>
> | | |
> |---|---|
> | KEY | determines the endpoints for the data in the key cell. The initial number of endpoints is based on the number of observations in the key cell by using the method of Terrell and Scott (1985). The procedure extends the endpoint list for the key cell in either direction as necessary until it spans the data in the remaining cells. |
> | UNIFORM | determines the endpoints by using all the observations as if there were no cells. In other words, the number of endpoints is based on the total sample size by using the method of Terrell and Scott (1985). |
>
> Neither KEY nor UNIFORM apply unless you use the CLASS statement.
>
> If you omit ENDPOINTS, the procedure uses the histogram midpoints as horizontal axis tick values. If you specify ENDPOINTS, the procedure computes the endpoints by using an algorithm (Terrell and Scott 1985) that is primarily applicable to continuous data that are approximately normally distributed.
>
> If you specify both MIDPOINTS= and ENDPOINTS, the procedure issues a warning message and uses the endpoints.

If you specify RTINCLUDE, the procedure includes the right endpoint of each histogram interval in that interval instead of including the left endpoint.

If you use a CLASS statement and specify ENDPOINTS, the procedure uses ENDPOINTS=KEY as the default. However if the key cell is empty, then the procedure uses ENDPOINTS=UNIFORM.

EXPONENTIAL < (*exponential-options*) >

EXP < (*exponential-options*) >

displays fitted exponential density curves on the histogram. The EXPONENTIAL option can occur only once in a HISTOGRAM statement, but it can request any number of exponential curves. The parameter θ must be less than or equal to the minimum data value. Use the THETA= *exponential-option* to specify θ. By default, THETA=0. You can specify THETA=EST to request the maximum likelihood estimate for θ. Use the SIGMA= *exponential-option* to specify σ. By default, the procedure computes a maximum likelihood estimate for σ. Table 4.6 lists options you can specify with the EXPONENTIAL option. See the section "Exponential Distribution" on page 417 for details.

FILL

fills areas under the fitted density curve or the kernel density estimate with colors and patterns. The FILL option can occur with only one fitted curve. Enclose the FILL option in parentheses after a density curve option or the KERNEL option. The CFILL= and PFILL= options specify the color and pattern for the area under the curve when producing traditional graphics. For a list of available colors and patterns, see *SAS/GRAPH: Help*.

† FRONTREF

draws reference lines requested with the HREF= and VREF= options in front of the histogram bars. When the NOGSTYLE system option is in effect for traditional graphics, reference lines are drawn behind the histogram bars by default, and they can be obscured by filled bars.

GAMMA < (*gamma-options*) >

displays fitted gamma density curves on the histogram. The GAMMA option can occur only once in a HISTOGRAM statement, but it can request any number of gamma curves. The parameter θ must be less than the minimum data value. Use the THETA= *gamma-option* to specify θ. By default, THETA=0. You can specify THETA=EST to request the maximum likelihood estimate for θ. Use the ALPHA= and the SIGMA= *gamma-options* to specify the shape parameter α and the scale parameter σ. By default, PROC UNIVARIATE computes maximum likelihood estimates for α and σ. The procedure calculates the maximum likelihood estimate of α iteratively by using the Newton-Raphson approximation. Table 4.6 lists options you can specify with the GAMMA option. See the section "Gamma Distribution" on page 418 for details, and see Example 4.22 for an example that uses the GAMMA option.

GAMMA=*value-list*

specifies the second shape parameter γ for Johnson S_B and Johnson S_U distribution functions requested with the SB and SU options. Enclose the GAMMA= option in parentheses after the SB or SU option. If you do not specify a value for γ, or if you specify the value *EST*, the procedure calculates an estimate.

GRID

displays a grid on the histogram. Grid lines are horizontal lines that are positioned at major tick marks on the vertical axis.

GUMBEL < (*Gumbel-options*) >

> displays fitted Gumbel density curves on the histogram. The GUMBEL option can occur only once in a HISTOGRAM statement, but it can request any number of Gumbel curves. Use the MU= and the SIGMA= *Gumbel-options* to specify the location parameter μ and the scale parameter σ. By default, PROC UNIVARIATE computes maximum likelihood estimates for μ and σ. Table 4.6 lists options you can specify with the GUMBEL option. See the section "Gumbel Distribution" on page 419 for details about the Gumbel distribution.

HANGING

HANG

> requests a hanging histogram, as illustrated in Figure 4.7.

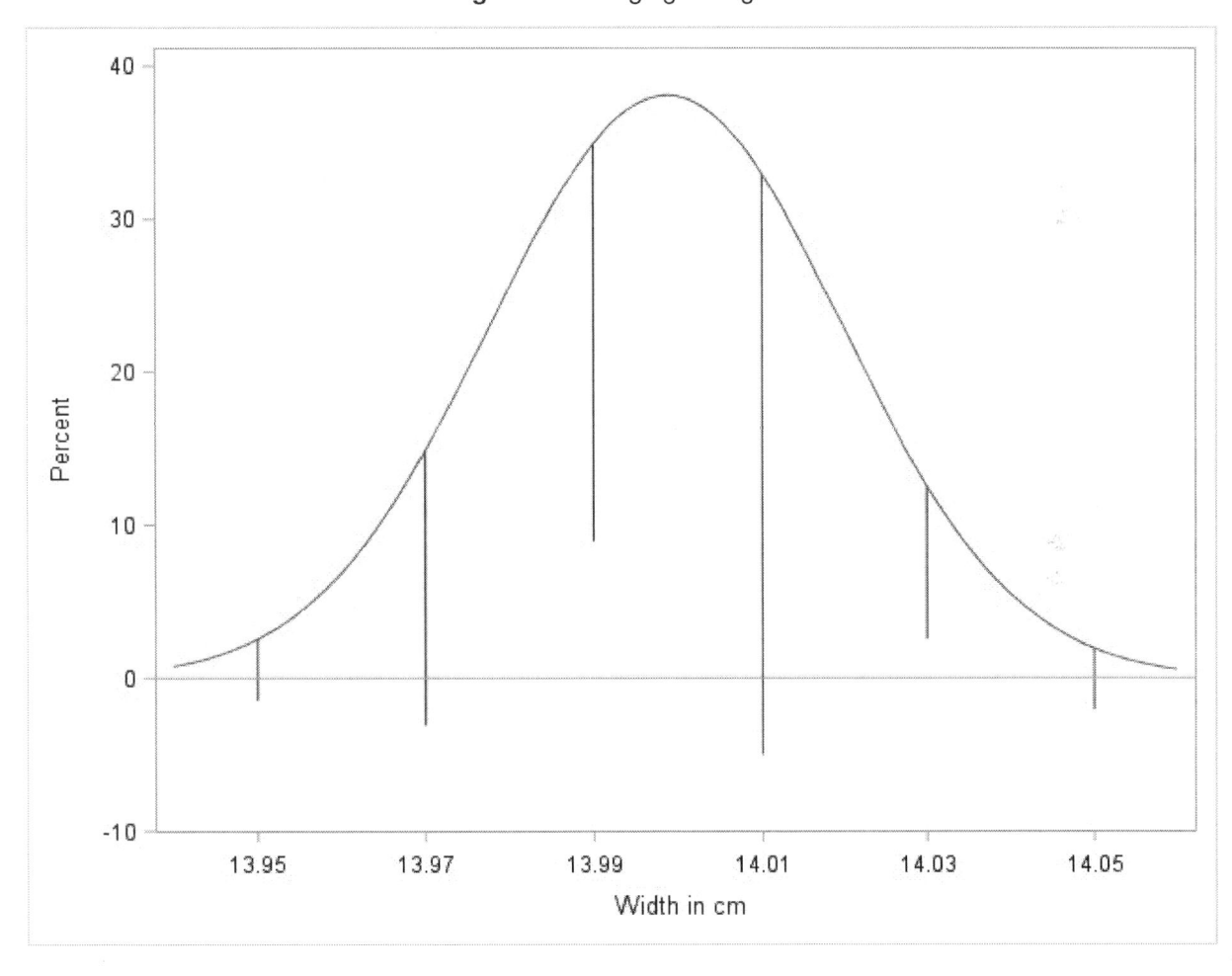

Figure 4.7 Hanging Histogram

You can use the HANGING option only when exactly one fitted density curve is requested. A hanging histogram aligns the tops of the histogram bars (displayed as lines) with the fitted curve. The lines are positioned at the midpoints of the histogram bins. A hanging histogram is a goodness-of-fit diagnostic in the sense that the closer the lines are to the horizontal axis, the better the fit. Hanging histograms are discussed by Tukey (1977), Wainer (1974), and Velleman and Hoaglin (1981).

† HOFFSET=*value*

specifies the offset, in percentage screen units, at both ends of the horizontal axis. You can use HOFFSET=0 to eliminate the default offset.

IGAUSS <(*iGauss-options*)>

displays fitted inverse Gaussian density curves on the histogram. The IGAUSS option can occur only once in a HISTOGRAM statement, but it can request any number of inverse Gaussian curves. Use the MU= and the LAMBDA= *iGauss-options* to specify the location parameter μ and the shape parameter λ. By default, PROC UNIVARIATE uses the sample mean for μ and computes a maximum likelihood estimate for λ. Table 4.6 lists options you can specify with the IGAUSS option. See the section "Inverse Gaussian Distribution" on page 419 for details.

† INTERBAR=*value*

specifies the space between histogram bars in percentage screen units. If both the INTERBAR= and BARWIDTH= options are specified, the INTERBAR= option takes precedence.

K=NORMAL | QUADRATIC | TRIANGULAR

specifies the kernel function (normal, quadratic, or triangular) used to compute a kernel density estimate. You can specify a list of values to request multiple estimates. You must enclose this option in parentheses after the KERNEL option. You can also use the K= *kernel-option* with the C= *kernel-option*, which specifies standardized bandwidths. If you specify more kernel functions than bandwidths, the procedure repeats the last bandwidth in the list for the remaining estimates. Similarly, if you specify more bandwidths than kernel functions, the procedure repeats the last kernel function for the remaining estimates. By default, K=NORMAL.

KERNEL<(*kernel-options*)>

superimposes kernel density estimates on the histogram. By default, the procedure uses the AMISE method to compute kernel density estimates. To request multiple kernel density estimates on the same histogram, specify a list of values for the C= *kernel-option* or K= *kernel-option*. Table 4.7 lists options you can specify with the KERNEL option. See the section "Kernel Density Estimates" on page 432 for more information about kernel density estimates, and see Example 4.23.

LAMBDA=*value*

specifies the shape parameter λ for fitted curves requested with the IGAUSS option. Enclose the LAMBDA= option in parentheses after the IGAUSS distribution keyword. If you do not specify a value for λ, the procedure calculates a maximum likelihood estimate.

† LGRID=*linetype*

specifies the line type for the grid when a grid displays on the histogram. This option also creates a grid if the GRID option is not specified.

LOGNORMAL<(*lognormal-options*)>

displays fitted lognormal density curves on the histogram. The LOGNORMAL option can occur only once in a HISTOGRAM statement, but it can request any number of lognormal curves. The parameter θ must be less than the minimum data value. Use the THETA= *lognormal-option* to specify θ. By default, THETA=0. You can specify THETA=EST to request the maximum likelihood estimate for θ. Use the SIGMA= and ZETA= *lognormal-options* to specify σ and ζ. By default, the procedure computes maximum likelihood estimates for σ and ζ. Table 4.6 lists options you can specify with the LOGNORMAL option. See the section "Lognormal Distribution" on page 420 for details, and see Example 4.22 and Example 4.24 for examples using the LOGNORMAL option.

LOWER=*value-list*

specifies lower bounds for kernel density estimates requested with the KERNEL option. Enclose the LOWER= option in parentheses after the KERNEL option. If you specify more kernel estimates than lower bounds, the last lower bound is repeated for the remaining estimates. The default is a missing value, indicating no lower bounds for fitted kernel density curves.

MAXNBIN=*n*

limits the number of bins displayed in the comparative histogram. This option is useful when the scales or ranges of the data distributions differ greatly from cell to cell. By default, the bin size and midpoints are determined for the key cell, and then the midpoint list is extended to accommodate the data ranges for the remaining cells. However, if the cell scales differ considerably, the resulting number of bins can be so great that each cell histogram is scaled into a narrow region. By using MAXNBIN= to limit the number of bins, you can narrow the window about the data distribution in the key cell. This option is not available unless you specify the CLASS statement. The MAXNBIN= option is an alternative to the MAXSIGMAS= option.

MAXSIGMAS=*value*

limits the number of bins displayed in the comparative histogram to a range of *value* standard deviations (of the data in the key cell) above and below the mean of the data in the key cell. This option is useful when the scales or ranges of the data distributions differ greatly from cell to cell. By default, the bin size and midpoints are determined for the key cell, and then the midpoint list is extended to accommodate the data ranges for the remaining cells. However, if the cell scales differ considerably, the resulting number of bins can be so great that each cell histogram is scaled into a narrow region. By using MAXSIGMAS= to limit the number of bins, you can narrow the window that surrounds the data distribution in the key cell. This option is not available unless you specify the CLASS statement.

MIDPERCENTS

requests a table listing the midpoints and percentage of observations in each histogram interval. If you specify MIDPERCENTS in parentheses after a density estimate option, the procedure displays a table that lists the midpoints, the observed percentage of observations, and the estimated percentage of the population in each interval (estimated from the fitted distribution). See Example 4.18.

MIDPOINTS=*values* | **KEY** | **UNIFORM**

specifies how to determine the midpoints for the histogram intervals, where *values* determines the width of the histogram bars as the difference between consecutive midpoints. The procedure uses the same values for all variables.

The range of midpoints, extended at each end by half of the bar width, must cover the range of the data. For example, if you specify

```
midpoints=2 to 10 by 0.5
```

then all of the observations should fall between 1.75 and 10.25. You must use evenly spaced midpoints listed in increasing order.

    KEY   determines the midpoints for the data in the key cell. The initial number of midpoints is based on the number of observations in the key cell that use the method of Terrell and Scott (1985). The procedure extends the midpoint list for the key cell in either direction as necessary until it spans the data in the remaining cells.

UNIFORM
: determines the midpoints by using all the observations as if there were no cells. In other words, the number of midpoints is based on the total sample size by using the method of Terrell and Scott (1985).

Neither KEY nor UNIFORM apply unless you use the CLASS statement. By default, if you use a CLASS statement, MIDPOINTS=KEY; however, if the key cell is empty then MIDPOINTS=UNIFORM. Otherwise, the procedure computes the midpoints by using an algorithm (Terrell and Scott 1985) that is primarily applicable to continuous data that are approximately normally distributed.

MU= *value-list*
: specifies the parameter μ for Gumbel, inverse Gaussian, and normal density curves requested with the GUMBEL, IGAUSS, and NORMAL options, respectively. Enclose the MU= option in parentheses after the distribution keyword. By default, or if you specify the value *EST*, the procedure uses the sample mean for μ for normal and inverse Gaussian distributions and computes a maximum likelihood estimate of μ for the Gumbel distribution. For more detail please see "Inverse Gaussian Distribution" on page 419 and "Gumbel Distribution" on page 419.

NENDPOINTS= *n*
: uses histogram interval endpoints as the tick mark values for the horizontal axis and determines the number of bins.

NMIDPOINTS= *n*
: specifies the number of histogram intervals.

NOBARS
: suppresses drawing of histogram bars, which is useful for viewing fitted curves only.

NOCURVELEGEND

NOLEGEND
: suppresses the legend for fitted curves. The NOCURVELEGEND option applies only to ODS Graphics output.

NOPLOT

NOCHART
: suppresses the creation of a plot. Use this option when you only want to tabulate summary statistics for a fitted density or create an OUTHISTOGRAM= data set.

NOPRINT
: suppresses tables summarizing the fitted curve. Enclose the NOPRINT option in parentheses following the distribution option.

NORMAL< *(normal-options)* **>**
: displays fitted normal density curves on the histogram. The NORMAL option can occur only once in a HISTOGRAM statement, but it can request any number of normal curves. Use the MU= and SIGMA= *normal-options* to specify μ and σ. By default, the procedure uses the sample mean and sample standard deviation for μ and σ. Table 4.6 lists options you can specify with the NORMAL option. See the section "Normal Distribution" on page 421 for details, and see Example 4.19 for an example that uses the NORMAL option.

NOTABCONTENTS
> suppresses the table of contents entries for tables produced by the HISTOGRAM statement.

OPTBOUNDRANGE=value
> defines the sampling range for each parameter during maximum likelihood estimation for the Johnson S_U distribution. PROC UNIVARIATE computes initial estimates for each parameter by using the method of percentiles. The *value* determines the range of parameter values around the initial estimate that can be sampled for local optimization starting values. The default is 100.

OPTMAXITER=value
> limits the number of iterations that are used by the optimizer in maximum likelihood estimation for the Johnson S_U distribution. The default is 500.

OPTMAXSTARTS=N
> defines the maximum number of starting points to be used for local optimization in maximum likelihood estimation for the Johnson S_U distribution. That is, no more than *N* local optimizations are used in the multistart algorithm. The default value is 100.

OPTPRINT
> prints the iteration history for the Johnson S_U distribution maximum likelihood estimation.

OPTSEED=value
> specifies a positive integer seed for generating random number sequences in Johnson S_U distribution maximum likelihood estimation. You can use this option to replicate results from different runs.

OPTTOLERANCE=value
> specifies the tolerance for declaring optimality in maximum likelihood estimation for the Johnson S_U distribution. The default value is 1E–8.

OUTHISTOGRAM=SAS-data-set

OUTHIST=SAS-data-set
> creates a SAS data set that contains information about histogram intervals. Specifically, the data set contains the midpoints of the histogram intervals (or the lower endpoints of the intervals if you specify the ENDPOINTS option), the observed percentage of observations in each interval, and the estimated percentage of observations in each interval (estimated from each of the specified fitted curves).

OUTKERNEL=SAS-data-set
> creates a SAS data set that contains information about kernel density estimates.

PARETO < (*Pareto-options*) >
> displays fitted generalized Pareto density curves on the histogram. The PARETO option can occur only once in a HISTOGRAM statement, but it can request any number of generalized Pareto curves. The parameter θ must be less than the minimum data value. Use the THETA= *Pareto-option* to specify θ. By default, THETA=0. Use the SIGMA= and the ALPHA= *Pareto-options* to specify the scale parameter σ and the shape parameter α. By default, PROC UNIVARIATE computes maximum likelihood estimates for σ and α. Table 4.6 lists options you can specify with the PARETO option. See the section "Generalized Pareto Distribution" on page 421 for details.

PERCENTS=_values_

PERCENT=_values_

specifies a list of percents for which quantiles calculated from the data and quantiles estimated from the fitted curve are tabulated. The percents must be between 0 and 100. Enclose the PERCENTS= option in parentheses after the curve option. The default percents are 1, 5, 10, 25, 50, 75, 90, 95, and 99.

† **PFILL=**_pattern_

specifies a pattern used to fill the bars of the histograms (or the areas under a fitted curve if you also specify the FILL option) when producing traditional graphics. See the entries for the CFILL= and FILL options for additional details. Refer to *SAS/GRAPH: Help* for a list of pattern values. The option does not apply to ODS Graphics output.

POWER < (_power-options_) >

displays fitted power function density curves on the histogram. The POWER option can occur only once in a HISTOGRAM statement, but it can request any number of power function curves. The parameter θ must be less than the minimum data value. Use the THETA= and SIGMA= _power-options_ to specify θ and σ. The default values are 0 and 1, respectively. Use the ALPHA= _power-option_ to specify the and the shape parameter, α. By default, PROC UNIVARIATE computes a maximum likelihood estimate for α. Table 4.6 lists options you can specify with the POWER option. See the section "Power Function Distribution" on page 422 for details.

RAYLEIGH < (_Rayleigh-options_) >

displays fitted Rayleigh density curves on the histogram. The RAYLEIGH option can occur only once in a HISTOGRAM statement, but it can request any number of Rayleigh curves. The parameter θ must be less than the minimum data value. Use the THETA= _Rayleigh-option_ to specify θ. By default, THETA=0. Use the SIGMA= _Rayleigh-option_ to specify the scale parameter σ. By default, PROC UNIVARIATE computes maximum likelihood estimates for σ. Table 4.6 lists options you can specify with the RAYLEIGH option. See the section "Rayleigh Distribution" on page 423 for details.

RTINCLUDE

includes the right endpoint of each histogram interval in that interval. By default, the left endpoint is included in the histogram interval.

SB< (_S_B-options_) >

displays fitted Johnson S_B density curves on the histogram. The SB option can occur only once in a HISTOGRAM statement, but it can request any number of Johnson S_B curves. Use the THETA= and SIGMA= _normal-options_ to specify θ and σ. By default, the procedure computes maximum likelihood estimates of θ and σ. Table 4.6 lists options you can specify with the SB option. See the section "Johnson S_B Distribution" on page 424 for details.

SIGMA=_value-list_

specifies the parameter σ for the fitted density curve when you request the BETA, EXPONENTIAL, GAMMA, GUMBEL, LOGNORMAL, NORMAL, PARETO, POWER, RAYLEIGH, SB, SU, or WEIBULL options.

See Table 4.9 for a summary of how to use the SIGMA= option. You must enclose this option in parentheses after the density curve option. You can specify the value *EST* to request a maximum likelihood estimate for σ.

Table 4.9 Uses of the SIGMA= Option

Distribution Keyword	SIGMA= Specifies	Default Value	Alias
BETA	scale parameter σ	1	SCALE=
EXPONENTIAL	scale parameter σ	maximum likelihood estimate	SCALE=
GAMMA	scale parameter σ	maximum likelihood estimate	SCALE=
GUMBEL	scale parameter σ	maximum likelihood estimate	
LOGNORMAL	shape parameter σ	maximum likelihood estimate	SHAPE=
NORMAL	scale parameter σ	standard deviation	
PARETO	scale parameter σ	1	
POWER	scale parameter σ	maximum likelihood estimate	
RAYLEIGH	scale parameter σ	maximum likelihood estimate	
SB	scale parameter σ	1	SCALE=
SU	scale parameter σ	percentile-based estimate	
WEIBULL	scale parameter σ	maximum likelihood estimate	SCALE=

SU< (S_U -options) >

displays fitted Johnson S_U density curves on the histogram. The SU option can occur only once in a HISTOGRAM statement, but it can request any number of Johnson S_U curves. Use the THETA= and SIGMA= *normal-options* to specify θ and σ. By default, the procedure computes maximum likelihood estimates of θ and σ. Table 4.6 lists options you can specify with the SU option. See the section "Johnson S_U Distribution" on page 426 for details.

THETA= *value-list*

THRESHOLD= *value-list*

specifies the lower threshold parameter θ for curves requested with the BETA, EXPONENTIAL, GAMMA, LOGNORMAL, PARETO, POWER, RAYLEIGH, SB, SU, and WEIBULL options. Enclose the THETA= option in parentheses after the curve option. By default, THETA=0. If you specify the value *EST*, an estimate is computed for θ.

UPPER= *value-list*

specifies upper bounds for kernel density estimates requested with the KERNEL option. Enclose the UPPER= option in parentheses after the KERNEL option. If you specify more kernel estimates than upper bounds, the last upper bound is repeated for the remaining estimates. The default is a missing value, indicating no upper bounds for fitted kernel density curves.

† VOFFSET= *value*

specifies the offset, in percentage screen units, at the upper end of the vertical axis.

VSCALE=COUNT | PERCENT | PROPORTION

specifies the scale of the vertical axis for a histogram. The value COUNT requests the data be scaled in units of the number of observations per data unit. The value PERCENT requests the data be scaled in units of percent of observations per data unit. The value PROPORTION requests the data be scaled in units of proportion of observations per data unit. The default is PERCENT.

† WBARLINE=n

specifies the width of bar outlines when producing traditional graphics. The option does not apply to ODS Graphics output.

WEIBULL<(Weibull-options)>

displays fitted Weibull density curves on the histogram. The WEIBULL option can occur only once in a HISTOGRAM statement, but it can request any number of Weibull curves. The parameter θ must be less than the minimum data value. Use the THETA= *Weibull-option* to specify θ. By default, THETA=0. You can specify THETA=EST to request the maximum likelihood estimate for θ. Use the C= and SIGMA= *Weibull-options* to specify the shape parameter c and the scale parameter σ. By default, the procedure computes the maximum likelihood estimates for c and σ. Table 4.6 lists options you can specify with the WEIBULL option. See the section "Weibull Distribution" on page 427 for details, and see Example 4.22 for an example that uses the WEIBULL option.

PROC UNIVARIATE calculates the maximum likelihood estimate of σ iteratively by using the Newton-Raphson approximation. See also the C=, SIGMA=, and THETA= *Weibull-options*.

† WGRID=n

specifies the line thickness for the grid when producing traditional graphics. The option does not apply to ODS Graphics output.

ZETA= value-list

specifies a value for the scale parameter ζ for lognormal density curves requested with the LOGNORMAL option. Enclose the ZETA= *lognormal-option* in parentheses after the LOGNORMAL option. By default, or if you specify the value *EST*, the procedure calculates a maximum likelihood estimate for ζ. You can specify the SCALE= option as an alias for the ZETA= option.

ID Statement

ID *variables* ;

The ID statement specifies one or more variables to include in the table of extreme observations. The corresponding values of the ID variables appear beside the n largest and n smallest observations, where n is the value of the NEXTROBS= option. See Example 4.3.

You can also include ID variables in the output data set created by an OUTPUT statement by specifying the IDOUT option in the PROC UNIVARIATE statement.

INSET Statement

INSET *keywords* < / *options* > ;

An INSET statement places a box or table of summary statistics, called an *inset*, directly in a graph created with a CDFPLOT, HISTOGRAM, PPPLOT, PROBPLOT, or QQPLOT statement. The INSET statement must follow the plot statement that creates the plot that you want to augment. The inset appears in all the graphs that the preceding plot statement produces.

You can use multiple INSET statements after a plot statement to add more than one inset to a plot. See Example 4.17.

In an INSET statement, you specify one or more keywords that identify the information to display in the inset. The information is displayed in the order that you request the keywords. Keywords can be any of the following:

- statistical keywords
- primary keywords
- secondary keywords

Statistical Keywords

The available statistical keywords are listed in Table 4.10.

Table 4.10 Statistical Keywords

Keyword	Description
Descriptive Statistic Keywords	
CSS	Corrected sum of squares
CV	Coefficient of variation
GEOMEAN	Geometric mean
KURTOSIS | KURT	Kurtosis
MAX	Largest value
MEAN	Sample mean
MIN	Smallest value
MODE	Most frequent value
N	Sample size
NEXCL	Number of observations excluded by MAXNBIN= or MAXSIGMAS= option
NMISS	Number of missing values
NOBS	Number of observations
RANGE	Range
SKEWNESS | SKEW	Skewness
STD | STDDEV	Standard deviation
STDMEAN | STDERR	Standard error of the mean
SUM	Sum of the observations
SUMWGT	Sum of the weights
USS	Uncorrected sum of squares
VAR	Variance
Percentile Statistic Keywords	
P1	1st percentile
P5	5th percentile
P10	10th percentile
Q1	
P25	Lower quartile (25th percentile)
MEDIAN	
Q2	
P50	Median (50th percentile)

Table 4.10 *continued*

Keyword	Description
Q3	
P75	Upper quartile (75th percentile)
P90	90th percentile
P95	95th percentile
P99	99th percentile
QRANGE	Interquartile range (Q3–Q1)
Keywords for Distribution-Free Confidence Limits for Percentiles (CIPCTLDF Option)	
P1_LCL_DF	1st percentile lower confidence limit
P1_UCL_DF	1st percentile upper confidence limit
P5_LCL_DF	5th percentile lower confidence limit
P5_UCL_DF	5th percentile upper confidence limit
P10_LCL_DF	10th percentile lower confidence limit
P10_UCL_DF	10th percentile upper confidence limit
Q1_LCL_DF	
P25_LCL_DF	Lower quartile (25th percentile) lower confidence limit
Q1_UCL_DF	
P25_UCL_DF	Lower quartile (25th percentile) upper confidence limit
MEDIAN_LCL_DF	
Q2_LCL_DF	
P50_LCL_DF	Median (50th percentile) lower confidence limit
MEDIAN_UCL_DF	
Q2_UCL_DF	
P50_UCL_DF	Median (50th percentile) upper confidence limit
Q3_LCL_DF	
P75_LCL_DF	Upper quartile (75th percentile) lower confidence limit
Q3_UCL_DF	
P75_UCL_DF	Upper quartile (75th percentile) upper confidence limit
P90_LCL_DF	90th percentile lower confidence limit
P90_UCL_DF	90th percentile upper confidence limit
P95_LCL_DF	95th percentile lower confidence limit
P95_UCL_DF	95th percentile upper confidence limit
P99_LCL_DF	99th percentile lower confidence limit
P99_UCL_DF	99th percentile upper confidence limit
Keywords Percentile Confidence Limits Assuming Normality (CIPCTLNORMAL Option)	
P1_LCL	1st percentile lower confidence limit
P1_UCL	1st percentile upper confidence limit
P5_LCL	5th percentile lower confidence limit
P5_UCL	5th percentile upper confidence limit
P10_LCL	10th percentile lower confidence limit
P10_UCL	10th percentile upper confidence limit
Q1_LCL	
P25_LCL	Lower quartile (25th percentile) lower confidence limit

Table 4.10 *continued*

Keyword	Description
Q1_UCL	
P25_UCL	Lower quartile (25th percentile) upper confidence limit
MEDIAN_LCL	
Q2_LCL	
P50_LCL	Median (50th percentile) lower confidence limit
MEDIAN_UCL	
Q2_UCL	
P50_UCL	Median (50th percentile) upper confidence limit
Q3_LCL	
P75_LCL	Upper quartile (75th percentile) lower confidence limit
Q3_UCL	
P75_UCL	Upper quartile (75th percentile) upper confidence limit
P90_LCL	90th percentile lower confidence limit
P90_UCL	90th percentile upper confidence limit
P95_LCL	95th percentile lower confidence limit
P95_UCL	95th percentile upper confidence limit
P99_LCL	99th percentile lower confidence limit
P99_UCL	99th percentile upper confidence limit

Robust Statistics Keywords

Keyword	Description
GINI	Gini's mean difference
MAD	Median absolute difference about the median
QN	Q_n, alternative to MAD
SN	S_n, alternative to MAD
STD_GINI	Gini's standard deviation
STD_MAD	MAD standard deviation
STD_QN	Q_n standard deviation
STD_QRANGE	Interquartile range standard deviation
STD_SN	S_n standard deviation

Hypothesis Testing Keywords

Keyword	Description
MSIGN	Sign statistic
NORMALTEST	Test statistic for normality
PNORMAL	Probability value for the test of normality
SIGNRANK	Signed rank statistic
PROBM	Probability of greater absolute value for the sign statistic
PROBN	Probability value for the test of normality
PROBS	Probability value for the signed rank test
PROBT	Probability value for the Student's t test
T	Statistics for Student's t test

Keyword for Reading an Input Data Set

Keyword	Description
DATA=	(label, value) pairs from input data set

334 ♦ *Chapter 4: The UNIVARIATE Procedure*

To create a completely customized inset, use a DATA= data set.

DATA=SAS-data-set
> requests that PROC UNIVARIATE display customized statistics from a SAS data set in the inset table. The data set must contain two variables:
>
> _LABEL_ is a character variable whose values provide labels for inset entries.
>
> _VALUE_ is a variable that is either character or numeric and whose values provide values for inset entries.
>
> The label and value from each observation in the data set occupy one line in the inset. The position of the DATA= keyword in the keyword list determines the position of its lines in the inset.

Primary and Secondary Keywords

A primary keyword specifies a fitted distribution, which is one of the parametric distributions or a kernel density estimate. You specify secondary keywords in parentheses after the primary keyword to request particular statistics associated with that distribution.

NOTE: When producing traditional graphics output, you can specify a primary keyword without secondary keywords to display a colored line and the distribution name as a key for the density curve.

In the HISTOGRAM statement you can request more than one fitted distribution from the same family (for example, two normal distributions). You can display inset statistics for individual curves by specifying the curve indices in square brackets immediately following the primary keyword.

The following statements produce a histogram with three fitted normal curves and an inset that contains goodness-of-fit statistics for the second curve only:

```
proc univariate data=score;
   histogram final / normal(sigma=1 2 3);
   inset normal[2](ad adpval);
run;
```

Table 4.11 lists the primary keywords and the plot statements with which they can be specified.

Table 4.11 Primary Keywords

Keyword	Distribution	Plot Statement Availability
BETA	Beta	All plot statements
EXPONENTIAL	Exponential	All plot statements
GAMMA	Gamma	All plot statements
GUMBEL	Gumbel	All plot statements
IGAUSS	Inverse Gaussian	CDFPLOT, HISTOGRAM, PPPLOT
KERNEL	Kernel density estimate	HISTOGRAM
LOGNORMAL	Lognormal	All plot statements
NORMAL	Normal	All plot statements
PARETO	Pareto	All plot statements
POWER	Power function	All plot statements
RAYLEIGH	Rayleigh	All plot statements

Table 4.11 *continued*

Keyword	Distribution	Plot Statement Availability
SB	Johnson S_B	HISTOGRAM
SU	Johnson S_U	HISTOGRAM
WEIBULL	Weibull(3-parameter)	All plot statements
WEIBULL2	Weibull(2-parameter)	PROBPLOT, QQPLOT

Table 4.12 lists the secondary keywords available with the primary keywords listed in Table 4.11.

Table 4.12 Secondary Keywords

Secondary Keyword	Alias	Description
BETA Secondary Keywords		
ALPHA	SHAPE1	First shape parameter α
BETA	SHAPE2	Second shape parameter β
MEAN		Mean of the fitted distribution
SIGMA	SCALE	Scale parameter σ
STD		Standard deviation of the fitted distribution
THETA	THRESHOLD	Lower threshold parameter θ
EXPONENTIAL Secondary Keywords		
MEAN		Mean of the fitted distribution
SIGMA	SCALE	Scale parameter σ
STD		Standard deviation of the fitted distribution
THETA	THRESHOLD	Threshold parameter θ
GAMMA Secondary Keywords		
ALPHA	SHAPE	Shape parameter α
MEAN		Mean of the fitted distribution
SIGMA	SCALE	Scale parameter σ
STD		Standard deviation of the fitted distribution
THETA	THRESHOLD	Threshold parameter θ
GUMBEL Secondary Keywords		
MEAN		Mean of the fitted distribution
MU		Location parameter μ
SIGMA	SCALE	Scale parameter σ
STD		Standard deviation of the fitted distribution
IGAUSS Secondary Keywords		
LAMBDA		Shape parameter λ
MEAN		Mean of the fitted distribution
MU		Mean parameter μ
STD		Standard deviation of the fitted distribution

Table 4.12 *continued*

Secondary Keyword	Alias	Description
KERNEL Secondary Keywords		
AMISE		Approximate mean integrated square error (MISE) for the kernel density
BANDWIDTH		Bandwidth λ for the density estimate
BWIDTH		Alias for BANDWIDTH
C		Standardized bandwidth for the density estimate
TYPE		Kernel type: normal, quadratic, or triangular
LOGNORMAL Secondary Keywords		
MEAN		Mean of the fitted distribution
SIGMA	SHAPE	Shape parameter σ
STD		Standard deviation of the fitted distribution
THETA	THRESHOLD	Threshold parameter θ
ZETA	SCALE	Scale parameter ζ
NORMAL Secondary Keywords		
MU	MEAN	Mean parameter μ
SIGMA	STD	Scale parameter σ
PARETO Secondary Keywords		
ALPHA		Shape parameter α
MEAN		Mean of the fitted distribution
SIGMA	SCALE	Scale parameter σ
STD		Standard deviation of the fitted distribution
THETA	THRESHOLD	Threshold parameter θ
POWER Secondary Keywords		
ALPHA		Shape parameter α
MEAN		Mean of the fitted distribution
SIGMA	SCALE	Scale parameter σ
STD		Standard deviation of the fitted distribution
THETA	THRESHOLD	Threshold parameter θ
RAYLEIGH Secondary Keywords		
MEAN		Mean of the fitted distribution
SIGMA	SCALE	Scale parameter σ
STD		Standard deviation of the fitted distribution
THETA	THRESHOLD	Threshold parameter θ
SB and SU Secondary Keywords		
DELTA	SHAPE1	First shape parameter δ
GAMMA	SHAPE2	Second shape parameter γ
MEAN		Mean of the fitted distribution
SIGMA	SCALE	Scale parameter σ

Table 4.12 *continued*

Secondary Keyword	Alias	Description
STD		Standard deviation of the fitted distribution
THETA	THRESHOLD	Lower threshold parameter θ

WEIBULL Secondary Keywords

C	SHAPE	Shape parameter c
MEAN		Mean of the fitted distribution
SIGMA	SCALE	Scale parameter σ
STD		Standard deviation of the fitted distribution
THETA	THRESHOLD	Threshold parameter θ

WEIBULL2 Secondary Keywords

C	SHAPE	Shape parameter c
MEAN		Mean of the fitted distribution
SIGMA	SCALE	Scale parameter σ
STD		Standard deviation of the fitted distribution
THETA	THRESHOLD	Known lower threshold θ_0

Keywords Available for All Parametric (non-KERNEL) Distributions

AD		Anderson-Darling EDF test statistic
ADPVAL		Anderson-Darling EDF test p-value
CVM		Cramér–von Mises EDF test statistic
CVMPVAL		Cramér–von Mises EDF test p-value
KSD		Kolmogorov-Smirnov EDF test statistic
KSDPVAL		Kolmogorov-Smirnov EDF test p-value

The inset statistics listed in Table 4.12 are not available unless you request a plot statement and options that calculate these statistics. For example, consider the following statements:

```
proc univariate data=score;
   histogram final / normal;
   inset mean std normal(ad adpval);
run;
```

The MEAN and STD keywords display the sample mean and standard deviation, respectively, of final. The NORMAL keyword with the secondary keywords AD and ADPVAL displays the Anderson-Darling goodness-of-fit test statistic and *p*-value, respectively. The statistics that are specified with the NORMAL *keyword* are available only because the NORMAL option is requested in the HISTOGRAM statement.

The KERNEL keyword is available only if you request a kernel density estimate in a HISTOGRAM statement. The WEIBULL2 keyword is available only if you request a two-parameter Weibull distribution in the PROBPLOT or QQPLOT statement.

INSET Statistic Labels and Formats

By default, PROC UNIVARIATE identifies inset statistics with appropriate labels and prints numeric values with appropriate formats. To customize the label, specify the *keyword* followed by an equal sign (=) and the desired label in quotes. To customize the format, specify a numeric format in parentheses after the *keyword*. Labels can have up to 24 characters. If you specify both a label and a format for a statistic, the label must appear before the format. For example, the following statement requests customized labels for two statistics and displays the standard deviation with a field width of 5 and two decimal places:

```
inset n='Sample Size' std='Std Dev' (5.2);
```

Summary of Options

Table 4.13 lists INSET statement *options*, which are specified after the slash (/) in the INSET statement. For complete descriptions, see the section "Dictionary of Options" on page 338.

Table 4.13 INSET Options

Option	Description
CFILL=*color*	BLANK	specifies color of inset background
CFILLH=*color*	specifies color of header background
CFRAME=*color*	specifies color of frame
CHEADER=*color*	specifies color of header text
CSHADOW=*color*	specifies color of drop shadow
CTEXT=*color*	specifies color of inset text
DATA	specifies data units for POSITION=(x, y) coordinates
FONT=*font*	specifies font of text
FORMAT=*format*	specifies format of values in inset
GUTTER=*value*	specifies gutter width for inset in top or bottom margin
HEADER='*string*'	specifies header text
HEIGHT=*value*	specifies height of inset text
NCOLS=	specifies number of columns for inset in top or bottom margin
NOFRAME	suppresses frame around inset
POSITION=*position*	specifies position of inset
REFPOINT=BR	BL	TR	TL	specifies reference point of inset positioned with POSITION=(x, y) coordinates

Dictionary of Options

The following entries provide detailed descriptions of options for the INSET statement. Options marked with † are applicable only when traditional graphics are produced.

† CFILL=*color* | BLANK

specifies the color of the background for traditional graphics. If you omit the CFILLH= option the header background is included. By default, the background is empty, which causes items that overlap the inset (such as curves or histogram bars) to show through the inset.

If you specify a value for CFILL= option, then overlapping items no longer show through the inset. Use CFILL=BLANK to leave the background uncolored and to prevent items from showing through the inset.

† CFILLH=_color_

specifies the color of the header background for traditional graphics. The default value is the CFILL= color.

† CFRAME=_color_

specifies the color of the frame for traditional graphics. The default value is the same color as the axis of the plot.

† CHEADER=_color_

specifies the color of the header text for traditional graphics. The default value is the CTEXT= color.

† CSHADOW=_color_

specifies the color of the drop shadow for traditional graphics. By default, if a CSHADOW= option is not specified, a drop shadow is not displayed.

† CTEXT=_color_

specifies the color of the text for traditional graphics. The default value is the same color as the other text on the plot.

DATA

specifies that data coordinates are to be used in positioning the inset with the POSITION= option. The DATA option is available only when you specify POSITION=(x,y). You must place DATA immediately after the coordinates (x,y). **NOTE:** Positioning insets with coordinates is not supported for ODS Graphics output.

† FONT=_font_

specifies the font of the text for traditional graphics. By default, if you locate the inset in the interior of the plot, then the font is SIMPLEX. If you locate the inset in the exterior of the plot, then the font is the same as the other text on the plot.

FORMAT=_format_

specifies a format for all the values in the inset. If you specify a format for a particular statistic, then that format overrides the one specified with the FORMAT= option. For more information about SAS formats, see _SAS Formats and Informats: Reference_.

GUTTER=_value_

specifies the gutter width in percent screen units for an inset located in the top or bottom margin. The gutter is the space between columns of (label, value) pairs in an inset. The default value is four. **NOTE:** The GUTTER= option applies only when ODS Graphics is enabled.

HEADER=_string_

specifies the header text. The _string_ cannot exceed 40 characters. By default, no header line appears in the inset. If all the keywords that you list in the INSET statement are secondary keywords that correspond to a fitted curve on a histogram, PROC UNIVARIATE displays a default header that indicates the distribution and identifies the curve.

† HEIGHT=value

specifies the height of the text for traditional graphics.

NCOLS=n

specifies the number of columns of (label, value) pairs displayed in an inset located in the top or bottom margin. The default value is three. **NOTE:** The NCOLS= option applies only when ODS Graphics is enabled.

NOFRAME

suppresses the frame drawn around the text.

POSITION=position

POS=position

determines the position of the inset. The position is a compass point keyword, a margin keyword, or a pair of coordinates (x,y). You can specify coordinates in axis percent units or axis data units. The default value is NW, which positions the inset in the upper left (northwest) corner of the display. See the section "Positioning Insets" on page 412.

NOTE: Positioning insets with coordinates is not supported for ODS Graphics output.

† REFPOINT=BR | BL | TR | TL

specifies the reference point for an inset that PROC UNIVARIATE positions by a pair of coordinates with the POSITION= option. The REFPOINT= option specifies which corner of the inset frame that you want to position at coordinates (x,y). The *keywords* are BL, BR, TL, and TR, which correspond to bottom left, bottom right, top left, and top right. The default value is BL. You must use REFPOINT= with POSITION=(x,y) coordinates. The option does not apply to ODS Graphics output.

OUTPUT Statement

OUTPUT < **OUT=***SAS-data-set*> < *keyword1=names . . . keywordk=names* > < *percentile-options* > ;

The OUTPUT statement saves statistics, BY variables, and CLASS variables in an output data set. When you specify a BY statement and/or a CLASS statement, there is an observation in the output data set that corresponds to each combination of BY groups and CLASS variable values. Otherwise, the output data set contains only one observation.

You can use any number of OUTPUT statements in the UNIVARIATE procedure. Each OUTPUT statement creates a new data set to contain the statistics specified in that statement. You must use the VAR statement with the OUTPUT statement. The OUTPUT statement must contain a specification of the form *keyword=names* or the PCTLPTS= and PCTLPRE= options. See Example 4.7 and Example 4.8.

You can use the OUT= option to specify the name of the output data set:

OUT=*SAS-data-set*

identifies the output data set. If *SAS-data-set* does not exist, PROC UNIVARIATE creates it. If you omit OUT=, the data set is named DATA*n*, where *n* is the smallest integer that makes the name unique.

A *keyword=names* specification selects a statistic to be included in the output data set and gives the names of new variables that contain the statistic. Specify a *keyword* for each desired statistic, followed by an equal sign, followed by the *names* of the variables to contain the statistic. In the output data set, the first variable listed after a keyword in the OUTPUT statement contains the statistic for the first variable listed in the VAR statement, the second variable contains the statistic for the second variable in the VAR statement, and so on. If the list of *names* following the equal sign is shorter than the list of variables in the VAR statement, the procedure uses the *names* in the order in which the variables are listed in the VAR statement. The available keywords are listed in Table 4.14.

Table 4.14 Statistical Keywords

Keyword	Description
Descriptive Statistic Keywords	
CSS	Corrected sum of squares
CV	Coefficient of variation
GEOMEAN	Geometric mean
KURTOSIS | KURT	Kurtosis
MAX	Largest value
MEAN	Sample mean
MIN	Smallest value
MODE	Most frequent value
N	Sample size
NMISS	Number of missing values
NOBS	Number of observations
RANGE	Range
SKEWNESS | SKEW	Skewness
STD | STDDEV	Standard deviation
STDMEAN | STDERR	Standard error of the mean
SUM	Sum of the observations
SUMWGT	Sum of the weights
USS	Uncorrected sum of squares
VAR	Variance
Quantile Statistic Keywords	
P1	1st percentile
P5	5th percentile
P10	10th percentile
Q1 | P25	Lower quartile (25th percentile)
MEDIAN | Q2 | P50	Median (50th percentile)
Q3 | P75	Upper quartile (75th percentile)
P90	90th percentile
P95	95th percentile
P99	99th percentile
QRANGE	Interquartile range (Q3–Q1)
Robust Statistic Keywords	
GINI	Gini's mean difference
MAD	Median absolute difference about the median
QN	Q_n, alternative to MAD
SN	S_n, alternative to MAD

Table 4.14 *continued*

Keyword	Description
STD_GINI	Gini's standard deviation
STD_MAD	MAD standard deviation
STD_QN	Q_n standard deviation
STD_QRANGE	Interquartile range standard deviation
STD_SN	S_n standard deviation
Hypothesis Testing Keywords	
MSIGN	Sign statistic
NORMALTEST	Test statistic for normality
SIGNRANK	Signed rank statistic
PROBM	Probability of a greater absolute value for the sign statistic
PROBN	Probability value for the test of normality
PROBS	Probability value for the signed rank test
PROBT	Probability value for the Student's t test
T	Statistic for the Student's t test

The UNIVARIATE procedure automatically computes the 1st, 5th, 10th, 25th, 50th, 75th, 90th, 95th, and 99th percentiles for the data. These can be saved in an output data set by using *keyword=names* specifications. You can request additional percentiles by using the PCTLPTS= option. The following *percentile-options* are related to these additional percentiles:

CIPCTLDF=(*cipctl-options*)

CIQUANTDF=(*cipctl-options*)

> requests distribution-free confidence limits for percentiles that are requested with the PCTLPTS= option. In other words, no specific parametric distribution such as the normal is assumed for the data. PROC UNIVARIATE uses order statistics (ranks) to compute the confidence limits as described by Hahn and Meeker (1991). This option does not apply if you use a WEIGHT statement. You can specify the following *cipctl-options*:
>
> **ALPHA=**α
>
> > specifies the level of significance α for $100(1-\alpha)\%$ confidence intervals. The value α must be between 0 and 1; the default value is 0.05, which results in 95% confidence intervals. The default value is the value of ALPHA= given in the PROC statement.
>
> **LOWERPRE=**ptofixes
>
> > specifies one or more prefixes that are used to create names for variables that contain the lower confidence limits. To save lower confidence limits for more than one analysis variable, specify a list of prefixes. The order of the prefixes corresponds to the order of the analysis variables in the VAR statement.
>
> **LOWERNAME=**suffixes
>
> > specifies one or more suffixes that are used to create names for variables that contain the lower confidence limits. PROC UNIVARIATE creates a variable name by combining the LOWERPRE=

value and suffix name. Because the suffixes are associated with the requested percentiles, list the suffixes in the same order as the PCTLPTS= percentiles.

TYPE=_keyword_

specifies the type of confidence limit, where _keyword_ is LOWER, UPPER, SYMMETRIC, or ASYMMETRIC. The default value is SYMMETRIC.

UPPERPRE=_prefixes_

specifies one or more prefixes that are used to create names for variables that contain the upper confidence limits. To save upper confidence limits for more than one analysis variable, specify a list of prefixes. The order of the prefixes corresponds to the order of the analysis variables in the VAR statement.

UPPERNAME=_suffixes_

specifies one or more suffixes that are used to create names for variables that contain the upper confidence limits. PROC UNIVARIATE creates a variable name by combining the UPPERPRE= value and suffix name. Because the suffixes are associated with the requested percentiles, list the suffixes in the same order as the PCTLPTS= percentiles.

NOTE: See the entries for the PCTLPTS=, PCTLPRE=, and PCTLNAME= options for a detailed description of how variable names are created using prefixes, percentile values, and suffixes.

CIPCTLNORMAL=(_cipctl-options_)

CIQUANTNORMAL=(_cipctl-options_)

requests confidence limits based on the assumption that the data are normally distributed for percentiles that are requested with the PCTLPTS= option. The computational method is described in Section 4.4.1 of Hahn and Meeker (1991) and uses the noncentral _t_ distribution as given by Odeh and Owen (1980). This option does not apply if you use a WEIGHT statement. You can specify the following _cipctl-options_:

ALPHA=α

specifies the level of significance α for $100(1 - \alpha)\%$ confidence intervals. The value α must be between 0 and 1; the default value is 0.05, which results in 95% confidence intervals. The default value is the value of ALPHA= given in the PROC statement.

LOWERPRE=_prefixes_

specifies one or more prefixes that are used to create names for variables that contain the lower confidence limits. To save lower confidence limits for more than one analysis variable, specify a list of prefixes. The order of the prefixes corresponds to the order of the analysis variables in the VAR statement.

LOWERNAME=_suffixes_

specifies one or more suffixes that are used to create names for variables that contain the lower confidence limits. PROC UNIVARIATE creates a variable name by combining the LOWERPRE= value and suffix name. Because the suffixes are associated with the requested percentiles, list the suffixes in the same order as the PCTLPTS= percentiles.

TYPE=_keyword_

specifies the type of confidence limit, where _keyword_ is LOWER, UPPER, or TWOSIDED. The default is TWOSIDED.

UPPERPRE=_prefixes_

specifies one or more prefixes that are used to create names for variables that contain the upper confidence limits. To save upper confidence limits for more than one analysis variable, specify a list of prefixes. The order of the prefixes corresponds to the order of the analysis variables in the VAR statement.

UPPERNAME=_suffixes_

specifies one or more suffixes that are used to create names for variables that contain the upper confidence limits. PROC UNIVARIATE creates a variable name by combining the UPPERPRE= value and suffix name. Because the suffixes are associated with the requested percentiles, list the suffixes in the same order as the PCTLPTS= percentiles.

NOTE: See the entries for the PCTLPTS=, PCTLPRE=, and PCTLNAME= options for a detailed description of how variable names are created using prefixes, percentile values, and suffixes.

PCTLGROUP=BYSTAT | BYVAR

specifies the order in which variables that you request with the PCTLPTS= option are added to the OUT= data set when the VAR statement lists more than one analysis variable. By default (or if you specify PCTLGROUP=BYSTAT), all variables that are associated with a percentile value are created consecutively. If you specify PCTLGROUP=BYVAR, all variables that are associated with an analysis variable are created consecutively.

Consider the following statements:

```
proc univariate data=Score;
   var PreTest PostTest;
   output out=ByStat pctlpts=20 40 pctlpre=Pre_ Post_;
   output out=ByVar pctlgroup=byvar pctlpts=20 40 pctlpre=Pre_ Post_;
run;
```

The order of variables in the data set ByStat is Pre_20, Post_20, Pre_40, Post_40. The order of variables in the data set ByVar is Pre_20, Pre_40, Post_20, Post_40.

PCTLNAME=_suffixes_

specifies one or more suffixes to create the names for the variables that contain the PCTLPTS= percentiles. PROC UNIVARIATE creates a variable name by combining the PCTLPRE= value and suffix name. Because the suffix names are associated with the percentiles that are requested, list the suffix names in the same order as the PCTLPTS= percentiles. If you specify _n suffixes_ with the PCTLNAME= option and _m_ percentile values with the PCTLPTS= option where $m > n$, the _suffixes_ are used to name the first _n_ percentiles and the default names are used for the remaining $m - n$ percentiles. For example, consider the following statements:

```
proc univariate;
   var Length Width Height;
   output pctlpts  = 20 40
          pctlpre  = pl pw ph
          pctlname = twenty;
run;
```

The value *twenty* in the PCTLNAME= option is used for only the first percentile in the PCTLPTS= list. This suffix is appended to the values in the PCTLPRE= option to generate the new variable names pltwenty, pwtwenty, and phtwenty, which contain the 20th percentiles for Length, Width, and Height, respectively. Because a second PCTLNAME= suffix is not specified, variable names for the 40th percentiles for Length, Width, and Height are generated using the prefixes and percentile values. Thus, the output data set contains the variables pltwenty, pl40, pwtwenty, pw40, phtwenty, and ph40.

You must specify PCTLPRE= to supply prefix names for the variables that contain the PCTLPTS= percentiles.

If the number of PCTLNAME= values is fewer than the number of percentiles or if you omit PCTLNAME=, PROC UNIVARIATE uses the percentile as the suffix to create the name of the variable that contains the percentile. For an integer percentile, PROC UNIVARIATE uses the percentile. Otherwise, PROC UNIVARIATE truncates decimal values of percentiles to two decimal places and replaces the decimal point with an underscore.

If either the prefix and suffix name combination or the prefix and percentile name combination is longer than 32 characters, PROC UNIVARIATE truncates the prefix name so that the variable name is 32 characters.

PCTLNDEC=*value*

specifies the number of decimal places in percentile values that are incorporated into percentile variable names. The default value is 2. For example, the following statements create two output data sets, each containing one percentile variable. The variable in data set short is named pwid85_12, while the one in data set long is named pwid85_125.

```
proc univariate;
   var width;
   output out=short pctlpts=85.125 pctlpre=pwid;
   output out=long  pctlpts=85.125 pctlpre=pwid pctlndec=3;
run;
```

PCTLPRE=*prefixes*

specifies one or more prefixes to create the variable names for the variables that contain the PCTLPTS= percentiles. To save the same percentiles for more than one analysis variable, specify a list of prefixes. The order of the prefixes corresponds to the order of the analysis variables in the VAR statement. The PCTLPRE= and PCTLPTS= options must be used together.

The procedure generates new variable names by using the *prefix* and the percentile values. If the specified percentile is an integer, the variable name is simply the *prefix* followed by the value. If the specified value is not an integer, an underscore replaces the decimal point in the variable name, and decimal values are truncated to one decimal place. For example, the following statements create the variables pwid20, pwid33_3, pwid66_6, and pwid80 for the 20th, 33.33rd, 66.67th, and 80th percentiles of Width, respectively:

```
proc univariate noprint;
   var Width;
   output pctlpts=20 33.33 66.67 80 pctlpre=pwid;
run;
```

If you request percentiles for more than one variable, you should list prefixes in the same order in which the variables appear in the VAR statement. If combining the *prefix* and percentile value results in a name longer than 32 characters, the prefix is truncated so that the variable name is 32 characters.

PCTLPTS=*percentiles*

specifies one or more percentiles that are not automatically computed by the UNIVARIATE procedure. The PCTLPRE= and PCTLPTS= options must be used together. You can specify percentiles with an expression of the form *start* TO *stop* BY *increment* where *start* is a starting number, *stop* is an ending number, and *increment* is a number to increment by. The PCTLPTS= option generates additional percentiles and outputs them to a data set. These additional percentiles are not printed.

To compute the 50th, 95th, 97.5th, and 100th percentiles, submit the statement

```
output pctlpre=P_ pctlpts=50,95 to 100 by 2.5;
```

PROC UNIVARIATE computes the requested percentiles based on the method that you specify with the PCTLDEF= option in the PROC UNIVARIATE statement. You must use PCTLPRE=, and optionally PCTLNAME=, to specify variable names for the percentiles. For example, the following statements create an output data set named Pctls that contains the 20th and 40th percentiles of the analysis variables PreTest and PostTest:

```
proc univariate data=Score;
   var PreTest PostTest;
   output out=Pctls pctlpts=20 40 pctlpre=PreTest_ PostTest_
          pctlname=P20 P40;
run;
```

PROC UNIVARIATE saves the 20th and 40th percentiles for PreTest and PostTest in the variables PreTest_P20, PostTest_P20, PreTest_P40, and PostTest_P40.

PPPLOT Statement

PPPLOT < *variables* > < / *options* > ;

The PPPLOT statement creates a probability-probability plot (also referred to as a P-P plot or percent plot), which compares the empirical cumulative distribution function (ECDF) of a variable with a specified theoretical cumulative distribution function such as the normal. If the two distributions match, the points on the plot form a linear pattern that passes through the origin and has unit slope. Thus, you can use a P-P plot to determine how well a theoretical distribution models a set of measurements.

You can specify one of the following theoretical distributions with the PPPLOT statement:

- beta
- exponential
- gamma
- Gumbel

- generalized Pareto
- inverse Gaussian
- lognormal
- normal
- power function
- Rayleigh
- Weibull

NOTE: Probability-probability plots should not be confused with probability plots, which compare a set of ordered measurements with *percentiles* from a specified distribution. You can create probability plots with the PROBPLOT statement.

You can use any number of PPPLOT statements in the UNIVARIATE procedure. The components of the PPPLOT statement are as follows.

variables

are the process variables for which P-P plots are created. If you specify a VAR statement, the *variables* must also be listed in the VAR statement. Otherwise, the *variables* can be any numeric variables in the input data set. If you do not specify a list of *variables*, then by default, the procedure creates a P-P plot for each variable listed in the VAR statement or for each numeric variable in the input data set if you do not specify a VAR statement. For example, if data set measures contains two numeric variables, length and width, the following two PPPLOT statements each produce a P-P plot for each of those variables:

```
proc univariate data=measures;
   var length width;
   ppplot;
run;

proc univariate data=measures;
   ppplot length width;
run;
```

options

specify the theoretical distribution for the plot or add features to the plot. If you specify more than one variable, the options apply equally to each variable. Specify all *options* after the slash (/) in the PPPLOT statement. You can specify only one option that names a distribution, but you can specify any number of other options. By default, the procedure produces a P-P plot based on the normal distribution.

In the following example, the NORMAL, MU=, and SIGMA= options request a P-P plot based on the normal distribution with mean 10 and standard deviation 0.3. The SQUARE option displays the plot in a square frame, and the CTEXT= option specifies the text color.

```
proc univariate data=measures;
   ppplot length width / normal(mu=10 sigma=0.3)
                         square
                         ctext=blue;
run;
```

Table 4.15 through Table 4.17 list the PPPLOT *options* by function. For complete descriptions, see the sections "Dictionary of Options" on page 351 and "Dictionary of Common Options" on page 384. *Options* can be any of the following:

- primary options
- secondary options
- general options

Distribution Options

Table 4.15 summarizes the options for requesting a specific theoretical distribution.

Table 4.15 Options for Specifying the Theoretical Distribution

Option	Description
BETA(*beta-options*)	specifies beta P-P plot
EXPONENTIAL(*exponential-options*)	specifies exponential P-P plot
GAMMA(*gamma-options*)	specifies gamma P-P plot
GUMBEL(*Gumbel-options*)	specifies Gumbel P-P plot
PARETO(*Pareto-options*)	specifies generalized Pareto P-P plot
IGAUSS(*iGauss-options*)	specifies inverse Gaussian P-P plot
LOGNORMAL(*lognormal-options*)	specifies lognormal P-P plot
NORMAL(*normal-options*)	specifies normal P-P plot
POWER(*power-options*)	specifies power function P-P plot
RAYLEIGH(*Rayleigh-options*)	specifies Rayleigh P-P plot
WEIBULL(*Weibull-options*)	specifies Weibull P-P plot

Table 4.16 summarizes options that specify distribution parameters and control the display of the diagonal distribution reference line. Specify these options in parentheses after the distribution option. For example, the following statements use the NORMAL option to request a normal P-P plot:

```
proc univariate data=measures;
   ppplot length / normal(mu=10 sigma=0.3 color=red);
run;
```

The MU= and SIGMA= *normal-options* specify μ and σ for the normal distribution, and the COLOR= *normal-option* specifies the color for the line.

Table 4.16 Secondary Distribution Reference Line Options

Option	Description
Options Used with All Distributions	
COLOR=	specifies color of distribution reference line
L=	specifies line type of distribution reference line
NOLINE	suppresses the distribution reference line
W=	specifies width of distribution reference line
Beta-Options	
ALPHA=	specifies shape parameter α
BETA=	specifies shape parameter β
SIGMA=	specifies scale parameter σ
THETA=	specifies lower threshold parameter θ
Exponential-Options	
SIGMA=	specifies scale parameter σ
THETA=	specifies threshold parameter θ
Gamma-Options	
ALPHA=	specifies shape parameter α
ALPHADELTA=	specifies change in successive estimates of α at which the Newton-Raphson approximation of $\hat{\alpha}$ terminates
ALPHAINITIAL=	specifies initial value for α in the Newton-Raphson approximation of $\hat{\alpha}$
MAXITER=	specifies maximum number of iterations in the Newton-Raphson approximation of $\hat{\alpha}$
SIGMA=	specifies scale parameter σ
THETA=	specifies threshold parameter θ
Gumbel-Options	
MU=	specifies location parameter μ
SIGMA=	specifies scale parameter σ
IGauss-Options	
LAMBDA=	specifies shape parameter λ
MU=	specifies mean μ
Lognormal-Options	
SIGMA=	specifies shape parameter σ
THETA=	specifies threshold parameter θ
ZETA=	specifies scale parameter ζ
Normal-Options	
MU=	specifies mean μ
SIGMA=	specifies standard deviation σ
Pareto-Options	
ALPHA=	specifies shape parameter α
SIGMA=	specifies scale parameter σ
THETA=	specifies threshold parameter θ
Power-Options	
ALPHA=	specifies shape parameter α
SIGMA=	specifies scale parameter σ
THETA=	specifies threshold parameter θ

Table 4.16 (continued)

Option	Description
Rayleigh-Options	
SIGMA=	specifies scale parameter σ
THETA=	specifies threshold parameter θ
Weibull-Options	
C=	specifies shape parameter c
ITPRINT	requests table of iteration history and optimizer details
MAXITER=	specifies maximum number of iterations in the Newton-Raphson approximation of $\hat{c}$
SIGMA=	specifies scale parameter σ
THETA=	specifies threshold parameter θ

General Options

Table 4.17 lists options that control the appearance of the plots. For complete descriptions, see the sections "Dictionary of Options" on page 351 and "Dictionary of Common Options" on page 384.

Table 4.17 General Graphics Options

Option	Description
General Graphics Options	
HREF=	specifies reference lines perpendicular to the horizontal axis
HREFLABELS=	specifies line labels for HREF= lines
HREFLABPOS=	specifies position for HREF= line labels
NOHLABEL	suppresses label for horizontal axis
NOVLABEL	suppresses label for vertical axis
NOVTICK	suppresses tick marks and tick mark labels for vertical axis
SQUARE	displays P-P plot in square format
VAXISLABEL=	specifies label for vertical axis
VREF=	specifies reference lines perpendicular to the vertical axis
VREFLABELS=	specifies line labels for VREF= lines
VREFLABPOS=	specifies position for VREF= line labels
Options for Traditional Graphics Output	
ANNOTATE=	provides an annotate data set
CAXIS=	specifies color for axis
CFRAME=	specifies color for frame
CHREF=	specifies colors for HREF= lines
CTEXT=	specifies color for text
CVREF=	specifies colors for VREF= lines
DESCRIPTION=	specifies description for plot in graphics catalog
FONT=	specifies software font for text
HAXIS=	specifies AXIS statement for horizontal axis
HEIGHT=	specifies height of text used outside framed areas
HMINOR=	specifies number of minor tick marks on horizontal axis

Table 4.17 continued

Option	Description
INFONT=	specifies software font for text inside framed areas
INHEIGHT=	specifies height of text inside framed areas
LHREF=	specifies line types for HREF= lines
LVREF=	specifies line types for VREF= lines
NAME=	specifies name for plot in graphics catalog
NOFRAME	suppresses frame around plotting area
TURNVLABELS	turns and vertically strings out characters in labels for vertical axis
VAXIS=	specifies AXIS statement for vertical axis
VMINOR=	specifies number of minor tick marks on vertical axis
WAXIS=	specifies line thickness for axes and frame
Options for ODS Graphics Output	
ODSFOOTNOTE=	specifies footnote displayed on plot
ODSFOOTNOTE2=	specifies secondary footnote displayed on plot
ODSTITLE=	specifies title displayed on plot
ODSTITLE2=	specifies secondary title displayed on plot
OVERLAY	overlays plots for different class levels (ODS Graphics only)
Options for Comparative Plots	
ANNOKEY	applies annotation requested in ANNOTATE= data set to key cell only
CFRAMESIDE=	specifies color for filling row label frames
CFRAMETOP=	specifies color for filling column label frames
CPROP=	specifies color for proportion of frequency bar
CTEXTSIDE=	specifies color for row labels
CTEXTTOP=	specifies color for column labels
INTERTILE=	specifies distance between tiles in comparative plot
NCOLS=	specifies number of columns in comparative plot
NROWS=	specifies number of rows in comparative plot
Miscellaneous Options	
CONTENTS=	specifies table of contents entry for P-P plot grouping

Dictionary of Options

The following entries provide detailed descriptions of options for the PPPLOT statement. See the section "Dictionary of Common Options" on page 384 for detailed descriptions of options common to all plot statements.

ALPHA=value

specifies the shape parameter α ($\alpha > 0$) for P-P plots requested with the BETA, GAMMA, PARETO, and POWER options.

BETA< (beta-options) >

creates a beta P-P plot. To create the plot, the n nonmissing observations are ordered from smallest to largest:

$$x_{(1)} \leq x_{(2)} \leq \cdots \leq x_{(n)}$$

The y-coordinate of the ith point is the empirical cdf value $\frac{i}{n}$. The x-coordinate is the theoretical beta cdf value

$$B_{\alpha\beta}\left(\frac{x_{(i)}-\theta}{\sigma}\right) = \int_{\theta}^{x_{(i)}} \frac{(t-\theta)^{\alpha-1}(\theta+\sigma-t)^{\beta-1}}{B(\alpha,\beta)\sigma^{(\alpha+\beta-1)}}dt$$

where $B_{\alpha\beta}(\cdot)$ is the normalized incomplete beta function, $B(\alpha,\beta) = \frac{\Gamma(\alpha)\Gamma(\beta)}{\Gamma(\alpha+\beta)}$, and

θ = lower threshold parameter
σ = scale parameter ($\sigma > 0$)
α = first shape parameter ($\alpha > 0$)
β = second shape parameter ($\beta > 0$)

You can specify α, β, σ, and θ with the ALPHA=, BETA=, SIGMA=, and THETA= *beta-options*, as illustrated in the following example:

```
proc univariate data=measures;
   ppplot width / beta(theta=1 sigma=2 alpha=3 beta=4);
run;
```

If you do not specify values for these parameters, then by default, $\theta = 0$, $\sigma = 1$, and maximum likelihood estimates are calculated for α and β.

IMPORTANT: If the default unit interval (0,1) does not adequately describe the range of your data, then you should specify THETA=θ and SIGMA=σ so that your data fall in the interval $(\theta, \theta + \sigma)$.

If the data are beta distributed with parameters α, β, σ, and θ, then the points on the plot for ALPHA=α, BETA=β, SIGMA=σ, and THETA=θ tend to fall on or near the diagonal line $y = x$, which is displayed by default. Agreement between the diagonal line and the point pattern is evidence that the specified beta distribution is a good fit. You can specify the SCALE= option as an alias for the SIGMA= option and the THRESHOLD= option as an alias for the THETA= option.

BETA=*value*

specifies the shape parameter β ($\beta > 0$) for P-P plots requested with the BETA distribution option. See the preceding entry for the BETA distribution option for an example.

C=*value*

specifies the shape parameter c ($c > 0$) for P-P plots requested with the WEIBULL option. See the entry for the WEIBULL option for examples.

EXPONENTIAL<(*exponential-options***)>**

EXP<(*exponential-options***)>**

creates an exponential P-P plot. To create the plot, the *n* nonmissing observations are ordered from smallest to largest:

$$x_{(1)} \leq x_{(2)} \leq \cdots \leq x_{(n)}$$

The y-coordinate of the ith point is the empirical cdf value $\frac{i}{n}$. The x-coordinate is the theoretical exponential cdf value

$$F(x_{(i)}) = 1 - \exp\left(-\frac{x_{(i)}-\theta}{\sigma}\right)$$

where

θ = threshold parameter

σ = scale parameter ($\sigma > 0$)

You can specify σ and θ with the SIGMA= and THETA= *exponential-options*, as illustrated in the following example:

```
proc univariate data=measures;
   ppplot width / exponential(theta=1 sigma=2);
run;
```

If you do not specify values for these parameters, then by default, $\theta = 0$ and a maximum likelihood estimate is calculated for σ.

IMPORTANT: Your data must be greater than or equal to the lower threshold θ. If the default $\theta = 0$ is not an adequate lower bound for your data, specify θ with the THETA= option.

If the data are exponentially distributed with parameters σ and θ, the points on the plot for SIGMA=σ and THETA=θ tend to fall on or near the diagonal line $y = x$, which is displayed by default. Agreement between the diagonal line and the point pattern is evidence that the specified exponential distribution is a good fit. You can specify the SCALE= option as an alias for the SIGMA= option and the THRESHOLD= option as an alias for the THETA= option.

GAMMA< (*gamma-options*) >

creates a gamma P-P plot. To create the plot, the n nonmissing observations are ordered from smallest to largest:

$$x_{(1)} \leq x_{(2)} \leq \cdots \leq x_{(n)}$$

The y-coordinate of the ith point is the empirical cdf value $\frac{i}{n}$. The x-coordinate is the theoretical gamma cdf value

$$G_\alpha\left(\frac{x_{(i)} - \theta}{\sigma}\right) = \int_\theta^{x_{(i)}} \frac{1}{\sigma \Gamma(\alpha)} \left(\frac{t - \theta}{\sigma}\right)^{\alpha-1} \exp\left(-\frac{t - \theta}{\sigma}\right) dt$$

where $G_\alpha(\cdot)$ is the normalized incomplete gamma function and

θ = threshold parameter

σ = scale parameter ($\sigma > 0$)

α = shape parameter ($\alpha > 0$)

You can specify α, σ, and θ with the ALPHA=, SIGMA=, and THETA= *gamma-options*, as illustrated in the following example:

```
proc univariate data=measures;
   ppplot width / gamma(alpha=1 sigma=2 theta=3);
run;
```

If you do not specify values for these parameters, then by default, $\theta = 0$ and maximum likelihood estimates are calculated for α and σ.

IMPORTANT: Your data must be greater than or equal to the lower threshold θ. If the default $\theta = 0$ is not an adequate lower bound for your data, specify θ with the THETA= option.

If the data are gamma distributed with parameters α, σ, and θ, the points on the plot for ALPHA=α, SIGMA=σ, and THETA=θ tend to fall on or near the diagonal line $y = x$, which is displayed by default. Agreement between the diagonal line and the point pattern is evidence that the specified gamma distribution is a good fit. You can specify the SHAPE= option as an alias for the ALPHA= option, the SCALE= option as an alias for the SIGMA= option, and the THRESHOLD= option as an alias for the THETA= option.

GUMBEL<(*Gumbel-options*)>

creates a Gumbel P-P plot. To create the plot, the n nonmissing observations are ordered from smallest to largest:

$$x_{(1)} \leq x_{(2)} \leq \cdots \leq x_{(n)}$$

The y-coordinate of the ith point is the empirical cdf value $\frac{i}{n}$. The x-coordinate is the theoretical Gumbel cdf value

$$F\left(x_{(i)}\right) = \exp(-e^{-(x_{(i)} - \mu)/\sigma})$$

where

μ = location parameter

σ = scale parameter ($\sigma > 0$)

You can specify μ and σ with the MU= and SIGMA= *Gumbel-options*, as illustrated in the following example:

```
proc univariate data=measures;
   ppplot width / gumbel(mu=1 sigma=2);
run;
```

If you do not specify values for these parameters, then by default, the maximum likelihood estimates are calculated for μ and σ.

If the data are Gumbel distributed with parameters μ and σ, the points on the plot for MU=μ and SIGMA=σ tend to fall on or near the diagonal line $y = x$, which is displayed by default. Agreement between the diagonal line and the point pattern is evidence that the specified Gumbel distribution is a good fit.

IGAUSS<(*iGauss-options*)>

creates an inverse Gaussian P-P plot. To create the plot, the n nonmissing observations are ordered from smallest to largest:

$$x_{(1)} \leq x_{(2)} \leq \cdots \leq x_{(n)}$$

The y-coordinate of the ith point is the empirical cdf value $\frac{i}{n}$. The x-coordinate is the theoretical inverse Gaussian cdf value

$$F(x_{(i)}) = \Phi\left\{\sqrt{\frac{\lambda}{x_{(i)}}}\left(\frac{x_{(i)}}{\mu} - 1\right)\right\} + e^{2\lambda/\mu}\Phi\left\{-\sqrt{\frac{\lambda}{x_{(i)}}}\left(\frac{x_{(i)}}{\mu} + 1\right)\right\}$$

where $\Phi(\cdot)$ is the standard normal distribution function and

μ = mean parameter ($\mu > 0$)

λ = shape parameter ($\lambda > 0$)

You can specify λ and μ with the LAMBDA= and MU= *IGauss-options*, as illustrated in the following example:

```
proc univariate data=measures;
   ppplot width / igauss(lambda=1 mu=2);
run;
```

If you do not specify values for these parameters, then by default, the maximum likelihood estimates are calculated for λ and μ.

If the data are inverse Gaussian distributed with parameters λ and μ, the points on the plot for LAMBDA=λ and MU=μ tend to fall on or near the diagonal line $y = x$, which is displayed by default. Agreement between the diagonal line and the point pattern is evidence that the specified inverse Gaussian distribution is a good fit.

LAMBDA=*value*

specifies the shape parameter λ for fitted curves requested with the IGAUSS option. Enclose the LAMBDA= option in parentheses after the IGAUSS distribution keyword. If you do not specify a value for λ, the procedure calculates a maximum likelihood estimate.

LOGNORMAL<(*lognormal-options***)>**

LNORM<(*lognormal-options***)>**

creates a lognormal P-P plot. To create the plot, the n nonmissing observations are ordered from smallest to largest:

$x_{(1)} \leq x_{(2)} \leq \cdots \leq x_{(n)}$

The y-coordinate of the ith point is the empirical cdf value $\frac{i}{n}$. The x-coordinate is the theoretical lognormal cdf value

$$\Phi\left(\frac{\log(x_{(i)} - \theta) - \zeta}{\sigma}\right)$$

where $\Phi(\cdot)$ is the cumulative standard normal distribution function and

θ = threshold parameter

ζ = scale parameter

σ = shape parameter ($\sigma > 0$)

You can specify θ, ζ, and σ with the THETA= ZETA=, and SIGMA= *lognormal-options*, as illustrated in the following example:

```
proc univariate data=measures;
   ppplot width / lognormal(theta=1 zeta=2);
run;
```

If you do not specify values for these parameters, then by default, $\theta = 0$ and maximum likelihood estimates are calculated for σ and ζ.

IMPORTANT: Your data must be greater than the lower threshold θ. If the default $\theta = 0$ is not an adequate lower bound for your data, specify θ with the THETA= option.

If the data are lognormally distributed with parameters σ, θ, and ζ, the points on the plot for SIGMA=σ, THETA=θ, and ZETA=ζ tend to fall on or near the diagonal line $y = x$, which is displayed by default. Agreement between the diagonal line and the point pattern is evidence that the specified lognormal distribution is a good fit. You can specify the SHAPE= option as an alias for the SIGMA=option, the SCALE= option as an alias for the ZETA= option, and the THRESHOLD= option as an alias for the THETA= option.

MU=value

specifies the parameter μ for P-P plots requested with the GUMBEL, IGAUSS, and NORMAL options. By default, the sample mean is used for μ with inverse Gaussian and normal distributions. A maximum likelihood estimate is computed by default with the Gumbel distribution. See Example 4.36.

NOLINE

suppresses the diagonal reference line.

NORMAL<(normal-options**)>**

NORM<(normal-options**)>**

creates a normal P-P plot. By default, if you do not specify a distribution option, the procedure displays a normal P-P plot. To create the plot, the n nonmissing observations are ordered from smallest to largest:

$$x_{(1)} \leq x_{(2)} \leq \cdots \leq x_{(n)}$$

The y-coordinate of the ith point is the empirical cdf value $\frac{i}{n}$. The x-coordinate is the theoretical normal cdf value

$$\Phi\left(\frac{x_{(i)} - \mu}{\sigma}\right) = \int_{-\infty}^{x_{(i)}} \frac{1}{\sigma\sqrt{2\pi}} \exp\left(-\frac{(t-\mu)^2}{2\sigma^2}\right) dt$$

where $\Phi(\cdot)$ is the cumulative standard normal distribution function and

μ = location parameter or mean

σ = scale parameter or standard deviation ($\sigma > 0$)

You can specify μ and σ with the MU= and SIGMA= *normal-options*, as illustrated in the following example:

```
proc univariate data=measures;
   ppplot width / normal(mu=1 sigma=2);
run;
```

By default, the sample mean and sample standard deviation are used for μ and σ.

If the data are normally distributed with parameters μ and σ, the points on the plot for MU=μ and SIGMA=σ tend to fall on or near the diagonal line $y = x$, which is displayed by default. Agreement between the diagonal line and the point pattern is evidence that the specified normal distribution is a good fit. See Example 4.36.

PARETO< (Pareto-options) >

creates a generalized Pareto P-P plot. To create the plot, the n nonmissing observations are ordered from smallest to largest:

$$x_{(1)} \leq x_{(2)} \leq \cdots \leq x_{(n)}$$

The y-coordinate of the ith point is the empirical cdf value $\frac{i}{n}$. The x-coordinate is the theoretical generalized Pareto cdf value

$$F(x_{(i)}) = 1 - \left(1 - \frac{\alpha(x_{(i)} - \theta)}{\sigma}\right)^{\frac{1}{\alpha}}$$

where

θ = threshold parameter
σ = scale parameter ($\sigma > 0$)
α = shape parameter

The parameter θ for the generalized Pareto distribution must be less than the minimum data value. You can specify θ with the THETA= *Pareto-option*. The default value for θ is 0. In addition, the generalized Pareto distribution has a shape parameter α and a scale parameter σ. You can specify these parameters with the ALPHA= and SIGMA= *Pareto-options*. By default, maximum likelihood estimates are computed for α and σ.

If the data are generalized Pareto distributed with parameters θ, σ, and α, the points on the plot for THETA=θ, SIGMA=σ, and ALPHA=α tend to fall on or near the diagonal line $y = x$, which is displayed by default. Agreement between the diagonal line and the point pattern is evidence that the specified generalized Pareto distribution is a good fit.

POWER< (Power-options) >

creates a power function P-P plot. To create the plot, the n nonmissing observations are ordered from smallest to largest:

$$x_{(1)} \leq x_{(2)} \leq \cdots \leq x_{(n)}$$

The y-coordinate of the ith point is the empirical cdf value $\frac{i}{n}$. The x-coordinate is the theoretical power function cdf value

$$F(x_{(i)}) = \left(\frac{x_{(i)} - \theta}{\sigma}\right)^{\alpha}$$

where

θ = lower threshold parameter (lower endpoint)
σ = scale parameter ($\sigma > 0$)
α = shape parameter ($\alpha > 0$)

The power function distribution is bounded below by the parameter θ and above by the value $\theta + \sigma$. You can specify θ and σ by using the THETA= and SIGMA= *power-options*. The default values for θ and σ are 0 and 1, respectively.

You can specify a value for the shape parameter, α, with the ALPHA= *power-option*. If you do not specify a value for α, the procedure calculates a maximum likelihood estimate.

The power function distribution is a special case of the beta distribution with its second shape parameter, $\beta = 1$.

If the data are power function distributed with parameters θ, σ, and α, the points on the plot for THETA=θ, SIGMA=σ, and ALPHA=α tend to fall on or near the diagonal line $y = x$, which is displayed by default. Agreement between the diagonal line and the point pattern is evidence that the specified power function distribution is a good fit.

RAYLEIGH< (Rayleigh-options)>

creates a Rayleigh P-P plot. To create the plot, the *n* nonmissing observations are ordered from smallest to largest:

$$x_{(1)} \leq x_{(2)} \leq \cdots \leq x_{(n)}$$

The *y*-coordinate of the *i*th point is the empirical cdf value $\frac{i}{n}$. The *x*-coordinate is the theoretical Rayleigh cdf value

$$F(x_{(i)}) = 1 - e^{-(x_{(i)}-\theta)^2/(2\sigma^2)}$$

where

θ = threshold parameter
σ = scale parameter ($\sigma > 0$)

The parameter θ for the Rayleigh distribution must be less than the minimum data value. You can specify θ with the THETA= *Rayleigh-option*. The default value for θ is 0. You can specify σ with the SIGMA= *Rayleigh-option*. By default, a maximum likelihood estimate is computed for σ.

If the data are Rayleigh distributed with parameters θ and σ, the points on the plot for THETA=θ and SIGMA=σ tend to fall on or near the diagonal line $y = x$, which is displayed by default. Agreement between the diagonal line and the point pattern is evidence that the specified Rayleigh distribution is a good fit.

SIGMA=*value*

specifies the parameter σ, where $\sigma > 0$. When used with the BETA, EXPONENTIAL, GAMMA, GUMBEL, NORMAL, PARETO, POWER, RAYLEIGH, and WEIBULL options, the SIGMA= option specifies the scale parameter. When used with the LOGNORMAL option, the SIGMA= option specifies the shape parameter. See Example 4.36.

SQUARE
> displays the P-P plot in a square frame. The default is a rectangular frame. See Example 4.36.

THETA=value

THRESHOLD=value
> specifies the lower threshold parameter θ for plots requested with the BETA, EXPONENTIAL, GAMMA, LOGNORMAL, PARETO, POWER, RAYLEIGH, and WEIBULL options.

WEIBULL<(Weibull-options)**>**

WEIB<(Weibull-options)**>**
> creates a Weibull P-P plot. To create the plot, the n nonmissing observations are ordered from smallest to largest:
>
> $$x_{(1)} \leq x_{(2)} \leq \cdots \leq x_{(n)}$$
>
> The y-coordinate of the ith point is the empirical cdf value $\frac{i}{n}$. The x-coordinate is the theoretical Weibull cdf value
>
> $$F(x_{(i)}) = 1 - \exp\left(-\left(\frac{x_{(i)} - \theta}{\sigma}\right)^c\right)$$
>
> where
>
> > θ = threshold parameter
> >
> > σ = scale parameter ($\sigma > 0$)
> >
> > c = shape parameter ($c > 0$)
>
> You can specify c, σ, and θ with the C=, SIGMA= and THETA= *Weibull-options*, as illustrated in the following example:
>
> ```
> proc univariate data=measures;
>    ppplot width / weibull(theta=1 sigma=2);
> run;
> ```
>
> If you do not specify values for these parameters, then by default $\theta = 0$ and maximum likelihood estimates are calculated for σ and c.
>
> **IMPORTANT:** Your data must be greater than or equal to the lower threshold θ. If the default $\theta = 0$ is not an adequate lower bound for your data, you should specify θ with the THETA= option.
>
> If the data are Weibull distributed with parameters c, σ, and θ, the points on the plot for C=c, SIGMA=σ, and THETA=θ tend to fall on or near the diagonal line $y = x$, which is displayed by default. Agreement between the diagonal line and the point pattern is evidence that the specified Weibull distribution is a good fit. You can specify the SHAPE= option as an alias for the C= option, the SCALE= option as an alias for the SIGMA= option, and the THRESHOLD= option as an alias for the THETA= option.

ZETA=*value*
> specifies a value for the scale parameter ζ for lognormal P-P plots requested with the LOGNORMAL option.

PROBPLOT Statement

> **PROBPLOT** < *variables* > < / *options* > ;

The PROBPLOT statement creates a probability plot, which compares ordered variable values with the percentiles of a specified theoretical distribution. If the data distribution matches the theoretical distribution, the points on the plot form a linear pattern. Consequently, you can use a probability plot to determine how well a theoretical distribution models a set of measurements.

Probability plots are similar to Q-Q plots, which you can create with the QQPLOT statement. Probability plots are preferable for graphical estimation of percentiles, whereas Q-Q plots are preferable for graphical estimation of distribution parameters.

You can use any number of PROBPLOT statements in the UNIVARIATE procedure. The components of the PROBPLOT statement are as follows.

variables
> are the variables for which probability plots are created. If you specify a VAR statement, the *variables* must also be listed in the VAR statement. Otherwise, the *variables* can be any numeric variables in the input data set. If you do not specify a list of *variables*, then by default the procedure creates a probability plot for each variable listed in the VAR statement, or for each numeric variable in the DATA= data set if you do not specify a VAR statement. For example, each of the following PROBPLOT statements produces two probability plots, one for Length and one for Width:

```
proc univariate data=Measures;
   var Length Width;
   probplot;

proc univariate data=Measures;
   probplot Length Width;
run;
```

options
> specify the theoretical distribution for the plot or add features to the plot. If you specify more than one variable, the *options* apply equally to each variable. Specify all *options* after the slash (/) in the PROBPLOT statement. You can specify only one *option* that names a distribution in each PROBPLOT statement, but you can specify any number of other *options*. The distributions available are the beta, exponential, gamma, generalized Pareto, Gumbel, lognormal, normal, Rayleigh, two-parameter Weibull, and three-parameter Weibull. By default, the procedure produces a plot for the normal distribution.

> In the following example, the NORMAL option requests a normal probability plot for each variable, while the MU= and SIGMA= *normal-options* request a distribution reference line corresponding to the normal distribution with $\mu = 10$ and $\sigma = 0.3$. The SQUARE option displays the plot in a square frame, and the CTEXT= option specifies the text color.

```
proc univariate data=Measures;
   probplot Length1 Length2 / normal(mu=10 sigma=0.3)
                              square ctext=blue;
run;
```

Table 4.18 through Table 4.20 list the PROBPLOT *options* by function. For complete descriptions, see the sections "Dictionary of Options" on page 365 and "Dictionary of Common Options" on page 384. *Options* can be any of the following:

- primary options
- secondary options
- general options

Distribution Options

Table 4.18 lists *options* for requesting a theoretical distribution.

Table 4.18 Primary Options for Theoretical Distributions

Option	Description
BETA(*beta-options*)	specifies beta probability plot for shape parameters α and β specified with mandatory ALPHA= and BETA= *beta-options*
EXPONENTIAL(*exponential-options*)	specifies exponential probability plot
GAMMA(*gamma-options*)	specifies gamma probability plot for shape parameter α specified with mandatory ALPHA= *gamma-option*
GUMBEL(*Gumbel-options*)	specifies Gumbel probability plot
LOGNORMAL(*lognormal-options*)	specifies lognormal probability plot for shape parameter σ specified with mandatory SIGMA= *lognormal-option*
NORMAL(*normal-options*)	specifies normal probability plot
PARETO(*Pareto-options*)	specifies generalized Pareto probability plot for shape parameter α specified with mandatory ALPHA= *Pareto-option*
POWER(*power-options*)	specifies power function probability plot for shape parameter α specified with mandatory ALPHA= *power-option*
RAYLEIGH(*Rayleigh-options*)	specifies Rayleigh probability plot
WEIBULL(*Weibull-options*)	specifies three-parameter Weibull probability plot for shape parameter c specified with mandatory C= *Weibull-option*

Table 4.18 (continued)

Option	Description
WEIBULL2(*Weibull2-options*)	specifies two-parameter Weibull probability plot

Table 4.19 lists secondary options that specify distribution parameters and control the display of a distribution reference line. Specify these options in parentheses after the distribution keyword. For example, you can request a normal probability plot with a distribution reference line by specifying the NORMAL option as follows:

```
proc univariate;
   probplot Length / normal(mu=10 sigma=0.3 color=red);
run;
```

The MU= and SIGMA= *normal-options* display a distribution reference line that corresponds to the normal distribution with mean $\mu_0 = 10$ and standard deviation $\sigma_0 = 0.3$, and the COLOR= *normal-option* specifies the color for the line.

Table 4.19 Secondary Distribution Options

Option	Description
Options Used with All Distributions	
COLOR=	specifies color of distribution reference line
L=	specifies line type of distribution reference line
W=	specifies width of distribution reference line
Beta-Options	
ALPHA=	specifies mandatory shape parameter α
BETA=	specifies mandatory shape parameter β
SIGMA=	specifies σ_0 for distribution reference line
THETA=	specifies θ_0 for distribution reference line
Exponential-Options	
SIGMA=	specifies σ_0 for distribution reference line
THETA=	specifies θ_0 for distribution reference line
Gamma-Options	
ALPHA=	specifies mandatory shape parameter α
ALPHADELTA=	specifies change in successive estimates of α at which the Newton-Raphson approximation of $\hat{\alpha}$ terminates
ALPHAINITIAL=	specifies initial value for α in the Newton-Raphson approximation of $\hat{\alpha}$
MAXITER=	specifies maximum number of iterations in the Newton-Raphson approximation of $\hat{\alpha}$
SIGMA=	specifies σ_0 for distribution reference line
THETA=	specifies θ_0 for distribution reference line
Gumbel-Options	
MU=	specifies μ_0 for distribution reference line
SIGMA=	specifies σ_0 for distribution reference line
Lognormal-Options	
SIGMA=	specifies mandatory shape parameter σ

Table 4.19 (continued)

Option	Description
SLOPE=	specifies slope of distribution reference line
THETA=	specifies θ_0 for distribution reference line
ZETA=	specifies ζ_0 for distribution reference line (slope is $\exp(\zeta_0)$)
Normal-Options	
MU=	specifies μ_0 for distribution reference line
SIGMA=	specifies σ_0 for distribution reference line
Pareto-Options	
ALPHA=	specifies mandatory shape parameter α
SIGMA=	specifies σ_0 for distribution reference line
THETA=	specifies θ_0 for distribution reference line
Power-Options	
ALPHA=	specifies mandatory shape parameter α
SIGMA=	specifies σ_0 for distribution reference line
THETA=	specifies θ_0 for distribution reference line
Rayleigh-Options	
SIGMA=	specifies σ_0 for distribution reference line
THETA=	specifies θ_0 for distribution reference line
Weibull-Options	
C=	specifies mandatory shape parameter c
ITPRINT	requests table of iteration history and optimizer details
MAXITER=	specifies maximum number of iterations in the Newton-Raphson approximation of $\hat{c}$
SIGMA=	specifies σ_0 for distribution reference line
THETA=	specifies θ_0 for distribution reference line
Weibull2-Options	
C=	specifies c_0 for distribution reference line (slope is $1/c_0$)
ITPRINT	requests table of iteration history and optimizer details
MAXITER=	specifies maximum number of iterations in the Newton-Raphson approximation of $\hat{c}$
SIGMA=	specifies σ_0 for distribution reference line (intercept is $\log(\sigma_0)$)
SLOPE=	specifies slope of distribution reference line
THETA=	specifies known lower threshold θ_0

General Graphics Options

Table 4.20 summarizes the general options for enhancing probability plots.

Table 4.20 General Graphics Options

Option	Description
General Graphics Options	
GRID	creates a grid
HREF=	specifies reference lines perpendicular to the horizontal axis

Table 4.20 (continued)

Option	Description
HREFLABELS=	specifies labels for HREF= lines
HREFLABPOS=	specifies position for HREF= line labels
NOHLABEL	suppresses label for horizontal axis
NOVLABEL	suppresses label for vertical axis
NOVTICK	suppresses tick marks and tick mark labels for vertical axis
PCTLORDER=	specifies tick mark labels for percentile axis
ROTATE	switches horizontal and vertical axes
SQUARE	displays plot in square format
VREF=	specifies reference lines perpendicular to the vertical axis
VREFLABELS=	specifies labels for VREF= lines
VREFLABPOS=	specifies horizontal position of labels for VREF= lines
VAXISLABEL=	specifies label for vertical axis
Options for Traditional Graphics Output	
ANNOTATE=	specifies annotate data set
CAXIS=	specifies color for axis
CFRAME=	specifies color for frame
CGRID=	specifies color for grid lines
CHREF=	specifies colors for HREF= lines
CSTATREF=	specifies colors for STATREF= lines
CTEXT=	specifies color for text
CVREF=	specifies colors for VREF= lines
DESCRIPTION=	specifies description for plot in graphics catalog
FONT=	specifies software font for text
HAXIS=	specifies AXIS statement for horizontal axis
HEIGHT=	specifies height of text used outside framed areas
HMINOR=	specifies number of horizontal minor tick marks
INFONT=	specifies software font for text inside framed areas
INHEIGHT=	specifies height of text inside framed areas
LGRID=	specifies a line type for grid lines
LHREF=	specifies line types for HREF= lines
LSTATREF=	specifies line types for STATREF= lines
LVREF=	specifies line types for VREF= lines
NAME=	specifies name for plot in graphics catalog
NOFRAME	suppresses frame around plotting area
PCTLMINOR	requests minor tick marks for percentile axis
WAXIS=	specifies line thickness for axes and frame
WGRID=	specifies line thickness for grid
TURNVLABELS	turns and vertically strings out characters in labels for vertical axis
VAXIS=	specifies AXIS statement for vertical axis
VMINOR=	specifies number of vertical minor tick marks
Options for ODS Graphics Output	
NOLINELEGEND	suppresses legend for distribution reference line
ODSFOOTNOTE=	specifies footnote displayed on plot
ODSFOOTNOTE2=	specifies secondary footnote displayed on plot

Table 4.20 (continued)

Option	Description
ODSTITLE=	specifies title displayed on plot
ODSTITLE2=	specifies secondary title displayed on plot
OVERLAY	overlays plots for different class levels (ODS Graphics only)
Options for Comparative Plots	
ANNOKEY	applies annotation requested in ANNOTATE= data set to key cell only
CFRAMESIDE=	specifies color for filling frame for row labels
CFRAMETOP=	specifies color for filling frame for column labels
CPROP=	specifies color for proportion of frequency bar
CTEXTSIDE=	specifies color for row labels
CTEXTTOP=	specifies color for column labels
INTERTILE=	specifies distance between tiles
NCOLS=	specifies number of columns in comparative probability plot
NROWS=	specifies number of rows in comparative probability plot
Miscellaneous Options	
CONTENTS=	specifies table of contents entry for probability plot grouping
NADJ=	adjusts sample size when computing percentiles
RANKADJ=	adjusts ranks when computing percentiles

Dictionary of Options

The following entries provide detailed descriptions of *options* in the PROBPLOT statement. Options marked with † are applicable only when traditional graphics are produced. See the section "Dictionary of Common Options" on page 384 for detailed descriptions of options common to all plot statements.

ALPHA=*value-list* | **EST**

specifies the mandatory shape parameter α for probability plots requested with the BETA, GAMMA, PARETO, and POWER options. Enclose the ALPHA= option in parentheses after the distribution keyword. If you specify ALPHA=EST, a maximum likelihood estimate is computed for α.

BETA(ALPHA=*value* | **EST BETA=***value* | **EST** < *beta-options* >**)**

creates a beta probability plot for each combination of the required shape parameters α and β specified by the required ALPHA= and BETA= *beta-options*. If you specify ALPHA=EST and BETA=EST, the procedure creates a plot based on maximum likelihood estimates for α and β. You can specify the SCALE= *beta-option* as an alias for the SIGMA= *beta-option* and the THRESHOLD= *beta-option* as an alias for the THETA= *beta-option*. To create a plot that is based on maximum likelihood estimates for α and β, specify ALPHA=EST and BETA=EST.

To obtain graphical estimates of α and β, specify lists of values in the ALPHA= and BETA= *beta-options*, and select the combination of α and β that most nearly linearizes the point pattern. To assess the point pattern, you can add a diagonal distribution reference line corresponding to lower threshold parameter θ_0 and scale parameter σ_0 with the THETA= and SIGMA= *beta-options*. Alternatively, you can add a line that corresponds to estimated values of θ_0 and σ_0 with the *beta-options* THETA=EST and SIGMA=EST. Agreement between the reference line and the point pattern indicates that the beta distribution with parameters α, β, θ_0, and σ_0 is a good fit.

BETA=value-list | **EST**
B=value-list | **EST**
> specifies the mandatory shape parameter β for probability plots requested with the BETA option. Enclose the BETA= option in parentheses after the BETA option. If you specify BETA=EST, a maximum likelihood estimate is computed for β.

C=value-list | **EST**
> specifies the shape parameter c for probability plots requested with the WEIBULL and WEIBULL2 options. Enclose this option in parentheses after the WEIBULL or WEIBULL2 option. C= is a required *Weibull-option* in the WEIBULL option; in this situation, it accepts a list of values, or if you specify C=EST, a maximum likelihood estimate is computed for c. You can optionally specify C=*value* or C=EST as a *Weibull2-option* with the WEIBULL2 option to request a distribution reference line; in this situation, you must also specify *Weibull2-option* SIGMA=*value* or SIGMA=EST.

† CGRID=color
> specifies the color for grid lines when a grid displays on the plot. This option also produces a grid.

EXPONENTIAL< (exponential-options) **>**
EXP< (exponential-options) **>**
> creates an exponential probability plot. To assess the point pattern, add a diagonal distribution reference line corresponding to θ_0 and σ_0 with the THETA= and SIGMA= *exponential-options*. Alternatively, you can add a line corresponding to estimated values of the threshold parameter θ_0 and the scale parameter σ with the *exponential-options* THETA=EST and SIGMA=EST. Agreement between the reference line and the point pattern indicates that the exponential distribution with parameters θ_0 and σ_0 is a good fit. You can specify the SCALE= *exponential-option* as an alias for the SIGMA= *exponential-option* and the THRESHOLD= *exponential-option* as an alias for the THETA= *exponential-option*.

GAMMA(ALPHA=value | **EST** < gamma-options >**)**
> creates a gamma probability plot for each value of the shape parameter α given by the mandatory ALPHA= *gamma-option*. If you specify ALPHA=EST, the procedure creates a plot based on a maximum likelihood estimate for α. To obtain a graphical estimate of α, specify a list of values for the ALPHA= *gamma-option* and select the value that most nearly linearizes the point pattern. To assess the point pattern, add a diagonal distribution reference line corresponding to θ_0 and σ_0 with the THETA= and SIGMA= *gamma-options*. Alternatively, you can add a line corresponding to estimated values of the threshold parameter θ_0 and the scale parameter σ with the *gamma-options* THETA=EST and SIGMA=EST. Agreement between the reference line and the point pattern indicates that the gamma distribution with parameters α, θ_0, and σ_0 is a good fit. You can specify the SCALE= *gamma-option* as an alias for the SIGMA= *gamma-option* and the THRESHOLD= *gamma-option* as an alias for the THETA= *gamma-option*.

GRID
> displays a grid. Grid lines are reference lines that are perpendicular to the percentile axis at major tick marks.

GUMBEL< (Gumbel-options) **>**
> creates a Gumbel probability plot. To assess the point pattern, add a diagonal distribution reference line corresponding to μ_0 and σ_0 with the MU= and SIGMA= *Gumbel-options*. Alternatively, you can add a line corresponding to estimated values of the location parameter μ_0 and the scale parameter σ with the *Gumbel-options* MU=EST and SIGMA=EST. Agreement between the reference line and the point pattern indicates that the exponential distribution with parameters μ_0 and σ_0 is a good fit.

† **LGRID=**_linetype_

specifies the line type for the grid requested by the GRID= option. By default, LGRID=1, which produces a solid line.

LOGNORMAL(SIGMA=_value_ | **EST** < _lognormal-options_ >**)**
LNORM(SIGMA=_value_ | **EST** < _lognormal-options_ >**)**

creates a lognormal probability plot for each value of the shape parameter σ given by the mandatory SIGMA= _lognormal-option_. If you specify SIGMA=EST, the procedure creates a plot based on a maximum likelihood estimate for σ. To obtain a graphical estimate of σ, specify a list of values for the SIGMA= _lognormal-option_ and select the value that most nearly linearizes the point pattern. To assess the point pattern, add a diagonal distribution reference line corresponding to θ_0 and ζ_0 with the THETA= and ZETA= _lognormal-options_. Alternatively, you can add a line corresponding to estimated values of the threshold parameter θ_0 and the scale parameter ζ_0 with the _lognormal-options_ THETA=EST and ZETA=EST. Agreement between the reference line and the point pattern indicates that the lognormal distribution with parameters σ, θ_0, and ζ_0 is a good fit. You can specify the THRESHOLD= _lognormal-option_ as an alias for the THETA= _lognormal-option_ and the SCALE= _lognormal-option_ as an alias for the ZETA= _lognormal-option_. See Example 4.26.

MU=_value_ | **EST**

specifies the mean μ_0 for a probability plot requested with the GUMBEL and NORMAL options. Enclose MU= in parentheses after the distribution keyword. You can specify MU=EST to request a distribution reference line with μ_0 equal to the sample mean with the normal distribution. If you specify MU=EST for the Gumbel distribution, the procedure computes a maximum likelihood estimate.

NADJ=_value_

specifies the adjustment value added to the sample size in the calculation of theoretical percentiles. By default, NADJ=$\frac{1}{4}$. Refer to Chambers et al. (1983).

NOLINELEGEND
NOLEGEND

suppresses the legend for the optional distribution reference line. The NOLINELEGEND option applies only to ODS Graphics output.

NORMAL< (_normal-options_) **>**

creates a normal probability plot. This is the default if you omit a distribution option. To assess the point pattern, you can add a diagonal distribution reference line corresponding to μ_0 and σ_0 with the MU= and SIGMA= _normal-options_. Alternatively, you can add a line corresponding to estimated values of μ_0 and σ_0 with the _normal-options_ MU=EST and SIGMA=EST; the estimates of the mean μ_0 and the standard deviation σ_0 are the sample mean and sample standard deviation. Agreement between the reference line and the point pattern indicates that the normal distribution with parameters μ_0 and σ_0 is a good fit.

PARETO(ALPHA=_value_ | **EST** < _Pareto-options_ >**)**

creates a generalized Pareto probability plot for each value of the shape parameter α given by the mandatory ALPHA= _Pareto-option_. If you specify ALPHA=EST, the procedure creates a plot based on a maximum likelihood estimate for α. To obtain a graphical estimate of α, specify a list of values for the ALPHA= _Pareto-option_ and select the value that most nearly linearizes the point pattern. To assess the point pattern, add a diagonal distribution reference line corresponding to θ_0 and σ_0 with the THETA= and SIGMA= _Pareto-options_. Alternatively, you can add a line corresponding to estimated values of the threshold parameter θ_0 and the scale parameter σ with the *Pareto-options* THETA=EST and SIGMA=EST. Agreement between the reference line and the point pattern indicates that the generalized Pareto distribution with parameters α, θ_0, and σ_0 is a good fit.

† PCTLMINOR

requests minor tick marks for the percentile axis. The HMINOR option overrides the minor tick marks requested by the PCTLMINOR option.

PCTLORDER=*values*

specifies the tick marks that are labeled on the theoretical percentile axis. Because the values are percentiles, the labels must be between 0 and 100, exclusive. The values must be listed in increasing order and must cover the plotted percentile range. Otherwise, the default values of 1, 5, 10, 25, 50, 75, 90, 95, and 99 are used.

POWER(ALPHA=*value* | EST <*power-options*>)

creates a power function probability plot for each value of the shape parameter α given by the mandatory ALPHA= *power-option*. If you specify ALPHA=EST, the procedure creates a plot based on a maximum likelihood estimate for α. To obtain a graphical estimate of α, specify a list of values for the ALPHA= *power-option* and select the value that most nearly linearizes the point pattern. To assess the point pattern, add a diagonal distribution reference line corresponding to θ_0 and σ_0 with the THETA= and SIGMA= *power-options*. Alternatively, you can add a line corresponding to estimated values of the threshold parameter θ_0 and the scale parameter σ with the *power-options* THETA=EST and SIGMA=EST. Agreement between the reference line and the point pattern indicates that the power function distribution with parameters α, θ_0, and σ_0 is a good fit.

RANKADJ=*value*

specifies the adjustment value added to the ranks in the calculation of theoretical percentiles. By default, RANKADJ=$-\frac{3}{8}$, as recommended by Blom (1958). Refer to Chambers et al. (1983) for additional information.

RAYLEIGH<(*Rayleigh-options*)>

creates an Rayleigh probability plot. To assess the point pattern, add a diagonal distribution reference line corresponding to θ_0 and σ_0 with the THETA= and SIGMA= *Rayleigh-options*. Alternatively, you can add a line corresponding to estimated values of the threshold parameter θ_0 and the scale parameter σ with the *Rayleigh-options* THETA=EST and SIGMA=EST. Agreement between the reference line and the point pattern indicates that the exponential distribution with parameters θ_0 and σ_0 is a good fit.

ROTATE

switches the horizontal and vertical axes so that the theoretical percentiles are plotted vertically while the data are plotted horizontally. Regardless of whether the plot has been rotated, horizontal axis options (such as HAXIS=) still refer to the horizontal axis, and vertical axis options (such as VAXIS=) still refer to the vertical axis. All other options that depend on axis placement adjust to the rotated axes.

SIGMA=*value-list* | EST

specifies the parameter σ, where $\sigma > 0$. Alternatively, you can specify SIGMA=EST to request a maximum likelihood estimate for σ_0. The interpretation and use of the SIGMA= option depend on the distribution option with which it is used. See Table 4.21 for a summary of how to use the SIGMA= option. You must enclose this option in parentheses after the distribution option.

Table 4.21 Uses of the SIGMA= Option

Distribution Option	Use of the SIGMA= Option
BETA EXPONENTIAL GAMMA PARETO POWER RAYLEIGH WEIBULL	THETA=θ_0 and SIGMA=σ_0 request a distribution reference line corresponding to θ_0 and σ_0.
GUMBEL	MU=μ_0 and SIGMA=σ_0 request a distribution reference line corresponding to μ_0 and σ_0.
LOGNORMAL	SIGMA=$\sigma_1 \ldots \sigma_n$ requests n probability plots with shape parameters $\sigma_1 \ldots \sigma_n$. The SIGMA= option must be specified.
NORMAL	MU=μ_0 and SIGMA=σ_0 request a distribution reference line corresponding to μ_0 and σ_0. SIGMA=EST requests a line with σ_0 equal to the sample standard deviation.
WEIBULL2	SIGMA=σ_0 and C=c_0 request a distribution reference line corresponding to σ_0 and c_0.

SLOPE=value | **EST**

specifies the slope for a distribution reference line requested with the LOGNORMAL and WEIBULL2 options. Enclose the SLOPE= option in parentheses after the distribution option. When you use the SLOPE= *lognormal-option* with the LOGNORMAL option, you must also specify a threshold parameter value θ_0 with the THETA= *lognormal-option* to request the line. The SLOPE= *lognormal-option* is an alternative to the ZETA= *lognormal-option* for specifying ζ_0, because the slope is equal to $\exp(\zeta_0)$.

When you use the SLOPE= *Weibull2-option* with the WEIBULL2 option, you must also specify a scale parameter value σ_0 with the SIGMA= *Weibull2-option* to request the line. The SLOPE= *Weibull2-option* is an alternative to the C= *Weibull2-option* for specifying c_0, because the slope is equal to $\frac{1}{c_0}$.

For example, the first and second PROBPLOT statements produce the same probability plots and the third and fourth PROBPLOT statements produce the same probability plots:

```
proc univariate data=Measures;
   probplot Width / lognormal(sigma=2 theta=0 zeta=0);
   probplot Width / lognormal(sigma=2 theta=0 slope=1);
   probplot Width / weibull2(sigma=2 theta=0 c=.25);
   probplot Width / weibull2(sigma=2 theta=0 slope=4);
run;
```

SQUARE
>displays the probability plot in a square frame. By default, the plot is in a rectangular frame.

THETA=value | **EST**

THRESHOLD=value | **EST**
>specifies the lower threshold parameter θ for plots requested with the BETA, EXPONENTIAL, GAMMA, PARETO, POWER, RAYLEIGH, LOGNORMAL, WEIBULL, and WEIBULL2 options. Enclose the THETA= option in parentheses after a distribution option. When used with the WEIBULL2 option, the THETA= option specifies the known lower threshold θ_0, for which the default is 0. When used with the other distribution options, the THETA= option specifies θ_0 for a distribution reference line; alternatively in this situation, you can specify THETA=EST to request a maximum likelihood estimate for θ_0. To request the line, you must also specify a scale parameter.

WEIBULL(C=value | **EST** < Weibull-options >**)**

WEIB(C=value | **EST** < Weibull-options >**)**
>creates a three-parameter Weibull probability plot for each value of the required shape parameter c specified by the mandatory C= *Weibull-option*. To create a plot that is based on a maximum likelihood estimate for c, specify C=EST. To obtain a graphical estimate of c, specify a list of values in the C= *Weibull-option* and select the value that most nearly linearizes the point pattern. To assess the point pattern, add a diagonal distribution reference line corresponding to θ_0 and σ_0 with the THETA= and SIGMA= *Weibull-options*. Alternatively, you can add a line corresponding to estimated values of θ_0 and σ_0 with the *Weibull-options* THETA=EST and SIGMA=EST. Agreement between the reference line and the point pattern indicates that the Weibull distribution with parameters c, θ_0, and σ_0 is a good fit. You can specify the SCALE= *Weibull-option* as an alias for the SIGMA= *Weibull-option* and the THRESHOLD= *Weibull-option* as an alias for the THETA= *Weibull-option*.

WEIBULL2<(Weibull2-options**)>**

W2<(Weibull2-options**)>**
>creates a two-parameter Weibull probability plot. You should use the WEIBULL2 option when your data have a *known* lower threshold θ_0, which is 0 by default. To specify the threshold value θ_0, use the THETA= *Weibull2-option*. By default, THETA=0. An advantage of the two-parameter Weibull plot over the three-parameter Weibull plot is that the parameters c and σ can be estimated from the slope and intercept of the point pattern. A disadvantage is that the two-parameter Weibull distribution applies only in situations where the threshold parameter is known. To obtain a graphical estimate of θ_0, specify a list of values for the THETA= *Weibull2-option* and select the value that most nearly linearizes the point pattern. To assess the point pattern, add a diagonal distribution reference line corresponding to σ_0 and c_0 with the SIGMA= and C= *Weibull2-options*. Alternatively, you can add a distribution reference line corresponding to estimated values of σ_0 and c_0 with the *Weibull2-options* SIGMA=EST and C=EST. Agreement between the reference line and the point pattern indicates that the Weibull distribution with parameters c_0, θ_0, and σ_0 is a good fit. You can specify the SCALE= *Weibull2-option* as an alias for the SIGMA= *Weibull2-option* and the SHAPE= *Weibull2-option* as an alias for the C= *Weibull2-option*.

† WGRID=n
>specifies the line thickness for the grid when producing traditional graphics. The option does not apply to ODS Graphics output.

ZETA=_value_ | **EST**

specifies a value for the scale parameter ζ for the lognormal probability plots requested with the LOGNORMAL option. Enclose the ZETA= _lognormal-option_ in parentheses after the LOGNORMAL option. To request a distribution reference line with intercept θ_0 and slope $\exp(\zeta_0)$, specify the THETA=θ_0 and ZETA=ζ_0.

QQPLOT Statement

QQPLOT < _variables_ > < / _options_ > ;

The QQPLOT statement creates quantile-quantile plots (Q-Q plots) and compares ordered variable values with quantiles of a specified theoretical distribution. If the data distribution matches the theoretical distribution, the points on the plot form a linear pattern. Thus, you can use a Q-Q plot to determine how well a theoretical distribution models a set of measurements.

Q-Q plots are similar to probability plots, which you can create with the PROBPLOT statement. Q-Q plots are preferable for graphical estimation of distribution parameters, whereas probability plots are preferable for graphical estimation of percentiles.

You can use any number of QQPLOT statements in the UNIVARIATE procedure. The components of the QQPLOT statement are as follows.

variables

are the variables for which Q-Q plots are created. If you specify a VAR statement, the _variables_ must also be listed in the VAR statement. Otherwise, the _variables_ can be any numeric variables in the input data set. If you do not specify a list of _variables_, then by default the procedure creates a Q-Q plot for each variable listed in the VAR statement, or for each numeric variable in the DATA= data set if you do not specify a VAR statement. For example, each of the following QQPLOT statements produces two Q-Q plots, one for Length and one for Width:

```
proc univariate data=Measures;
   var Length Width;
   qqplot;

proc univariate data=Measures;
   qqplot Length Width;
run;
```

options

specify the theoretical distribution for the plot or add features to the plot. If you specify more than one variable, the _options_ apply equally to each variable. Specify all _options_ after the slash (/) in the QQPLOT statement. You can specify only one _option_ that names the distribution in each QQPLOT statement, but you can specify any number of other _options_. The distributions available are the beta, exponential, gamma, lognormal, normal, two-parameter Weibull, and three-parameter Weibull. By default, the procedure produces a plot for the normal distribution.

In the following example, the NORMAL option requests a normal Q-Q plot for each variable. The MU= and SIGMA= _normal-options_ request a distribution reference line with intercept 10 and slope 0.3 for each plot, corresponding to a normal distribution with mean $\mu = 10$ and standard deviation

$\sigma = 0.3$. The SQUARE option displays the plot in a square frame, and the CTEXT= option specifies the text color.

```
proc univariate data=measures;
   qqplot length1 length2 / normal(mu=10 sigma=0.3)
                            square ctext=blue;
run;
```

Table 4.22 through Table 4.24 list the QQPLOT *options* by function. For complete descriptions, see the sections "Dictionary of Options" on page 376 and "Dictionary of Common Options" on page 384.

Options can be any of the following:

- primary options
- secondary options
- general options

Distribution Options

Table 4.22 lists primary options for requesting a theoretical distribution. See the section "Distributions for Probability and Q-Q Plots" on page 436 for detailed descriptions of these distributions.

Table 4.22 Primary Options for Theoretical Distributions

Option	Description
BETA(*beta-options*)	specifies beta Q-Q plot for shape parameters α and β specified with mandatory ALPHA= and BETA= *beta-options*
EXPONENTIAL(*exponential-options*)	specifies exponential Q-Q plot
GAMMA(*gamma-options*)	specifies gamma Q-Q plot for shape parameter α specified with mandatory ALPHA= *gamma-option*
GUMBEL(*Gumbel-options*)	specifies gumbel Q-Q plot
LOGNORMAL(*lognormal-options*)	specifies lognormal Q-Q plot for shape parameter σ specified with mandatory SIGMA= *lognormal-option*
NORMAL(*normal-options*)	specifies normal Q-Q plot
PARETO(*Pareto-options*)	specifies generalized Pareto Q-Q plot for shape parameter α specified with mandatory ALPHA= *Pareto-option*
POWER(*power-options*)	specifies power function Q-Q plot for shape parameter α specified with mandatory ALPHA= *power-option*
RAYLEIGH(*Rayleigh-options*)	specifies Rayleigh Q-Q plot

QQPLOT Statement ◆ 373

Table 4.22 (*continued*)

Option	Description
WEIBULL(*Weibull-options*)	specifies three-parameter Weibull Q-Q plot for shape parameter c specified with mandatory C= *Weibull-option*
WEIBULL2(*Weibull2-options*)	specifies two-parameter Weibull Q-Q plot

Table 4.23 lists secondary options that specify distribution parameters and control the display of a distribution reference line. Specify these options in parentheses after the distribution keyword. For example, you can request a normal Q-Q plot with a distribution reference line by specifying the NORMAL option as follows:

```
proc univariate;
   qqplot Length / normal(mu=10 sigma=0.3 color=red);
run;
```

The MU= and SIGMA= *normal-options* display a distribution reference line that corresponds to the normal distribution with mean $\mu_0 = 10$ and standard deviation $\sigma_0 = 0.3$, and the COLOR= *normal-option* specifies the color for the line.

Table 4.23 Secondary Distribution Reference Line Options

Option	Description
Options Used with All Distributions	
COLOR=	specifies color of distribution reference line
L=	specifies line type of distribution reference line
W=	specifies width of distribution reference line
Beta-Options	
ALPHA=	specifies mandatory shape parameter α
BETA=	specifies mandatory shape parameter β
SIGMA=	specifies σ_0 for distribution reference line
THETA=	specifies θ_0 for distribution reference line
Exponential-Options	
SIGMA=	specifies σ_0 for distribution reference line
THETA=	specifies θ_0 for distribution reference line
Gamma-Options	
ALPHA=	specifies mandatory shape parameter α
ALPHADELTA=	specifies change in successive estimates of α at which the Newton-Raphson approximation of $\hat{\alpha}$ terminates
ALPHAINITIAL=	specifies initial value for α in the Newton-Raphson approximation of $\hat{\alpha}$
MAXITER=	specifies maximum number of iterations in the Newton-Raphson approximation of $\hat{\alpha}$
SIGMA=	specifies σ_0 for distribution reference line
THETA=	specifies θ_0 for distribution reference line
Gumbel-Options	
MU=	specifies μ_0 for distribution reference line

Table 4.23 (continued)

Option	Description
SIGMA=	specifies σ_0 for distribution reference line
Lognormal-Options	
SIGMA=	specifies mandatory shape parameter σ
SLOPE=	specifies slope of distribution reference line
THETA=	specifies θ_0 for distribution reference line
ZETA=	specifies ζ_0 for distribution reference line (slope is $\exp(\zeta_0)$)
Normal-Options	
MU=	specifies μ_0 for distribution reference line
SIGMA=	specifies σ_0 for distribution reference line
Pareto-Options	
ALPHA=	specifies mandatory shape parameter α
SIGMA=	specifies σ_0 for distribution reference line
THETA=	specifies θ_0 for distribution reference line
Power-Options	
ALPHA=	specifies mandatory shape parameter α
SIGMA=	specifies σ_0 for distribution reference line
THETA=	specifies θ_0 for distribution reference line
Rayleigh-Options	
SIGMA=	specifies σ_0 for distribution reference line
THETA=	specifies θ_0 for distribution reference line
Weibull-Options	
C=	specifies mandatory shape parameter c
SIGMA=	specifies σ_0 for distribution reference line
THETA=	specifies θ_0 for distribution reference line
Weibull2-Options	
C=	specifies c_0 for distribution reference line (slope is $1/c_0$)
SIGMA=	specifies σ_0 for distribution reference line (intercept is $\log(\sigma_0)$)
SLOPE=	specifies slope of distribution reference line
THETA=	specifies known lower threshold θ_0
Weibull-Options	
C=	specifies mandatory shape parameter c
ITPRINT	requests table of iteration history and optimizer details
MAXITER=	specifies maximum number of iterations in the Newton-Raphson approximation of $\hat{c}$
SIGMA=	specifies σ_0 for distribution reference line
THETA=	specifies θ_0 for distribution reference line
Weibull2-Options	
C=	specifies c_0 for distribution reference line (slope is $1/c_0$)
ITPRINT	requests table of iteration history and optimizer details
MAXITER=	specifies maximum number of iterations in the Newton-Raphson approximation of $\hat{c}$
SIGMA=	specifies σ_0 for distribution reference line (intercept is $\log(\sigma_0)$)
SLOPE=	specifies slope of distribution reference line
THETA=	specifies known lower threshold θ_0

General Options

Table 4.24 summarizes general options for enhancing Q-Q plots.

Table 4.24 General Graphics Options

Option	Description
General Graphics Options	
GRID	creates a grid
HREF=	specifies reference lines perpendicular to the horizontal axis
HREFLABELS=	specifies labels for HREF= lines
HREFLABPOS=	specifies vertical position of labels for HREF= lines
NOHLABEL	suppresses label for horizontal axis
NOVLABEL	suppresses label for vertical axis
NOVTICK	suppresses tick marks and tick mark labels for vertical axis
PCTLAXIS	displays a nonlinear percentile axis
PCTLSCALE	replaces theoretical quantiles with percentiles
ROTATE	switches horizontal and vertical axes
SQUARE	displays plot in square format
VAXISLABEL=	specifies label for vertical axis
VREF=	specifies reference lines perpendicular to the vertical axis
VREFLABELS=	specifies labels for VREF= lines
VREFLABPOS=	specifies horizontal position of labels for VREF= lines
Options for Traditional Graphics Output	
ANNOTATE=	specifies annotate data set
CAXIS=	specifies color for axis
CFRAME=	specifies color for frame
CGRID=	specifies color for grid lines
CHREF=	specifies colors for HREF= lines
CSTATREF=	specifies colors for STATREF= lines
CTEXT=	specifies color for text
CVREF=	specifies colors for VREF= lines
DESCRIPTION=	specifies description for plot in graphics catalog
FONT=	specifies software font for text
HEIGHT=	specifies height of text used outside framed areas
HMINOR=	specifies number of horizontal minor tick marks
INFONT=	specifies software font for text inside framed areas
INHEIGHT=	specifies height of text inside framed areas
LGRID=	specifies a line type for grid lines
LHREF=	specifies line types for HREF= lines
LSTATREF=	specifies line types for STATREF= lines
LVREF=	specifies line types for VREF= lines
NAME=	specifies name for plot in graphics catalog
NOFRAME	suppresses frame around plotting area
PCTLMINOR	requests minor tick marks for percentile axis
VAXIS=	specifies AXIS statement for vertical axis
VMINOR=	specifies number of vertical minor tick marks
WAXIS=	specifies line thickness for axes and frame

Table 4.24 (continued)

Option	Description
WGRID=	specifies line thickness for grid
Options for ODS Graphics Output	
NOLINELEGEND	suppresses legend for distribution reference line
ODSFOOTNOTE=	specifies footnote displayed on plot
ODSFOOTNOTE2=	specifies secondary footnote displayed on plot
ODSTITLE=	specifies title displayed on plot
ODSTITLE2=	specifies secondary title displayed on plot
OVERLAY	overlays plots for different class levels (ODS Graphics only)
Options for Comparative Plots	
ANNOKEY	applies annotation requested in ANNOTATE= data set to key cell only
CFRAMESIDE=	specifies color for filling frame for row labels
CFRAMETOP=	specifies color for filling frame for column labels
CPROP=	specifies color for proportion of frequency bar
INTERTILE=	specifies distance between tiles
NCOLS=	specifies number of columns in comparative Q-Q plot
NROWS=	specifies number of rows in comparative Q-Q plot
Miscellaneous Options	
CONTENTS=	specifies table of contents entry for Q-Q plot grouping
NADJ=	adjusts sample size when computing percentiles
RANKADJ=	adjusts ranks when computing percentiles

Dictionary of Options

The following entries provide detailed descriptions of *options* in the QQPLOT statement. Options marked with † are applicable only when traditional graphics are produced. See the section "Dictionary of Common Options" on page 384 for detailed descriptions of options common to all plot statements.

ALPHA=*value-list* | **EST**

specifies the mandatory shape parameter α for quantile plots requested with the BETA, GAMMA, PARETO, and POWER options. Enclose the ALPHA= option in parentheses after the distribution keyword. If you specify ALPHA=EST, a maximum likelihood estimate is computed for α.

BETA(ALPHA=*value* | **EST BETA=***value* | **EST** < *beta-options* >**)**

creates a beta quantile plot for each combination of the required shape parameters α and β specified by the required ALPHA= and BETA= *beta-options*. If you specify ALPHA=EST and BETA=EST, the procedure creates a plot based on maximum likelihood estimates for α and β. You can specify the SCALE= *beta-option* as an alias for the SIGMA= *beta-option* and the THRESHOLD= *beta-option* as an alias for the THETA= *beta-option*. To create a plot that is based on maximum likelihood estimates for α and β, specify ALPHA=EST and BETA=EST. See the section "Beta Distribution" on page 437 for details.

To obtain graphical estimates of α and β, specify lists of values in the ALPHA= and BETA= *beta-options* and select the combination of α and β that most nearly linearizes the point pattern. To assess the point pattern, you can add a diagonal distribution reference line corresponding to lower threshold parameter θ_0 and scale parameter σ_0 with the THETA= and SIGMA= *beta-options*. Alternatively, you can add a line that corresponds to estimated values of θ_0 and σ_0 with the *beta-options* THETA=EST and SIGMA=EST. Agreement between the reference line and the point pattern indicates that the beta distribution with parameters α, β, θ_0, and σ_0 is a good fit.

BETA=*value-list* **| EST**

B=*value* **| EST**

specifies the mandatory shape parameter β for quantile plots requested with the BETA option. Enclose the BETA= option in parentheses after the BETA option. If you specify BETA=EST, a maximum likelihood estimate is computed for β.

C=*value-list* **| EST**

specifies the shape parameter c for quantile plots requested with the WEIBULL and WEIBULL2 options. Enclose this option in parentheses after the WEIBULL or WEIBULL2 option. C= is a required *Weibull-option* in the WEIBULL option; in this situation, it accepts a list of values, or if you specify C=EST, a maximum likelihood estimate is computed for c. You can optionally specify C=*value* or C=EST as a *Weibull2-option* with the WEIBULL2 option to request a distribution reference line; in this situation, you must also specify *Weibull2-option* SIGMA=*value* or SIGMA=EST.

† CGRID=*color*

specifies the color for grid lines when a grid displays on the plot. This option also produces a grid.

EXPONENTIAL<(*exponential-options***)>**

EXP<(*exponential-options***)>**

creates an exponential quantile plot. To assess the point pattern, add a diagonal distribution reference line corresponding to θ_0 and σ_0 with the THETA= and SIGMA= *exponential-options*. Alternatively, you can add a line corresponding to estimated values of the threshold parameter θ_0 and the scale parameter σ with the *exponential-options* THETA=EST and SIGMA=EST. Agreement between the reference line and the point pattern indicates that the exponential distribution with parameters θ_0 and σ_0 is a good fit. You can specify the SCALE= *exponential-option* as an alias for the SIGMA= *exponential-option* and the THRESHOLD= *exponential-option* as an alias for the THETA= *exponential-option*. See the section "Exponential Distribution" on page 438 for details.

GAMMA(ALPHA=*value* **| EST <** *gamma-options* **>)**

creates a gamma quantile plot for each value of the shape parameter α given by the mandatory ALPHA= *gamma-option*. If you specify ALPHA=EST, the procedure creates a plot based on a maximum likelihood estimate for α. To obtain a graphical estimate of α, specify a list of values for the ALPHA= *gamma-option* and select the value that most nearly linearizes the point pattern. To assess the point pattern, add a diagonal distribution reference line corresponding to θ_0 and σ_0 with the THETA= and SIGMA= *gamma-options*. Alternatively, you can add a line corresponding to estimated values of the threshold parameter θ_0 and the scale parameter σ with the *gamma-options* THETA=EST and SIGMA=EST. Agreement between the reference line and the point pattern indicates that the gamma distribution with parameters α, θ_0, and σ_0 is a good fit. You can specify the SCALE= *gamma-option* as an alias for the SIGMA= *gamma-option* and the THRESHOLD= *gamma-option* as an alias for the THETA= *gamma-option*. See the section "Gamma Distribution" on page 438 for details.

GRID
> displays a grid of horizontal lines positioned at major tick marks on the vertical axis.

GUMBEL< (*Gumbel-options***) >**
> creates a Gumbel quantile plot. To assess the point pattern, add a diagonal distribution reference line corresponding to μ_0 and σ_0 with the MU= and SIGMA= *Gumbel-options*. Alternatively, you can add a line corresponding to estimated values of the location parameter μ_0 and the scale parameter σ with the *Gumbel-options* MU=EST and SIGMA=EST. Agreement between the reference line and the point pattern indicates that the exponential distribution with parameters μ_0 and σ_0 is a good fit. See the section "Gumbel Distribution" on page 438 for details.

† LGRID=*linetype*
> specifies the line type for the grid requested by the GRID option. By default, LGRID=1, which produces a solid line. The LGRID= option also produces a grid.

LOGNORMAL(SIGMA=*value* **| EST** *< lognormal-options >***)**
LNORM(SIGMA=*value* **| EST** *< lognormal-options >***)**
> creates a lognormal quantile plot for each value of the shape parameter σ given by the mandatory SIGMA= *lognormal-option*. If you specify SIGMA=EST, the procedure creates a plot based on a maximum likelihood estimate for σ. To obtain a graphical estimate of σ, specify a list of values for the SIGMA= *lognormal-option* and select the value that most nearly linearizes the point pattern. To assess the point pattern, add a diagonal distribution reference line corresponding to θ_0 and ζ_0 with the THETA= and ZETA= *lognormal-options*. Alternatively, you can add a line corresponding to estimated values of the threshold parameter θ_0 and the scale parameter ζ_0 with the *lognormal-options* THETA=EST and ZETA=EST. Agreement between the reference line and the point pattern indicates that the lognormal distribution with parameters σ, θ_0, and ζ_0 is a good fit. You can specify the THRESHOLD= *lognormal-option* as an alias for the THETA= *lognormal-option* and the SCALE= *lognormal-option* as an alias for the ZETA= *lognormal-option*. See the section "Lognormal Distribution" on page 439 for details, and see Example 4.31 through Example 4.33 for examples that use the LOGNORMAL option.

MU=*value* **| EST**
> specifies the mean μ_0 for a quantile plot requested with the GUMBEL and NORMAL options. Enclose MU= in parentheses after the distribution keyword. You can specify MU=EST to request a distribution reference line with μ_0 equal to the sample mean with the normal distribution. If you specify MU=EST for the Gumbel distribution, the procedure computes a maximum likelihood estimate.

NADJ=*value*
> specifies the adjustment value added to the sample size in the calculation of theoretical percentiles. By default, NADJ=$\frac{1}{4}$. Refer to Chambers et al. (1983) for additional information.

NOLINELEGEND
NOLEGEND
> suppresses the legend for the optional distribution reference line. The NOLINELEGEND option applies only to ODS Graphics output.

NORMAL< (*normal-options*) >

creates a normal quantile plot. This is the default if you omit a distribution option. To assess the point pattern, you can add a diagonal distribution reference line corresponding to μ_0 and σ_0 with the MU= and SIGMA= *normal-options*. Alternatively, you can add a line corresponding to estimated values of μ_0 and σ_0 with the *normal-options* MU=EST and SIGMA=EST; the estimates of the mean μ_0 and the standard deviation σ_0 are the sample mean and sample standard deviation. Agreement between the reference line and the point pattern indicates that the normal distribution with parameters μ_0 and σ_0 is a good fit. See the section "Normal Distribution" on page 439 for details, and see Example 4.28 and Example 4.30 for examples that use the NORMAL option.

PARETO(ALPHA=*value* | **EST** < *Pareto-options* >)

creates a generalized Pareto quantile plot for each value of the shape parameter α given by the mandatory ALPHA= *Pareto-option*. If you specify ALPHA=EST, the procedure creates a plot based on a maximum likelihood estimate for α. To obtain a graphical estimate of α, specify a list of values for the ALPHA= *Pareto-option* and select the value that most nearly linearizes the point pattern. To assess the point pattern, add a diagonal distribution reference line corresponding to θ_0 and σ_0 with the THETA= and SIGMA= *Pareto-options*. Alternatively, you can add a line corresponding to estimated values of the threshold parameter θ_0 and the scale parameter σ with the *Pareto-options* THETA=EST and SIGMA=EST. Agreement between the reference line and the point pattern indicates that the generalized Pareto distribution with parameters α, θ_0, and σ_0 is a good fit. See the section "Generalized Pareto Distribution" on page 439 for details.

PCTLAXIS< (*axis-options*) >

adds a nonlinear percentile axis along the frame of the Q-Q plot opposite the theoretical quantile axis. The added axis is identical to the axis for probability plots produced with the PROBPLOT statement. When using the PCTLAXIS option, you must specify HREF= values in quantile units, and you cannot use the NOFRAME option. You can specify the following *axis-options*:

Table 4.25 PCTLAXIS Axis Options

Option	Description
CGRID=	specifies color for grid lines
GRID	draws grid lines at major percentiles
LABEL='*string*'	specifies label for percentile axis
LGRID=*linetype*	specifies line type for grid
WGRID=*n*	specifies line thickness for grid

† PCTLMINOR

requests minor tick marks for the percentile axis when you specify PCTLAXIS. The HMINOR option overrides the PCTLMINOR option.

PCTLSCALE

requests scale labels for the theoretical quantile axis in percentile units, resulting in a nonlinear axis scale. Tick marks are drawn uniformly across the axis based on the quantile scale. In all other respects, the plot remains the same, and you must specify HREF= values in quantile units. For a true nonlinear axis, use the PCTLAXIS option or use the PROBPLOT statement.

POWER(ALPHA=*value* | EST <*power-options*>)

creates a power function quantile plot for each value of the shape parameter α given by the mandatory ALPHA= *power-option*. If you specify ALPHA=EST, the procedure creates a plot based on a maximum likelihood estimate for α. To obtain a graphical estimate of α, specify a list of values for the ALPHA= *power-option* and select the value that most nearly linearizes the point pattern. To assess the point pattern, add a diagonal distribution reference line corresponding to θ_0 and σ_0 with the THETA= and SIGMA= *power-options*. Alternatively, you can add a line corresponding to estimated values of the threshold parameter θ_0 and the scale parameter σ with the *power-options* THETA=EST and SIGMA=EST. Agreement between the reference line and the point pattern indicates that the power function distribution with parameters α, θ_0, and σ_0 is a good fit. See the section "Power Function Distribution" on page 440 for details.

RAYLEIGH<(*Rayleigh-options*)>

creates a Rayleigh quantile plot. To assess the point pattern, add a diagonal distribution reference line corresponding to θ_0 and σ_0 with the THETA= and SIGMA= *Rayleigh-options*. Alternatively, you can add a line corresponding to estimated values of the threshold parameter θ_0 and the scale parameter σ with the *Rayleigh-options* THETA=EST and SIGMA=EST. Agreement between the reference line and the point pattern indicates that the exponential distribution with parameters θ_0 and σ_0 is a good fit. See the section "Rayleigh Distribution" on page 440 for details.

RANKADJ=*value*

specifies the adjustment value added to the ranks in the calculation of theoretical percentiles. By default, RANKADJ=$-\frac{3}{8}$, as recommended by Blom (1958). Refer to Chambers et al. (1983) for additional information.

ROTATE

switches the horizontal and vertical axes so that the theoretical quantiles are plotted vertically while the data are plotted horizontally. Regardless of whether the plot has been rotated, horizontal axis options (such as HAXIS=) still refer to the horizontal axis, and vertical axis options (such as VAXIS=) still refer to the vertical axis. All other options that depend on axis placement adjust to the rotated axes.

SIGMA=*value* | EST

specifies the parameter σ, where $\sigma > 0$. Alternatively, you can specify SIGMA=EST to request a maximum likelihood estimate for σ_0. The interpretation and use of the SIGMA= option depend on the distribution option with which it is used, as summarized in Table 4.26. Enclose this option in parentheses after the distribution option.

Table 4.26 Uses of the SIGMA= Option

Distribution Option	Use of the SIGMA= Option
BETA EXPONENTIAL GAMMA PARETO POWER RAYLEIGH WEIBULL	THETA=θ_0 and SIGMA=σ_0 request a distribution reference line corresponding to θ_0 and σ_0.
GUMBEL	MU=μ_0 and SIGMA=σ_0 request a distribution reference line corresponding to μ_0 and σ_0.
LOGNORMAL	SIGMA=$\sigma_1 \ldots \sigma_n$ requests n quantile plots with shape parameters $\sigma_1 \ldots \sigma_n$. The SIGMA= option must be specified.
NORMAL	MU=μ_0 and SIGMA=σ_0 request a distribution reference line corresponding to μ_0 and σ_0. SIGMA=EST requests a line with σ_0 equal to the sample standard deviation.
WEIBULL2	SIGMA=σ_0 and C=c_0 request a distribution reference line corresponding to σ_0 and c_0.

SLOPE=value | EST

specifies the slope for a distribution reference line requested with the LOGNORMAL and WEIBULL2 options. Enclose the SLOPE= option in parentheses after the distribution option. When you use the SLOPE= *lognormal-option* with the LOGNORMAL option, you must also specify a threshold parameter value θ_0 with the THETA= *lognormal-option* to request the line. The SLOPE= *lognormal-option* is an alternative to the ZETA= *lognormal-option* for specifying ζ_0, because the slope is equal to $\exp(\zeta_0)$.

When you use the SLOPE= *Weibull2-option* with the WEIBULL2 option, you must also specify a scale parameter value σ_0 with the SIGMA= *Weibull2-option* to request the line. The SLOPE= *Weibull2-option* is an alternative to the C= *Weibull2-option* for specifying c_0, because the slope is equal to $\frac{1}{c_0}$.

For example, the first and second QQPLOT statements produce the same quantile plots and the third and fourth QQPLOT statements produce the same quantile plots:

```
proc univariate data=Measures;
   qqplot Width / lognormal(sigma=2 theta=0 zeta=0);
   qqplot Width / lognormal(sigma=2 theta=0 slope=1);
   qqplot Width / weibull2(sigma=2 theta=0 c=.25);
   qqplot Width / weibull2(sigma=2 theta=0 slope=4);
```

SQUARE

displays the quantile plot in a square frame. By default, the frame is rectangular.

THETA=value | **EST**
THRESHOLD=value | **EST**

specifies the lower threshold parameter θ for plots requested with the BETA, EXPONENTIAL, GAMMA, PARETO, POWER, RAYLEIGH, LOGNORMAL, WEIBULL, and WEIBULL2 options. Enclose the THETA= option in parentheses after a distribution option. When used with the WEIBULL2 option, the THETA= option specifies the known lower threshold θ_0, for which the default is 0. When used with the other distribution options, the THETA= option specifies θ_0 for a distribution reference line; alternatively in this situation, you can specify THETA=EST to request a maximum likelihood estimate for θ_0. To request the line, you must also specify a scale parameter.

WEIBULL(C=value | **EST** < Weibull-options >**)**
WEIB(C=value | **EST** < Weibull-options >**)**

creates a three-parameter Weibull quantile plot for each value of the required shape parameter c specified by the mandatory C= *Weibull-option*. To create a plot that is based on a maximum likelihood estimate for c, specify C=EST. To obtain a graphical estimate of c, specify a list of values in the C= *Weibull-option* and select the value that most nearly linearizes the point pattern. To assess the point pattern, add a diagonal distribution reference line corresponding to θ_0 and σ_0 with the THETA= and SIGMA= *Weibull-options*. Alternatively, you can add a line corresponding to estimated values of θ_0 and σ_0 with the *Weibull-options* THETA=EST and SIGMA=EST. Agreement between the reference line and the point pattern indicates that the Weibull distribution with parameters c, θ_0, and σ_0 is a good fit. You can specify the SCALE= *Weibull-option* as an alias for the SIGMA= *Weibull-option* and the THRESHOLD= *Weibull-option* as an alias for the THETA= *Weibull-option*. See Example 4.34.

WEIBULL2<(Weibull2-options**) >**
W2<(Weibull2-options**) >**

creates a two-parameter Weibull quantile plot. You should use the WEIBULL2 option when your data have a *known* lower threshold θ_0, which is 0 by default. To specify the threshold value θ_0, use the THETA= *Weibull2-option*. By default, THETA=0. An advantage of the two-parameter Weibull plot over the three-parameter Weibull plot is that the parameters c and σ can be estimated from the slope and intercept of the point pattern. A disadvantage is that the two-parameter Weibull distribution applies only in situations where the threshold parameter is known. To obtain a graphical estimate of θ_0, specify a list of values for the THETA= *Weibull2-option* and select the value that most nearly linearizes the point pattern. To assess the point pattern, add a diagonal distribution reference line corresponding to σ_0 and c_0 with the SIGMA= and C= *Weibull2-options*. Alternatively, you can add a distribution reference line corresponding to estimated values of σ_0 and c_0 with the *Weibull2-options* SIGMA=EST and C=EST. Agreement between the reference line and the point pattern indicates that the Weibull distribution with parameters c_0, θ_0, and σ_0 is a good fit. You can specify the SCALE= *Weibull2-option* as an alias for the SIGMA= *Weibull2-option* and the SHAPE= *Weibull2-option* as an alias for the C= *Weibull2-option*. See Example 4.34.

† **WGRID=**n

specifies the line thickness for the grid when producing traditional graphics. The option does not apply to ODS Graphics output.

ZETA=value | **EST**

specifies a value for the scale parameter ζ for the lognormal quantile plots requested with the LOGNORMAL option. Enclose the ZETA= *lognormal-option* in parentheses after the LOGNORMAL option. To request a distribution reference line with intercept θ_0 and slope $\exp(\zeta_0)$, specify the THETA=θ_0 and ZETA=ζ_0.

VAR Statement

VAR *variables* ;

The VAR statement specifies the analysis variables and their order in the results. By default, if you omit the VAR statement, PROC UNIVARIATE analyzes all numeric variables that are not listed in the other statements.

Using the OUTPUT Statement with the VAR Statement

You must provide a VAR statement when you use an OUTPUT statement. To store the same statistic for several analysis variables in the OUT= data set, you specify a list of names in the OUTPUT statement. PROC UNIVARIATE makes a one-to-one correspondence between the order of the analysis variables in the VAR statement and the list of names that follow a statistic keyword.

WEIGHT Statement

WEIGHT *variable* ;

The WEIGHT statement specifies numeric weights for analysis variables in the statistical calculations. The UNIVARIATE procedure uses the values w_i of the WEIGHT variable to modify the computation of a number of summary statistics by assuming that the variance of the ith value x_i of the analysis variable is equal to σ^2/w_i, where σ is an unknown parameter. The values of the WEIGHT variable do not have to be integers and are typically positive. By default, observations with nonpositive or missing values of the WEIGHT variable are handled as follows:

- If the value is zero, the observation is counted in the total number of observations.
- If the value is negative, it is converted to zero, and the observation is counted in the total number of observations.
- If the value is missing, the observation is excluded from the analysis.

To exclude observations that contain negative and zero weights from the analysis, use EXCLNPWGT. Note that most SAS/STAT procedures, such as PROC GLM, exclude negative and zero weights by default. The weight variable does not change how the procedure determines the range, mode, extreme values, extreme observations, or number of missing values. When you specify a WEIGHT statement, the procedure also computes a weighted standard error and a weighted version of Student's t test. The Student's t test is the only test of location that PROC UNIVARIATE computes when you weight the analysis variables.

When you specify a WEIGHT variable, the procedure uses its values, w_i, to compute weighted versions of the statistics provided in the Moments table. For example, the procedure computes a weighted mean $\bar{x}_w$ and a weighted variance s_w^2 as

$$\bar{x}_w = \frac{\sum_i w_i x_i}{\sum_i w_i}$$

and

$$s_w^2 = \frac{1}{d} \sum_i w_i (x_i - \overline{x}_w)^2$$

where x_i is the ith variable value. The divisor d is controlled by the VARDEF= option in the PROC UNIVARIATE statement.

The WEIGHT statement does not affect the determination of the mode, extreme values, extreme observations, or the number of missing values of the analysis variables. However, the weights w_i are used to compute weighted percentiles. The WEIGHT variable has no effect on graphical displays produced with the plot statements.

The CIPCTLDF, CIPCTLNORMAL, LOCCOUNT, NORMAL, ROBUSTSCALE, TRIMMED=, and WINSORIZED= options are not available with the WEIGHT statement.

To compute weighted skewness or kurtosis, use VARDEF=DF or VARDEF=N in the PROC statement.

You cannot specify the HISTOGRAM, PROBPLOT, or QQPLOT statements with the WEIGHT statement.

When you use the WEIGHT statement, consider which value of the VARDEF= option is appropriate. See VARDEF= and the calculation of weighted statistics for more information.

Dictionary of Common Options

The following entries provide detailed descriptions of *options* that are common to all the plot statements: CDFPLOT, HISTOGRAM, PPPLOT, PROBPLOT, and QQPLOT. Options marked with † are applicable only when traditional graphics are produced.

ALPHADELTA=*value*

specifies the change in successive estimates of $\hat{\alpha}$ at which iteration terminates in the Newton-Raphson approximation of the maximum likelihood estimate of α for gamma distributions requested with the GAMMA option. Enclose the ALPHADELTA= option in parentheses after the GAMMA keyword. Iteration continues until the change in α is less than the value specified or the number of iterations exceeds the value of the MAXITER= option. The default value is 0.00001.

ALPHAINITIAL=*value*

specifies the initial value for $\hat{\alpha}$ in the Newton-Raphson approximation of the maximum likelihood estimate of α for gamma distributions requested with the GAMMA option. Enclose the ALPHAINITIAL= option in parentheses after the GAMMA keyword. The default value is Thom's approximation of the estimate of α (refer to Johnson, Kotz, and Balakrishnan (1995).

† ANNOKEY

applies the annotation requested with the ANNOTATE= option only to the key cell of a comparative plot. By default, the procedure applies annotation to all of the cells. This option is not available unless you use the CLASS statement. You can use the KEYLEVEL= option in the CLASS statement to specify the key cell.

† ANNOTATE=*SAS-data-set*
† ANNO=*SAS-data-set*
specifies an input data set that contains annotate variables as described in *SAS/GRAPH: Help*. The ANNOTATE= data set you specify in the plot statement is used for all plots created by the statement. You can also specify an ANNOTATE= data set in the PROC UNIVARIATE statement to enhance all plots created by the procedure (see the section "ANNOTATE= Data Sets" on page 444).

† CAXIS=*color*
† CAXES=*color*
† CA=*color*
specifies the color for the axes and tick marks. This option overrides any COLOR= specifications in an AXIS statement.

† CFRAME=*color*
specifies the color for the area that is enclosed by the axes and frame. The area is not filled by default.

† CFRAMESIDE=*color*
specifies the color to fill the frame area for the row labels that display along the left side of a comparative plot. This color also fills the frame area for the label of the corresponding CLASS variable (if you associate a label with the variable). By default, these areas are not filled. This option is not available unless you use the CLASS statement.

† CFRAMETOP=*color*
specifies the color to fill the frame area for the column labels that display across the top of a comparative plot. This color also fills the frame area for the label of the corresponding CLASS variable (if you associate a label with the variable). By default, these areas are not filled. This option is not available unless you use the CLASS statement.

† CHREF=*color* | *(color-list)*
† CH=*color* | *(color-list)*
specifies the colors for horizontal axis reference lines requested by the HREF= option. If you specify a single color, it is used for all HREF= lines. Otherwise, if there are fewer colors specified than reference lines requested, the remaining lines are displayed with the default reference line color. You can also specify the value _*default* in the color list to request the default color.

† COLOR=*color*
† COLOR=*color-list*
specifies the color of the curve or reference line associated with a distribution or kernel density estimate. Enclose the COLOR= option in parentheses after a distribution option or the KERNEL option. In a HISTOGRAM statement, you can specify a list of colors in parentheses for multiple density curves.

CONTENTS='*string*'
specifies the table of contents grouping entry for output produced by the plot statement. You can specify CONTENTS='' to suppress the grouping entry.

† CPROP=color | EMPTY
CPROP
specifies the color for a horizontal bar whose length (relative to the width of the tile) indicates the proportion of the total frequency that is represented by the corresponding cell in a comparative plot. By default, no proportion bars are displayed. This option is not available unless you use the CLASS statement. You can specify the keyword EMPTY to display empty bars. See Example 4.20.

For ODS Graphics and traditional graphics with the GSTYLE system option in effect, you can specify CPROP with no argument to produce proportion bars using an appropriate color from the ODS style.

† CSTATREF=color | (color-list)
specifies the colors for reference lines that you request with the STATREF= option. If you specify a single color, it is used for all STATREF= lines. Otherwise, if there are fewer colors specified than reference lines requested, the remaining lines are displayed with the default reference line color. You can also specify the value _default in the color list to request the default color.

† CTEXT=color
† CT=color
specifies the color for tick mark values and axis labels. The default is the color specified for the CTEXT= option in the GOPTIONS statement.

† CTEXTSIDE=color
specifies the color for the row labels that display along the left side of a comparative plot. By default, the color specified by the CTEXT= option is used. If you omit the CTEXT= option, the color specified in the GOPTIONS statement is used. This option is not available unless you use the CLASS statement. You can specify the CFRAMESIDE= option to change the background color for the row labels.

† CTEXTTOP=color
specifies the color for the column labels that display along the left side of a comparative plot. By default, the color specified by the CTEXT= option is used. If you omit the CTEXT= option, the color specified in the GOPTIONS statement is used. This option is not available unless you specify the CLASS statement. You can use the CFRAMETOP= option to change the background color for the column labels.

† CVREF=color | (color-list)
† CV=color | (color-list)
specifies the colors for lines requested with the VREF= option. If you specify a single color, it is used for all VREF= lines. Otherwise, if there are fewer colors specified than reference lines requested, the remaining lines are displayed with the default reference line color. You can also specify the value _default in the color list to request the default color.

† DESCRIPTION='string'
† DES='string'
specifies a description, up to 256 characters long, that appears in the PROC GREPLAY master menu for a traditional graphics chart. The default value is the analysis variable name.

FITINTERVAL=value
specifies the value of z for the method of percentiles when this method is used to fit a Johnson S_B or Johnson S_U distribution. The FITINTERVAL= option is specified in parentheses after the SB or SU option. The default of z is 0.524.

FITMETHOD=PERCENTILE | MLE | MOMENTS

specifies the method used to estimate the parameters of a Johnson S_B or Johnson S_U distribution. The FITMETHOD= option is specified in parentheses after the SB or SU option. By default, the method of percentiles is used. You can specify the MLE keyword to request maximum likelihood estimation. The OPTBOUNDRANGE=, OPTMAXITER=, OPTMAXSTARTS=, OPTPRINT, OPTSEED=, and OPTTOLERANCE= options control the optimizer that performs the maximum likelihood calculation.

FITTOLERANCE=value

specifies the tolerance value for the ratio criterion when the method of percentiles is used to fit a Johnson S_B or Johnson S_U distribution. The FITTOLERANCE= option is specified in parentheses after the SB or SU option. The default value is 0.01.

† FONT=font

specifies a software font for reference line and axis labels. You can also specify fonts for axis labels in an AXIS statement. The FONT= font takes precedence over the FTEXT= font specified in the GOPTIONS statement.

HAXIS=value

specifies the name of an AXIS statement describing the horizontal axis.

† HEIGHT=value

specifies the height, in percentage screen units, of text for axis labels, tick mark labels, and legends. This option takes precedence over the HTEXT= option in the GOPTIONS statement.

† HMINOR=n

† HM=n

specifies the number of minor tick marks between each major tick mark on the horizontal axis. Minor tick marks are not labeled. By default, HMINOR=0.

HREF=values

draws reference lines that are perpendicular to the horizontal axis at the values that you specify. Also see the CHREF= and LHREF= options.

HREFLABELS='label1' ... 'labeln'

HREFLABEL='label1' ... 'labeln'

HREFLAB='label1' ... 'labeln'

specifies labels for the lines requested by the HREF= option. The number of labels must equal the number of lines. Enclose each label in quotes. Labels can have up to 16 characters.

HREFLABPOS=n

specifies the vertical position of HREFLABELS= labels, as described in the following table.

n	Position
1	along top of plot
2	staggered from top to bottom of plot
3	along bottom of plot
4	staggered from bottom to top of plot

By default, HREFLABPOS=1. **NOTE:** HREFLABPOS=2 and HREFLABPOS=4 are not supported for ODS Graphics output.

† INFONT=_font_

specifies a software font to use for text inside the framed areas of the plot. The INFONT= option takes precedence over the FTEXT= option in the GOPTIONS statement. For a list of fonts, see *SAS/GRAPH: Help*.

† INHEIGHT=_value_

specifies the height, in percentage screen units, of text used inside the framed areas of the histogram. By default, the height specified by the HEIGHT= option is used. If you do not specify the HEIGHT= option, the height specified with the HTEXT= option in the GOPTIONS statement is used.

† INTERTILE=_value_

specifies the distance in horizontal percentage screen units between the framed areas, called *tiles*, of a comparative plot. By default, INTERTILE=0.75 percentage screen units. This option is not available unless you use the CLASS statement. You can specify INTERTILE=0 to create contiguous tiles.

ITPRINT

requests a table that shows the iteration history and optimizer details of the maximum likelihood parameter estimation of a Weibull distribution requested with the WEIBULL or WEIBULL2 option.

† L=_linetype_

† L=_linetype-list_

specifies the line type of the curve or reference line associated with a distribution or kernel density estimate. Enclose the L= option in parentheses after the distribution option or the KERNEL option. In a HISTOGRAM statement, you can specify a list of line types in parentheses for multiple density curves.

† LHREF=_linetype_ | _linetype-list_

† LH=_linetype_ | _linetype-list_

specifies the line types for the reference lines that you request with the HREF= option. If you specify a single line type, it is used for all HREF= lines. Otherwise, if there are fewer line types specified than reference lines requested, the remaining lines are displayed with the default reference line type. You can also specify line type 0 to request the default line type.

† LSTATREF=_linetype_ | _linetype-list_

specifies the line types for the reference lines that you request with the STATREF= option. If you specify a single line type, it is used for all STATREF= lines. Otherwise, if there are fewer line types specified than reference lines requested, the remaining lines are displayed with the default reference line type. You can also specify line type 0 to request the default line type.

† LVREF=_linetype_ | _linetype-list_

† LV=_linetype_ | _linetype-list_

specifies the line types for lines requested with the VREF= option. If you specify a single line type, it is used for all VREF= lines. Otherwise, if there are fewer line types specified than reference lines requested, the remaining lines are displayed with the default reference line type. You can also specify line type 0 to request the default line type.

MAXITER=_n_

specifies the maximum number of iterations in the Newton-Raphson approximation of the maximum likelihood estimate of α for gamma distributions requested with the GAMMA option and c for Weibull distributions requested with the WEIBULL and WEIBULL2 options. Enclose the MAXITER= option in parentheses after the GAMMA, WEIBULL, or WEIBULL2 keywords. The default value of n is 20.

† NAME='*string*'

specifies a name for the plot, up to eight characters long, that appears in the PROC GREPLAY master menu for a traditional graphics chart. The default value is 'UNIVAR'.

NCOLS=*n*
NCOL=*n*

specifies the number of columns per panel in a comparative plot. This option is not available unless you use the CLASS statement. By default, NCOLS=1 if you specify only one CLASS variable, and NCOLS=2 if you specify two CLASS variables. If you specify two CLASS variables, you can use the NCOLS= option with the NROWS= option.

NOFRAME

suppresses the frame around the subplot area.

NOHLABEL

suppresses the label for the horizontal axis. You can use this option to reduce clutter.

NOVLABEL

suppresses the label for the vertical axis. You can use this option to reduce clutter.

NOVTICK

suppresses the tick marks and tick mark labels for the vertical axis. This option also suppresses the label for the vertical axis.

NROWS=*n*
NROW=*n*

specifies the number of rows per panel in a comparative plot. This option is not available unless you use the CLASS statement. By default, NROWS=2. If you specify two CLASS variables, you can use the NCOLS= option with the NROWS= option.

ODSFOOTNOTE=FOOTNOTE | FOOTNOTE1 | '*string*'

adds a footnote to ODS Graphics output. If you specify the FOOTNOTE (or FOOTNOTE1) keyword, the value of the SAS FOOTNOTE statement is used as the graph footnote. If you specify a quoted string, that is used as the footnote. The quoted string can contain either of the following escaped characters, which are replaced with the appropriate values from the analysis:

\n   is replaced by the analysis variable name.

\l   is replaced by the analysis variable label (or name if the analysis variable has no label).

ODSFOOTNOTE2=FOOTNOTE2 | '*string*'

adds a secondary footnote to ODS Graphics output. If you specify the FOOTNOTE2 keyword, the value of the SAS FOOTNOTE2 statement is used as the secondary graph footnote. If you specify a quoted string, that is used as the secondary footnote. The quoted string can contain any of the following escaped characters, which are replaced with the appropriate values from the analysis:

\n   is replaced by the analysis variable name.

\l   is replaced by the analysis variable label (or name if the analysis variable has no label).

ODSTITLE=TITLE | TITLE1 | NONE | DEFAULT | LABELFMT | '*string*'

specifies a title for ODS Graphics output. You can specify the following values:

TITLE (or TITLE1)   uses the value of the SAS TITLE statement as the graph title.

NONE   suppresses all titles from the graph.

DEFAULT   uses the default ODS Graphics title (a descriptive title that consists of the plot type and the analysis variable name).

LABELFMT   uses the default ODS Graphics title with the variable label instead of the variable name.

If you specify a quoted string, that is used as the graph title. The quoted string can contain the following escaped characters, which are replaced with the appropriate values from the analysis:

\n   is replaced by the analysis variable name.

\l   is replaced by the analysis variable label (or name if the analysis variable has no label).

ODSTITLE2=TITLE2 | '*string*'

specifies a secondary title for ODS Graphics output. If you specify the TITLE2 keyword, the value of the SAS TITLE2 statement is used as the secondary graph title. If you specify a quoted string, that is used as the secondary title. The quoted string can contain the following escaped characters, which are replaced with the appropriate values from the analysis:

\n   is replaced by the analysis variable name.

\l   is replaced by the analysis variable label (or name if the analysis variable has no label).

OVERLAY

specifies that plots associated with different levels of a CLASS variable be overlaid onto a single plot, rather than displayed as separate cells in a comparative plot. If you specify the OVERLAY option with one CLASS variable, the output associated with each level of the CLASS variable is overlaid on a single plot. If you specify the OVERLAY option with two CLASS variables, a comparative plot based on the first CLASS variable's levels is produced. Each cell in this comparative plot contains overlaid output associated with the levels of the second CLASS variable.

The OVERLAY option applies only to ODS Graphics output.

SCALE=*value*

is an alias for the SIGMA= option for distributions requested by the BETA, EXPONENTIAL, GAMMA, SB, SU, WEIBULL, and WEIBULL2 options and for the ZETA= option for distributions requested by the LOGNORMAL option.

SHAPE=*value*

is an alias for the ALPHA= option for distributions requested by the GAMMA option, for the SIGMA= option for distributions requested by the LOGNORMAL option, and for the C= option for distributions requested by the WEIBULL and WEIBULL2 options.

STATREF=_keyword-list_

draws reference lines at the values of the statistics that are requested in the space-delimited *keyword-list*. These reference lines are perpendicular to the horizontal axis in a histogram or cdf plot, and perpendicular to the vertical axis in a probability or Q-Q plot (unless the ROTATE option is specified). The STATREF= option does not apply to the PPPLOT statement.

Valid *keywords* are listed in the following table.

Keyword	Statistic
MAX	Largest value
MEAN	Sample mean
MEDIAN | Q2	Median (50th percentile)
MIN	Smallest value
MODE	Most frequent value
P *p*	*p*th percentile
Q1	Lower quartile (25th percentile)
Q3	Upper quartile (75th percentile)
factor STD	*factor* standard deviations from the mean

Note that the *factor* that is specified with the STD keyword can be positive (which puts a reference line above the mean) or negative (which puts a reference line below the mean).

Also see the CSTATREF=, LSTATREF=, STATREFLABELS=, and STATREFSUBCHAR= options.

STATREFLABELS='*label1*' ... '*labeln*'

STATREFLABEL='*label1*' ... '*labeln*'

STATREFLAB='*label1*' ... '*labeln*'

specifies labels for the lines that you request with the STATREF= option. The number of labels must equal the number of lines. Enclose each label in quotes. Labels can be up to 16 characters long.

STATREFSUBCHAR='*character*'

specifies a substitution character (such as #) for labels that you specify with the STATREFLABELS= option. When the labels are displayed in a graph, the first occurrence of the specified character in each label is replaced with the value of the corresponding STATREF= statistic.

For example, suppose the mean of variable Weight is 155. The following statement creates a histogram with a vertical reference line at 155 with the label "Average=155":

```
histogram Weight / statref=mean statreflabel='Average=#' statrefsubchar='#';
```

† TURNVLABELS

† TURNVLABEL

turns the characters in the vertical axis labels so that they display vertically. This happens by default when you use a hardware font.

VAXIS=_name_

VAXIS=_value-list_

specifies the name of an AXIS statement describing the vertical axis. In a HISTOGRAM statement, you can alternatively specify a *value-list* for the vertical axis.

VAXISLABEL='*label*'

specifies a label for the vertical axis. Labels can have up to 40 characters.

† VMINOR=*n*

† VM=*n*

specifies the number of minor tick marks between each major tick mark on the vertical axis. Minor tick marks are not labeled. The default is zero.

VREF=*value-list*

draws reference lines perpendicular to the vertical axis at the values specified. Also see the CVREF= and LVREF= options.

VREFLABELS='*label1*'...'*labeln*'

VREFLABEL='*label1*'...'*labeln*'

VREFLAB='*label1*'...'*labeln*'

specifies labels for the lines requested by the VREF= option. The number of labels must equal the number of lines. Enclose each label in quotes. Labels can have up to 16 characters.

VREFLABPOS=*n*

specifies the horizontal position of VREFLABELS= labels. If you specify VREFLABPOS=1, the labels are positioned at the left of the plot. If you specify VREFLABPOS=2, the labels are positioned at the right of the plot. By default, VREFLABPOS=1.

† W=*value*

† W=*value-list*

specifies the width in pixels of the curve or reference line associated with a distribution or kernel density estimate. Enclose the W= option in parentheses after the distribution option or the KERNEL option. In a HISTOGRAM statement, you can specify a list of widths in parentheses for multiple density curves.

† WAXIS=*n*

specifies the line thickness, in pixels, for the axes and frame.

Details: UNIVARIATE Procedure

Missing Values

PROC UNIVARIATE excludes missing values for an analysis variable before calculating statistics. Each analysis variable is treated individually; a missing value for an observation in one variable does not affect the calculations for other variables. The statements handle missing values as follows:

- If a BY or an ID variable value is missing, PROC UNIVARIATE treats it like any other BY or ID variable value. The missing values form a separate BY group.

- If the FREQ variable value is missing or nonpositive, PROC UNIVARIATE excludes the observation from the analysis.

- If the WEIGHT variable value is missing, PROC UNIVARIATE excludes the observation from the analysis.

PROC UNIVARIATE tabulates the number of missing values and reports this information in the ODS table named "Missing Values." See the section "ODS Table Names" on page 451. Before the number of missing values is tabulated, PROC UNIVARIATE excludes observations when either of the following conditions exist:

- you use the FREQ statement and the frequencies are nonpositive

- you use the WEIGHT statement and the weights are missing or nonpositive (you must specify the EXCLNPWGT option)

Rounding

When you specify ROUND=u, PROC UNIVARIATE rounds a variable by using the rounding unit to divide the number line into intervals with midpoints of the form ui, where u is the nonnegative rounding unit and i is an integer. The interval width is u. Any variable value that falls in an interval is rounded to the midpoint of that interval. A variable value that is midway between two midpoints, and is therefore on the boundary of two intervals, rounds to the even midpoint. Even midpoints occur when i is an even integer $(0, \pm 2, \pm 4, \ldots)$.

When ROUND=1 and the analysis variable values are between -2.5 and 2.5, the intervals are as follows:

Table 4.27 Intervals for Rounding When ROUND=1

i	Interval	Midpoint	Left endpt rounds to	Right endpt rounds to
−2	[−2.5, −1.5]	−2	−2	−2
−1	[−1.5, −0.5]	−1	−2	0
0	[−0.5, 0.5]	0	0	0
1	[0.5, 1.5]	1	0	2
2	[1.5, 2.5]	2	2	2

When ROUND=0.5 and the analysis variable values are between -1.25 and 1.25, the intervals are as follows:

Table 4.28 Intervals for Rounding When ROUND=0.5

i	Interval	Midpoint	Left endpt rounds to	Right endpt rounds to
−2	[−1.25, −0.75]	−1.0	−1	−1
−1	[−0.75, −0.25]	−0.5	−1	0
0	[−0.25, 0.25]	0.0	0	0
1	[0.25, 0.75]	0.5	0	1
2	[0.75, 1.25]	1.0	1	1

As the rounding unit increases, the interval width also increases. This reduces the number of unique values and decreases the amount of memory that PROC UNIVARIATE needs.

Descriptive Statistics

This section provides computational details for the descriptive statistics that are computed with the PROC UNIVARIATE statement. These statistics can also be saved in an OUT= data set by specifying keywords listed in Table 4.14 in the OUTPUT statement.

Standard algorithms (Fisher 1973) are used to compute the moment statistics. The computational methods used by the UNIVARIATE procedure are consistent with those used by other SAS procedures for calculating descriptive statistics.

The following sections give specific details on a number of statistics calculated by the UNIVARIATE procedure.

Mean

The sample mean is calculated as

$$\bar{x}_w = \frac{\sum_{i=1}^{n} w_i x_i}{\sum_{i=1}^{n} w_i}$$

where n is the number of nonmissing values for a variable, x_i is the ith value of the variable, and w_i is the weight associated with the ith value of the variable. If there is no WEIGHT variable, the formula reduces to

$$\bar{x} = \frac{1}{n} \sum_{i=1}^{n} x_i$$

Sum

The sum is calculated as $\sum_{i=1}^{n} w_i x_i$, where n is the number of nonmissing values for a variable, x_i is the ith value of the variable, and w_i is the weight associated with the ith value of the variable. If there is no WEIGHT variable, the formula reduces to $\sum_{i=1}^{n} x_i$.

Sum of the Weights

The sum of the weights is calculated as $\sum_{i=1}^{n} w_i$, where n is the number of nonmissing values for a variable and w_i is the weight associated with the ith value of the variable. If there is no WEIGHT variable, the sum of the weights is n.

Variance

The variance is calculated as

$$\frac{1}{d} \sum_{i=1}^{n} w_i (x_i - \bar{x}_w)^2$$

where n is the number of nonmissing values for a variable, x_i is the ith value of the variable, $\bar{x}_w$ is the weighted mean, w_i is the weight associated with the ith value of the variable, and d is the divisor controlled by the VARDEF= option in the PROC UNIVARIATE statement:

$$d = \begin{cases} n-1 & \text{if VARDEF=DF (default)} \\ n & \text{if VARDEF=N} \\ (\sum_i w_i) - 1 & \text{if VARDEF=WDF} \\ \sum_i w_i & \text{if VARDEF=WEIGHT | WGT} \end{cases}$$

If there is no WEIGHT variable, the formula reduces to

$$\frac{1}{d} \sum_{i=1}^{n} (x_i - \bar{x})^2$$

Standard Deviation

The standard deviation is calculated as

$$s_w = \sqrt{\frac{1}{d} \sum_{i=1}^{n} w_i (x_i - \bar{x}_w)^2}$$

where n is the number of nonmissing values for a variable, x_i is the ith value of the variable, $\bar{x}_w$ is the weighted mean, w_i is the weight associated with the ith value of the variable, and d is the divisor controlled by the VARDEF= option in the PROC UNIVARIATE statement. If there is no WEIGHT variable, the formula reduces to

$$s = \sqrt{\frac{1}{d} \sum_{i=1}^{n} (x_i - \bar{x})^2}$$

Skewness

The sample skewness, which measures the tendency of the deviations to be larger in one direction than in the other, is calculated as follows depending on the VARDEF= option:

Table 4.29 Formulas for Skewness

VARDEF	Formula
DF (default)	$\dfrac{n}{(n-1)(n-2)} \sum_{i=1}^{n} w_i^{3/2} \left(\dfrac{x_i - \bar{x}_w}{s_w}\right)^3$
N	$\dfrac{1}{n} \sum_{i=1}^{n} w_i^{3/2} \left(\dfrac{x_i - \bar{x}_w}{s_w}\right)^3$
WDF	missing
WEIGHT	WGT	missing

where n is the number of nonmissing values for a variable, x_i is the ith value of the variable, $\bar{x}_w$ is the sample average, s is the sample standard deviation, and w_i is the weight associated with the ith value of the variable. If VARDEF=DF, then n must be greater than 2. If there is no WEIGHT variable, then $w_i = 1$ for all $i = 1, \ldots, n$.

The sample skewness can be positive or negative; it measures the asymmetry of the data distribution and estimates the theoretical skewness $\sqrt{\beta_1} = \mu_3 \mu_2^{-\frac{3}{2}}$, where μ_2 and μ_3 are the second and third central moments. Observations that are normally distributed should have a skewness near zero.

Kurtosis

The sample kurtosis, which measures the heaviness of tails, is calculated as follows depending on the VARDEF= option:

Table 4.30 Formulas for Kurtosis

VARDEF	Formula
DF (default)	$\dfrac{n(n+1)}{(n-1)(n-2)(n-3)} \sum_{i=1}^{n} w_i^2 \left(\dfrac{x_i - \bar{x}_w}{s_w} \right)^4 - \dfrac{3(n-1)^2}{(n-2)(n-3)}$
N	$\dfrac{1}{n} \sum_{i=1}^{n} w_i^2 \left(\dfrac{x_i - \bar{x}_w}{s_w} \right)^4 - 3$
WDF	missing
WEIGHT | WGT	missing

where n is the number of nonmissing values for a variable, x_i is the ith value of the variable, $\bar{x}_w$ is the sample average, s_w is the sample standard deviation, and w_i is the weight associated with the ith value of the variable. If VARDEF=DF, then n must be greater than 3. If there is no WEIGHT variable, then $w_i = 1$ for all $i = 1, \ldots, n$.

The sample kurtosis measures the heaviness of the tails of the data distribution. It estimates the adjusted theoretical kurtosis denoted as $\beta_2 - 3$, where $\beta_2 = \frac{\mu_4}{\mu_2^2}$, and μ_4 is the fourth central moment. Observations that are normally distributed should have a kurtosis near zero.

Coefficient of Variation (CV)

The coefficient of variation is calculated as

$$CV = \frac{100 \times s_w}{\bar{x}_w}$$

Geometric Mean

The geometric mean is calculated as

$$\left(\prod_{i=1}^{n} x_i^{w_i}\right)^{1/\sum_{i=1}^{n} w_i}$$

where n is the number of nonmissing values for a variable, x_i is the ith value of the variable, and w_i is the weight associated with the ith value of the variable.

If there is no WEIGHT variable, the formula reduces to

$$\left(\prod_{i=1}^{n} x_i\right)^{1/n}$$

If any x_i is negative, the geometric mean is set to missing.

Calculating the Mode

The mode is the value that occurs most often in the data. PROC UNIVARIATE counts repetitions of the values of the analysis variables or, if you specify the ROUND= option, the rounded values. If a tie occurs for the most frequent value, the procedure reports the lowest mode in the table labeled "Basic Statistical Measures" in the statistical output. To list all possible modes, use the MODES option in the PROC UNIVARIATE statement. When no repetitions occur in the data (as with truly continuous data), the procedure does not report the mode. The WEIGHT statement has no effect on the mode. See Example 4.2.

Calculating Percentiles

The UNIVARIATE procedure automatically computes the 1st, 5th, 10th, 25th, 50th, 75th, 90th, 95th, and 99th percentiles (quantiles), as well as the minimum and maximum of each analysis variable. To compute percentiles other than these default percentiles, use the PCTLPTS= and PCTLPRE= options in the OUTPUT statement.

You can specify one of five definitions for computing the percentiles with the PCTLDEF= option. Let n be the number of nonmissing values for a variable, and let $x_1, x_2, \ldots, x_n$ represent the ordered values of the variable. Let the tth percentile be y, set $p = \frac{t}{100}$, and let

$$\begin{aligned} np &= j + g \quad \text{when PCTLDEF=1, 2, 3, or 5} \\ (n+1)p &= j + g \quad \text{when PCTLDEF=4} \end{aligned}$$

where j is the integer part of np, and g is the fractional part of np. Then the PCTLDEF= option defines the tth percentile, y, as described in the following table.

Table 4.31 Percentile Definitions

PCTLDEF	Description	Formula
1	weighted average at x_{np}	$y = (1-g)x_j + gx_{j+1}$ where x_0 is taken to be x_1
2	observation numbered closest to np	$y = x_j$ if $g < \frac{1}{2}$ $y = x_j$ if $g = \frac{1}{2}$ and j is even $y = x_{j+1}$ if $g = \frac{1}{2}$ and j is odd $y = x_{j+1}$ if $g > \frac{1}{2}$
3	empirical distribution function	$y = x_j$ if $g = 0$ $y = x_{j+1}$ if $g > 0$
4	weighted average aimed at $x_{(n+1)p}$	$y = (1-g)x_j + gx_{j+1}$ where x_{n+1} is taken to be x_n
5	empirical distribution function with averaging	$y = \frac{1}{2}(x_j + x_{j+1})$ if $g = 0$ $y = x_{j+1}$ if $g > 0$

Weighted Percentiles

When you use a WEIGHT statement, the percentiles are computed differently. The $100p$th weighted percentile y is computed from the empirical distribution function with averaging:

$$y = \begin{cases} x_1 & \text{if } w_1 > pW \\ \frac{1}{2}(x_i + x_{i+1}) & \text{if } \sum_{j=1}^{i} w_j = pW \\ x_{i+1} & \text{if } \sum_{j=1}^{i} w_j < pW < \sum_{j=1}^{i+1} w_j \end{cases}$$

where w_i is the weight associated with x_i and $W = \sum_{i=1}^{n} w_i$ is the sum of the weights.

Note that the PCTLDEF= option is not applicable when a WEIGHT statement is used. However, in this case, if all the weights are identical, the weighted percentiles are the same as the percentiles that would be computed without a WEIGHT statement and with PCTLDEF=5.

Confidence Limits for Percentiles

You can use the CIPCTLNORMAL option to request confidence limits for percentiles, assuming the data are normally distributed. These limits are described in Section 4.4.1 of Hahn and Meeker (1991). When $0 < p < \frac{1}{2}$, the two-sided $100(1-\alpha)\%$ confidence limits for the $100p$th percentile are

$$\text{lower limit} = \bar{X} - g'(\tfrac{\alpha}{2}; 1-p, n)s$$
$$\text{upper limit} = \bar{X} - g'(1-\tfrac{\alpha}{2}; p, n)s$$

where n is the sample size. When $\frac{1}{2} \leq p < 1$, the two-sided $100(1-\alpha)\%$ confidence limits for the $100p$th percentile are

$$\text{lower limit} = \bar{X} + g'(\tfrac{\alpha}{2}; 1-p, n)s$$
$$\text{upper limit} = \bar{X} + g'(1-\tfrac{\alpha}{2}; p, n)s$$

One-sided $100(1-\alpha)\%$ confidence bounds are computed by replacing $\frac{\alpha}{2}$ by α in the appropriate preceding equation. The factor $g'(\gamma, p, n)$ is related to the noncentral t distribution and is described in Owen and Hua (1977) and Odeh and Owen (1980). See Example 4.10.

You can use the CIPCTLDF option to request distribution-free confidence limits for percentiles. In particular, it is not necessary to assume that the data are normally distributed. These limits are described in Section 5.2 of Hahn and Meeker (1991). The two-sided $100(1-\alpha)\%$ confidence limits for the $100p$th percentile are

$$\text{lower limit} = X_{(l)}$$
$$\text{upper limit} = X_{(u)}$$

where $X_{(j)}$ is the jth order statistic when the data values are arranged in increasing order:

$$X_{(1)} \leq X_{(2)} \leq \cdots \leq X_{(n)}$$

The lower rank l and upper rank u are integers that are symmetric (or nearly symmetric) around $\lfloor np \rfloor + 1$, where $\lfloor np \rfloor$ is the integer part of np and n is the sample size. Furthermore, l and u are chosen so that $X_{(l)}$ and $X_{(u)}$ are as close to $X_{\lfloor np \rfloor + 1}$ as possible while satisfying the coverage probability requirement,

$$Q(u-1; n, p) - Q(l-1; n, p) \geq 1 - \alpha$$

where $Q(k; n, p)$ is the cumulative binomial probability,

$$Q(k; n, p) = \sum_{i=0}^{k} \binom{n}{i} p^i (1-p)^{n-i}$$

In some cases, the coverage requirement cannot be met, particularly when n is small and p is near 0 or 1. To relax the requirement of symmetry, you can specify CIPCTLDF(TYPE = ASYMMETRIC). This option requests symmetric limits when the coverage requirement can be met, and asymmetric limits otherwise.

If you specify CIPCTLDF(TYPE = LOWER), a one-sided $100(1-\alpha)\%$ lower confidence bound is computed as $X_{(l)}$, where l is the largest integer that satisfies the inequality

$$1 - Q(l-1; n, p) \geq 1 - \alpha \quad \text{where } 0 < u \leq n$$

If you specify CIPCTLDF(TYPE = UPPER), a one-sided $100(1-\alpha)\%$ upper confidence bound is computed as $X_{(u)}$, where u is the smallest integer that satisfies the inequality

$$Q(u-1; n, p) \geq 1 - \alpha \quad \text{where } 0 < u \leq n$$

Note that confidence limits for percentiles are not computed when a WEIGHT statement is specified. See Example 4.10.

Tests for Location

PROC UNIVARIATE provides three tests for location: Student's t test, the sign test, and the Wilcoxon signed rank test. All three tests produce a test statistic for the null hypothesis that the mean or median is equal to a given value μ_0 against the two-sided alternative that the mean or median is not equal to μ_0. By default, PROC UNIVARIATE sets the value of μ_0 to zero. You can use the MU0= option in the PROC UNIVARIATE statement to specify the value of μ_0. Student's t test is appropriate when the data are from an approximately normal population; otherwise, use nonparametric tests such as the sign test or the signed rank test. For large sample situations, the t test is asymptotically equivalent to a z test. If you use the WEIGHT statement, PROC UNIVARIATE computes only one weighted test for location, the t test. You must use the default value for the VARDEF= option in the PROC statement (VARDEF=DF). See Example 4.12.

You can also use these tests to compare means or medians of *paired data*. Data are said to be paired when subjects or units are matched in pairs according to one or more variables, such as pairs of subjects with the same age and gender. Paired data also occur when each subject or unit is measured at two times or under two conditions. To compare the means or medians of the two times, create an analysis variable that is the difference between the two measures. The test that the mean or the median difference of the variables equals zero is equivalent to the test that the means or medians of the two original variables are equal. Note that you can also carry out these tests by using the PAIRED statement in the TTEST procedure; see Chapter 122, "The TTEST Procedure" (*SAS/STAT User's Guide*). Also see Example 4.13.

Student's *t* Test

PROC UNIVARIATE calculates the t statistic as

$$t = \frac{\bar{x} - \mu_0}{s/\sqrt{n}}$$

where $\bar{x}$ is the sample mean, n is the number of nonmissing values for a variable, and s is the sample standard deviation. The null hypothesis is that the population mean equals μ_0. When the data values are approximately normally distributed, the probability under the null hypothesis of a t statistic that is as extreme, or more extreme, than the observed value (the p-value) is obtained from the t distribution with $n - 1$ degrees of freedom. For large n, the t statistic is asymptotically equivalent to a z test. When you use the WEIGHT statement and the default value of VARDEF=, which is DF, the t statistic is calculated as

$$t_w = \frac{\bar{x}_w - \mu_0}{s_w/\sqrt{\sum_{i=1}^{n} w_i}}$$

where $\bar{x}_w$ is the weighted mean, s_w is the weighted standard deviation, and w_i is the weight for ith observation. The t_w statistic is treated as having a Student's t distribution with $n - 1$ degrees of freedom. If you specify the EXCLNPWGT option in the PROC statement, n is the number of nonmissing observations when the value of the WEIGHT variable is positive. By default, n is the number of nonmissing observations for the WEIGHT variable.

Sign Test

PROC UNIVARIATE calculates the sign test statistic as

$$M = (n^+ - n^-)/2$$

where n^+ is the number of values that are greater than μ_0, and n^- is the number of values that are less than μ_0. Values equal to μ_0 are discarded. Under the null hypothesis that the population median is equal to μ_0, the p-value for the observed statistic M_{obs} is

$$\Pr(|M_{obs}| \geq |M|) = 0.5^{(n_t-1)} \sum_{j=0}^{min(n^+,n^-)} \binom{n_t}{i}$$

where $n_t = n^+ + n^-$ is the number of x_i values not equal to μ_0.

NOTE: If n^+ and n^- are equal, the p-value is equal to one.

Wilcoxon Signed Rank Test

The signed rank statistic S is computed as

$$S = \sum_{i:x_i>\mu_0} r_i^+ - \frac{n_t(n_t+1)}{4}$$

where r_i^+ is the rank of $|x_i - \mu_0|$ after discarding values of $x_i = \mu_0$, and n_t is the number of x_i values not equal to μ_0. Average ranks are used for tied values.

If $n_t \leq 20$, the significance of S is computed from the exact distribution of S, where the distribution is a convolution of scaled binomial distributions. When $n_t > 20$, the significance of S is computed by treating

$$S\sqrt{\frac{n_t-1}{n_t V - S^2}}$$

as a Student's t variate with $n_t - 1$ degrees of freedom. V is computed as

$$V = \frac{1}{24}n_t(n_t+1)(2n_t+1) - \frac{1}{48}\sum t_i(t_i+1)(t_i-1)$$

where the sum is over groups tied in absolute value and where t_i is the number of values in the ith group (Iman 1974; Conover 1980). The null hypothesis tested is that the mean (or median) is μ_0, assuming that the distribution is symmetric. Refer to Lehmann and D'Abrera (1975).

Confidence Limits for Parameters of the Normal Distribution

The two-sided $100(1-\alpha)\%$ confidence interval for the mean has upper and lower limits

$$\bar{x} \pm t_{1-\frac{\alpha}{2};n-1}\frac{s}{\sqrt{n}}$$

where $s^2 = \frac{1}{n-1}\sum(x_i - \bar{x})^2$ and $t_{1-\frac{\alpha}{2};n-1}$ is the $(1-\frac{\alpha}{2})$ percentile of the t distribution with $n-1$ degrees of freedom. The one-sided upper $100(1-\alpha)\%$ confidence limit is computed as $\bar{x} + \frac{s}{\sqrt{n}}t_{1-\alpha;n-1}$ and the one-sided lower $100(1-\alpha)\%$ confidence limit is computed as $\bar{x} - \frac{s}{\sqrt{n}}t_{1-\alpha;n-1}$. See Example 4.9.

The two-sided $100(1-\alpha)\%$ confidence interval for the standard deviation has lower and upper limits,

$$s\sqrt{\frac{n-1}{\chi^2_{1-\frac{\alpha}{2};n-1}}} \quad \text{and} \quad s\sqrt{\frac{n-1}{\chi^2_{\frac{\alpha}{2};n-1}}}$$

respectively, where $\chi^2_{1-\frac{\alpha}{2};n-1}$ and $\chi^2_{\frac{\alpha}{2};n-1}$ are the $(1-\frac{\alpha}{2})$ and $\frac{\alpha}{2}$ percentiles of the chi-square distribution with $n-1$ degrees of freedom. A one-sided $100(1-\alpha)\%$ confidence limit has lower and upper limits,

$$s\sqrt{\frac{n-1}{\chi^2_{1-\alpha;n-1}}} \quad \text{and} \quad s\sqrt{\frac{n-1}{\chi^2_{\alpha;n-1}}}$$

respectively. The $100(1-\alpha)\%$ confidence interval for the variance has upper and lower limits equal to the squares of the corresponding upper and lower limits for the standard deviation.

When you use the WEIGHT statement and specify VARDEF=DF in the PROC statement, the $100(1-\alpha)\%$ confidence interval for the weighted mean is

$$\bar{x}_w \pm t_{1-\frac{\alpha}{2}} \frac{s_w}{\sqrt{\sum_{i=1}^n w_i}}$$

where $\bar{x}_w$ is the weighted mean, s_w is the weighted standard deviation, w_i is the weight for ith observation, and $t_{1-\frac{\alpha}{2}}$ is the $(1-\frac{\alpha}{2})$ percentile for the t distribution with $n-1$ degrees of freedom.

Confidence intervals for the weighted standard deviation are computed by substituting s_w for s in the preceding formulas for confidence limits for the standard deviation.

Robust Estimators

A statistical method is robust if it is insensitive to moderate or even large departures from the assumptions that justify the method. PROC UNIVARIATE provides several methods for robust estimation of location and scale. See Example 4.11.

Winsorized Means

The Winsorized mean is a robust estimator of the location that is relatively insensitive to outliers. The k-times Winsorized mean is calculated as

$$\bar{x}_{wk} = \frac{1}{n}\left((k+1)x_{(k+1)} + \sum_{i=k+2}^{n-k-1} x_{(i)} + (k+1)x_{(n-k)}\right)$$

where n is the number of observations and $x_{(i)}$ is the ith order statistic when the observations are arranged in increasing order:

$$x_{(1)} \leq x_{(2)} \leq \cdots \leq x_{(n)}$$

The Winsorized mean is computed as the ordinary mean after the k smallest observations are replaced by the $(k+1)$st smallest observation and the k largest observations are replaced by the $(k+1)$st largest observation.

For data from a symmetric distribution, the Winsorized mean is an unbiased estimate of the population mean. However, the Winsorized mean does not have a normal distribution even if the data are from a normal population.

The Winsorized sum of squared deviations is defined as

$$s_{wk}^2 = (k+1)(x_{(k+1)} - \bar{x}_{wk})^2 + \sum_{i=k+2}^{n-k-1} (x_{(i)} - \bar{x}_{wk})^2 + (k+1)(x_{(n-k)} - \bar{x}_{wk})^2$$

The Winsorized t statistic is given by

$$t_{wk} = \frac{\bar{x}_{wk} - \mu_0}{\text{SE}(\bar{x}_{wk})}$$

where μ_0 denotes the location under the null hypothesis and the standard error of the Winsorized mean is

$$\text{SE}(\bar{x}_{wk}) = \frac{n-1}{n-2k-1} \times \frac{s_{wk}}{\sqrt{n(n-1)}}$$

When the data are from a symmetric distribution, the distribution of t_{wk} is approximated by a Student's t distribution with $n - 2k - 1$ degrees of freedom (Tukey and McLaughlin 1963; Dixon and Tukey 1968).

The Winsorized $100(1 - \frac{\alpha}{2})\%$ confidence interval for the location parameter has upper and lower limits

$$\bar{x}_{wk} \pm t_{1-\frac{\alpha}{2};n-2k-1} \text{SE}(\bar{x}_{wk})$$

where $t_{1-\frac{\alpha}{2};n-2k-1}$ is the $(100(1 - \frac{\alpha}{2}))$th percentile of the Student's t distribution with $n - 2k - 1$ degrees of freedom.

Trimmed Means

Like the Winsorized mean, the trimmed mean is a robust estimator of the location that is relatively insensitive to outliers. The k-times trimmed mean is calculated as

$$\bar{x}_{tk} = \frac{1}{n-2k} \sum_{i=k+1}^{n-k} x_{(i)}$$

where n is the number of observations and $x_{(i)}$ is the ith order statistic when the observations are arranged in increasing order:

$$x_{(1)} \leq x_{(2)} \leq \cdots \leq x_{(n)}$$

The trimmed mean is computed after the k smallest and k largest observations are deleted from the sample. In other words, the observations are trimmed at each end.

For a symmetric distribution, the symmetrically trimmed mean is an unbiased estimate of the population mean. However, the trimmed mean does not have a normal distribution even if the data are from a normal population.

A robust estimate of the variance of the trimmed mean t_{tk} can be based on the Winsorized sum of squared deviations s_{wk}^2, which is defined in the section "Winsorized Means" on page 402; see Tukey and McLaughlin (1963). This can be used to compute a trimmed t test which is based on the test statistic

$$t_{tk} = \frac{(\bar{x}_{tk} - \mu_0)}{\text{SE}(\bar{x}_{tk})}$$

where the standard error of the trimmed mean is

$$\text{SE}(\bar{x}_{tk}) = \frac{s_{wk}}{\sqrt{(n-2k)(n-2k-1)}}$$

When the data are from a symmetric distribution, the distribution of t_{tk} is approximated by a Student's t distribution with $n - 2k - 1$ degrees of freedom (Tukey and McLaughlin 1963; Dixon and Tukey 1968).

The "trimmed" $100(1 - \alpha)\%$ confidence interval for the location parameter has upper and lower limits

$$\bar{x}_{tk} \pm t_{1-\frac{\alpha}{2};n-2k-1}\text{SE}(\bar{x}_{tk})$$

where $t_{1-\frac{\alpha}{2};n-2k-1}$ is the $(100(1 - \frac{\alpha}{2}))$th percentile of the Student's t distribution with $n - 2k - 1$ degrees of freedom.

Robust Estimates of Scale

The sample standard deviation, which is the most commonly used estimator of scale, is sensitive to outliers. Robust scale estimators, on the other hand, remain bounded when a single data value is replaced by an arbitrarily large or small value. The UNIVARIATE procedure computes several robust measures of scale, including the interquartile range, Gini's mean difference G, the median absolute deviation about the median (MAD), Q_n, and S_n. In addition, the procedure computes estimates of the normal standard deviation σ derived from each of these measures.

The interquartile range (IQR) is simply the difference between the upper and lower quartiles. For a normal population, σ can be estimated as IQR/1.34898.

Gini's mean difference is computed as

$$G = \frac{1}{\binom{n}{2}} \sum_{i<j} |x_i - x_j|$$

For a normal population, the expected value of G is $2\sigma/\sqrt{\pi}$. Thus $G\sqrt{\pi}/2$ is a robust estimator of σ when the data are from a normal sample. For the normal distribution, this estimator has high efficiency relative to the usual sample standard deviation, and it is also less sensitive to the presence of outliers.

A very robust scale estimator is the MAD, the median absolute deviation from the median (Hampel 1974), which is computed as

$$\text{MAD} = \text{med}_i(|x_i - \text{med}_j(x_j)|)$$

where the inner median, $\text{med}_j(x_j)$, is the median of the n observations, and the outer median (taken over i) is the median of the n absolute values of the deviations about the inner median. For a normal population, $1.4826 \times \text{MAD}$ is an estimator of σ.

The MAD has low efficiency for normal distributions, and it may not always be appropriate for symmetric distributions. Rousseeuw and Croux (1993) proposed two statistics as alternatives to the MAD. The first is

$$S_n = 1.1926 \times \text{med}_i(\text{med}_j(|x_i - x_j|))$$

where the outer median (taken over i) is the median of the n medians of $|x_i - x_j|$, $j = 1, 2, \ldots, n$. To reduce small-sample bias, $c_{sn}S_n$ is used to estimate σ, where c_{sn} is a correction factor; see Croux and Rousseeuw (1992).

The second statistic proposed by Rousseeuw and Croux (1993) is

$$Q_n = 2.2219\{|x_i - x_j|; i < j\}_{(k)}$$

where

$$k = \binom{\left[\frac{n}{2}\right] + 1}{2}$$

In other words, Q_n is 2.2219 times the kth order statistic of the $\binom{n}{2}$ distances between the data points. The bias-corrected statistic $c_{qn}Q_n$ is used to estimate σ, where c_{qn} is a correction factor; see Croux and Rousseeuw (1992).

Creating Line Printer Plots

When ODS Graphics is disabled, the PLOTS option in the PROC UNIVARIATE statement provides up to four diagnostic line printer plots to examine the data distribution. These plots are the stem-and-leaf plot or horizontal bar chart, the box plot, the normal probability plot, and the side-by-side box plots. If you specify the WEIGHT statement, PROC UNIVARIATE provides a weighted histogram, a weighted box plot based on the weighted quantiles, and a weighted normal probability plot.

Note that these plots are a legacy feature of the UNIVARIATE procedure in earlier versions of SAS. They predate the addition of the CDFPLOT, HISTOGRAM, PPPLOT, PROBPLOT, and QQPLOT statements, which provide high-resolution graphics displays. Also note that line printer plots requested with the PLOTS option are mainly intended for use with the ODS LISTING destination. See Example 4.5.

Stem-and-Leaf Plot

The first plot in the output is either a stem-and-leaf plot (Tukey 1977) or a horizontal bar chart. If any single interval contains more than 49 observations, the horizontal bar chart appears. Otherwise, the stem-and-leaf plot appears. The stem-and-leaf plot is like a horizontal bar chart in that both plots provide a method to visualize the overall distribution of the data. The stem-and-leaf plot provides more detail because each point in the plot represents an individual data value.

To change the number of stems that the plot displays, use PLOTSIZE= to increase or decrease the number of rows. Instructions that appear below the plot explain how to determine the values of the variable. If no instructions appear, you multiply *Stem.Leaf* by 1 to determine the values of the variable. For example, if the stem value is 10 and the leaf value is 1, then the variable value is approximately 10.1. For the stem-and-leaf plot, the procedure rounds a variable value to the nearest leaf. If the variable value is exactly halfway between two leaves, the value rounds to the nearest leaf with an even integer value. For example, a variable value of 3.15 has a stem value of 3 and a leaf value of 2.

Box Plot

The box plot, also known as a schematic box plot, appears beside the stem-and-leaf plot. Both plots use the same vertical scale. The box plot provides a visual summary of the data and identifies outliers. The bottom and top edges of the box correspond to the sample 25th (Q1) and 75th (Q3) percentiles. The box length is one *interquartile range* (Q3 – Q1). The center horizontal line with asterisk endpoints corresponds to the sample median. The central plus sign (+) corresponds to the sample mean. If the mean and median are equal, the plus sign falls on the line inside the box. The vertical lines that project out from the box, called *whiskers*, extend as far as the data extend, up to and including a distance of 1.5 interquartile ranges. Values farther away are potential outliers. The procedure identifies the extreme values with a zero or an asterisk (*). If zero appears, the value is between 1.5 and 3 interquartile ranges from the top or bottom edge of the box. If an asterisk appears, the value is more extreme.

NOTE: To produce box plots that use high-resolution graphics, use the BOXPLOT procedure in SAS/STAT software. See Chapter 28, "The BOXPLOT Procedure" (*SAS/STAT User's Guide*).

Normal Probability Plot

The normal probability plot plots the empirical quantiles against the quantiles of a standard normal distribution. Asterisks (*) indicate the data values. The plus signs (+) provide a straight reference line that is drawn by using the sample mean and standard deviation. If the data are from a normal distribution, the asterisks tend to fall along the reference line. The vertical coordinate is the data value, and the horizontal coordinate is $\Phi^{-1}(v_i)$ where

$$v_i = \frac{r_i - \frac{3}{8}}{n + \frac{1}{4}}$$

$\Phi^{-1}(\cdot)$ = inverse of the standard normal distribution function

r_i = rank of the ith data value when ordered from smallest to largest

n = number of nonmissing observations

For a weighted normal probability plot, the ith ordered observation is plotted against $\Phi^{-1}(v_i)$ where

$$v_i = \frac{(1 - \frac{3}{8i}) \sum_{j=1}^{i} w_{(j)}}{(1 + \frac{1}{4n}) \sum_{i=1}^{n} w_i}$$

$w_{(j)}$ = weight associated with the jth ordered observation

When each observation has an identical weight, $w_j = w$, the formula for v_i reduces to the expression for v_i in the unweighted normal probability plot:

$$v_i = \frac{i - \frac{3}{8}}{n + \frac{1}{4}}$$

When the value of VARDEF= is WDF or WEIGHT, a reference line with intercept $\hat{\mu}$ and slope $\hat{\sigma}$ is added to the plot. When the value of VARDEF= is DF or N, the slope is $\frac{\hat{\sigma}}{\sqrt{\bar{w}}}$ where $\bar{w} = \frac{\sum_{i=1}^{n} w_i}{n}$ is the average weight.

When each observation has an identical weight and the value of VARDEF= is DF, N, or WEIGHT, the reference line reduces to the usual reference line with intercept $\hat{\mu}$ and slope $\hat{\sigma}$ in the unweighted normal probability plot.

If the data are normally distributed with mean μ and standard deviation σ, and each observation has an identical weight w, then the points on the plot should lie approximately on a straight line. The intercept for this line is μ. The slope is σ when VARDEF= is WDF or WEIGHT, and the slope is $\frac{\sigma}{\sqrt{w}}$ when VARDEF= is DF or N.

NOTE: To produce high-resolution probability plots, use the PROBPLOT statement in PROC UNIVARIATE; see the section "PROBPLOT Statement" on page 360.

Side-by-Side Box Plots

When you use a BY statement with the PLOTS option, PROC UNIVARIATE produces side-by-side box plots, one for each BY group. The box plots (also known as schematic plots) use a common scale that enables you to compare the data distribution across BY groups. This plot appears after the univariate analyses of all BY groups. Use the NOBYPLOT option to suppress this plot.

NOTE: To produce high-resolution side-by-side box plots, use the BOXPLOT procedure in SAS/STAT software. See Chapter 28, "The BOXPLOT Procedure" (*SAS/STAT User's Guide*).

Creating High-Resolution Graphics

You can use the CDFPLOT, HISTOGRAM, PPPLOT, PROBPLOT, and QQPLOT statements to create high-resolution graphs.

The CDFPLOT statement plots the observed cumulative distribution function of a variable. You can optionally superimpose a fitted theoretical distribution on the plot.

The HISTOGRAM statement creates histograms that enable you to examine the data distribution. You can optionally fit families of density curves and superimpose kernel density estimates on the histograms. For additional information about the fitted distributions and kernel density estimates, see the sections "Formulas for Fitted Continuous Distributions" on page 416 and "Kernel Density Estimates" on page 432.

The PPPLOT statement creates a probability-probability (P-P) plot, which compares the empirical cumulative distribution function (ECDF) of a variable with a specified theoretical cumulative distribution function. You can use a P-P plot to determine how well a theoretical distribution models a set of measurements.

The PROBPLOT statement creates a probability plot, which compares ordered values of a variable with percentiles of a specified theoretical distribution. Probability plots are useful for graphical estimation of percentiles.

The QQPLOT statement creates a quantile-quantile plot, which compares ordered values of a variable with quantiles of a specified theoretical distribution. Q-Q plots are useful for graphical estimation of distribution parameters.

NOTE: You can use the CLASS statement with any of these plot statements to produce comparative versions of the plots.

Alternatives for Producing Graphics

The UNIVARIATE procedure supports two kinds of graphical output.

- ODS Statistical Graphics output is produced if ODS Graphics is enabled, for example by specifying the ODS GRAPHICS ON statement prior to the PROC statement.

- Otherwise, traditional graphics are produced if SAS/GRAPH is licensed.

Traditional graphics are saved in graphics catalogs. Their appearance is controlled by the SAS/GRAPH GOPTIONS, AXIS, and SYMBOL statements (as described in *SAS/GRAPH: Help*) and numerous specialized plot statement options.

ODS Statistical Graphics (or ODS Graphics for short) is an extension to the Output Delivery System (ODS) that can be enabled by specifying the ODS GRAPHICS statement prior to your procedure statements. An ODS graph is produced in ODS output (not a graphics catalog), and the details of its appearance and layout are controlled by ODS styles and templates rather than by SAS/GRAPH statements and procedure options. See Chapter 21, "Statistical Graphics Using ODS" (*SAS/STAT User's Guide*), for a thorough discussion of ODS Graphics.

Prior to SAS 9.2, the plots produced by PROC UNIVARIATE were extremely basic by default. Producing attractive graphical output required the careful selection of colors, fonts, and other elements, which were specified via SAS/GRAPH statements and plot statement options. Beginning with SAS 9.2, the default appearance of traditional graphs is governed by the prevailing ODS style, which automatically produces attractive, consistent output. The SAS/GRAPH statements and procedure options for controlling graph appearance continue to be honored for traditional graphics. You can specify the NOGSTYLE system option to prevent the ODS style from affecting the appearance of traditional graphs. This enables existing PROC UNIVARIATE programs to produce customized graphs that appear as they did under previous SAS releases.

The appearance of ODS Graphics output is also controlled by the ODS style, but it is not affected by SAS/GRAPH statements or plot statement options that govern traditional graphics, For example, the CAXIS= option used to specify the color of graph axes in traditional graphics is ignored when producing ODS Graphics output. **NOTE:** Some features available with traditional graphics are not supported in ODS Graphics.

The traditional graphics system enables you to control every detail of a graph through convenient procedure syntax. ODS Graphics provides the highest quality output with minimal syntax and full compatibility with graphics produced by SAS/STAT and SAS/ETS procedures.

The following code produces a histogram with a fitted lognormal distribution of the LoanToValueRatio data introduced in the section "Summarizing a Data Distribution" on page 281:

```
options nogstyle;
ods graphics off;
proc univariate data=HomeLoans noprint;
   histogram LoanToValueRatio / lognormal;
   inset lognormal(theta sigma zeta) / position=ne;
run;
```

The NOGSTYLE system option keeps the ODS style from influencing the output, and no SAS/GRAPH statements or procedure options affecting the appearance of the plot are specified. Figure 4.8 shows the resulting histogram, which is essentially identical to the default output produced under releases prior to SAS 9.2.

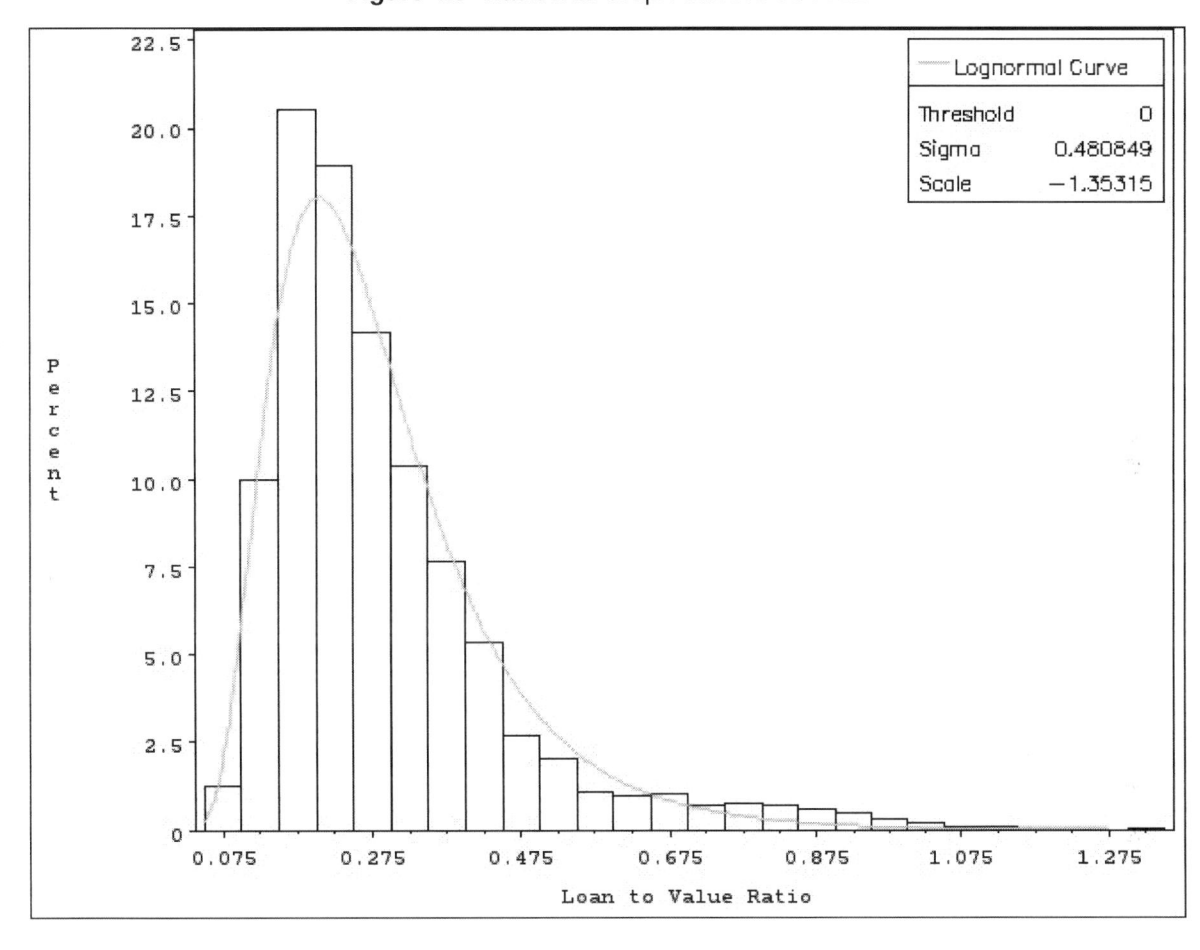

Figure 4.8 Traditional Graph with NOGSTYLE

410 ✦ *Chapter 4: The UNIVARIATE Procedure*

Figure 4.9 shows the result of executing the same code with the GSTYLE system option turned on (the default). Note the influence of the ODS style on the histogram's appearance. For example, the quality of the text is improved and histogram bars are filled by default.

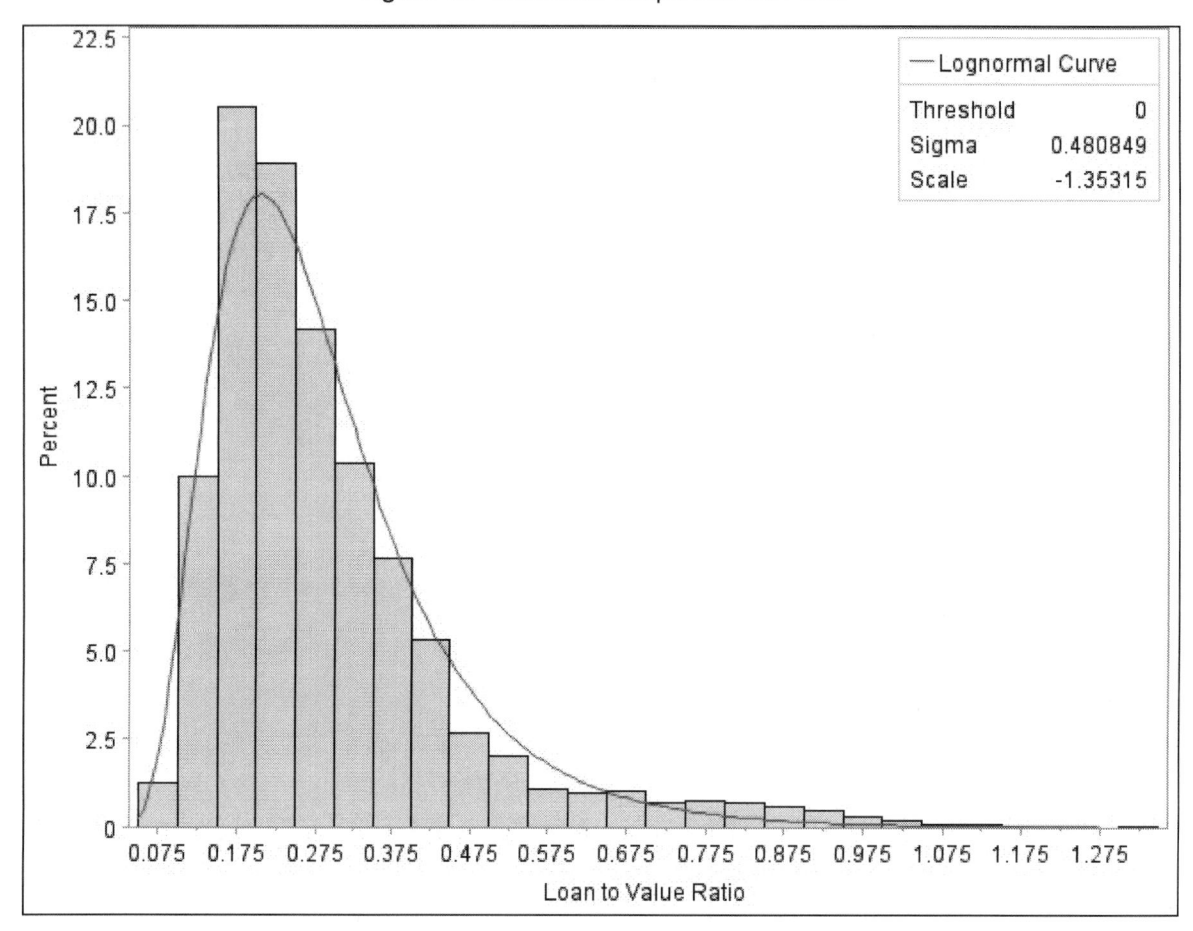

Figure 4.9 Traditional Graph with GSTYLE

Figure 4.10 shows the same histogram produced using ODS Graphics. The histogram's appearance is governed by the same style elements as in Figure 4.9, but the plots are not identical. Note, for example, the title incorporated in the ODS Graphics output and the smoother appearance of the fitted curve.

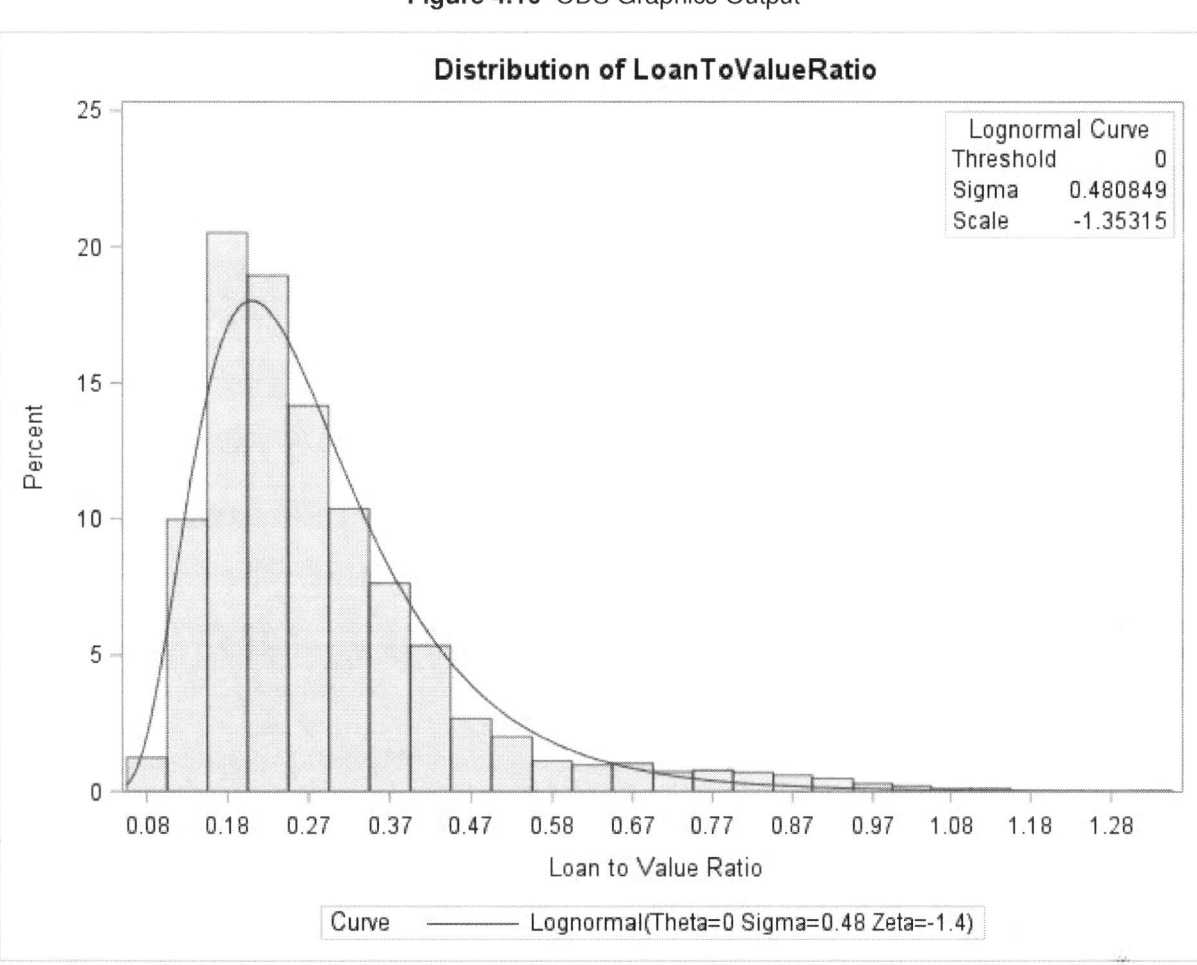

Figure 4.10 ODS Graphics Output

Distribution of LoanToValueRatio

Lognormal Curve
Threshold 0
Sigma 0.480849
Scale -1.35315

Curve —— Lognormal(Theta=0 Sigma=0.48 Zeta=-1.4)

Using the CLASS Statement to Create Comparative Plots

When you use the CLASS statement with the CDFPLOT, HISTOGRAM, PPPLOT, PROBPLOT, or QQPLOT statements, PROC UNIVARIATE creates comparative versions of the plots. You can use these plot statements with the CLASS statement to create one-way and two-way comparative plots. When you use one CLASS variable, PROC UNIVARIATE displays an array of component plots (stacked or side-by-side), one for each level of the classification variable. When you use two CLASS variables, PROC UNIVARIATE displays a matrix of component plots, one for each combination of levels of the classification variables. The observations in a given level are referred to collectively as a *cell*.

When you create a one-way comparative plot, the observations in the input data set are sorted by the method specified in the ORDER= option. PROC UNIVARIATE creates a separate plot for the analysis variable values in each level and arranges these component plots in an array to form the comparative plot with uniform horizontal and vertical axes. See Example 4.15.

When you create a two-way comparative plot, the observations in the input data set are cross-classified according to the values (levels) of these variables. PROC UNIVARIATE creates a separate plot for the analysis variable values in each cell of the cross-classification and arranges these component plots in a matrix to form the comparative plot with uniform horizontal and vertical axes. The levels of the first CLASS variable are the labels for the rows of the matrix, and the levels of the second CLASS variable are the labels for the columns of the matrix. See Example 4.16.

PROC UNIVARIATE determines the layout of a two-way comparative plot by using the order for the first CLASS variable to obtain the order of the rows from top to bottom. Then it applies the order for the second CLASS variable to the observations that correspond to the first row to obtain the order of the columns from left to right. If any columns remain unordered (that is, the categories are unbalanced), PROC UNIVARIATE applies the order for the second CLASS variable to the observations in the second row, and so on, until all the columns have been ordered.

If you associate a label with a CLASS variable, PROC UNIVARIATE displays the variable label in the comparative plot and this label is parallel to the column (or row) labels.

Use the MISSING option to treat missing values as valid levels.

To reduce the number of classification levels, use a FORMAT statement to combine variable values.

Positioning Insets

Positioning an Inset Using Compass Point Values

To position an inset by using a compass point position, specify the value N, NE, E, SE, S, SW, W, or NW with the POSITION= option. The default position of the inset is NW. The following statements produce a histogram to show the position of the inset for the eight compass points:

```
data Score;
   input Student $ PreTest PostTest @@;
   label ScoreChange = 'Change in Test Scores';
   ScoreChange = PostTest - PreTest;
datalines;
Capalleti  94 91  Dubose     51 65
Engles     95 97  Grant      63 75
Krupski    80 75  Lundsford  92 55
Mcbane     75 78  Mullen     89 82
Nguyen     79 76  Patel      71 77
Si         75 70  Tanaka     87 73
;

title 'Test Scores for a College Course';
ods graphics off;
proc univariate data=Score noprint;
   histogram PreTest / midpoints = 45 to 95 by 10;
   inset n      / cfill=blank
                  header='Position = NW' pos=nw;
   inset mean   / cfill=blank
                  header='Position = N ' pos=n ;
   inset sum    / cfill=blank
                  header='Position = NE' pos=ne;
   inset max    / cfill=blank
                  header='Position = E ' pos=e ;
   inset min    / cfill=blank
```

```
                      header='Position = SE' pos=se;
     inset nobs   / cfill=blank
                      header='Position = S ' pos=s ;
     inset range  / cfill=blank
                      header='Position = SW' pos=sw;
     inset mode   / cfill=blank
                      header='Position = W ' pos=w ;
     label PreTest = 'Pretest Score';
run;
```

Figure 4.11 Compass Positions for Inset

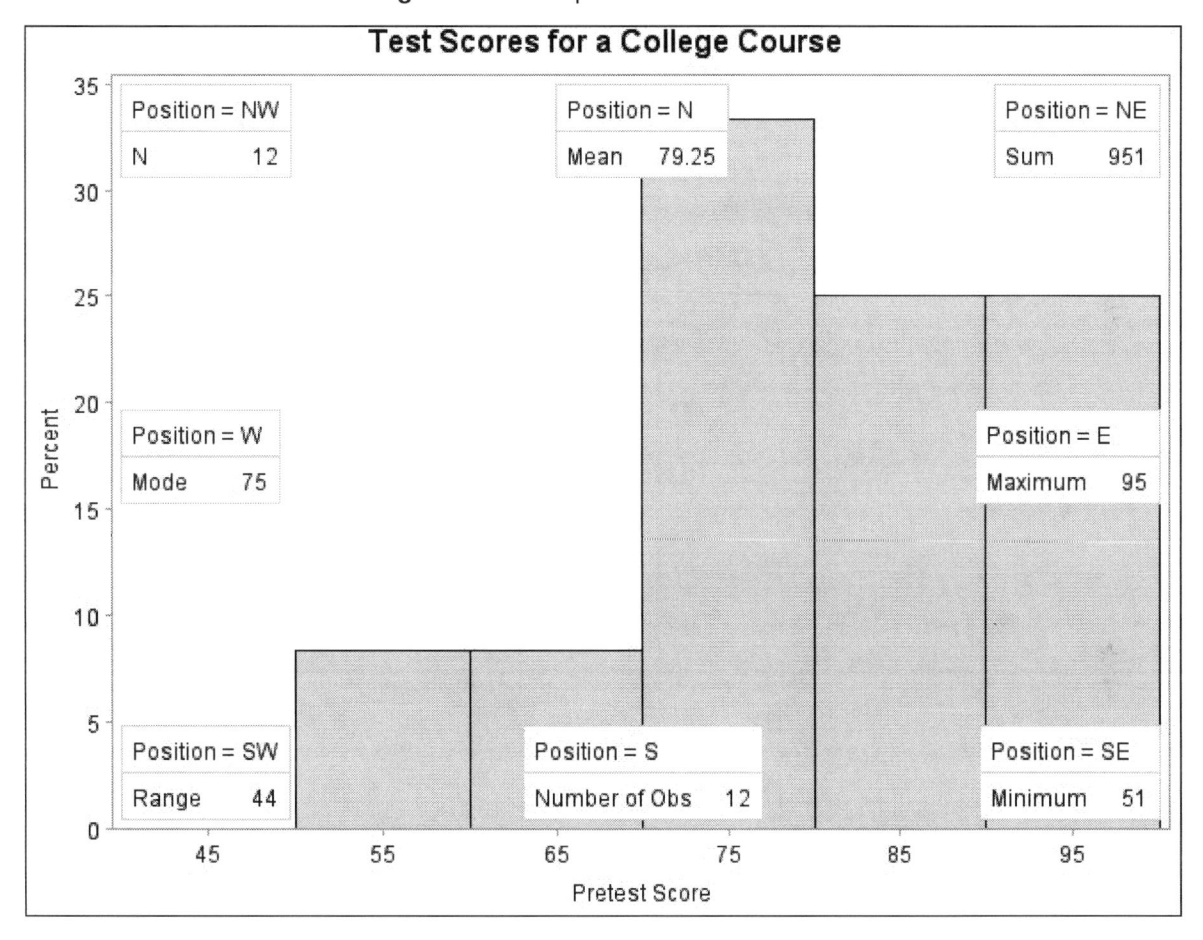

Positioning Insets in the Margins

To position an inset in one of the four margins that surround the plot area, specify the value LM, RM, TM, or BM with the POSITION= option. Margin positions are recommended if you list a large number of statistics in the INSET statement. If you attempt to display a lengthy inset in the interior of the plot, the inset is likely to collide with the data display.

414 ✦ *Chapter 4: The UNIVARIATE Procedure*

Positioning an Inset Using Coordinates

To position an inset with coordinates, use POSITION=(x,y). You specify the coordinates in axis data units or in axis percentage units (the default). **NOTE:** You cannot position an inset with coordinates when producing ODS Graphics output.

If you specify the DATA option immediately following the coordinates, PROC UNIVARIATE positions the inset by using axis data units. For example, the following statements place the bottom left corner of the inset at 45 on the horizontal axis and 10 on the vertical axis:

```
title 'Test Scores for a College Course';
proc univariate data=Score noprint;
   histogram PreTest / midpoints = 45 to 95 by 10;
   inset n / header   = 'Position=(45,10)'
             position = (45,10) data;
run;
```

Figure 4.12 Coordinate Position for Inset

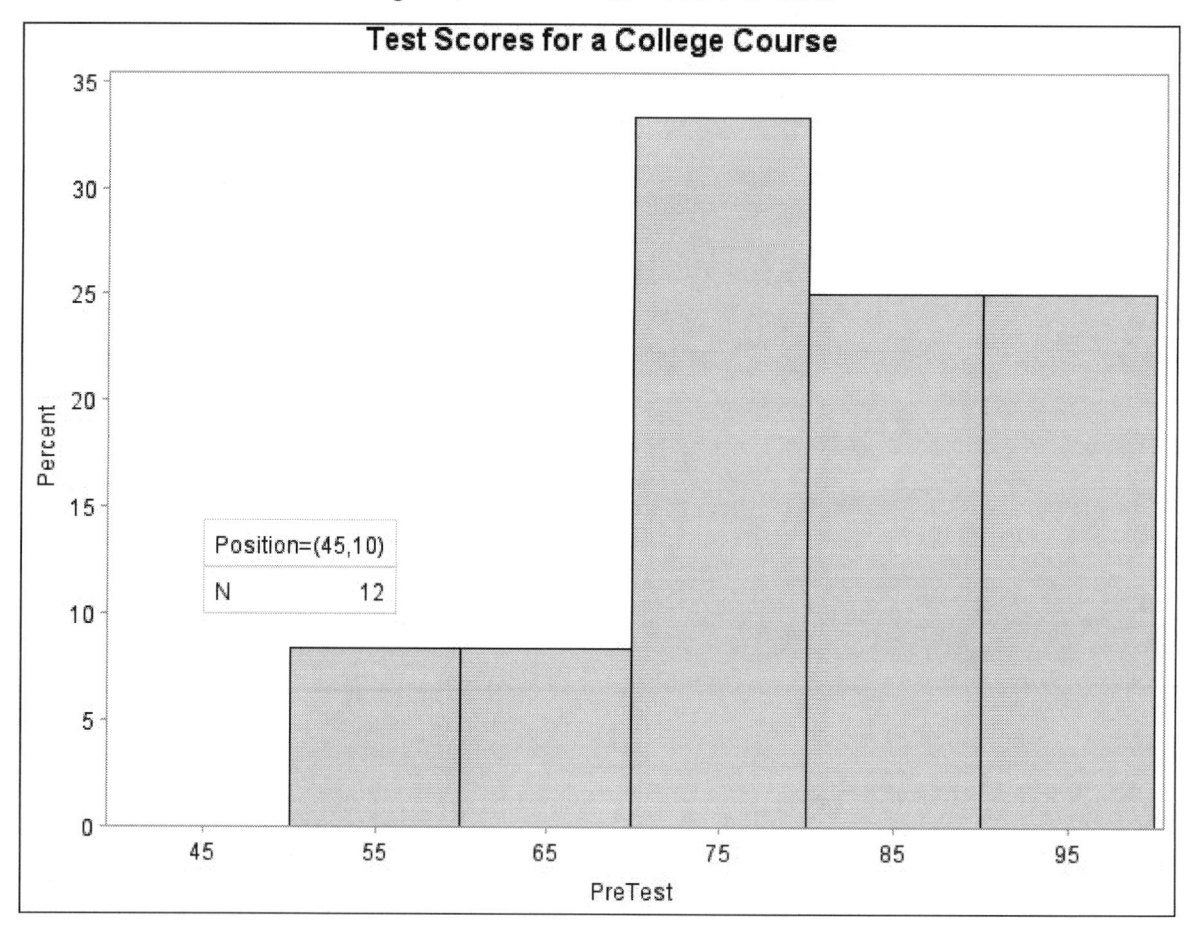

By default, the specified coordinates determine the position of the bottom left corner of the inset. To change this reference point, use the REFPOINT= option (see below).

If you omit the DATA option, PROC UNIVARIATE positions the inset by using axis percentage units. The coordinates in axis percentage units must be between 0 and 100. The coordinates of the bottom left corner of the display are (0,0), while the upper right corner is (100, 100). For example, the following statements create a histogram and use coordinates in axis percentage units to position the two insets:

```
title 'Test Scores for a College Course';
proc univariate data=Score noprint;
   histogram PreTest / midpoints = 45 to 95 by 10;
   inset min / position = (5,25)
               header   = 'Position=(5,25)'
               refpoint = tl;
   inset max / position = (95,95)
               header   = 'Position=(95,95)'
               refpoint = tr;
run;
```

The REFPOINT= option determines which corner of the inset to place at the coordinates that are specified with the POSITION= option. The first inset uses REFPOINT=TL, so that the top left corner of the inset is positioned 5% of the way across the horizontal axis and 25% of the way up the vertical axis. The second inset uses REFPOINT=TR, so that the top right corner of the inset is positioned 95% of the way across the horizontal axis and 95% of the way up the vertical axis.

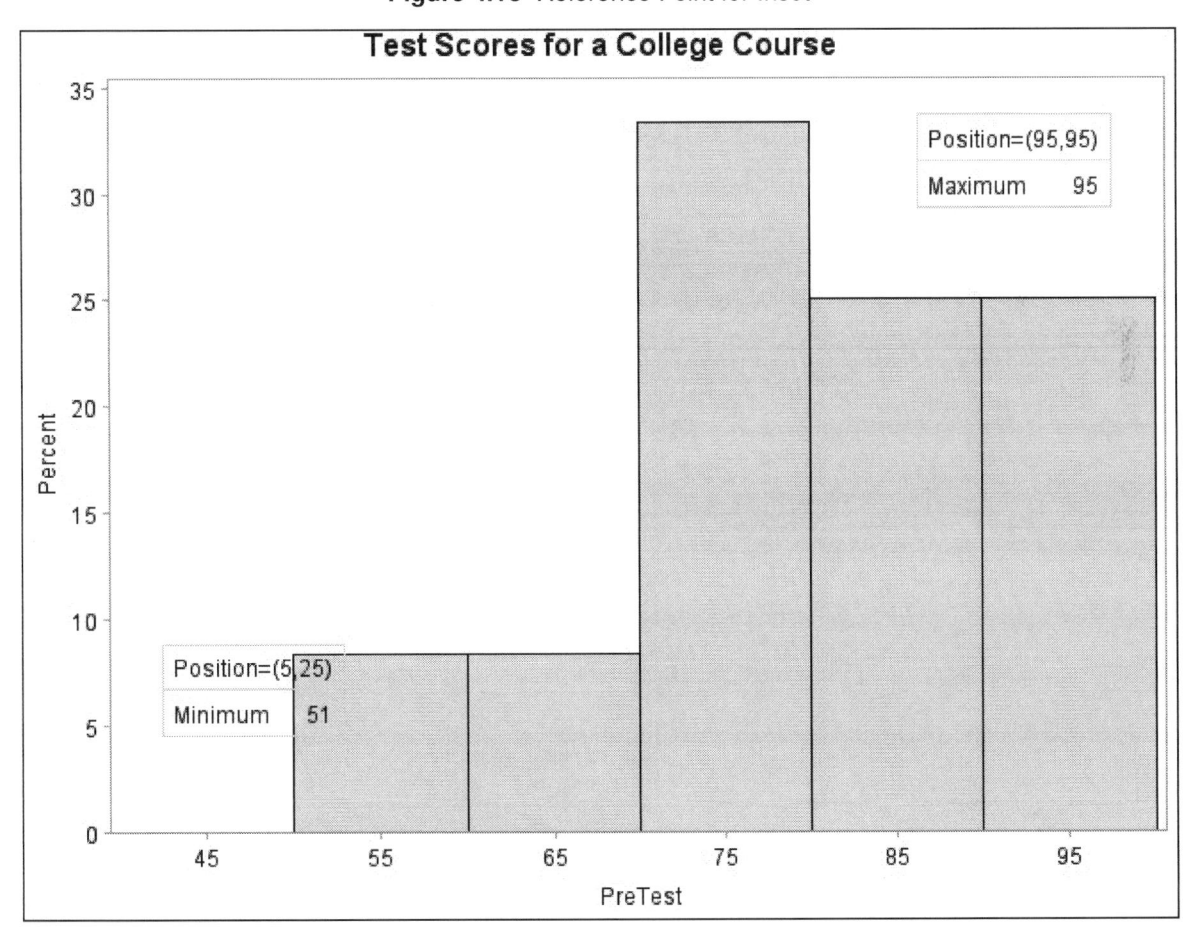

Figure 4.13 Reference Point for Inset

A sample program for these examples, *univar3.sas*, is available in the SAS Sample Library for Base SAS software.

Formulas for Fitted Continuous Distributions

The following sections provide information about the families of parametric distributions that you can fit with the HISTOGRAM statement. Properties of these distributions are discussed by Johnson, Kotz, and Balakrishnan (1994, 1995).

Beta Distribution

The fitted density function is

$$p(x) = \begin{cases} hv \frac{(x-\theta)^{\alpha-1}(\sigma+\theta-x)^{\beta-1}}{B(\alpha,\beta)\sigma^{(\alpha+\beta-1)}} & \text{for } \theta < x < \theta + \sigma \\ 0 & \text{for } x \leq \theta \text{ or } x \geq \theta + \sigma \end{cases}$$

where $B(\alpha, \beta) = \frac{\Gamma(\alpha)\Gamma(\beta)}{\Gamma(\alpha+\beta)}$ and

θ = lower threshold parameter (lower endpoint parameter)

σ = scale parameter ($\sigma > 0$)

α = shape parameter ($\alpha > 0$)

β = shape parameter ($\beta > 0$)

h = width of histogram interval

v = vertical scaling factor

and

$$v = \begin{cases} n & \text{the sample size, for VSCALE=COUNT} \\ 100 & \text{for VSCALE=PERCENT} \\ 1 & \text{for VSCALE=PROPORTION} \end{cases}$$

NOTE: This notation is consistent with that of other distributions that you can fit with the HISTOGRAM statement. However, many texts, including Johnson, Kotz, and Balakrishnan (1995), write the beta density function as

$$p(x) = \begin{cases} \frac{(x-a)^{p-1}(b-x)^{q-1}}{B(p,q)(b-a)^{p+q-1}} & \text{for } a < x < b \\ 0 & \text{for } x \leq a \text{ or } x \geq b \end{cases}$$

The two parameterizations are related as follows:

$\sigma = b - a$

$\theta = a$

$\alpha = p$

$\beta = q$

The range of the beta distribution is bounded below by a threshold parameter $\theta = a$ and above by $\theta + \sigma = b$. If you specify a fitted beta curve by using the BETA option, θ must be less than the minimum data value and $\theta + \sigma$ must be greater than the maximum data value. You can specify θ and σ with the THETA= and SIGMA= *beta-options* in parentheses after the keyword BETA. By default, $\sigma = 1$ and $\theta = 0$. If you specify THETA=EST and SIGMA=EST, maximum likelihood estimates are computed for θ and σ. However, three- and four-parameter maximum likelihood estimation does not always converge.

In addition, you can specify α and β with the ALPHA= and BETA= *beta-options*, respectively. By default, the procedure calculates maximum likelihood estimates for α and β. For example, to fit a beta density curve to a set of data bounded below by 32 and above by 212 with maximum likelihood estimates for α and β, use the following statement:

```
histogram Length / beta(theta=32 sigma=180);
```

The beta distributions are also referred to as Pearson Type I or II distributions. These include the power function distribution ($\beta = 1$), the arc sine distribution ($\alpha = \beta = \frac{1}{2}$), and the generalized arc sine distributions ($\alpha + \beta = 1, \beta \neq \frac{1}{2}$).

You can use the DATA step function QUANTILE to compute beta quantiles and the DATA step function CDF to compute beta probabilities.

Exponential Distribution

The fitted density function is

$$p(x) = \begin{cases} \frac{hv}{\sigma} \exp(-(\frac{x-\theta}{\sigma})) & \text{for } x \geq \theta \\ 0 & \text{for } x < \theta \end{cases}$$

where

θ = threshold parameter

σ = scale parameter ($\sigma > 0$)

h = width of histogram interval

v = vertical scaling factor

and

$$v = \begin{cases} n & \text{the sample size, for VSCALE=COUNT} \\ 100 & \text{for VSCALE=PERCENT} \\ 1 & \text{for VSCALE=PROPORTION} \end{cases}$$

The threshold parameter θ must be less than or equal to the minimum data value. You can specify θ with the THRESHOLD= *exponential-option*. By default, $\theta = 0$. If you specify THETA=EST, a maximum likelihood estimate is computed for θ. In addition, you can specify σ with the SCALE= *exponential-option*. By default, the procedure calculates a maximum likelihood estimate for σ. Note that some authors define the scale parameter as $\frac{1}{\sigma}$.

The exponential distribution is a special case of both the gamma distribution (with $\alpha = 1$) and the Weibull distribution (with $c = 1$). A related distribution is the extreme value distribution. If $Y = \exp(-X)$ has an exponential distribution, then X has an extreme value distribution.

You can use the DATA step function QUANTILE to compute exponential quantiles and the DATA step function CDF to compute exponential probabilities.

Gamma Distribution

The fitted density function is

$$p(x) = \begin{cases} \frac{hv}{\Gamma(\alpha)\sigma}(\frac{x-\theta}{\sigma})^{\alpha-1}\exp(-(\frac{x-\theta}{\sigma})) & \text{for } x > \theta \\ 0 & \text{for } x \leq \theta \end{cases}$$

where

θ = threshold parameter

σ = scale parameter ($\sigma > 0$)

α = shape parameter ($\alpha > 0$)

h = width of histogram interval

v = vertical scaling factor

and

$$v = \begin{cases} n & \text{the sample size, for VSCALE=COUNT} \\ 100 & \text{for VSCALE=PERCENT} \\ 1 & \text{for VSCALE=PROPORTION} \end{cases}$$

The threshold parameter θ must be less than the minimum data value. You can specify θ with the THRESHOLD= *gamma-option*. By default, $\theta = 0$. If you specify THETA=EST, a maximum likelihood estimate is computed for θ. In addition, you can specify σ and α with the SCALE= and ALPHA= *gamma-options*. By default, the procedure calculates maximum likelihood estimates for σ and α.

The gamma distributions are also referred to as Pearson Type III distributions, and they include the chi-square, exponential, and Erlang distributions. The probability density function for the chi-square distribution is

$$p(x) = \begin{cases} \frac{1}{2\Gamma(\frac{v}{2})}(\frac{x}{2})^{\frac{v}{2}-1}\exp(-\frac{x}{2}) & \text{for } x > 0 \\ 0 & \text{for } x \leq 0 \end{cases}$$

Notice that this is a gamma distribution with $\alpha = \frac{v}{2}$, $\sigma = 2$, and $\theta = 0$. The exponential distribution is a gamma distribution with $\alpha = 1$, and the Erlang distribution is a gamma distribution with α being a positive integer. A related distribution is the Rayleigh distribution. If $R = \frac{\max(X_1,\dots,X_n)}{\min(X_1,\dots,X_n)}$ where the X_i's are independent χ_v^2 variables, then $\log R$ is distributed with a χ_v distribution having a probability density function of

$$p(x) = \begin{cases} \left[2^{\frac{v}{2}-1}\Gamma(\frac{v}{2})\right]^{-1} x^{v-1}\exp(-\frac{x^2}{2}) & \text{for } x > 0 \\ 0 & \text{for } x \leq 0 \end{cases}$$

If $v = 2$, the preceding distribution is referred to as the Rayleigh distribution.

You can use the DATA step function QUANTILE to compute gamma quantiles and the DATA step function CDF to compute gamma probabilities.

Gumbel Distribution

The fitted density function is

$$p(x) = \frac{hv}{\sigma} e^{-(x-\mu)/\sigma} \exp\left(-e^{-(x-\mu)/\sigma}\right)$$

where

μ = location parameter

σ = scale parameter ($\sigma > 0$)

h = width of histogram interval

v = vertical scaling factor

and

$$v = \begin{cases} n & \text{the sample size, for VSCALE=COUNT} \\ 100 & \text{for VSCALE=PERCENT} \\ 1 & \text{for VSCALE=PROPORTION} \end{cases}$$

You can specify μ and σ with the MU= and SIGMA= *Gumbel-options*, respectively. By default, the procedure calculates maximum likelihood estimates for these parameters.

NOTE: The Gumbel distribution is also referred to as Type 1 extreme value distribution.

NOTE: The random variable X has Gumbel (Type 1 extreme value) distribution if and only if e^X has Weibull distribution and $\exp((X - \mu)/\sigma)$ has standard exponential distribution.

Inverse Gaussian Distribution

The fitted density function is

$$p(x) = \begin{cases} hv \left(\frac{\lambda}{2\pi x^3}\right)^{1/2} \exp(-\frac{\lambda}{2\mu^2 x}(x - \mu)^2) & \text{for } x > 0 \\ 0 & \text{for } x \leq 0 \end{cases}$$

where

μ = location parameter ($\mu > 0$)

λ = shape parameter ($\lambda > 0$)

h = width of histogram interval

v = vertical scaling factor

and

$$v = \begin{cases} n & \text{the sample size, for VSCALE=COUNT} \\ 100 & \text{for VSCALE=PERCENT} \\ 1 & \text{for VSCALE=PROPORTION} \end{cases}$$

The location parameter μ has to be greater then zero. You can specify μ with the MU= *iGauss-option*. In addition, you can specify shape parameter λ with LAMBDA= *iGauss-option*. By default, the procedure calculates maximum likelihood estimates for μ and λ.

NOTE: The special case where $\mu = 1$ and $\lambda = \phi$ corresponds to the Wald distribution.

You can use the DATA step function QUANTILE to compute inverse Gaussian quantiles and the DATA step function CDF to compute inverse Gaussian probabilities.

Lognormal Distribution

The fitted density function is

$$p(x) = \begin{cases} \frac{hv}{\sigma\sqrt{2\pi}(x-\theta)} \exp\left(-\frac{(\log(x-\theta)-\zeta)^2}{2\sigma^2}\right) & \text{for } x > \theta \\ 0 & \text{for } x \leq \theta \end{cases}$$

where

θ = threshold parameter

ζ = scale parameter ($-\infty < \zeta < \infty$)

σ = shape parameter ($\sigma > 0$)

h = width of histogram interval

v = vertical scaling factor

and

$$v = \begin{cases} n & \text{the sample size, for VSCALE=COUNT} \\ 100 & \text{for VSCALE=PERCENT} \\ 1 & \text{for VSCALE=PROPORTION} \end{cases}$$

The threshold parameter θ must be less than the minimum data value. You can specify θ with the THRESHOLD= *lognormal-option*. By default, $\theta = 0$. If you specify THETA=EST, a maximum likelihood estimate is computed for θ. You can specify ζ and σ with the SCALE= and SHAPE= *lognormal-options*, respectively. By default, the procedure calculates maximum likelihood estimates for these parameters.

NOTE: The lognormal distribution is also referred to as the S_L distribution in the Johnson system of distributions.

NOTE: This book uses σ to denote the shape parameter of the lognormal distribution, whereas σ is used to denote the scale parameter of the other distributions. The use of σ to denote the lognormal shape parameter is based on the fact that $\frac{1}{\sigma}(\log(X - \theta) - \zeta)$ has a standard normal distribution if X is lognormally distributed. Based on this relationship, you can use the DATA step function PROBIT to compute lognormal quantiles and the DATA step function PROBNORM to compute probabilities.

Normal Distribution

The fitted density function is

$$p(x) = \frac{hv}{\sigma\sqrt{2\pi}} \exp\left(-\frac{1}{2}\left(\frac{x-\mu}{\sigma}\right)^2\right) \quad \text{for } -\infty < x < \infty$$

where

μ = mean

σ = standard deviation ($\sigma > 0$)

h = width of histogram interval

v = vertical scaling factor

and

$$v = \begin{cases} n & \text{the sample size, for VSCALE=COUNT} \\ 100 & \text{for VSCALE=PERCENT} \\ 1 & \text{for VSCALE=PROPORTION} \end{cases}$$

You can specify μ and σ with the MU= and SIGMA= *normal-options*, respectively. By default, the procedure estimates μ with the sample mean and σ with the sample standard deviation.

You can use the DATA step function QUANTILE to compute beta quantiles and the DATA step function CDF to compute normal probabilities.

NOTE: The normal distribution is also referred to as the S_N distribution in the Johnson system of distributions.

Generalized Pareto Distribution

The fitted density function is

$$p(x) = \begin{cases} \frac{hv}{\sigma}(1 - \alpha(x-\theta)/\sigma)^{1/\alpha - 1} & \text{if } \alpha \neq 0 \\ \frac{hv}{\sigma} \exp(-x/\sigma) & \text{if } \alpha = 0 \end{cases}$$

where

θ = threshold parameter

α = shape parameter

σ = shape parameter ($\sigma > 0$)

h = width of histogram interval

v = vertical scaling factor

and

$$v = \begin{cases} n & \text{the sample size, for VSCALE=COUNT} \\ 100 & \text{for VSCALE=PERCENT} \\ 1 & \text{for VSCALE=PROPORTION} \end{cases}$$

The support of the distribution is $x > \theta$ for $\alpha \leq 0$ and $\theta < x < \sigma/\alpha$ for $\alpha > 0$.

NOTE: Special cases of Pareto distribution with $\alpha = 0$ and $\alpha = 1$ correspond respectively to the exponential distribution with mean σ and uniform distribution on the interval (θ, σ).

The threshold parameter θ must be less than the minimum data value. You can specify θ with the THETA= *Pareto-option*. By default, $\theta = 0$. You can also specify α and σ with the ALPHA= and SIGMA= *Pareto-options*, respectively. By default, the procedure calculates maximum likelihood estimates for these parameters.

NOTE: Maximum likelihood estimation of the parameters works well if $\alpha < \frac{1}{2}$, but not otherwise. In this case the estimators are asymptotically normal and asymptotically efficient. The asymptotic normal distribution of the maximum likelihood estimates has mean (α, σ) and variance-covariance matrix

$$\frac{1}{n} \begin{pmatrix} (1-\alpha)^2 & \sigma(1-\alpha) \\ \sigma(1-\alpha) & 2\sigma^2(1-\alpha) \end{pmatrix}.$$

NOTE: If no local minimum is found in the region

$$\{\alpha < 0, \sigma > 0\} \cup \{0 < \alpha \leq 1, \sigma/\alpha > \max(X_i)\},$$

there is no maximum likelihood estimator. More details on how to find maximum likelihood estimators and suggested algorithm can be found in Grimshaw(1993).

Power Function Distribution

The fitted density function is

$$p(x) = \begin{cases} hv\frac{\alpha}{\sigma}\left(\frac{x-\theta}{\sigma}\right)^{\alpha-1} & \text{for } \theta < x < \theta + \sigma \\ 0 & \text{for } x \leq \theta \text{ or } x \geq \theta + \sigma \end{cases}$$

where

θ = lower threshold parameter (lower endpoint parameter)

σ = scale parameter ($\sigma > 0$)

α = shape parameter ($\alpha > 0$)

h = width of histogram interval

v = vertical scaling factor

and

$$v = \begin{cases} n & \text{the sample size, for VSCALE=COUNT} \\ 100 & \text{for VSCALE=PERCENT} \\ 1 & \text{for VSCALE=PROPORTION} \end{cases}$$

NOTE: This notation is consistent with that of other distributions that you can fit with the HISTOGRAM statement. However, many texts, including Johnson, Kotz, and Balakrishnan (1995), write the density function of power function distribution as

$$p(x) = \begin{cases} \frac{p}{b-a}\left(\frac{x-a}{b-a}\right)^{p-1} & \text{for } a < x < b \\ 0 & \text{for } x \leq a \text{ or } x \geq b \end{cases}$$

The two parameterizations are related as follows:

$\sigma = b - a$

$\theta = a$

$\alpha = p$

NOTE: The family of power function distributions is subclass of beta distribution with density function

$$p(x) = \begin{cases} hv \frac{(x-\theta)^{\alpha-1}(\sigma+\theta-x)^{\beta-1}}{B(\alpha,\beta)\sigma^{(\alpha+\beta-1)}} & \text{for } \theta < x < \theta + \sigma \\ 0 & \text{for } x \leq \theta \text{ or } x \geq \theta + \sigma \end{cases}$$

where $B(\alpha, \beta) = \frac{\Gamma(\alpha)\Gamma(\beta)}{\Gamma(\alpha+\beta)}$ with parameter $\beta = 1$. Therefore, all properties and estimation procedures of beta distribution apply.

The range of the power function distribution is bounded below by a threshold parameter $\theta = a$ and above by $\theta + \sigma = b$. If you specify a fitted power function curve by using the POWER option, θ must be less than the minimum data value and $\theta + \sigma$ must be greater than the maximum data value. You can specify θ and σ with the THETA= and SIGMA= *power-options* in parentheses after the keyword POWER. By default, $\sigma = 1$ and $\theta = 0$. If you specify THETA=EST and SIGMA=EST, maximum likelihood estimates are computed for θ and σ. However, three-parameter maximum likelihood estimation does not always converge.

In addition, you can specify α with the ALPHA= *power-option*. By default, the procedure calculates maximum likelihood estimate for α. For example, to fit a power function density curve to a set of data bounded below by 32 and above by 212 with maximum likelihood estimate for α, use the following statement:

```
histogram Length / power(theta=32 sigma=180);
```

Rayleigh Distribution

The fitted density function is

$$p(x) = \begin{cases} hv \frac{x-\theta}{\sigma^2} e^{-(x-\theta)^2/(2\sigma^2)} & \text{for } x \geq \theta \\ 0 & \text{for } x < \theta \end{cases}$$

where

θ = lower threshold parameter (lower endpoint parameter)

σ = scale parameter ($\sigma > 0$)

h = width of histogram interval

v = vertical scaling factor

and

$$v = \begin{cases} n & \text{the sample size, for VSCALE=COUNT} \\ 100 & \text{for VSCALE=PERCENT} \\ 1 & \text{for VSCALE=PROPORTION} \end{cases}$$

NOTE: The Rayleigh distribution is Weibull distribution with density function

$$p(x) = \begin{cases} hv\frac{k}{\lambda}\left(\frac{x-\theta}{\lambda}\right)^{k-1} \exp(-(\frac{x-\theta}{\lambda})^k) & \text{for } x \geq \theta \\ 0 & \text{for } x < \theta \end{cases}$$

and with shape parameter $k = 2$ and scale parameter $\lambda = \sqrt{2}\sigma$.

The threshold parameter θ must be less than the minimum data value. You can specify θ with the THETA= *Rayleigh-option*. By default, $\theta = 0$. In addition you can specify σ with the SIGMA= *Rayleigh-option*. By default, the procedure calculates maximum likelihood estimate for σ.

For example, to fit a Rayleigh density curve to a set of data bounded below by 32 with maximum likelihood estimate for σ, use the following statement:

```
histogram Length / rayleigh(theta=32);
```

Johnson S_B Distribution

The fitted density function is

$$p(x) = \begin{cases} \frac{\delta h v}{\sigma \sqrt{2\pi}} \left[\left(\frac{x-\theta}{\sigma}\right)\left(1 - \frac{x-\theta}{\sigma}\right)\right]^{-1} \times \\ \exp\left[-\frac{1}{2}\left(\gamma + \delta \log(\frac{x-\theta}{\theta+\sigma-x})\right)^2\right] & \text{for } \theta < x < \theta + \sigma \\ 0 & \text{for } x \leq \theta \text{ or } x \geq \theta + \sigma \end{cases}$$

where

θ = threshold parameter ($-\infty < \theta < \infty$)

σ = scale parameter ($\sigma > 0$)

δ = shape parameter ($\delta > 0$)

γ = shape parameter ($-\infty < \gamma < \infty$)

h = width of histogram interval

v = vertical scaling factor

and

$$v = \begin{cases} n & \text{the sample size, for VSCALE=COUNT} \\ 100 & \text{for VSCALE=PERCENT} \\ 1 & \text{for VSCALE=PROPORTION} \end{cases}$$

The S_B distribution is bounded below by the parameter θ and above by the value $\theta + \sigma$. The parameter θ must be less than the minimum data value. You can specify θ with the THETA= S_B-option, or you can request that θ be estimated with the THETA = EST S_B-option. The default value for θ is zero. The sum $\theta + \sigma$ must be greater than the maximum data value. The default value for σ is one. You can specify σ with the SIGMA= S_B-option, or you can request that σ be estimated with the SIGMA = EST S_B-option.

By default, the method of percentiles given by Slifker and Shapiro (1980) is used to estimate the parameters. This method is based on four data percentiles, denoted by x_{-3z}, x_{-z}, x_z, and x_{3z}, which correspond to the four equally spaced percentiles of a standard normal distribution, denoted by $-3z, -z, z$, and $3z$, under the transformation

$$z = \gamma + \delta \log\left(\frac{x - \theta}{\theta + \sigma - x}\right)$$

The default value of z is 0.524. The results of the fit are dependent on the choice of z, and you can specify other values with the FITINTERVAL= option (specified in parentheses after the SB option). If you use the method of percentiles, you should select a value of z that corresponds to percentiles which are critical to your application.

The following values are computed from the data percentiles:

$$\begin{aligned} m &= x_{3z} - x_z \\ n &= x_{-z} - x_{-3z} \\ p &= x_z - x_{-z} \end{aligned}$$

It was demonstrated by Slifker and Shapiro (1980) that

$$\begin{aligned} \frac{mn}{p^2} &> 1 \quad \text{for any } S_U \text{ distribution} \\ \frac{mn}{p^2} &< 1 \quad \text{for any } S_B \text{ distribution} \\ \frac{mn}{p^2} &= 1 \quad \text{for any } S_L \text{ (lognormal) distribution} \end{aligned}$$

A tolerance interval around one is used to discriminate among the three families with this ratio criterion. You can specify the tolerance with the FITTOLERANCE= option (specified in parentheses after the SB option). The default tolerance is 0.01. Assuming that the criterion satisfies the inequality

$$\frac{mn}{p^2} < 1 - \text{tolerance}$$

the parameters of the S_B distribution are computed using the explicit formulas derived by Slifker and Shapiro (1980).

If you specify FITMETHOD = MOMENTS (in parentheses after the SB option), the method of moments is used to estimate the parameters. If you specify FITMETHOD = MLE (in parentheses after the SB option), the method of maximum likelihood is used to estimate the parameters. Note that maximum likelihood estimates may not always exist. Refer to Bowman and Shenton (1983) for discussion of methods for fitting Johnson distributions.

Johnson S_U Distribution

The fitted density function is

$$p(x) = \begin{cases} \frac{\delta h v}{\sigma\sqrt{2\pi}} \frac{1}{\sqrt{1+((x-\theta)/\sigma)^2}} \times \\ \exp\left[-\frac{1}{2}\left(\gamma + \delta \sinh^{-1}\left(\frac{x-\theta}{\sigma}\right)\right)^2\right] & \text{for } x > \theta \\ 0 & \text{for } x \leq \theta \end{cases}$$

where

θ = location parameter $(-\infty < \theta < \infty)$

σ = scale parameter $(\sigma > 0)$

δ = shape parameter $(\delta > 0)$

γ = shape parameter $(-\infty < \gamma < \infty)$

h = width of histogram interval

v = vertical scaling factor

and

$$v = \begin{cases} n & \text{the sample size, for VSCALE=COUNT} \\ 100 & \text{for VSCALE=PERCENT} \\ 1 & \text{for VSCALE=PROPORTION} \end{cases}$$

You can specify the parameters with the THETA=, SIGMA=, DELTA=, and GAMMA= S_U-options, which are enclosed in parentheses after the SU option. If you do not specify these parameters, they are estimated.

By default, the method of percentiles given by Slifker and Shapiro (1980) is used to estimate the parameters. This method is based on four data percentiles, denoted by x_{-3z}, x_{-z}, x_z, and x_{3z}, which correspond to the four equally spaced percentiles of a standard normal distribution, denoted by $-3z$, $-z$, z, and $3z$, under the transformation

$$z = \gamma + \delta \sinh^{-1}\left(\frac{x-\theta}{\sigma}\right)$$

The default value of z is 0.524. The results of the fit are dependent on the choice of z, and you can specify other values with the FITINTERVAL= option (specified in parentheses after the SU option). If you use the method of percentiles, you should select a value of z that corresponds to percentiles that are critical to your application.

The following values are computed from the data percentiles:

$m = x_{3z} - x_z$
$n = x_{-z} - x_{-3z}$
$p = x_z - x_{-z}$

It was demonstrated by Slifker and Shapiro (1980) that

$\frac{mn}{p^2} > 1$ for any S_U distribution

$\frac{mn}{p^2} < 1$ for any S_B distribution

$\frac{mn}{p^2} = 1$ for any S_L (lognormal) distribution

A tolerance interval around one is used to discriminate among the three families with this ratio criterion. You can specify the tolerance with the FITTOLERANCE= option (specified in parentheses after the SU option). The default tolerance is 0.01. Assuming that the criterion satisfies the inequality

$$\frac{mn}{p^2} > 1 + \text{tolerance}$$

the parameters of the S_U distribution are computed using the explicit formulas derived by Slifker and Shapiro (1980).

If you specify FITMETHOD = MOMENTS (in parentheses after the SU option), the method of moments is used to estimate the parameters. If you specify FITMETHOD = MLE (in parentheses after the SU option), the method of maximum likelihood is used to estimate the parameters. Note that maximum likelihood estimates do not always exist. Refer to Bowman and Shenton (1983) for discussion of methods for fitting Johnson distributions.

Weibull Distribution

The fitted density function is

$$p(x) = \begin{cases} hv\frac{c}{\sigma}(\frac{x-\theta}{\sigma})^{c-1} \exp(-(\frac{x-\theta}{\sigma})^c) & \text{for } x > \theta \\ 0 & \text{for } x \leq \theta \end{cases}$$

where

θ = threshold parameter

σ = scale parameter ($\sigma > 0$)

c = shape parameter ($c > 0$)

h = width of histogram interval

v = vertical scaling factor

and

$$v = \begin{cases} n & \text{the sample size, for VSCALE=COUNT} \\ 100 & \text{for VSCALE=PERCENT} \\ 1 & \text{for VSCALE=PROPORTION} \end{cases}$$

The threshold parameter θ must be less than the minimum data value. You can specify θ with the THRESHOLD= *Weibull-option*. By default, $\theta = 0$. If you specify THETA=EST, a maximum likelihood estimate is computed for θ. You can specify σ and c with the SCALE= and SHAPE= *Weibull-options*, respectively. By default, the procedure calculates maximum likelihood estimates for σ and c.

The exponential distribution is a special case of the Weibull distribution where $c = 1$.

You can use the DATA step function QUANTILE to compute Weibull quantiles and the DATA step function CDF to compute Weibull probabilities.

Goodness-of-Fit Tests

When you specify the NORMAL option in the PROC UNIVARIATE statement or you request a fitted parametric distribution in the HISTOGRAM statement, the procedure computes goodness-of-fit tests for the null hypothesis that the values of the analysis variable are a random sample from the specified theoretical distribution. See Example 4.22.

When you specify the NORMAL option, these tests, which are summarized in the output table labeled "Tests for Normality," include the following:

- Shapiro-Wilk test
- Kolmogorov-Smirnov test
- Anderson-Darling test
- Cramér–von Mises test

The Kolmogorov-Smirnov D statistic, the Anderson-Darling statistic, and the Cramér–von Mises statistic are based on the empirical distribution function (EDF). However, some EDF tests are not supported when certain combinations of the parameters of a specified distribution are estimated. See Table 4.32 for a list of the EDF tests available. You determine whether to reject the null hypothesis by examining the p-value that is associated with a goodness-of-fit statistic. When the p-value is less than the predetermined critical value (α), you reject the null hypothesis and conclude that the data did not come from the specified distribution.

If you want to test the normality assumptions for analysis of variance methods, beware of using a statistical test for normality alone. A test's ability to reject the null hypothesis (known as the *power* of the test) increases with the sample size. As the sample size becomes larger, increasingly smaller departures from normality can be detected. Because small deviations from normality do not severely affect the validity of analysis of variance tests, it is important to examine other statistics and plots to make a final assessment of normality. The skewness and kurtosis measures and the plots that are provided by the PLOTS option, the HISTOGRAM statement, the PROBPLOT statement, and the QQPLOT statement can be very helpful. For small sample sizes, power is low for detecting larger departures from normality that may be important. To increase the test's ability to detect such deviations, you may want to declare significance at higher levels, such as 0.15 or 0.20, rather than the often-used 0.05 level. Again, consulting plots and additional statistics can help you assess the severity of the deviations from normality.

Shapiro-Wilk Statistic

If the sample size is less than or equal to 2000 and you specify the NORMAL option, PROC UNIVARIATE computes the Shapiro-Wilk statistic, W (also denoted as W_n to emphasize its dependence on the sample size n). The W statistic is the ratio of the best estimator of the variance (based on the square of a linear combination of the order statistics) to the usual corrected sum of squares estimator of the variance (Shapiro and Wilk 1965). When n is greater than three, the coefficients to compute the linear combination of the order statistics are approximated by the method of Royston (1992). The statistic W is always greater than zero and less than or equal to one ($0 < W \leq 1$).

Small values of W lead to the rejection of the null hypothesis of normality. The distribution of W is highly skewed. Seemingly large values of W (such as 0.90) may be considered small and lead you to reject the null hypothesis. The method for computing the *p*-value (the probability of obtaining a *W* statistic less than or equal to the observed value) depends on *n*. For *n* = 3, the probability distribution of *W* is known and is used to determine the *p*-value. For *n* > 4, a normalizing transformation is computed:

$$Z_n = \begin{cases} (-\log(\gamma - \log(1 - W_n)) - \mu)/\sigma & \text{if } 4 \leq n \leq 11 \\ (\log(1 - W_n) - \mu)/\sigma & \text{if } 12 \leq n \leq 2000 \end{cases}$$

The values of σ, γ, and μ are functions of *n* obtained from simulation results. Large values of Z_n indicate departure from normality, and because the statistic Z_n has an approximately standard normal distribution, this distribution is used to determine the *p*-values for *n* > 4.

EDF Goodness-of-Fit Tests

When you fit a parametric distribution, PROC UNIVARIATE provides a series of goodness-of-fit tests based on the empirical distribution function (EDF). The EDF tests offer advantages over traditional chi-square goodness-of-fit test, including improved power and invariance with respect to the histogram midpoints. For a thorough discussion, refer to D'Agostino and Stephens (1986).

The empirical distribution function is defined for a set of *n* independent observations $X_1, \ldots, X_n$ with a common distribution function $F(x)$. Denote the observations ordered from smallest to largest as $X_{(1)}, \ldots, X_{(n)}$. The empirical distribution function, $F_n(x)$, is defined as

$$F_n(x) = 0, \quad x < X_{(1)}$$
$$F_n(x) = \frac{i}{n}, \quad X_{(i)} \leq x < X_{(i+1)} \quad i = 1, \ldots, n-1$$
$$F_n(x) = 1, \quad X_{(n)} \leq x$$

Note that $F_n(x)$ is a step function that takes a step of height $\frac{1}{n}$ at each observation. This function estimates the distribution function $F(x)$. At any value *x*, $F_n(x)$ is the proportion of observations less than or equal to *x*, while $F(x)$ is the probability of an observation less than or equal to *x*. EDF statistics measure the discrepancy between $F_n(x)$ and $F(x)$.

The computational formulas for the EDF statistics make use of the probability integral transformation $U = F(X)$. If $F(X)$ is the distribution function of *X*, the random variable *U* is uniformly distributed between 0 and 1.

Given *n* observations $X_{(1)}, \ldots, X_{(n)}$, the values $U_{(i)} = F(X_{(i)})$ are computed by applying the transformation, as discussed in the next three sections.

PROC UNIVARIATE provides three EDF tests:

- Kolmogorov-Smirnov
- Anderson-Darling
- Cramér–von Mises

The following sections provide formal definitions of these EDF statistics.

Kolmogorov D Statistic

The Kolmogorov-Smirnov statistic (D) is defined as

$$D = \sup_x |F_n(x) - F(x)|$$

The Kolmogorov-Smirnov statistic belongs to the supremum class of EDF statistics. This class of statistics is based on the largest vertical difference between $F(x)$ and $F_n(x)$.

The Kolmogorov-Smirnov statistic is computed as the maximum of D^+ and D^-, where D^+ is the largest vertical distance between the EDF and the distribution function when the EDF is greater than the distribution function, and D^- is the largest vertical distance when the EDF is less than the distribution function.

$$\begin{aligned} D^+ &= \max_i \left(\frac{i}{n} - U_{(i)} \right) \\ D^- &= \max_i \left(U_{(i)} - \frac{i-1}{n} \right) \\ D &= \max(D^+, D^-) \end{aligned}$$

PROC UNIVARIATE uses a modified Kolmogorov D statistic to test the data against a normal distribution with mean and variance equal to the sample mean and variance.

Anderson-Darling Statistic

The Anderson-Darling statistic and the Cramér–von Mises statistic belong to the quadratic class of EDF statistics. This class of statistics is based on the squared difference $(F_n(x) - F(x))^2$. Quadratic statistics have the following general form:

$$Q = n \int_{-\infty}^{+\infty} (F_n(x) - F(x))^2 \psi(x) dF(x)$$

The function $\psi(x)$ weights the squared difference $(F_n(x) - F(x))^2$.

The Anderson-Darling statistic (A^2) is defined as

$$A^2 = n \int_{-\infty}^{+\infty} (F_n(x) - F(x))^2 \left[F(x)(1 - F(x)) \right]^{-1} dF(x)$$

Here the weight function is $\psi(x) = [F(x)(1 - F(x))]^{-1}$.

The Anderson-Darling statistic is computed as

$$A^2 = -n - \frac{1}{n} \sum_{i=1}^{n} \left[(2i - 1) \log U_{(i)} + (2n + 1 - 2i) \log(1 - U_{(i)}) \right]$$

Cramér–von Mises Statistic

The Cramér–von Mises statistic (W^2) is defined as

$$W^2 = n \int_{-\infty}^{+\infty} (F_n(x) - F(x))^2 dF(x)$$

Here the weight function is $\psi(x) = 1$.

The Cramér–von Mises statistic is computed as

$$W^2 = \sum_{i=1}^{n} \left(U_{(i)} - \frac{2i - 1}{2n} \right)^2 + \frac{1}{12n}$$

Goodness-of-Fit Tests ✦ 431

Probability Values of EDF Tests

Once the EDF test statistics are computed, PROC UNIVARIATE computes the associated probability values (*p*-values).

For the Gumbel, inverse Gaussian, generalized Pareto, and Rayleigh distributions, PROC UNIVARIATE computes associated probability values (*p*-values) by resampling from the estimated distribution. By default 500 EDF test statistics are computed and then compared to the EDF test statistic for the specified (fitted) distribution. The number of samples can be controlled by setting EDFNSAMPLES=*n*. For example, to request Gumbel distribution Goodness-of-Fit test *p*-values based on 5000 simulations, use the following statement:

```
proc univariate data=test;
   histogram / gumbel(edfnsamples=5000);
run;
```

For the beta, exponential, gamma, lognormal, normal, power function, and Weibull distributions the UNIVARIATE procedure uses internal tables of probability levels similar to those given by D'Agostino and Stephens (1986). If the value is between two probability levels, then linear interpolation is used to estimate the probability value.

The probability value depends upon the parameters that are known and the parameters that are estimated for the distribution. Table 4.32 summarizes different combinations fitted for which EDF tests are available.

Table 4.32 Availability of EDF Tests

Distribution	Threshold	Scale	Shape	Tests Available
beta	θ known	σ known	α, β known	all
	θ known	σ known	$\alpha, \beta < 5$ unknown	all
exponential	θ known,	σ known		all
	θ known	σ unknown		all
	θ unknown	σ known		all
	θ unknown	σ unknown		all
gamma	θ known	σ known	α known	all
	θ known	σ unknown	α known	all
	θ known	σ known	α unknown	all
	θ known	σ unknown	$\alpha > 1$ unknown	all
	θ unknown	σ known	$\alpha > 1$ known	all
	θ unknown	σ unknown	$\alpha > 1$ known	all
	θ unknown	σ known	$\alpha > 1$ unknown	all
	θ unknown	σ unknown	$\alpha > 1$ unknown	all
lognormal	θ known	ζ known	σ known	all
	θ known	ζ known	σ unknown	A^2 and W^2
	θ known	ζ unknown	σ known	A^2 and W^2
	θ known	ζ unknown	σ unknown	all
	θ unknown	ζ known	$\sigma < 3$ known	all
	θ unknown	ζ known	$\sigma < 3$ unknown	all
	θ unknown	ζ unknown	$\sigma < 3$ known	all
	θ unknown	ζ unknown	$\sigma < 3$ unknown	all

Table 4.32 (continued)

Distribution	Threshold	Scale	Shape	Tests Available
normal	θ known	σ known		all
	θ known	σ unknown		A^2 and W^2
	θ unknown	σ known		A^2 and W^2
	θ unknown	σ unknown		all
power function	θ known	σ known	α known	all
	θ known	σ known	$\alpha < 5$ unknown	all
Weibull	θ known	σ known	c known	all
	θ known	σ unknown	c known	A^2 and W^2
	θ known	σ known	c unknown	A^2 and W^2
	θ known	σ unknown	c unknown	A^2 and W^2
	θ unknown	σ known	$c > 2$ known	all
	θ unknown	σ unknown	$c > 2$ known	all
	θ unknown	σ known	$c > 2$ unknown	all
	θ unknown	σ unknown	$c > 2$ unknown	all

Kernel Density Estimates

You can use the KERNEL option to superimpose kernel density estimates on histograms. Smoothing the data distribution with a kernel density estimate can be more effective than using a histogram to identify features that might be obscured by the choice of histogram bins or sampling variation. A kernel density estimate can also be more effective than a parametric curve fit when the process distribution is multi-modal. See Example 4.23.

The general form of the kernel density estimator is

$$\hat{f}_\lambda(x) = \frac{hv}{n\lambda} \sum_{i=1}^{n} K_0 \left(\frac{x - x_i}{\lambda} \right)$$

where

$K_0(\cdot)$ is the kernel function

λ is the bandwidth

n is the sample size

x_i is the ith observation

v = vertical scaling factor

and

$$v = \begin{cases} n & \text{for VSCALE=COUNT} \\ 100 & \text{for VSCALE=PERCENT} \\ 1 & \text{for VSCALE=PROPORTION} \end{cases}$$

The KERNEL option provides three kernel functions (K_0): normal, quadratic, and triangular. You can specify the function with the K= *kernel-option* in parentheses after the KERNEL option. Values for the K= option are NORMAL, QUADRATIC, and TRIANGULAR (with aliases of N, Q, and T, respectively). By default, a normal kernel is used. The formulas for the kernel functions are

Normal $\quad K_0(t) = \frac{1}{\sqrt{2\pi}} \exp(-\frac{1}{2}t^2) \quad$ for $-\infty < t < \infty$
Quadratic $\quad K_0(t) = \frac{3}{4}(1 - t^2) \quad$ for $|t| \leq 1$
Triangular $\quad K_0(t) = 1 - |t| \quad$ for $|t| \leq 1$

The value of λ, referred to as the bandwidth parameter, determines the degree of smoothness in the estimated density function. You specify λ indirectly by specifying a standardized bandwidth c with the C= *kernel-option*. If Q is the interquartile range and n is the sample size, then c is related to λ by the formula

$$\lambda = cQn^{-\frac{1}{5}}$$

For a specific kernel function, the discrepancy between the density estimator $\hat{f}_\lambda(x)$ and the true density $f(x)$ is measured by the mean integrated square error (MISE):

$$\text{MISE}(\lambda) = \int_x \{E(\hat{f}_\lambda(x)) - f(x)\}^2 dx + \int_x var(\hat{f}_\lambda(x))dx$$

The MISE is the sum of the integrated squared bias and the variance. An approximate mean integrated square error (AMISE) is:

$$\text{AMISE}(\lambda) = \frac{1}{4}\lambda^4 \left(\int_t t^2 K(t)dt\right)^2 \int_x \left(f''(x)\right)^2 dx + \frac{1}{n\lambda}\int_t K(t)^2 dt$$

A bandwidth that minimizes AMISE can be derived by treating $f(x)$ as the normal density that has parameters μ and σ estimated by the sample mean and standard deviation. If you do not specify a bandwidth parameter or if you specify C=MISE, the bandwidth that minimizes AMISE is used. The value of AMISE can be used to compare different density estimates. You can also specify C=SJPI to select the bandwidth by using a plug-in formula of Sheather and Jones (Jones, Marron, and Sheather 1996). For each estimate, the bandwidth parameter c, the kernel function type, and the value of AMISE are reported in the SAS log.

The general kernel density estimates assume that the domain of the density to estimate can take on all values on a real line. However, sometimes the domain of a density is an interval bounded on one or both sides. For example, if a variable Y is a measurement of only positive values, then the kernel density curve should be bounded so that is zero for negative Y values. You can use the LOWER= and UPPER= *kernel-options* to specify the bounds.

The UNIVARIATE procedure uses a reflection technique to create the bounded kernel density curve, as described in Silverman (1986, pp. 30-31). It adds the reflections of the kernel density that are outside the boundary to the bounded kernel estimates. The general form of the bounded kernel density estimator is computed by replacing $K_0\left(\frac{x-x_i}{\lambda}\right)$ in the original equation with

$$\left\{K_0\left(\frac{x-x_i}{\lambda}\right) + K_0\left(\frac{(x-x_l)+(x_i-x_l)}{\lambda}\right) + K_0\left(\frac{(x_u-x)+(x_u-x_i)}{\lambda}\right)\right\}$$

where x_l is the lower bound and x_u is the upper bound.

Without a lower bound, $x_l = -\infty$ and $K_0\left(\frac{(x-x_l)+(x_i-x_l)}{\lambda}\right) = 0$. Similarly, without an upper bound, $x_u = \infty$ and $K_0\left(\frac{(x_u-x)+(x_u-x_i)}{\lambda}\right) = 0$.

When C=MISE is used with a bounded kernel density, the UNIVARIATE procedure uses a bandwidth that minimizes the AMISE for its corresponding unbounded kernel.

Construction of Quantile-Quantile and Probability Plots

Figure 4.14 illustrates how a Q-Q plot is constructed for a specified theoretical distribution. First, the n nonmissing values of the variable are ordered from smallest to largest:

$$x_{(1)} \leq x_{(2)} \leq \cdots \leq x_{(n)}$$

Then the ith ordered value $x_{(i)}$ is plotted as a point whose y-coordinate is $x_{(i)}$ and whose x-coordinate is $F^{-1}\left(\frac{i-0.375}{n+0.25}\right)$, where $F(\cdot)$ is the specified distribution with zero location parameter and unit scale parameter.

You can modify the adjustment constants –0.375 and 0.25 with the RANKADJ= and NADJ= options. This default combination is recommended by Blom (1958). For additional information, see Chambers et al. (1983). Because $x_{(i)}$ is a quantile of the empirical cumulative distribution function (ECDF), a Q-Q plot compares quantiles of the ECDF with quantiles of a theoretical distribution. Probability plots (see the section "PROBPLOT Statement" on page 360) are constructed the same way, except that the x-axis is scaled nonlinearly in percentiles.

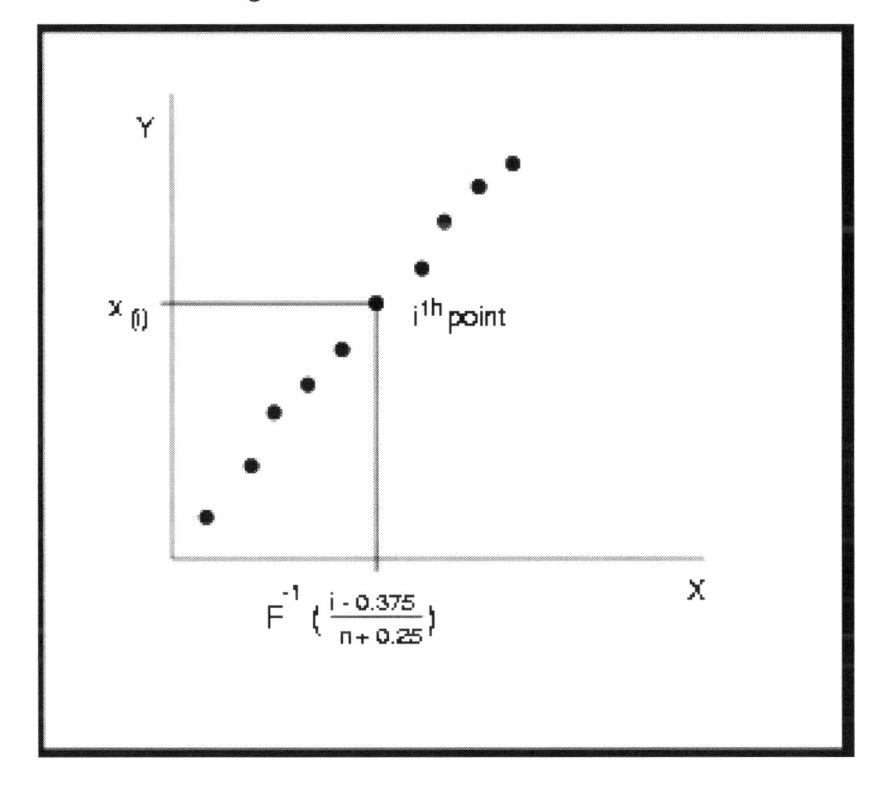

Figure 4.14 Construction of a Q-Q Plot

Interpretation of Quantile-Quantile and Probability Plots

The following properties of Q-Q plots and probability plots make them useful diagnostics of how well a specified theoretical distribution fits a set of measurements:

- If the quantiles of the theoretical and data distributions agree, the plotted points fall on or near the line $y = x$.

- If the theoretical and data distributions differ only in their location or scale, the points on the plot fall on or near the line $y = ax + b$. The slope a and intercept b are visual estimates of the scale and location parameters of the theoretical distribution.

Q-Q plots are more convenient than probability plots for graphical estimation of the location and scale parameters because the x-axis of a Q-Q plot is scaled linearly. On the other hand, probability plots are more convenient for estimating percentiles or probabilities.

There are many reasons why the point pattern in a Q-Q plot may not be linear. Chambers et al. (1983) and Fowlkes (1987) discuss the interpretations of commonly encountered departures from linearity, and these are summarized in Table 4.33.

In some applications, a nonlinear pattern may be more revealing than a linear pattern. However, Chambers et al. (1983) note that departures from linearity can also be due to chance variation.

Table 4.33 Quantile-Quantile Plot Diagnostics

Description of Point Pattern	Possible Interpretation
all but a few points fall on a line	outliers in the data
left end of pattern is below the line; right end of pattern is above the line	long tails at both ends of the data distribution
left end of pattern is above the line; right end of pattern is below the line	short tails at both ends of the data distribution
curved pattern with slope increasing from left to right	data distribution is skewed to the right
curved pattern with slope decreasing from left to right	data distribution is skewed to the left
staircase pattern (plateaus and gaps)	data have been rounded or are discrete

When the pattern is linear, you can use Q-Q plots to estimate shape, location, and scale parameters and to estimate percentiles. See Example 4.26 through Example 4.34.

Distributions for Probability and Q-Q Plots

You can use the PROBPLOT and QQPLOT statements to request probability and Q-Q plots that are based on the theoretical distributions summarized in Table 4.34.

Table 4.34 Distributions and Parameters

Distribution	Density Function $p(x)$	Range	Location	Scale	Shape
Beta	$\frac{(x-\theta)^{\alpha-1}(\theta+\sigma-x)^{\beta-1}}{B(\alpha,\beta)\sigma^{(\alpha+\beta-1)}}$	$\theta < x < \theta+\sigma$	θ	σ	α, β
Exponential	$\frac{1}{\sigma} \exp\left(-\frac{x-\theta}{\sigma}\right)$	$x \geq \theta$	θ	σ	
Gamma	$\frac{1}{\sigma\Gamma(\alpha)} \left(\frac{x-\theta}{\sigma}\right)^{\alpha-1} \exp\left(-\frac{x-\theta}{\sigma}\right)$	$x > \theta$	θ	σ	α
Gumbel	$\frac{e^{-(x-\mu)/\sigma}}{\sigma} \exp\left(-e^{-(x-\mu)/\sigma}\right)$	all x	μ	σ	
Lognormal (3-parameter)	$\frac{1}{\sigma\sqrt{2\pi}(x-\theta)} \exp\left(-\frac{(\log(x-\theta)-\zeta)^2}{2\sigma^2}\right)$	$x > \theta$	θ	ζ	σ
Normal	$\frac{1}{\sigma\sqrt{2\pi}} \exp\left(-\frac{(x-\mu)^2}{2\sigma^2}\right)$	all x	μ	σ	
Generalized Pareto	$\alpha \neq 0$: $\frac{1}{\sigma}(1-\alpha(x-\theta)/\sigma)^{1/\alpha-1}$ $\alpha = 0$: $\frac{1}{\sigma}\exp(-(x-\theta)/\sigma)$	$x > \theta$	θ	σ	α
Power Function	$\frac{\alpha}{\sigma}\left(\frac{x-\theta}{\sigma}\right)^{\alpha-1}$	$x > \theta$	θ	σ	α

Table 4.34 (continued)

Distribution	Density Function $p(x)$	Range	Location	Scale	Shape
Rayleigh	$\frac{x-\theta}{\sigma^2} \exp(-(x-\theta)^2/(2\sigma^2))$	$x \geq \theta$	θ	σ	
Weibull (3-parameter)	$\frac{c}{\sigma}\left(\frac{x-\theta}{\sigma}\right)^{c-1} \exp\left(-\left(\frac{x-\theta}{\sigma}\right)^c\right)$	$x > \theta$	θ	σ	c
Weibull (2-parameter)	$\frac{c}{\sigma}\left(\frac{x-\theta_0}{\sigma}\right)^{c-1} \exp\left(-\left(\frac{x-\theta_0}{\sigma}\right)^c\right)$	$x > \theta_0$ (known)	θ_0	σ	c

You can request these distributions with the BETA, EXPONENTIAL, GAMMA, PARETO, GUMBEL, LOGNORMAL, NORMAL, POWER, RAYLEIGH, WEIBULL, and WEIBULL2 options, respectively. If you do not specify a distribution option, a normal probability plot or a normal Q-Q plot is created.

The following sections provide details for constructing Q-Q plots that are based on these distributions. Probability plots are constructed similarly except that the horizontal axis is scaled in percentile units.

Beta Distribution

To create the plot, the observations are ordered from smallest to largest, and the *i*th ordered observation is plotted against the quantile $B_{\alpha\beta}^{-1}\left(\frac{i-0.375}{n+0.25}\right)$, where $B_{\alpha\beta}^{-1}(\cdot)$ is the inverse normalized incomplete beta function, *n* is the number of nonmissing observations, and α and β are the shape parameters of the beta distribution. In a probability plot, the horizontal axis is scaled in percentile units.

The pattern on the plot for ALPHA=α and BETA=β tends to be linear with intercept θ and slope σ if the data are beta distributed with the specific density function

$$p(x) = \begin{cases} \frac{(x-\theta)^{\alpha-1}(\theta+\sigma-x)^{\beta-1}}{B(\alpha,\beta)\sigma^{(\alpha+\beta-1)}} & \text{for } \theta < x < \theta+\sigma \\ 0 & \text{for } x \leq \theta \text{ or } x \geq \theta+\sigma \end{cases}$$

where $B(\alpha,\beta) = \frac{\Gamma(\alpha)\Gamma(\beta)}{\Gamma(\alpha+\beta)}$ and

θ = lower threshold parameter

σ = scale parameter ($\sigma > 0$)

α = first shape parameter ($\alpha > 0$)

β = second shape parameter ($\beta > 0$)

Exponential Distribution

To create the plot, the observations are ordered from smallest to largest, and the *i*th ordered observation is plotted against the quantile $-\log\left(1 - \frac{i-0.375}{n+0.25}\right)$, where n is the number of nonmissing observations. In a probability plot, the horizontal axis is scaled in percentile units.

The pattern on the plot tends to be linear with intercept θ and slope σ if the data are exponentially distributed with the specific density function

$$p(x) = \begin{cases} \frac{1}{\sigma} \exp\left(-\frac{x-\theta}{\sigma}\right) & \text{for } x \geq \theta \\ 0 & \text{for } x < \theta \end{cases}$$

where θ is a threshold parameter, and σ is a positive scale parameter.

Gamma Distribution

To create the plot, the observations are ordered from smallest to largest, and the *i*th ordered observation is plotted against the quantile $G_\alpha^{-1}\left(\frac{i-0.375}{n+0.25}\right)$, where $G_\alpha^{-1}(\cdot)$ is the inverse normalized incomplete gamma function, n is the number of nonmissing observations, and α is the shape parameter of the gamma distribution. In a probability plot, the horizontal axis is scaled in percentile units.

The pattern on the plot for ALPHA=α tends to be linear with intercept θ and slope σ if the data are gamma distributed with the specific density function

$$p(x) = \begin{cases} \frac{1}{\sigma \Gamma(\alpha)} \left(\frac{x-\theta}{\sigma}\right)^{\alpha-1} \exp\left(-\frac{x-\theta}{\sigma}\right) & \text{for } x > \theta \\ 0 & \text{for } x \leq \theta \end{cases}$$

where

θ = threshold parameter

σ = scale parameter ($\sigma > 0$)

α = shape parameter ($\alpha > 0$)

Gumbel Distribution

To create the plot, the observations are ordered from smallest to largest, and the *i*th ordered observation is plotted against the quantile $-\log\left(-\log\left(\frac{i-0.375}{n+0.25}\right)\right)$, where n is the number of nonmissing observations. In a probability plot, the horizontal axis is scaled in percentile units.

The pattern on the plot tends to be linear with intercept μ and slope σ if the data are Gumbel distributed with the specific density function

$$p(x) = \frac{e^{-(x-\mu)/\sigma}}{\sigma} \exp\left(-e^{-(x-\mu)/\sigma}\right)$$

μ = location parameter

σ = scale parameter ($\sigma > 0$)

Lognormal Distribution

To create the plot, the observations are ordered from smallest to largest, and the ith ordered observation is plotted against the quantile $\exp\left(\sigma \Phi^{-1}\left(\frac{i-0.375}{n+0.25}\right)\right)$, where $\Phi^{-1}(\cdot)$ is the inverse cumulative standard normal distribution, n is the number of nonmissing observations, and σ is the shape parameter of the lognormal distribution. In a probability plot, the horizontal axis is scaled in percentile units.

The pattern on the plot for SIGMA=σ tends to be linear with intercept θ and slope $\exp(\zeta)$ if the data are lognormally distributed with the specific density function

$$p(x) = \begin{cases} \frac{1}{\sigma\sqrt{2\pi}(x-\theta)} \exp\left(-\frac{(\log(x-\theta)-\zeta)^2}{2\sigma^2}\right) & \text{for } x > \theta \\ 0 & \text{for } x \leq \theta \end{cases}$$

where

θ = threshold parameter

ζ = scale parameter

σ = shape parameter ($\sigma > 0$)

See Example 4.26 and Example 4.33.

Normal Distribution

To create the plot, the observations are ordered from smallest to largest, and the ith ordered observation is plotted against the quantile $\Phi^{-1}\left(\frac{i-0.375}{n+0.25}\right)$, where $\Phi^{-1}(\cdot)$ is the inverse cumulative standard normal distribution and n is the number of nonmissing observations. In a probability plot, the horizontal axis is scaled in percentile units.

The point pattern on the plot tends to be linear with intercept μ and slope σ if the data are normally distributed with the specific density function

$$p(x) = \frac{1}{\sigma\sqrt{2\pi}} \exp\left(-\frac{(x-\mu)^2}{2\sigma^2}\right) \quad \text{for all } x$$

where μ is the mean and σ is the standard deviation ($\sigma > 0$).

Generalized Pareto Distribution

To create the plot, the observations are ordered from smallest to largest, and the ith ordered observation is plotted against the quantile $(1 - (1 - \frac{i-0.375}{n+0.25})^\alpha)/\alpha$ ($\alpha \neq 0$) or $-\log(1 - \frac{i-0.375}{n+0.25})$ ($\alpha = 0$), where n is the number of nonmissing observations and α is the shape parameter of the generalized Pareto distribution. The horizontal axis is scaled in percentile units.

The point pattern on the plot for ALPHA=α tends to be linear with intercept θ and slope σ if the data are generalized Pareto distributed with the specific density function

$$p(x) = \begin{cases} \frac{1}{\sigma}(1 - \alpha(x-\theta)/\sigma)^{1/\alpha - 1} & \text{if } \alpha \neq 0 \\ \frac{1}{\sigma} \exp(-(x-\theta)/\sigma) & \text{if } \alpha = 0 \end{cases}$$

where

θ = threshold parameter
σ = scale parameter ($\sigma > 0$)
α = shape parameter ($\alpha > 0$)

Power Function Distribution

To create the plot, the observations are ordered from smallest to largest, and the ith ordered observation is plotted against the quantile $B_{\alpha(1)}^{-1}\left(\frac{i-0.375}{n+0.25}\right)$, where $B_{\alpha(1)}^{-1}(\cdot)$ is the inverse normalized incomplete beta function, n is the number of nonmissing observations, α is one shape parameter of the beta distribution, and the second shape parameter, $\beta = 1$. The horizontal axis is scaled in percentile units.

The point pattern on the plot for ALPHA=α tends to be linear with intercept θ and slope σ if the data are power function distributed with the specific density function

$$p(x) = \begin{cases} \frac{\alpha}{\sigma}\left(\frac{x-\theta}{\sigma}\right)^{\alpha-1} & \text{for } \theta < x < \theta + \sigma \\ 0 & \text{for } x \leq \theta \text{ or } x \geq \theta + \sigma \end{cases}$$

where

θ = threshold parameter
σ = shape parameter ($\sigma > 0$)
α = shape parameter ($\alpha > 0$)

Rayleigh Distribution

To create the plot, the observations are ordered from smallest to largest, and the ith ordered observation is plotted against the quantile $\sqrt{-2\log\left(1 - \frac{i-0.375}{n+0.25}\right)}$, where n is the number of nonmissing observations. The horizontal axis is scaled in percentile units.

The point pattern on the plot tends to be linear with intercept θ and slope σ if the data are Rayleigh distributed with the specific density function

$$p(x) = \begin{cases} \frac{x-\theta}{\sigma^2}\exp(-(x-\theta)^2/(2\sigma^2)) & \text{for } x \geq \theta \\ 0 & \text{for } x < \theta \end{cases}$$

where θ is a threshold parameter, and σ is a positive scale parameter.

Three-Parameter Weibull Distribution

To create the plot, the observations are ordered from smallest to largest, and the ith ordered observation is plotted against the quantile $\left(-\log\left(1 - \frac{i-0.375}{n+0.25}\right)\right)^{\frac{1}{c}}$, where n is the number of nonmissing observations, and c is the Weibull distribution shape parameter. In a probability plot, the horizontal axis is scaled in percentile units.

The pattern on the plot for C=c tends to be linear with intercept θ and slope σ if the data are Weibull distributed with the specific density function

$$p(x) = \begin{cases} \frac{c}{\sigma}\left(\frac{x-\theta}{\sigma}\right)^{c-1} \exp\left(-\left(\frac{x-\theta}{\sigma}\right)^{c}\right) & \text{for } x > \theta \\ 0 & \text{for } x \leq \theta \end{cases}$$

where

θ = threshold parameter

σ = scale parameter ($\sigma > 0$)

c = shape parameter ($c > 0$)

See Example 4.34.

Two-Parameter Weibull Distribution

To create the plot, the observations are ordered from smallest to largest, and the log of the shifted ith ordered observation $x_{(i)}$, denoted by $\log(x_{(i)} - \theta_0)$, is plotted against the quantile $\log\left(-\log\left(1 - \frac{i-0.375}{n+0.25}\right)\right)$, where n is the number of nonmissing observations. In a probability plot, the horizontal axis is scaled in percentile units.

Unlike the three-parameter Weibull quantile, the preceding expression is free of distribution parameters. Consequently, the C= shape parameter is not mandatory with the WEIBULL2 distribution option.

The pattern on the plot for THETA=θ_0 tends to be linear with intercept $\log(\sigma)$ and slope $\frac{1}{c}$ if the data are Weibull distributed with the specific density function

$$p(x) = \begin{cases} \frac{c}{\sigma}\left(\frac{x-\theta_0}{\sigma}\right)^{c-1} \exp\left(-\left(\frac{x-\theta_0}{\sigma}\right)^{c}\right) & \text{for } x > \theta_0 \\ 0 & \text{for } x \leq \theta_0 \end{cases}$$

where

θ_0 = known lower threshold

σ = scale parameter ($\sigma > 0$)

c = shape parameter ($c > 0$)

See Example 4.34.

Estimating Shape Parameters Using Q-Q Plots

Some of the distribution options in the PROBPLOT or QQPLOT statements require you to specify one or two shape parameters in parentheses after the distribution keyword. These are summarized in Table 4.35.

You can visually estimate the value of a shape parameter by specifying a list of values for the shape parameter option. A separate plot is produced for each value, and you can then select the value of the shape parameter that produces the most nearly linear point pattern. Alternatively, you can request that the plot be created using an estimated shape parameter. See the entries for the distribution options in the section "Dictionary of Options" on page 365 (for the PROBPLOT statement) and in the section "Dictionary of Options" on page 376 (for the QQPLOT statement).

NOTE: For Q-Q plots created with the WEIBULL2 option, you can estimate the shape parameter c from a linear pattern by using the fact that the slope of the pattern is $\frac{1}{c}$.

Table 4.35 Shape Parameter Options

Distribution Keyword	Mandatory Shape Parameter Option	Range
BETA	ALPHA=α, BETA=β	$\alpha > 0, \beta > 0$
EXPONENTIAL	None	
GAMMA	ALPHA=α	$\alpha > 0$
GUMBEL	None	
LOGNORMAL	SIGMA=σ	$\sigma > 0$
NORMAL	None	
PARETO	ALPHA=α	$\alpha > 0$
POWER	ALPHA=α	$\alpha > 0$
RAYLEIGH	None	
WEIBULL	C=c	$c > 0$
WEIBULL2	None	

Estimating Location and Scale Parameters Using Q-Q Plots

If you specify location and scale parameters for a distribution in a PROBPLOT or QQPLOT statement (or if you request estimates for these parameters), a diagonal distribution reference line is displayed on the plot. (An exception is the two-parameter Weibull distribution, for which a line is displayed when you specify or estimate the scale and shape parameters.) Agreement between this line and the point pattern indicates that the distribution with these parameters is a good fit.

When the point pattern on a Q-Q plot is linear, its intercept and slope provide estimates of the location and scale parameters. (An exception to this rule is the two-parameter Weibull distribution, for which the intercept and slope are related to the scale and shape parameters.)

Table 4.36 shows how the specified parameters determine the intercept and slope of the line. The intercept and slope are based on the quantile scale for the horizontal axis, which is used in Q-Q plots.

Table 4.36 Intercept and Slope of Distribution Reference Line

Distribution	Location	Scale	Shape	Intercept	Slope
Beta	θ	σ	α, β	θ	σ
Exponential	θ	σ		θ	σ
Gamma	θ	σ	α	θ	σ
Gumbel	μ	σ		μ	σ
Lognormal	θ	ζ	σ	θ	$\exp(\zeta)$
Normal	μ	σ		μ	σ
Generalized Pareto	θ	σ	α	θ	σ
Power Function	θ	σ	α	θ	σ
Rayleigh	θ	σ		θ	σ
Weibull (3-parameter)	θ	σ	c	θ	σ
Weibull (2-parameter)	θ_0 (known)	σ	c	$\log(\sigma)$	$\frac{1}{c}$

For instance, specifying MU=3 and SIGMA=2 with the NORMAL option requests a line with intercept 3 and slope 2. Specifying SIGMA=1 and C=2 with the WEIBULL2 option requests a line with intercept $\log(1) = 0$ and slope $\frac{1}{2}$. On a probability plot with the LOGNORMAL and WEIBULL2 options, you can specify the slope directly with the SLOPE= option. That is, for the LOGNORMAL option, specifying THETA= θ_0 and SLOPE=$\exp(\zeta_0)$ displays the same line as specifying THETA= θ_0 and ZETA= ζ_0. For the WEIBULL2 option, specifying SIGMA= σ_0 and SLOPE= $\frac{1}{c_0}$ displays the same line as specifying SIGMA= σ_0 and C= c_0.

Estimating Percentiles Using Q-Q Plots

There are two ways to estimate percentiles from a Q-Q plot:

- Specify the PCTLAXIS option, which adds a percentile axis opposite the theoretical quantile axis. The scale for the percentile axis ranges between 0 and 100 with tick marks at percentile values such as 1, 5, 10, 25, 50, 75, 90, 95, and 99.

- Specify the PCTLSCALE option, which relabels the horizontal axis tick marks with their percentile equivalents but does not alter their spacing. For example, on a normal Q-Q plot, the tick mark labeled "0" is relabeled as "50" because the 50th percentile corresponds to the zero quantile.

You can also estimate percentiles by using probability plots created with the PROBPLOT statement. See Example 4.32.

Input Data Sets

DATA= Data Set

The DATA= data set provides the set of variables that are analyzed. The UNIVARIATE procedure must have a DATA= data set. If you do not specify one with the DATA= option in the PROC UNIVARIATE statement, the procedure uses the last data set created.

ANNOTATE= Data Sets

You can add features to plots by specifying ANNOTATE= data sets either in the PROC UNIVARIATE statement or in individual plot statements.

Information contained in an ANNOTATE= data set specified in the PROC UNIVARIATE statement is used for all plots produced in a given PROC step; this is a "global" ANNOTATE= data set. By using this global data set, you can keep information common to all high-resolution plots in one data set.

Information contained in the ANNOTATE= data set specified in a plot statement is used only for plots produced by that statement; this is a "local" ANNOTATE= data set. By using this data set, you can add statement-specific features to plots. For example, you can add different features to plots produced by the HISTOGRAM and QQPLOT statements by specifying an ANNOTATE= data set in each plot statement.

You can specify an ANNOTATE= data set in the PROC UNIVARIATE statement and in plot statements. This enables you to add some features to all plots and also add statement-specific features to plots. See Example 4.25.

OUT= Output Data Set in the OUTPUT Statement

PROC UNIVARIATE creates an OUT= data set for each OUTPUT statement. This data set contains an observation for each combination of levels of the variables in the BY and CLASS statements, or a single observation if you do not specify a BY or CLASS statement. Thus the number of observations in the new data set corresponds to the number of groups for which statistics are calculated. Without a BY or CLASS statement, the procedure computes statistics and percentiles by using all the observations in the input data set. With a BY statement, the procedure computes statistics and percentiles by using the observations within each BY group. With a CLASS statement, the procedure computes statistics and percentiles by using the observations that correspond to each level of the CLASS variables within each BY group.

The variables in the OUT= data set are as follows:

- BY statement variables. The values of these variables match the values in the corresponding BY group in the DATA= data set and indicate which BY group each observation summarizes.

- CLASS statement variables. The values of these variables match the CLASS levels within a BY group that each observation summarizes.

- variables created by selecting statistics in the OUTPUT statement. The statistics are computed using all the nonmissing data, or they are computed for each CLASS level within each BY group if you specify BY and/or CLASS statements.

- variables created by requesting new percentiles with the PCTLPTS= option. The names of these new variables depend on the values of the PCTLPRE= and PCTLNAME= options.

If the output data set contains a percentile variable or a quartile variable, the percentile definition assigned with the PCTLDEF= option in the PROC UNIVARIATE statement is recorded in the output data set label. See Example 4.8.

The following table lists variables available in the OUT= data set.

Table 4.37 Variables Available in the OUT= Data Set

Variable Name	Description
Descriptive Statistics	
CSS	sum of squares corrected for the mean
CV	percent coefficient of variation
KURTOSIS	KURT	measurement of the heaviness of tails
MAX	largest (maximum) value
MEAN	arithmetic mean
MIN	smallest (minimum) value
MODE	most frequent value (if not unique, the smallest mode)
N	number of observations on which calculations are based
NMISS	number of missing observations
NOBS	total number of observations
RANGE	difference between the maximum and minimum values
SKEWNESS	SKEW	measurement of the tendency of the deviations to be larger in one direction than in the other
STD	STDDEV	standard deviation
STDMEAN	STDERR	standard error of the mean
SUM	sum
SUMWGT	sum of the weights
USS	uncorrected sum of squares
VAR	variance
Quantile Statistics	
MEDIAN	Q2	P50	middle value (50th percentile)
P1	1st percentile
P5	5th percentile
P10	10th percentile
P90	90th percentile
P95	95th percentile
P99	99th percentile
Q1	P25	lower quartile (25th percentile)
Q3	P75	upper quartile (75th percentile)
QRANGE	difference between the upper and lower quartiles (also known as the inner quartile range)
Robust Statistics	
GINI	Gini's mean difference
MAD	median absolute difference
QN	2nd variation of median absolute difference

Table 4.37 (continued)

Variable Name	Description
SN	1st variation of median absolute difference
STD_GINI	standard deviation for Gini's mean difference
STD_MAD	standard deviation for median absolute difference
STD_QN	standard deviation for the second variation of the median absolute difference
STD_QRANGE	estimate of the standard deviation, based on interquartile range
STD_SN	standard deviation for the first variation of the median absolute difference
Hypothesis Test Statistics	
MSIGN	sign statistic
NORMAL	test statistic for normality. If the sample size is less than or equal to 2000, this is the Shapiro-Wilk W statistic. Otherwise, it is the Kolmogorov D statistic.
PROBM	probability of a greater absolute value for the sign statistic
PROBN	probability that the data came from a normal distribution
PROBS	probability of a greater absolute value for the signed rank statistic
PROBT	two-tailed p-value for Student's t statistic with $n-1$ degrees of freedom
SIGNRANK	signed rank statistic
T	Student's t statistic to test the null hypothesis that the population mean is equal to μ_0

OUTHISTOGRAM= Output Data Set

You can create an OUTHISTOGRAM= data set with the HISTOGRAM statement. This data set contains information about histogram intervals. Because you can specify multiple HISTOGRAM statements with the UNIVARIATE procedure, you can create multiple OUTHISTOGRAM= data sets.

An OUTHISTOGRAM= data set contains a group of observations for each variable in the HISTOGRAM statement. The group contains an observation for each interval of the histogram, beginning with the leftmost interval that contains a value of the variable and ending with the rightmost interval that contains a value of the variable. These intervals do not necessarily coincide with the intervals displayed in the histogram because the histogram might be padded with empty intervals at either end. If you superimpose one or more fitted curves on the histogram, the OUTHISTOGRAM= data set contains multiple groups of observations for each variable (one group for each curve). If you use BY and/or CLASS statement, the OUTHISTOGRAM= data set contains groups of observations for each CLASS level within each BY group. ID variables are not saved in an OUTHISTOGRAM= data set.

By default, an OUTHISTOGRAM= data set contains the _MIDPT_ variable, whose values identify histogram intervals by their midpoints. When the ENDPOINTS= or NENDPOINTS option is specified, intervals are identified by endpoint values instead. If the RTINCLUDE option is specified, the _MAXPT_ variable contains upper endpoint values. Otherwise, the _MINPT_ variable contains lower endpoint values. See Example 4.18.

Table 4.38 Variables in the OUTHISTOGRAM= Data Set

Variable	Description
COUNT	number of variable values in histogram interval
CURVE	name of fitted distribution (if requested in HISTOGRAM statement)
EXPPCT	estimated percent of population in histogram interval determined from optional fitted distribution
MAXPT	upper endpoint of histogram interval
MIDPT	midpoint of histogram interval
MINPT	lower endpoint of histogram interval
OBSPCT	percent of variable values in histogram interval
VAR	variable name

OUTKERNEL= Output Data Set

You can create an OUTKERNEL= data set with the HISTOGRAM statement. This data set contains information about histogram intervals. Because you can specify multiple HISTOGRAM statements with the UNIVARIATE procedure, you can create multiple OUTKERNEL= data sets.

An OUTKERNEL= data set contains a group of observations for each kernel density estimate requested with the HISTOGRAM statement. These observations span a range of analysis variable values recorded in the _VALUE_ variable. The procedure determines the increment between values, and therefore the number of observations in the group. The variable _DENSITY_ contains the kernel density calculated for the corresponding analysis variable value.

When a density curve is overlaid on a histogram, the curve is scaled so that the area under the curve equals the total area of the histogram bars. The scaled density values are saved in the variable _COUNT_, _PERCENT_, or _PROPORTION_, depending on the histogram's vertical axis scale, determined by the VSCALE= option. Only one of these variables appears in a given OUTKERNEL= data set.

Table 4.39 lists the variables in an OUTKERNEL= data set.

Table 4.39 Variables in the OUTKERNEL= Data Set

Variable	Description
_C_	standardized bandwidth parameter
COUNT	kernel density scaled for VSCALE=COUNT
DENSITY	kernel density
PERCENT	kernel density scaled for VSCALE=PERCENT (default)
PROPORTION	kernel density scaled for VSCALE=PROPORTION
TYPE	kernel function
VALUE	variable value at which kernel function is calculated
VAR	variable name

OUTTABLE= Output Data Set

The OUTTABLE= data set saves univariate statistics in a data set that contains one observation per analysis variable. The following variables are saved:

Table 4.40 Variables in the OUTTABLE= Data Set

Variable	Description
CSS	corrected sum of squares
_CV_	coefficient of variation
GEOMEAN	geometric mean
GINI	Gini's mean difference
KURT	kurtosis
MAD	median absolute difference about the median
MAX	maximum
MEAN	mean
MEDIAN	median
MIN	minimum
MODE	mode
MSIGN	sign statistic
NMISS	number of missing observations
NOBS	number of nonmissing observations
NORMAL	test statistic for normality
_P1_	1st percentile
_P5_	5th percentile
_P10_	10th percentile
_P90_	90th percentile
_P95_	95th percentile
_P99_	99th percentile
PROBM	p-value of sign statistic
PROBN	p-value of test for normality
PROBS	p-value of signed rank test
PROBT	p-value of t statistic
_Q1_	25th percentile (lower quartile)
_Q3_	75th percentile (upper quartile)
_QN_	Q_n
QRANGE	interquartile range (upper quartile minus lower quartile)
RANGE	range
SGNRNK	centered sign rank
SKEW	skewness
_SN_	S_n (see "Robust Estimates of Scale" on page 404)
STD	standard deviation
STDGINI	Gini's standard deviation
STDMAD	MAD standard deviation
STDMEAN	standard error of the mean
STDQN	Q_n standard deviation
STDQRANGE	interquartile range standard deviation

Table 4.40 *(continued)*

Variable	Description
STDSN	S_n standard deviation
SUMWGT	sum of the weights
SUM	sum
_T_	Student's t statistic
USS	uncorrected sum of squares
VARI	variance
VAR	variable name

The OUTTABLE= data set and the OUT= data set (see the section "OUT= Output Data Set in the OUTPUT Statement" on page 444) contain essentially the same information. However, the structure of the OUTTABLE= data set may be more appropriate when you are computing summary statistics for more than one analysis variable in the same invocation of the UNIVARIATE procedure. Each observation in the OUTTABLE= data set corresponds to a different analysis variable, and the variables in the data set correspond to summary statistics and indices.

For example, suppose you have 10 analysis variables (P1-P10). The following statements create an OUTTABLE= data set named Table, which contains summary statistics for each of these variables:

```
data Analysis;
   input A1-A10;
   datalines;
 72  223  332  138  110  145   23  293  353  458
 97   54   61  196  275  171  117   72   81  141
 56  170  140  400  371   72   60   20  484  138
124    6  332  493  214   43  125   55  372   30
152  236  222   76  187  126  192  334  109  546
  5  260  194  277  176   96  109  184  240  261
161  253  153  300   37  156  282  293  451  299
128  121  254  297  363  132  209  257  429  295
116  152  331   27  442  103   80  393  383   94
 43  178  278  159   25  180  253  333   51  225
 34  128  182  415  524  112   13  186  145  131
142  236  234  255  211   80  281  135  179   11
108  215  335   66  254  196  190  363  226  379
 62  232  219  474   31  139   15   56  429  298
177  218  275  171  457  146  163   18  155  129
  0  235   83  239  398   99  226  389  498   18
147  199  324  258  504    2  218  295  422  287
 39  161  156  198  214   58  238   19  231  548
120   42  372  420  232  112  157   79  197  166
178   83  238  492  463   68   46  386   45   81
161  267  372  296  501   96   11  288  330   74
 14    2   52   81  169   63  194  161  173   54
 22  181   92  272  417   94  188  180  367  342
 55  248  214  422  133  193  144  318  271  479
 56   83  169   30  379    5  296  320  396  597
;
```

```
proc univariate data=Analysis outtable=Table noprint;
   var A1-A10;
run;
```

The following statements create the table shown in Figure 4.15, which contains the mean, standard deviation, and so on, for each analysis variable:

```
proc print data=Table label noobs;
   var _VAR_ _MIN_ _MEAN_ _MAX_ _STD_;
   label _VAR_='Analysis';
run;
```

Figure 4.15 Tabulating Results for Multiple Process Variables

Test Scores for a College Course

Analysis	Minimum	Mean	Maximum	Standard Deviation
A1	0	90.76	178	57.024
A2	2	167.32	267	81.628
A3	52	224.56	372	96.525
A4	27	258.08	493	145.218
A5	25	283.48	524	157.033
A6	2	107.48	196	52.437
A7	11	153.20	296	90.031
A8	18	217.08	393	130.031
A9	45	280.68	498	140.943
A10	11	243.24	597	178.799

Tables for Summary Statistics

By default, PROC UNIVARIATE produces ODS tables of moments, basic statistical measures, tests for location, quantiles, and extreme observations. You must specify options in the PROC UNIVARIATE statement to request other statistics and tables. The CIBASIC option produces a table that displays confidence limits for the mean, standard deviation, and variance. The CIPCTLDF and CIPCTLNORMAL options request tables of confidence limits for the quantiles. The LOCCOUNT option requests a table that shows the number of values greater than, not equal to, and less than the value of MU0=. The FREQ option requests a table of frequencies counts. The NEXTRVAL= option requests a table of extreme values. The NORMAL option requests a table with tests for normality.

The TRIMMED=, WINSORIZED=, and ROBUSTSCALE options request tables with robust estimators. The table of trimmed or Winsorized means includes the percentage and the number of observations that are trimmed or Winsorized at each end, the mean and standard error, confidence limits, and the Student's t test. The table with robust measures of scale includes interquartile range, Gini's mean difference G, MAD, Q_n, and S_n, with their corresponding estimates of σ.

See the section "ODS Table Names" on page 451 for the names of ODS tables created by PROC UNIVARIATE.

ODS Table Names

PROC UNIVARIATE assigns a name to each table that it creates. You can use these names to reference the table when you use the Output Delivery System (ODS) to select tables and create output data sets.

Table 4.41   ODS Tables Produced with the PROC UNIVARIATE Statement

ODS Table Name	Description	Option
BasicIntervals	confidence intervals for mean, standard deviation, variance	CIBASIC
BasicMeasures	measures of location and variability	default
ExtremeObs	extreme observations	default
ExtremeValues	extreme values	NEXTRVAL=
Frequencies	frequencies	FREQ
LocationCounts	counts used for sign test and signed rank test	LOCCOUNT
MissingValues	missing values	default, if missing values exist
Modes	modes	MODES
Moments	sample moments	default
Plots	line printer plots	PLOTS
Quantiles	quantiles	default
RobustScale	robust measures of scale	ROBUSTSCALE
SSPlots	line printer side-by-side box plots	PLOTS (with BY statement)
TestsForLocation	tests for location	default
TestsForNormality	tests for normality	NORMALTEST
TrimmedMeans	trimmed means	TRIMMED=
WinsorizedMeans	Winsorized means	WINSORIZED=

Table 4.42   ODS Tables Produced with the HISTOGRAM Statement

ODS Table Name	Description	Option
Bins	histogram bins	MIDPERCENTS secondary option
FitQuantiles	quantiles of fitted distribution	any distribution option
GoodnessOfFit	goodness-of-fit tests for fitted distribution	any distribution option
HistogramBins	histogram bins	MIDPERCENTS option
ParameterEstimates	parameter estimates for fitted distribution	any distribution option

ODS Tables for Fitted Distributions

If you request a fitted parametric distribution with a HISTOGRAM statement, PROC UNIVARIATE creates a summary that is organized into the ODS tables described in this section.

Parameters

The ParameterEstimates table lists the estimated (or specified) parameters for the fitted curve as well as the estimated mean and estimated standard deviation. See "Formulas for Fitted Continuous Distributions" on page 416.

EDF Goodness-of-Fit Tests

When you fit a parametric distribution, the HISTOGRAM statement provides a series of goodness-of-fit tests based on the empirical distribution function (EDF). See "EDF Goodness-of-Fit Tests" on page 429. These are displayed in the GoodnessOfFit table.

Histogram Intervals

The Bins table is included in the summary only if you specify the MIDPERCENTS option in parentheses after the distribution option. This table lists the midpoints for the histogram bins along with the observed and estimated percentages of the observations that lie in each bin. The estimated percentages are based on the fitted distribution.

If you specify the MIDPERCENTS option without requesting a fitted distribution, the HistogramBins table is included in the summary. This table lists the interval midpoints with the observed percent of observations that lie in the interval. See the entry for the MIDPERCENTS option.

Quantiles

The FitQuantiles table lists observed and estimated quantiles. You can use the PERCENTS= option to specify the list of quantiles in this table. By default, the table lists observed and estimated quantiles for the 1, 5, 10, 25, 50, 75, 90, 95, and 99 percent of a fitted parametric distribution.

ODS Graphics

Statistical procedures use ODS Graphics to create graphs as part of their output. ODS Graphics is described in detail in Chapter 21, "Statistical Graphics Using ODS" (*SAS/STAT User's Guide*).

Before you create graphs, ODS Graphics must be enabled (for example, with the ODS GRAPHICS ON statement). For more information about enabling and disabling ODS Graphics, see the section "Enabling and Disabling ODS Graphics" in that chapter.

The overall appearance of graphs is controlled by ODS styles. Styles and other aspects of using ODS Graphics are discussed in the section "A Primer on ODS Statistical Graphics" in that chapter.

PROC UNIVARIATE assigns a name to each graph it creates by using ODS Graphics. You can use these names to reference the graphs when you use ODS. The names are listed in Table 4.43.

Table 4.43 ODS Graphics Produced by PROC UNIVARIATE

ODS Graph Name	Plot Description	Statement
CDFPlot	cdf plot	CDFPLOT
Histogram	histogram	HISTOGRAM
PPPlot	P-P plot	PPPLOT
ProbPlot	probability plot	PROBPLOT
QQPlot	Q-Q plot	QQPLOT

Computational Resources

Because the UNIVARIATE procedure computes quantile statistics, it requires additional memory to store a copy of the data in memory. By default, the MEANS, SUMMARY, and TABULATE procedures require less memory because they do not automatically compute quantiles. These procedures also provide an option to use a new fixed-memory quantiles estimation method that is usually less memory-intensive.

In the UNIVARIATE procedure, the only factor that limits the number of variables that you can analyze is the computer resources that are available. The amount of temporary storage and CPU time required depends on the statements and the options that you specify. To calculate the computer resources the procedure needs, let

- N be the number of observations in the data set
- V be the number of variables in the VAR statement
- U_i be the number of unique values for the ith variable

Then the minimum memory requirement in bytes to process all variables is $M = 24 \sum_i U_i$. If M bytes are not available, PROC UNIVARIATE must process the data multiple times to compute all the statistics. This reduces the minimum memory requirement to $M = 24 \max(U_i)$.

Using the ROUND= option reduces the number of unique values (U_i), thereby reducing memory requirements. The ROBUSTSCALE option requires $40 U_i$ bytes of temporary storage.

Several factors affect the CPU time:

- The time to create V tree structures to internally store the observations is proportional to $NV \log(N)$.

- The time to compute moments and quantiles for the ith variable is proportional to U_i.

- The time to compute the NORMAL option test statistics is proportional to N.

- The time to compute the ROBUSTSCALE option test statistics is proportional to $U_i \log(U_i)$.

- The time to compute the exact significance level of the sign rank statistic can increase when the number of nonzero values is less than or equal to 20.

Each of these factors has a different constant of proportionality. For additional information about optimizing CPU performance and memory usage, see the SAS documentation for your operating environment.

Examples: UNIVARIATE Procedure

Example 4.1: Computing Descriptive Statistics for Multiple Variables

This example computes univariate statistics for two variables. The following statements create the data set BPressure, which contains the systolic (Systolic) and diastolic (Diastolic) blood pressure readings for 22 patients:

```
data BPressure;
   length PatientID $2;
   input PatientID $ Systolic Diastolic @@;
   datalines;
CK 120  50    SS  96  60   FR 100  70
CP 120  75    BL 140  90   ES 120  70
CP 165 110    JI 110  40   MC 119  66
FC 125  76    RW 133  60   KD 108  54
DS 110  50    JW 130  80   BH 120  65
JW 134  80    SB 118  76   NS 122  78
GS 122  70    AB 122  78   EC 112  62
HH 122  82
;
```

The following statements produce descriptive statistics and quantiles for the variables Systolic and Diastolic:

```
title 'Systolic and Diastolic Blood Pressure';
ods select BasicMeasures Quantiles;
proc univariate data=BPressure;
   var Systolic Diastolic;
run;
```

The ODS SELECT statement restricts the output, which is shown in Output 4.1.1, to the "BasicMeasures" and "Quantiles" tables; see the section "ODS Table Names" on page 451. You use the PROC UNIVARIATE statement to request univariate statistics for the variables listed in the VAR statement, which specifies the analysis variables and their order in the output. Formulas for computing the statistics in the "BasicMeasures" table are provided in the section "Descriptive Statistics" on page 394. The quantiles are calculated using *Definition 5*, which is the default definition; see the section "Calculating Percentiles" on page 397.

A sample program for this example, *uniex01.sas*, is available in the SAS Sample Library for Base SAS software.

Output 4.1.1 Display Basic Measures and Quantiles

Systolic and Diastolic Blood Pressure

The UNIVARIATE Procedure
Variable: Systolic

Basic Statistical Measures			
Location		Variability	
Mean	121.2727	Std Deviation	14.28346
Median	120.0000	Variance	204.01732
Mode	120.0000	Range	69.00000
		Interquartile Range	13.00000

Note: The mode displayed is the smallest of 2 modes with a count of 4.

Quantiles (Definition 5)	
Level	Quantile
100% Max	165
99%	165
95%	140
90%	134
75% Q3	125
50% Median	120
25% Q1	112
10%	108
5%	100
1%	96
0% Min	96

Systolic and Diastolic Blood Pressure

The UNIVARIATE Procedure
Variable: Diastolic

Basic Statistical Measures			
Location		Variability	
Mean	70.09091	Std Deviation	15.16547
Median	70.00000	Variance	229.99134
Mode	70.00000	Range	70.00000
		Interquartile Range	18.00000

Output 4.1.1 *continued*

Quantiles (Definition 5)	
Level	Quantile
100% Max	110
99%	110
95%	90
90%	82
75% Q3	78
50% Median	70
25% Q1	60
10%	50
5%	50
1%	40
0% Min	40

Example 4.2: Calculating Modes

An instructor is interested in calculating all the modes of the scores on a recent exam. The following statements create a data set named Exam, which contains the exam scores in the variable Score:

```
data Exam;
   label Score = 'Exam Score';
   input Score @@;
   datalines;
81 97 78 99 77 81 84 86 86 97
85 86 94 76 75 42 91 90 88 86
97 97 89 69 72 82 83 81 80 81
;
```

The following statements use the MODES option to request a table of all possible modes:

```
title 'Table of Modes for Exam Scores';
ods select Modes;
proc univariate data=Exam modes;
   var Score;
run;
```

The ODS SELECT statement restricts the output to the "Modes" table; see the section "ODS Table Names" on page 451.

Output 4.2.1 Table of Modes Display

Table of Modes for Exam Scores

The UNIVARIATE Procedure
Variable: Score (Exam Score)

Modes	
Mode	Count
81	4
86	4
97	4

By default, when the MODES option is used and there is more than one mode, the lowest mode is displayed in the "BasicMeasures" table. The following statements illustrate the default behavior:

```
title 'Default Output';
ods select BasicMeasures;
proc univariate data=Exam;
   var Score;
run;
```

Output 4.2.2 Default Output (Without MODES Option)

Default Output

The UNIVARIATE Procedure
Variable: Score (Exam Score)

Basic Statistical Measures			
Location		Variability	
Mean	83.66667	Std Deviation	11.08069
Median	84.50000	Variance	122.78161
Mode	81.00000	Range	57.00000
		Interquartile Range	10.00000

Note: The mode displayed is the smallest of 3 modes with a count of 4.

The default output displays a mode of 81 and includes a note regarding the number of modes; the modes 86 and 97 are not displayed. The ODS SELECT statement restricts the output to the "BasicMeasures" table; see the section "ODS Table Names" on page 451.

A sample program for this example, *uniex02.sas*, is available in the SAS Sample Library for Base SAS software.

Example 4.3: Identifying Extreme Observations and Extreme Values

This example, which uses the data set BPressure introduced in Example 4.1, illustrates how to produce a table of the extreme observations and a table of the extreme values in a data set. The following statements generate the "Extreme Observations" tables for Systolic and Diastolic, which enable you to identify the extreme observations for each variable:

```
title 'Extreme Blood Pressure Observations';
ods select ExtremeObs;
proc univariate data=BPressure;
   var Systolic Diastolic;
   id PatientID;
run;
```

The ODS SELECT statement restricts the output to the "ExtremeObs" table; see the section "ODS Table Names" on page 451. The ID statement requests that the extreme observations are to be identified using the value of PatientID as well as the observation number. By default, the five lowest and five highest observations are displayed. You can use the NEXTROBS= option to request a different number of extreme observations.

Output 4.3.1 shows that the patient identified as 'CP' (Observation 7) has the highest values for both Systolic and Diastolic. To visualize extreme observations, you can create histograms; see Example 4.14.

Output 4.3.1 Blood Pressure Extreme Observations

Extreme Blood Pressure Observations

The UNIVARIATE Procedure
Variable: Systolic

Extreme Observations

| \multicolumn{3}{c}{Lowest} | \multicolumn{3}{c}{Highest} |
Value	PatientID	Obs	Value	PatientID	Obs
96	SS	2	130	JW	14
100	FR	3	133	RW	11
108	KD	12	134	JW	16
110	DS	13	140	BL	5
110	JI	8	165	CP	7

Extreme Blood Pressure Observations

The UNIVARIATE Procedure
Variable: Diastolic

Extreme Observations

| \multicolumn{3}{c}{Lowest} | \multicolumn{3}{c}{Highest} |
Value	PatientID	Obs	Value	PatientID	Obs
40	JI	8	80	JW	14
50	DS	13	80	JW	16
50	CK	1	82	HH	22
54	KD	12	90	BL	5
60	RW	11	110	CP	7

The following statements generate the "Extreme Values" tables for Systolic and Diastolic, which tabulate the tails of the distributions:

```
title 'Extreme Blood Pressure Values';
ods select ExtremeValues;
proc univariate data=BPressure nextrval=5;
   var Systolic Diastolic;
run;
```

The ODS SELECT statement restricts the output to the "ExtremeValues" table; see the section "ODS Table Names" on page 451. The NEXTRVAL= option specifies the number of extreme values at each end of the distribution to be shown in the tables in Output 4.3.2.

Output 4.3.2 shows that the values 78 and 80 occurred twice for Diastolic and the maximum of Diastolic is 110. Note that Output 4.3.1 displays the value of 80 twice for Diastolic because there are two observations with that value. In Output 4.3.2, the value 80 is only displayed once.

Output 4.3.2 Blood Pressure Extreme Values

Extreme Blood Pressure Values

The UNIVARIATE Procedure
Variable: Systolic

\multicolumn{6}{c}{Extreme Values}
\multicolumn{3}{c}{Lowest}	\multicolumn{3}{c}{Highest}
Order	Value	Freq	Order	Value	Freq
1	96	1	11	130	1
2	100	1	12	133	1
3	108	1	13	134	1
4	110	2	14	140	1
5	112	1	15	165	1

Extreme Blood Pressure Values

The UNIVARIATE Procedure
Variable: Diastolic

\multicolumn{6}{c}{Extreme Values}
\multicolumn{3}{c}{Lowest}	\multicolumn{3}{c}{Highest}
Order	Value	Freq	Order	Value	Freq
1	40	1	11	78	2
2	50	2	12	80	2
3	54	1	13	82	1
4	60	2	14	90	1
5	62	1	15	110	1

A sample program for this example, *uniex01.sas*, is available in the SAS Sample Library for Base SAS software.

Example 4.4: Creating a Frequency Table

An instructor is interested in creating a frequency table of score changes between a pair of tests given in one of his college courses. The data set Score contains test scores for his students who took a pretest and a posttest on the same material. The variable ScoreChange contains the difference between the two test scores. The following statements create the data set:

```
data Score;
   input Student $ PreTest PostTest @@;
   label ScoreChange = 'Change in Test Scores';
   ScoreChange = PostTest - PreTest;
   datalines;
Capalleti  94 91  Dubose     51 65
Engles     95 97  Grant      63 75
Krupski    80 75  Lundsford  92 55
Mcbane     75 78  Mullen     89 82
Nguyen     79 76  Patel      71 77
Si         75 70  Tanaka     87 73
;
```

The following statements produce a frequency table for the variable ScoreChange:

```
title 'Analysis of Score Changes';
ods select Frequencies;
proc univariate data=Score freq;
   var ScoreChange;
run;
```

The ODS SELECT statement restricts the output to the "Frequencies" table; see the section "ODS Table Names" on page 451. The FREQ option in the PROC UNIVARIATE statement requests the table of frequencies shown in Output 4.4.1.

Output 4.4.1 Table of Frequencies

Analysis of Score Changes

The UNIVARIATE Procedure
Variable: ScoreChange (Change in Test Scores)

Frequency Counts

Value	Count	Cell Percent	Cum Percent
-37	1	8.3	8.3
-14	1	8.3	16.7
-7	1	8.3	25.0
-5	2	16.7	41.7
-3	2	16.7	58.3
2	1	8.3	66.7
3	1	8.3	75.0
6	1	8.3	83.3
12	1	8.3	91.7
14	1	8.3	100.0

From Output 4.4.1, the instructor sees that only score changes of -3 and -5 occurred more than once.

A sample program for this example, *uniex03.sas*, is available in the SAS Sample Library for Base SAS software.

Example 4.5: Creating Basic Summary Plots

The PLOTS option in the PROC UNIVARIATE statement requests several basic summary plots. For more information about plots created by the PLOTS option, see the section "Creating Line Printer Plots" on page 405. This example illustrates the use of the PLOT option as well as BY processing in PROC UNIVARIATE.

A researcher is analyzing a data set consisting of air pollution data from three different measurement sites. The data set AirPoll, created by the following statements, contains the variables Site and Ozone, which are the site number and ozone level, respectively.

```
data AirPoll (keep = Site Ozone);
   label Site  = 'Site Number'
         Ozone = 'Ozone level (in ppb)';
   do i = 1 to 3;
      input Site @@;
      do j = 1 to 15;
         input Ozone @@;
         output;
      end;
   end;
   datalines;
102 4 6 3 4 7 8 2 3 4 1 3 8 9 5 6
134 5 3 6 2 1 2 4 3 2 4 6 4 6 3 1
137 8 9 7 8 6 7 6 7 9 8 9 8 7 8 5
;
```

The following statements produce basic plots for each site in the AirPoll data set:

```
ods graphics off;
ods select Plots SSPlots;
proc univariate data=AirPoll plot;
   by Site;
   var Ozone;
run;
```

The ODS GRAPHICS OFF statement specified before the PROC statement disables ODS Graphics, which causes the PLOTS option to produce legacy line printer plots. The PLOTS option produces a stem-and-leaf plot, a box plot, and a normal probability plot for the Ozone variable at each site. Because a BY statement is specified, a side-by-side box plot is also created to compare the ozone levels across sites. Note that AirPoll is sorted by Site; in general, the data set should be sorted by the BY variable by using the SORT procedure. The ODS SELECT statement restricts the output to the "Plots" and "SSPlots" tables; see the section "ODS Table Names" on page 451. Optionally, you can specify the PLOTSIZE=*n* option to control the approximate number of rows (between 8 and the page size) that the plots occupy.

Output 4.5.1 through Output 4.5.3 show the plots produced for each BY group. Output 4.5.4 shows the side-by-side box plot for comparing Ozone values across sites.

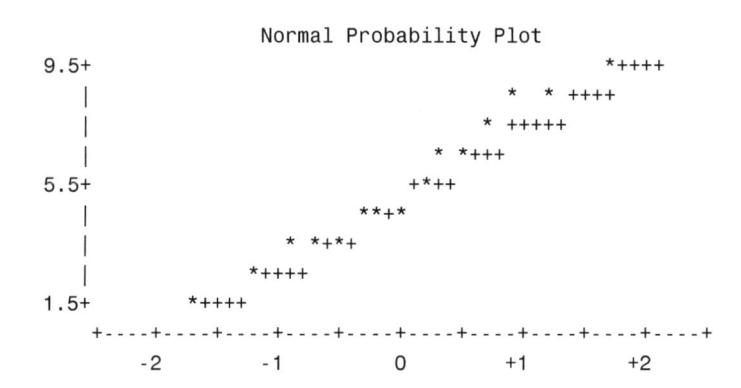

Output 4.5.1 Ozone Plots for BY Group Site = 102

Analysis of Score Changes

The UNIVARIATE Procedure
Variable: Ozone (Ozone level (in ppb))

Site Number=102

```
Stem Leaf                    #        Boxplot
  9 0                        1           |
  8 00                       2           |
  7 0                        1        +-----+
  6 00                       2        |     |
  5 0                        1        |     |
  4 000                      3        *--+--*
  3 000                      3        +-----+
  2 0                        1           |
  1 0                        1           |
    ----+----+----+----+
```

```
                    Normal Probability Plot
   9.5+                                      *++++
      |                                  *  * ++++
      |                                *  +++++
      |                             *  *+++
   5.5+                         +*++
      |                      **+*
      |                  *  *+*+
      |               *++++
   1.5+         *++++
      +----+----+----+----+----+----+----+----+----+----+
         -2        -1        0        +1        +2
```

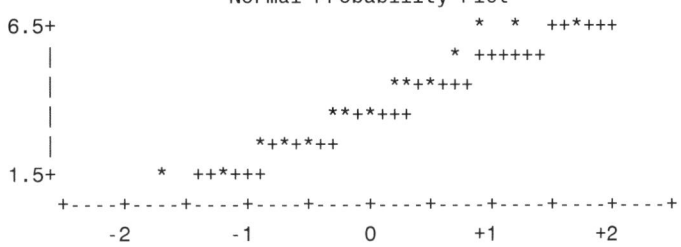

Output 4.5.2 Ozone Plots for BY Group Site = 134

Analysis of Score Changes

The UNIVARIATE Procedure
Variable: Ozone (Ozone level (in ppb))

Site Number=134

```
Stem Leaf                    #      Boxplot
   6 000                      3        |
   5 0                        1     +-----+
   4 000                      3     |     |
   3 000                      3     *--+--*
   2 000                      3     +-----+
   1 00                       2        |
     ----+----+----+----+
```

```
                    Normal Probability Plot
   6.5+                                  *   *  ++*+++
      |                                    * ++++++
      |                                 **+*+++
      |                            **+*+++
      |                        *+*+*++
   1.5+          *    ++*+++
      +----+----+----+----+----+----+----+----+----+----+
         -2        -1         0        +1        +2
```

Output 4.5.3 Ozone Plots for BY Group Site = 137

Analysis of Score Changes

The UNIVARIATE Procedure
Variable: Ozone (Ozone level (in ppb))

Site Number=137

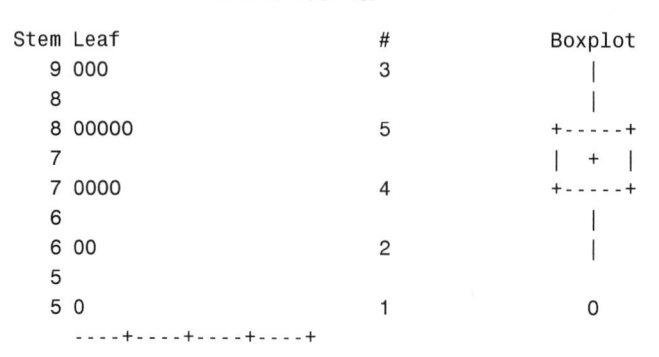

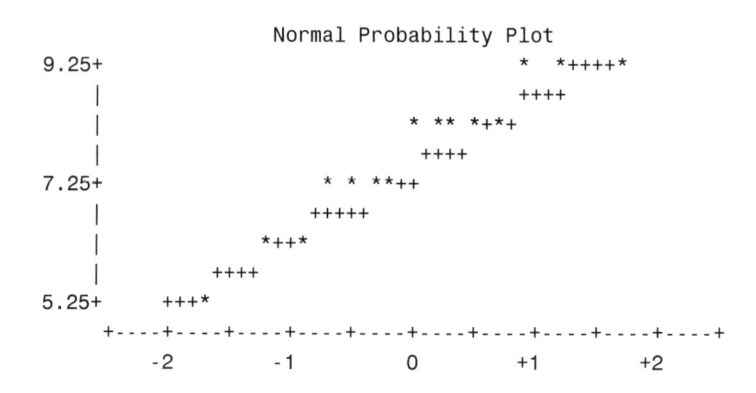

```
Stem Leaf                       #             Boxplot
   9 000                        3                |
   8                                             |
   8 00000                      5             +-----+
   7                                          |  +  |
   7 0000                       4             +-----+
   6                                             |
   6 00                         2                |
   5
   5 0                          1                0
     ----+----+----+----+
```

```
                         Normal Probability Plot
    9.25+                                    *  *++++*
        |                                        ++++
        |                                * ** *+*+
        |                                 ++++
    7.25+                           * *  **++
        |                             +++++
        |                         *++*
        |                       ++++
    5.25+          +++*
         +----+----+----+----+----+----+----+----+----+
            -2       -1        0       +1       +2
```

Output 4.5.4 Ozone Side-by-Side Boxplot for All BY Groups

Analysis of Score Changes

The UNIVARIATE Procedure
Variable: Ozone (Ozone level (in ppb))

Schematic Plots

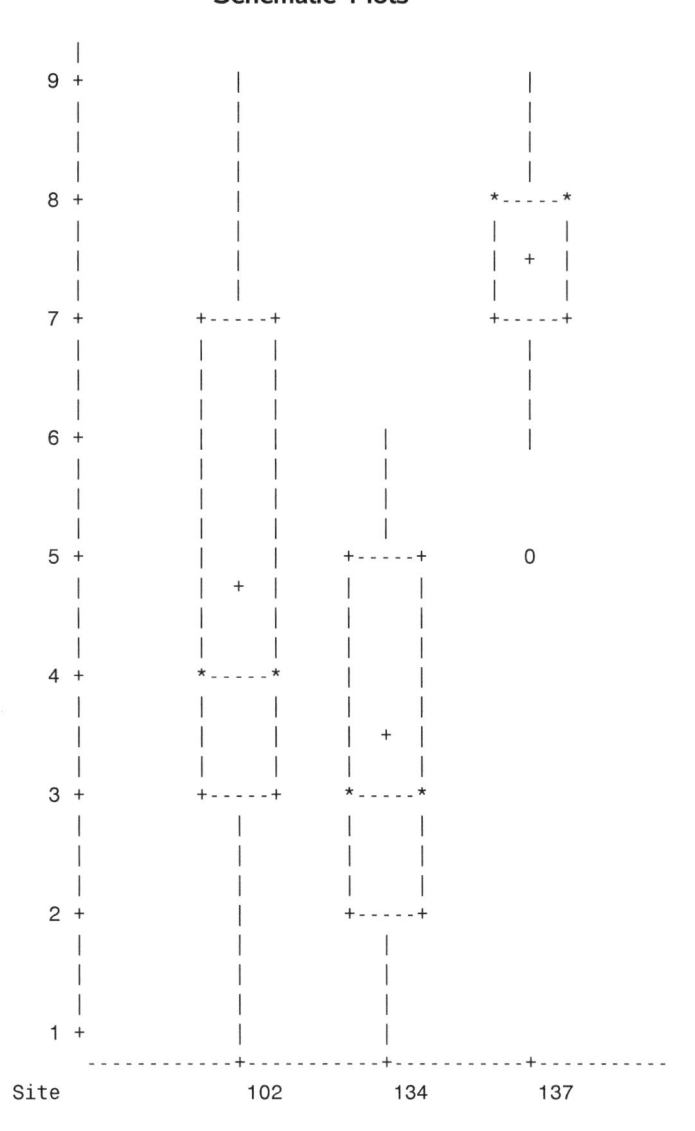

```
      |
   9 +         |                        |
      |         |                        |
      |         |                        |
      |         |                        |
   8 +         |                       *-----*
      |         |                        |   |
      |         |                        | + |
      |         |                        |   |
   7 +      +-----+                     +-----+
      |      |     |                        |
      |      |     |                        |
      |      |     |                        |
   6 +      |     |        |               |
      |      |     |        |
      |      |     |        |
      |      |     |        |
   5 +      |     |     +-----+             0
      |      |  +  |     |     |
      |      |     |     |     |
      |      |     |     |     |
   4 +      *-----*     |     |
      |      |     |     |     |
      |      |     |     |  +  |
      |      |     |     |     |
   3 +      +-----+     *-----*
      |         |           |
      |         |           |
      |         |           |
   2 +         |        +-----+
      |         |           |
      |         |           |
      |         |           |
   1 +         |           |
      ------------+-----------+-----------+------------
 Site         102         134         137
```

The following statements produce basic plots by using ODS Graphics:

```
ods graphics on;
ods select Plots SSPlots;
proc univariate data=AirPoll plot;
   by Site;
   var Ozone;
run;
```

466 ✦ *Chapter 4: The UNIVARIATE Procedure*

Output 4.5.5 through Output 4.5.8 show the plots produced by using ODS Graphics. Note that the line printer stem-and-leaf plots are replaced by horizontal histograms in ODS Graphics output.

Output 4.5.5 Ozone Plots for BY Group Site = 102

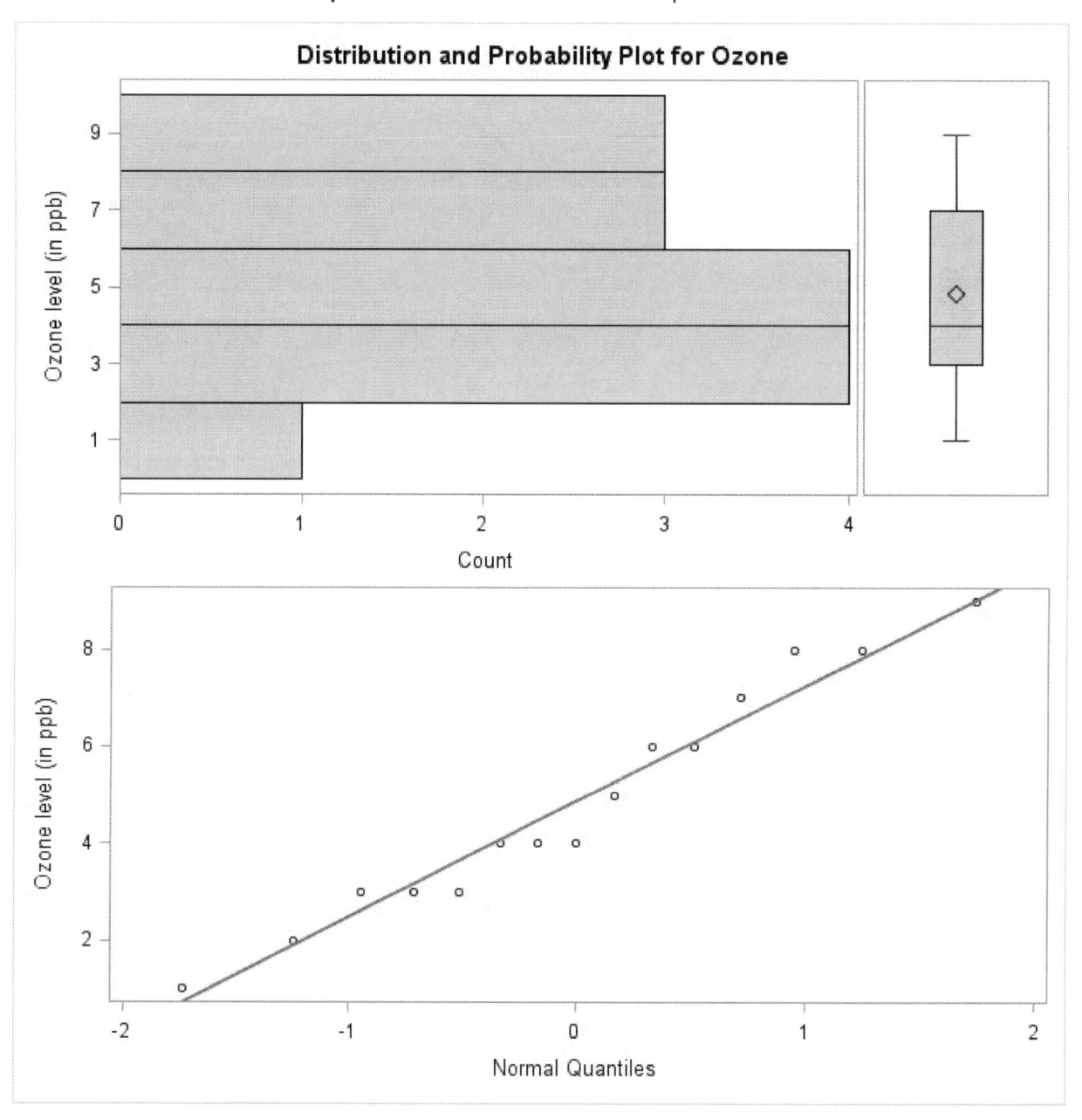

Example 4.5: Creating Basic Summary Plots ✦ 467

Output 4.5.6 Ozone Plots for BY Group Site = 134

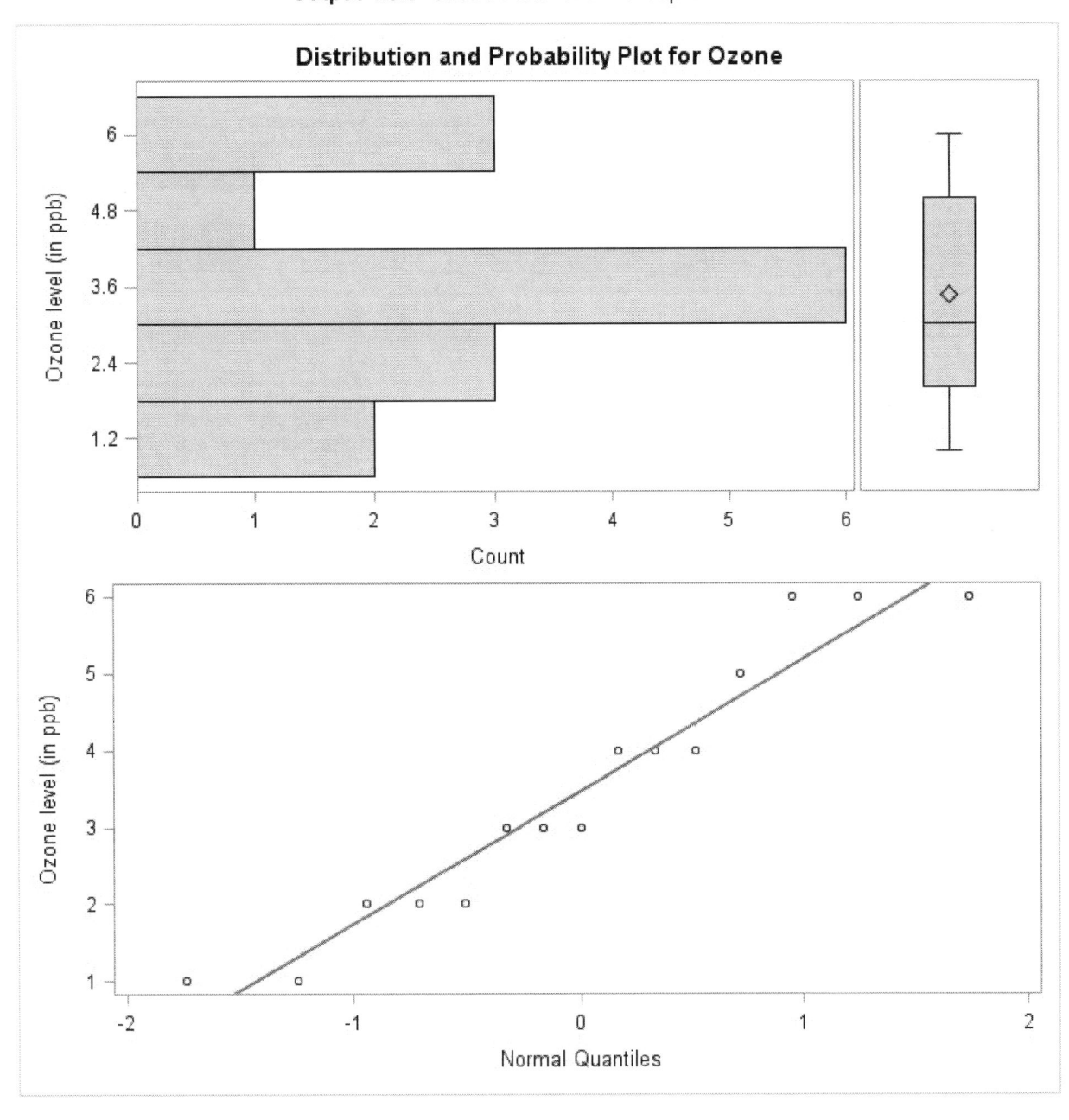

Output 4.5.7 Ozone Plots for BY Group Site = 137

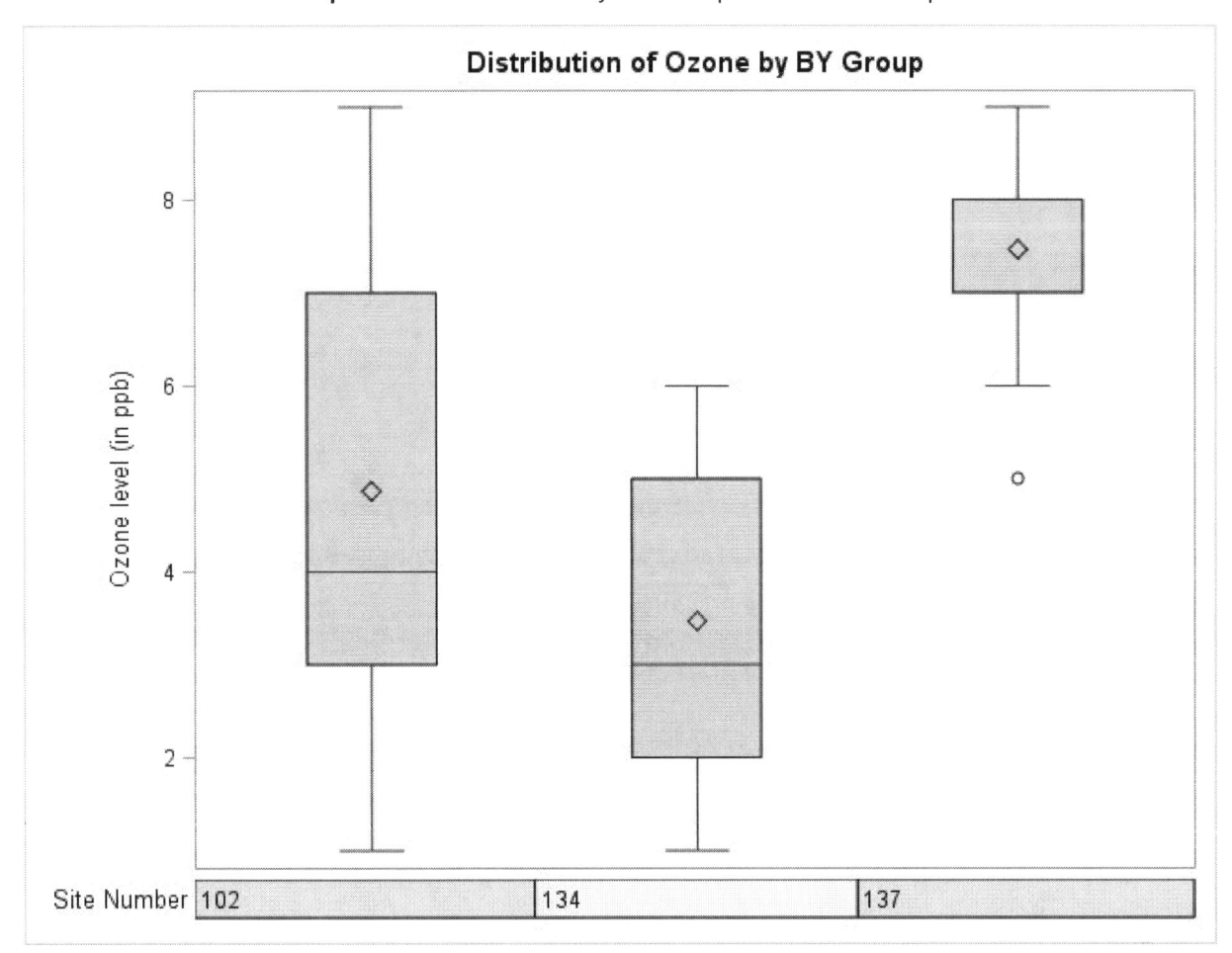

Output 4.5.8 Ozone Side-by-Side Boxplot for All BY Groups

[Figure: Distribution of Ozone by BY Group — side-by-side boxplots of Ozone level (in ppb) for Site Numbers 102, 134, and 137]

NOTE: You can use the PROBPLOT statement with the NORMAL option to produce high-resolution normal probability plots; see the section "Modeling a Data Distribution" on page 286. You can use the BOXPLOT procedure to produce box plots that use high-resolution graphics. See Chapter 28, "The BOXPLOT Procedure" (*SAS/STAT User's Guide*).

A sample program for this example, *uniex04.sas*, is available in the SAS Sample Library for Base SAS software.

Example 4.6: Analyzing a Data Set With a FREQ Variable

This example illustrates how to use PROC UNIVARIATE to analyze a data set with a variable that contains the frequency of each observation. The data set Speeding contains data on the number of cars pulled over for speeding on a stretch of highway with a 65 mile per hour speed limit. Speed is the speed at which the cars were traveling, and Number is the number of cars at each speed. The following statements create the data set:

```
data Speeding;
   label Speed = 'Speed (in miles per hour)';
   do Speed = 66 to 85;
      input Number @@;
      output;
   end;
   datalines;
 2  3  2  1  3  6  8  9 10 13
12 14  6  2  0  0  1  1  0  1
;
```

The following statements create a table of moments for the variable Speed:

```
title 'Analysis of Speeding Data';
ods select Moments;
proc univariate data=Speeding;
   freq Number;
   var Speed;
run;
```

The ODS SELECT statement restricts the output, which is shown in Output 4.6.1, to the "Moments" table; see the section "ODS Table Names" on page 451. The FREQ statement specifies that the value of the variable Number represents the frequency of each observation.

For the formulas used to compute these moments, see the section "Descriptive Statistics" on page 394. A sample program for this example, *uniex05.sas*, is available in the SAS Sample Library for Base SAS software.

Output 4.6.1 Table of Moments

Analysis of Speeding Data

The UNIVARIATE Procedure
Variable: Speed (Speed (in miles per hour))

Freq: Number

Moments			
N	94	Sum Weights	94
Mean	74.3404255	Sum Observations	6988
Std Deviation	3.44403237	Variance	11.861359
Skewness	-0.1275543	Kurtosis	0.92002287
Uncorrected SS	520594	Corrected SS	1103.10638
Coeff Variation	4.63278538	Std Error Mean	0.35522482

Example 4.7: Saving Summary Statistics in an OUT= Output Data Set

This example illustrates how to save summary statistics in an output data set. The following statements create a data set named Belts, which contains the breaking strengths (Strength) and widths (Width) of a sample of 50 automotive seat belts:

```
data Belts;
   label Strength = 'Breaking Strength (lb/in)'
         Width    = 'Width in Inches';
   input Strength Width @@;
   datalines;
1243.51  3.036  1221.95  2.995  1131.67  2.983  1129.70  3.019
1198.08  3.106  1273.31  2.947  1250.24  3.018  1225.47  2.980
1126.78  2.965  1174.62  3.033  1250.79  2.941  1216.75  3.037
1285.30  2.893  1214.14  3.035  1270.24  2.957  1249.55  2.958
1166.02  3.067  1278.85  3.037  1280.74  2.984  1201.96  3.002
1101.73  2.961  1165.79  3.075  1186.19  3.058  1124.46  2.929
1213.62  2.984  1213.93  3.029  1289.59  2.956  1208.27  3.029
1247.48  3.027  1284.34  3.073  1209.09  3.004  1146.78  3.061
1224.03  2.915  1200.43  2.974  1183.42  3.033  1195.66  2.995
1258.31  2.958  1136.05  3.022  1177.44  3.090  1246.13  3.022
1183.67  3.045  1206.50  3.024  1195.69  3.005  1223.49  2.971
1147.47  2.944  1171.76  3.005  1207.28  3.065  1131.33  2.984
1215.92  3.003  1202.17  3.058
;
```

The following statements produce two output data sets containing summary statistics:

```
proc univariate data=Belts noprint;
   var Strength Width;
   output out=Means         mean=StrengthMean WidthMean;
   output out=StrengthStats mean=StrengthMean std=StrengthSD
                            min=StrengthMin   max=StrengthMax;
run;
```

When you specify an OUTPUT statement, you must also specify a VAR statement. You can use multiple OUTPUT statements with a single procedure statement. Each OUTPUT statement creates a new data set with the name specified by the OUT= option. In this example, two data sets, Means and StrengthStats, are created. See Output 4.7.1 for a listing of Means and Output 4.7.2 for a listing of StrengthStats.

Output 4.7.1 Listing of Output Data Set Means

Analysis of Speeding Data

Obs	StrengthMean	WidthMean
1	1205.75	3.00584

Output 4.7.2 Listing of Output Data Set StrengthStats

Analysis of Speeding Data

Obs	StrengthMean	StrengthSD	StrengthMax	StrengthMin
1	1205.75	48.3290	1289.59	1101.73

Summary statistics are saved in an output data set by specifying *keyword=names* after the OUT= option. In the preceding statements, the first OUTPUT statement specifies the *keyword* MEAN followed by the *names* StrengthMean and WidthMean. The second OUTPUT statement specifies the *keywords* MEAN, STD, MAX, and MIN, for which the *names* StrengthMean, StrengthSD, StrengthMax, and StrengthMin are given.

The *keyword* specifies the statistic to be saved in the output data set, and the *names* determine the names for the new variables. The first *name* listed after a keyword contains that statistic for the first variable listed in the VAR statement; the second *name* contains that statistic for the second variable in the VAR statement, and so on.

The data set Means contains the mean of Strength in a variable named StrengthMean and the mean of Width in a variable named WidthMean. The data set StrengthStats contains the mean, standard deviation, maximum value, and minimum value of Strength in the variables StrengthMean, StrengthSD, StrengthMax, and StrengthMin, respectively.

See the section "OUT= Output Data Set in the OUTPUT Statement" on page 444 for more information about OUT= output data sets.

A sample program for this example, *uniex06.sas*, is available in the SAS Sample Library for Base SAS software.

Example 4.8: Saving Percentiles in an Output Data Set

This example, which uses the Belts data set from the previous example, illustrates how to save percentiles in an output data set. The UNIVARIATE procedure automatically computes the 1st, 5th, 10th, 25th, 75th, 90th, 95th, and 99th percentiles for each variable. You can save these percentiles in an output data set by specifying the appropriate keywords. For example, the following statements create an output data set named PctlStrength, which contains the 5th and 95th percentiles of the variable Strength:

```
proc univariate data=Belts noprint;
   var Strength Width;
   output out=PctlStrength p5=p5str p95=p95str;
run;
```

The output data set PctlStrength is listed in Output 4.8.1.

Output 4.8.1 Listing of Output Data Set PctlStrength

Analysis of Speeding Data

Obs	p95str	p5str
1	1284.34	1126.78

You can use the PCTLPTS=, PCTLPRE=, and PCTLNAME= options to save percentiles not automatically computed by the UNIVARIATE procedure. For example, the following statements create an output data set named Pctls, which contains the 20th and 40th percentiles of the variables Strength and Width:

```
proc univariate data=Belts noprint;
   var Strength Width;
   output out=Pctls pctlpts  = 20 40
                    pctlpre  = Strength Width
                    pctlname = pct20 pct40;
run;
```

The PCTLPTS= option specifies the percentiles to compute (in this case, the 20th and 40th percentiles). The PCTLPRE= and PCTLNAME= options build the names for the variables containing the percentiles. The

PCTLPRE= option gives prefixes for the new variables, and the PCTLNAME= option gives a suffix to add to the prefix. When you use the PCTLPTS= specification, you must also use the PCTLPRE= specification.

The OUTPUT statement saves the 20th and 40th percentiles of Strength and Width in the variables Strengthpct20, Widthpct20, Strengthpct40, and Weightpct40. The output data set Pctls is listed in Output 4.8.2.

Output 4.8.2 Listing of Output Data Set Pctls

Analysis of Speeding Data

Obs	Strengthpct20	Widthpct20	Strengthpct40	Widthpct40
1	1165.91	2.9595	1199.26	2.995

A sample program for this example, *uniex06.sas*, is available in the SAS Sample Library for Base SAS software.

Example 4.9: Computing Confidence Limits for the Mean, Standard Deviation, and Variance

This example illustrates how to compute confidence limits for the mean, standard deviation, and variance of a population. A researcher is studying the heights of a certain population of adult females. She has collected a random sample of heights of 75 females, which are saved in the data set Heights:

```
data Heights;
   label Height = 'Height (in)';
   input Height @@;
   datalines;
64.1 60.9 64.1 64.7 66.7 65.0 63.7 67.4 64.9 63.7
64.0 67.5 62.8 63.9 65.9 62.3 64.1 60.6 68.6 68.6
63.7 63.0 64.7 68.2 66.7 62.8 64.0 64.1 62.1 62.9
62.7 60.9 61.6 64.6 65.7 66.6 66.7 66.0 68.5 64.4
60.5 63.0 60.0 61.6 64.3 60.2 63.5 64.7 66.0 65.1
63.6 62.0 63.6 65.8 66.0 65.4 63.5 66.3 66.2 67.5
65.8 63.1 65.8 64.4 64.0 64.9 65.7 61.0 64.1 65.5
68.6 66.6 65.7 65.1 70.0
;
```

The following statements produce confidence limits for the mean, standard deviation, and variance of the population of heights:

```
title 'Analysis of Female Heights';
ods select BasicIntervals;
proc univariate data=Heights cibasic;
   var Height;
run;
```

The CIBASIC option requests confidence limits for the mean, standard deviation, and variance. For example, Output 4.9.1 shows that the 95% confidence interval for the population mean is (64.06, 65.07). The ODS SELECT statement restricts the output to the "BasicIntervals" table; see the section "ODS Table Names" on page 451.

The confidence limits in Output 4.9.1 assume that the heights are normally distributed, so you should check this assumption before using these confidence limits. See the section "Shapiro-Wilk Statistic" on page 428 for information about the Shapiro-Wilk test for normality in PROC UNIVARIATE. See Example 4.19 for an example that uses the test for normality.

Output 4.9.1 Default 95% Confidence Limits

Analysis of Female Heights

The UNIVARIATE Procedure
Variable: Height (Height (in))

Basic Confidence Limits Assuming Normality			
Parameter	Estimate	95% Confidence Limits	
Mean	64.56667	64.06302	65.07031
Std Deviation	2.18900	1.88608	2.60874
Variance	4.79171	3.55731	6.80552

By default, the confidence limits produced by the CIBASIC option produce 95% confidence intervals. You can request different level confidence limits by using the ALPHA= option in parentheses after the CIBASIC option. The following statements produce 90% confidence limits:

```
title 'Analysis of Female Heights';
ods select BasicIntervals;
proc univariate data=Heights cibasic(alpha=.1);
   var Height;
run;
```

The 90% confidence limits are displayed in Output 4.9.2.

Output 4.9.2 90% Confidence Limits

Analysis of Female Heights

The UNIVARIATE Procedure
Variable: Height (Height (in))

Basic Confidence Limits Assuming Normality			
Parameter	Estimate	90% Confidence Limits	
Mean	64.56667	64.14564	64.98770
Std Deviation	2.18900	1.93114	2.53474
Variance	4.79171	3.72929	6.42492

For the formulas used to compute these limits, see the section "Confidence Limits for Parameters of the Normal Distribution" on page 401.

A sample program for this example, *uniex07.sas*, is available in the SAS Sample Library for Base SAS software.

Example 4.10: Computing Confidence Limits for Quantiles and Percentiles

This example, which is a continuation of Example 4.9, illustrates how to compute confidence limits for quantiles and percentiles. A second researcher is more interested in summarizing the heights with quantiles than the mean and standard deviation. He is also interested in computing 90% confidence intervals for the quantiles. The following statements produce estimated quantiles and confidence limits for the population quantiles:

```
title 'Analysis of Female Heights';
ods select Quantiles;
proc univariate data=Heights ciquantnormal(alpha=.1);
   var Height;
run;
```

The ODS SELECT statement restricts the output to the "Quantiles" table; see the section "ODS Table Names" on page 451. The CIQUANTNORMAL option produces confidence limits for the quantiles. As noted in Output 4.10.1, these limits assume that the data are normally distributed. You should check this assumption before using these confidence limits. See the section "Shapiro-Wilk Statistic" on page 428 for information about the Shapiro-Wilk test for normality in PROC UNIVARIATE; see Example 4.19 for an example that uses the test for normality.

Output 4.10.1 Normal-Based Quantile Confidence Limits

Analysis of Female Heights

The UNIVARIATE Procedure
Variable: Height (Height (in))

Quantiles (Definition 5)

Level	Quantile	90% Confidence Limits Assuming Normality	
100% Max	70.0		
99%	70.0	68.94553	70.58228
95%	68.6	67.59184	68.89311
90%	67.5	66.85981	68.00273
75% Q3	66.0	65.60757	66.54262
50% Median	64.4	64.14564	64.98770
25% Q1	63.1	62.59071	63.52576
10%	61.6	61.13060	62.27352
5%	60.6	60.24022	61.54149
1%	60.0	58.55106	60.18781
0% Min	60.0		

It is also possible to use PROC UNIVARIATE to compute confidence limits for quantiles without assuming normality. The following statements use the CIQUANTDF option to request distribution-free confidence limits for the quantiles of the population of heights:

```
title 'Analysis of Female Heights';
ods select Quantiles;
proc univariate data=Heights ciquantdf(alpha=.1);
   var Height;
run;
```

The distribution-free confidence limits are shown in Output 4.10.2.

Output 4.10.2 Distribution-Free Quantile Confidence Limits

Analysis of Female Heights

The UNIVARIATE Procedure
Variable: Height (Height (in))

Quantiles (Definition 5)

Level	Quantile	90% Confidence Limits Distribution Free		LCL Rank	UCL Rank	Coverage
100% Max	70.0					
99%	70.0	68.6	70.0	73	75	48.97
95%	68.6	67.5	70.0	68	75	94.50
90%	67.5	66.6	68.6	63	72	91.53
75% Q3	66.0	65.7	66.6	50	63	91.77
50% Median	64.4	64.1	65.1	31	46	91.54
25% Q1	63.1	62.7	63.7	13	26	91.77
10%	61.6	60.6	62.7	4	13	91.53
5%	60.6	60.0	61.6	1	8	94.50
1%	60.0	60.0	60.5	1	3	48.97
0% Min	60.0					

The table in Output 4.10.2 includes the ranks from which the confidence limits are computed. For more information about how these confidence limits are calculated, see the section "Confidence Limits for Percentiles" on page 398. Note that confidence limits for quantiles are not produced when the WEIGHT statement is used.

A sample program for this example, *uniex07.sas*, is available in the SAS Sample Library for Base SAS software.

Example 4.11: Computing Robust Estimates

This example illustrates how you can use the UNIVARIATE procedure to compute robust estimates of location and scale. The following statements compute these estimates for the variable Systolic in the data set BPressure, which was introduced in Example 4.1:

```
title 'Robust Estimates for Blood Pressure Data';
ods select TrimmedMeans WinsorizedMeans RobustScale;
proc univariate data=BPressure trimmed=1 .1
                winsorized=.1  robustscale;
   var Systolic;
run;
```

The ODS SELECT statement restricts the output to the "TrimmedMeans," "WinsorizedMeans," and "RobustScale" tables; see the section "ODS Table Names" on page 451. The TRIMMED= option computes two trimmed means, the first after removing one observation and the second after removing 10% of the observations. If the value of TRIMMED= is greater than or equal to one, it is interpreted as the number of observations to be trimmed. The WINSORIZED= option computes a Winsorized mean that replaces three observations from the tails with the next closest observations. (Three observations are replaced because $np = (22)(.1) = 2.2$, and three is the smallest integer greater than 2.2.) The trimmed and Winsorized means for Systolic are displayed in Output 4.11.1.

Output 4.11.1 Computation of Trimmed and Winsorized Means

Robust Estimates for Blood Pressure Data

The UNIVARIATE Procedure
Variable: Systolic

Trimmed Means

Percent Trimmed in Tail	Number Trimmed in Tail	Trimmed Mean	Std Error Trimmed Mean	95% Confidence Limits	DF	t for H0: Mu0=0.00	Pr > |t|
4.55	1	120.3500	2.573536	114.9635 125.7365	19	46.76446	<.0001
13.64	3	120.3125	2.395387	115.2069 125.4181	15	50.22675	<.0001

Winsorized Means

Percent Winsorized in Tail	Number Winsorized in Tail	Winsorized Mean	Std Error Winsorized Mean	95% Confidence Limits	DF	t for H0: Mu0=0.00	Pr > |t|
13.64	3	120.6364	2.417065	115.4845 125.7882	15	49.91027	<.0001

Output 4.11.1 shows the trimmed mean for Systolic is 120.35 after one observation has been trimmed, and 120.31 after 3 observations are trimmed. The Winsorized mean for Systolic is 120.64. For details on trimmed and Winsorized means, see the section "Robust Estimators" on page 402. The trimmed means can be compared with the means shown in Output 4.1.1 (from Example 4.1), which displays the mean for Systolic as 121.273.

The ROBUSTSCALE option requests a table, displayed in Output 4.11.2, which includes the interquartile range, Gini's mean difference, the median absolute deviation about the median, Q_n, and S_n.

Output 4.11.2 shows the robust estimates of scale for Systolic. For instance, the interquartile range is 13. The estimates of σ range from 9.54 to 13.32. See the section "Robust Estimators" on page 402.

A sample program for this example, *uniex01.sas*, is available in the SAS Sample Library for Base SAS software.

Output 4.11.2 Computation of Robust Estimates of Scale

Robust Measures of Scale

Measure	Value	Estimate of Sigma
Interquartile Range	13.00000	9.63691
Gini's Mean Difference	15.03030	13.32026
MAD	6.50000	9.63690
Sn	9.54080	9.54080
Qn	13.33140	11.36786

Example 4.12: Testing for Location

This example, which is a continuation of Example 4.9, illustrates how to carry out three tests for location: the Student's *t* test, the sign test, and the Wilcoxon signed rank test. These tests are discussed in the section "Tests for Location" on page 399.

The following statements demonstrate the tests for location by using the Heights data set introduced in Example 4.9. Because the data consists of adult female heights, the researchers are not interested in testing whether the mean of the population is equal to zero inches, which is the default μ_0 value. Instead, they are interested in testing whether the mean is equal to 66 inches. The following statements test the null hypothesis $H_0: \mu_0 = 66$:

```
title 'Analysis of Female Height Data';
ods select TestsForLocation LocationCounts;
proc univariate data=Heights mu0=66 loccount;
   var Height;
run;
```

The ODS SELECT statement restricts the output to the "TestsForLocation" and "LocationCounts" tables; see the section "ODS Table Names" on page 451. The MU0= option specifies the null hypothesis value of μ_0 for the tests for location; by default, $\mu_0 = 0$. The LOCCOUNT option produces the table of the number of observations greater than, not equal to, and less than 66 inches.

Output 4.12.1 contains the results of the tests for location. All three tests are highly significant, causing the researchers to reject the hypothesis that the mean is 66 inches.

A sample program for this example, *uniex07.sas*, is available in the SAS Sample Library for Base SAS software.

Output 4.12.1 Tests for Location with MU0=66 and LOCCOUNT

Analysis of Female Height Data

The UNIVARIATE Procedure
Variable: Height (Height (in))

Tests for Location: Mu0=66				
Test		Statistic	p Value	
Student's t	t	-5.67065	Pr > |t|	<.0001
Sign	M	-20	Pr >= |M|	<.0001
Signed Rank	S	-849	Pr >= |S|	<.0001

Location Counts: Mu0=66.00	
Count	Value
Num Obs > Mu0	16
Num Obs ^= Mu0	72
Num Obs < Mu0	56

Example 4.13: Performing a Sign Test Using Paired Data

This example demonstrates a sign test for paired data, which is a specific application of the tests for location discussed in Example 4.12.

The instructor from Example 4.4 is now interested in performing a sign test for the pairs of test scores in his college course. The following statements request basic statistical measures and tests for location:

```
title 'Test Scores for a College Course';
ods select BasicMeasures TestsForLocation;
proc univariate data=Score;
   var ScoreChange;
run;
```

The ODS SELECT statement restricts the output to the "BasicMeasures" and "TestsForLocation" tables; see the section "ODS Table Names" on page 451. The instructor is not willing to assume that the ScoreChange variable is normal or even symmetric, so he decides to examine the sign test. The large *p*-value (0.7744) of the sign test provides insufficient evidence of a difference in test score medians.

Output 4.13.1 Sign Test for ScoreChange

Test Scores for a College Course

The UNIVARIATE Procedure
Variable: ScoreChange (Change in Test Scores)

Basic Statistical Measures			
Location		Variability	
Mean	-3.08333	Std Deviation	13.33797
Median	-3.00000	Variance	177.90152
Mode	-5.00000	Range	51.00000
		Interquartile Range	10.50000

Note: The mode displayed is the smallest of 2 modes with a count of 2.

Tests for Location: Mu0=0				
Test		Statistic		p Value
Student's t	t	-0.80079	Pr > |t|	0.4402
Sign	M	-1	Pr >= |M|	0.7744
Signed Rank	S	-8.5	Pr >= |S|	0.5278

A sample program for this example, *uniex03.sas*, is available in the SAS Sample Library for Base SAS software.

Example 4.14: Creating a Histogram

This example illustrates how to create a histogram. A semiconductor manufacturer produces printed circuit boards that are sampled to determine the thickness of their copper plating. The following statements create a data set named Trans, which contains the plating thicknesses (Thick) of 100 boards:

```
data Trans;
   input Thick @@;
   label Thick = 'Plating Thickness (mils)';
   datalines;
3.468 3.428 3.509 3.516 3.461 3.492 3.478 3.556 3.482 3.512
3.490 3.467 3.498 3.519 3.504 3.469 3.497 3.495 3.518 3.523
3.458 3.478 3.443 3.500 3.449 3.525 3.461 3.489 3.514 3.470
3.561 3.506 3.444 3.479 3.524 3.531 3.501 3.495 3.443 3.458
3.481 3.497 3.461 3.513 3.528 3.496 3.533 3.450 3.516 3.476
3.512 3.550 3.441 3.541 3.569 3.531 3.468 3.564 3.522 3.520
3.505 3.523 3.475 3.470 3.457 3.536 3.528 3.477 3.536 3.491
3.510 3.461 3.431 3.502 3.491 3.506 3.439 3.513 3.496 3.539
3.469 3.481 3.515 3.535 3.460 3.575 3.488 3.515 3.484 3.482
3.517 3.483 3.467 3.467 3.502 3.471 3.516 3.474 3.500 3.466
;
```

The following statements create the histogram shown in Output 4.14.1.

```
title 'Analysis of Plating Thickness';
ods graphics on;
proc univariate data=Trans noprint;
   histogram Thick / odstitle = title;
run;

title 'Enhancing a Histogram';
proc univariate data=Trans noprint;
   histogram Thick / midpoints     = 3.4375 to 3.5875 by .025
                     rtinclude
                     outhistogram = OutMdpts
                     odstitle     = title;
run;

proc print data=OutMdpts;
run;
```

The NOPRINT option in the PROC UNIVARIATE statement suppresses tables of summary statistics for the variable Thick that would be displayed by default. A histogram is created for each variable listed in the HISTOGRAM statement.

Output 4.14.1 Histogram for Plating Thickness

A sample program for this example, *uniex08.sas*, is available in the SAS Sample Library for Base SAS software.

Example 4.15: Creating a One-Way Comparative Histogram

This example illustrates how to create a comparative histogram. The effective channel length (in microns) is measured for 1225 field effect transistors. The channel lengths (Length) are stored in a data set named Channel, which is partially listed in Output 4.15.1:

Output 4.15.1 Partial Listing of Data Set Channel

The Data Set Channel

Lot	Length
Lot 1	0.91
.	.
Lot 1	1.17
Lot 2	1.47
.	.
Lot 2	1.39
Lot 3	2.04
.	.
Lot 3	1.91

The following statements request a histogram of Length ignoring the lot source:

```
title 'Histogram of Length Ignoring Lot Source';
ods graphics on;
proc univariate data=Channel noprint;
   histogram Length / odstitle = title;
run;
```

The resulting histogram is shown in Output 4.15.2.

Output 4.15.2 Histogram for Length Ignoring Lot Source

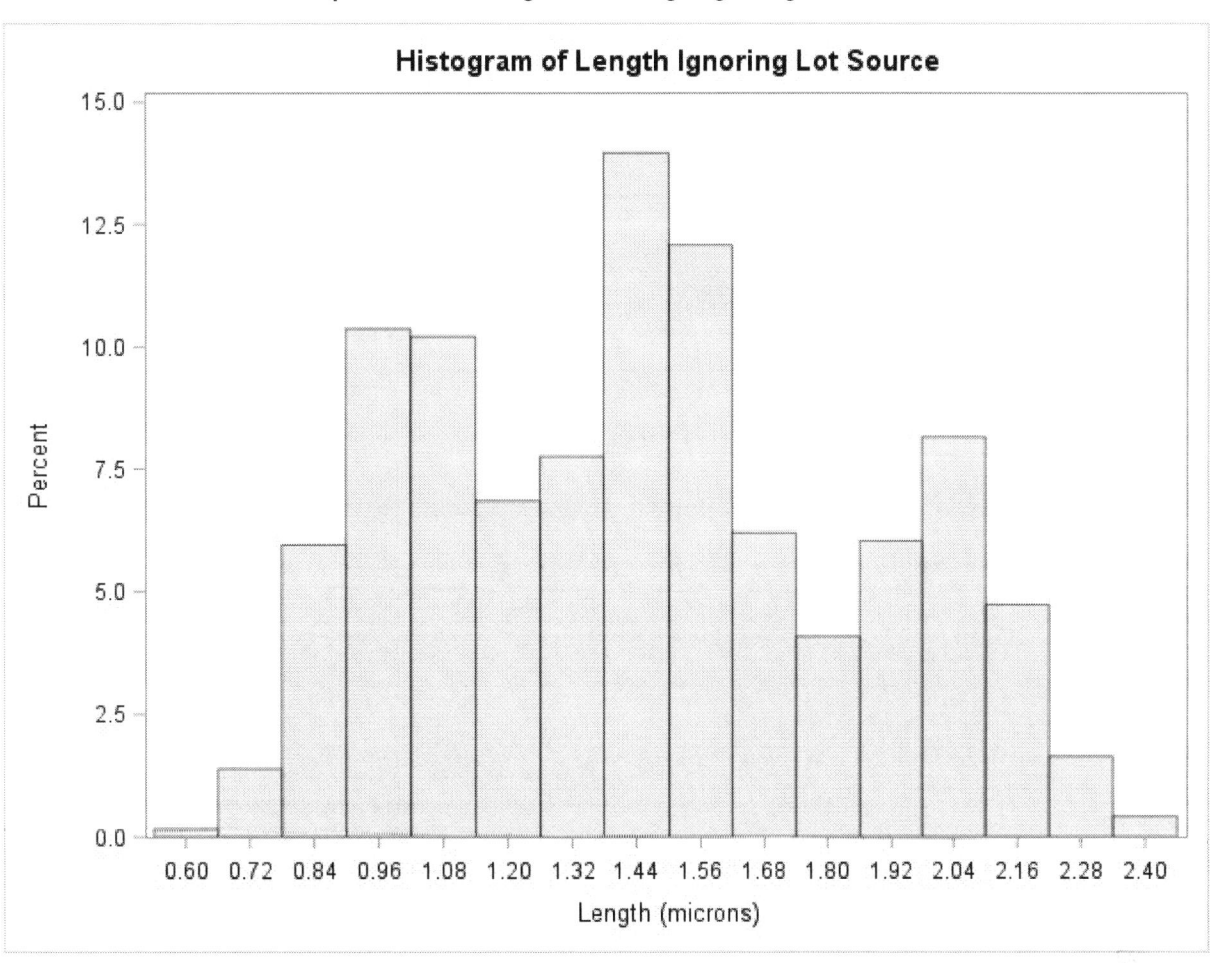

To investigate whether the peaks (modes) in Output 4.15.2 are related to the lot source, you can create a comparative histogram by using Lot as a classification variable. The following statements create the histogram shown in Output 4.15.3:

```
title 'Comparative Analysis of Lot Source';
proc univariate data=Channel noprint;
   class Lot;
   histogram Length / nrows    = 3
                     odstitle = title;
run;
```

The CLASS statement requests comparisons for each level (distinct value) of the classification variable Lot. The HISTOGRAM statement requests a comparative histogram for the variable Length. The NROWS= option specifies the number of rows per panel in the comparative histogram. By default, comparative histograms are displayed in two rows per panel.

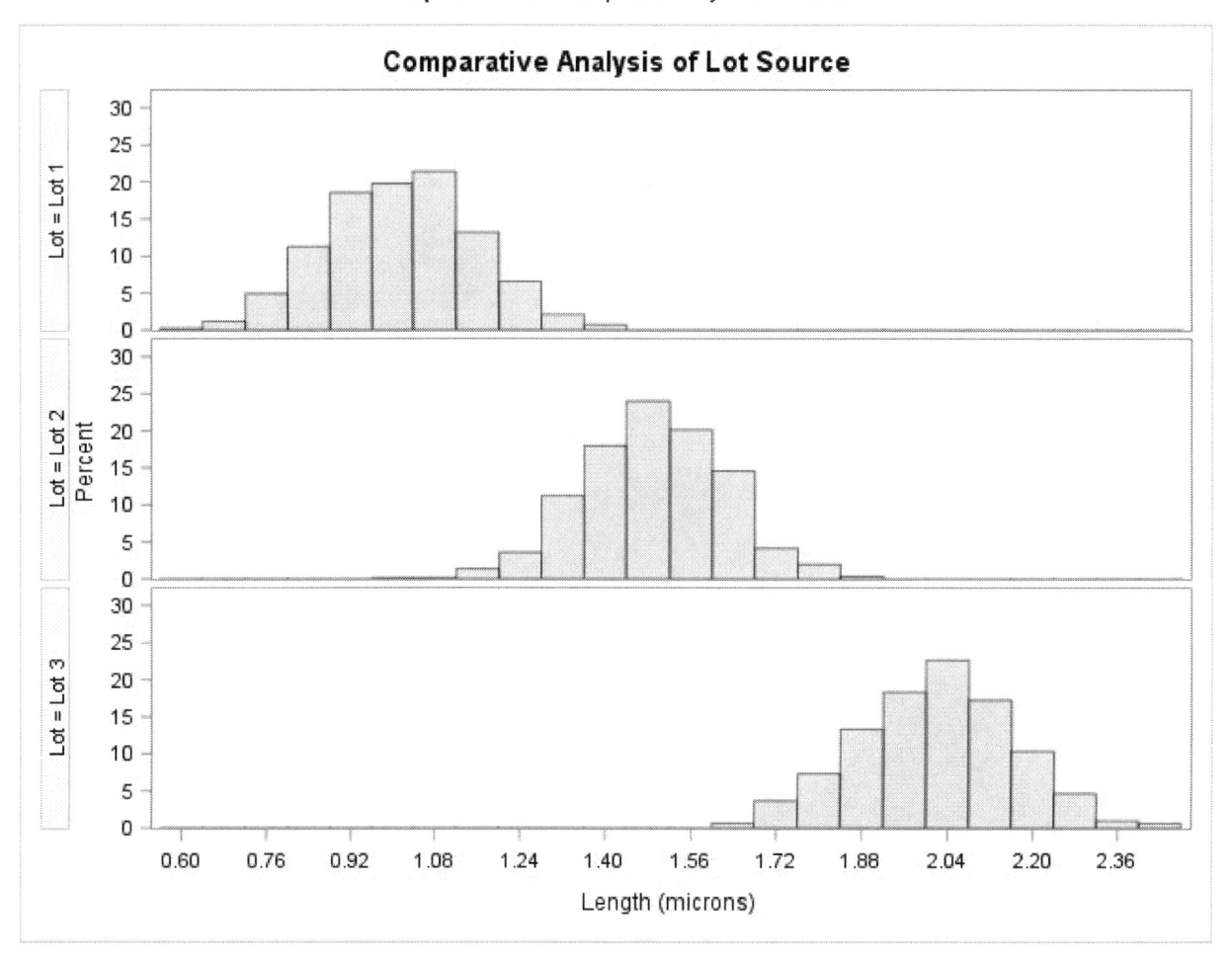

Output 4.15.3 Comparison by Lot Source

Output 4.15.3 reveals that the distributions of Length are similarly distributed except for shifts in mean.

A sample program for this example, *uniex09.sas*, is available in the SAS Sample Library for Base SAS software.

Example 4.16: Creating a Two-Way Comparative Histogram

This example illustrates how to create a two-way comparative histogram. Two suppliers (A and B) provide disk drives for a computer manufacturer. The manufacturer measures the disk drive opening width to determine whether there has been a change in variability from 2002 to 2003 for each supplier.

The following statements save the measurements in a data set named Disk. There are two classification variables, Supplier and Year, and a user-defined format is associated with Year.

```
proc format ;
   value mytime  1 = '2002' 2 = '2003';

data Disk;
   input @1 Supplier $10. Year Width;
   label Width = 'Opening Width (inches)';
   format Year mytime.;
datalines;
Supplier A    1    1.8932
   .          .       .
Supplier B    1    1.8986
Supplier A    2    1.8978
   .          .       .
Supplier B    2    1.8997
;
```

The following statements create the comparative histogram in Output 4.16.1:

```
title 'Results of Supplier Training Program';
ods graphics on;
proc univariate data=Disk noprint;
   class Supplier Year / keylevel = ('Supplier A' '2003');
   histogram Width / vaxis     = 0 10 20 30
                     ncols     = 2
                     nrows     = 2
                     odstitle  = title;
run;
```

The KEYLEVEL= option specifies the key cell as the cell for which Supplier is equal to 'SUPPLIER A' and Year is equal to '2003.' This cell determines the binning for the other cells, and the columns are arranged so that this cell is displayed in the upper left corner. Without the KEYLEVEL= option, the default key cell would be the cell for which Supplier is equal to 'SUPPLIER A' and Year is equal to '2002'; the column labeled '2002' would be displayed to the left of the column labeled '2003.'

The VAXIS= option specifies the tick mark labels for the vertical axis. The NROWS=2 and NCOLS=2 options specify a 2 × 2 arrangement for the tiles. Output 4.16.1 provides evidence that both suppliers have reduced variability from 2002 to 2003.

Output 4.16.1 Two-Way Comparative Histogram

A sample program for this example, *uniex10.sas*, is available in the SAS Sample Library for Base SAS software.

Example 4.17: Adding Insets with Descriptive Statistics

This example illustrates how to add insets with descriptive statistics to a comparative histogram; see Output 4.17.1. Three similar machines are used to attach a part to an assembly. One hundred assemblies are sampled from the output of each machine, and a part position is measured in millimeters. The following statements create the data set Machines, which contains the measurements in a variable named Position:

```
data Machines;
   input Position @@;
   label Position = 'Position in Millimeters';
   if       (_n_ <= 100) then Machine = 'Machine 1';
   else if (_n_ <= 200) then Machine = 'Machine 2';
   else                       Machine = 'Machine 3';
   datalines;
-0.17  -0.19  -0.24  -0.24  -0.12   0.07  -0.61   0.22   1.91  -0.08
-0.59   0.05  -0.38   0.82  -0.14   0.32   0.12  -0.02   0.26   0.19
-0.07   0.13  -0.49   0.07   0.65   0.94  -0.51  -0.61  -0.57  -0.51

 ... more lines ...

 0.48   0.41   0.78   0.58   0.43   0.07   0.27   0.49   0.79   0.92
 0.79   0.66   0.22   0.71   0.53   0.57   0.90   0.48   1.17   1.03
;
```

The following statements create the comparative histogram in Output 4.17.1:

```
title 'Machine Comparison Study';
ods graphics on;
proc univariate data=Machines noprint;
   class Machine;
   histogram Position / nrows     = 3
                        midpoints = -1.2 to 2.2 by 0.1
                        vaxis     =  0 to 16 by 4
                        odstitle  = title;
   inset mean std="Std Dev" / pos = ne format = 6.3;
run;
```

The INSET statement requests insets that contain the sample mean and standard deviation for each machine in the corresponding tile. The MIDPOINTS= option specifies the midpoints of the histogram bins.

488 ✦ *Chapter 4: The UNIVARIATE Procedure*

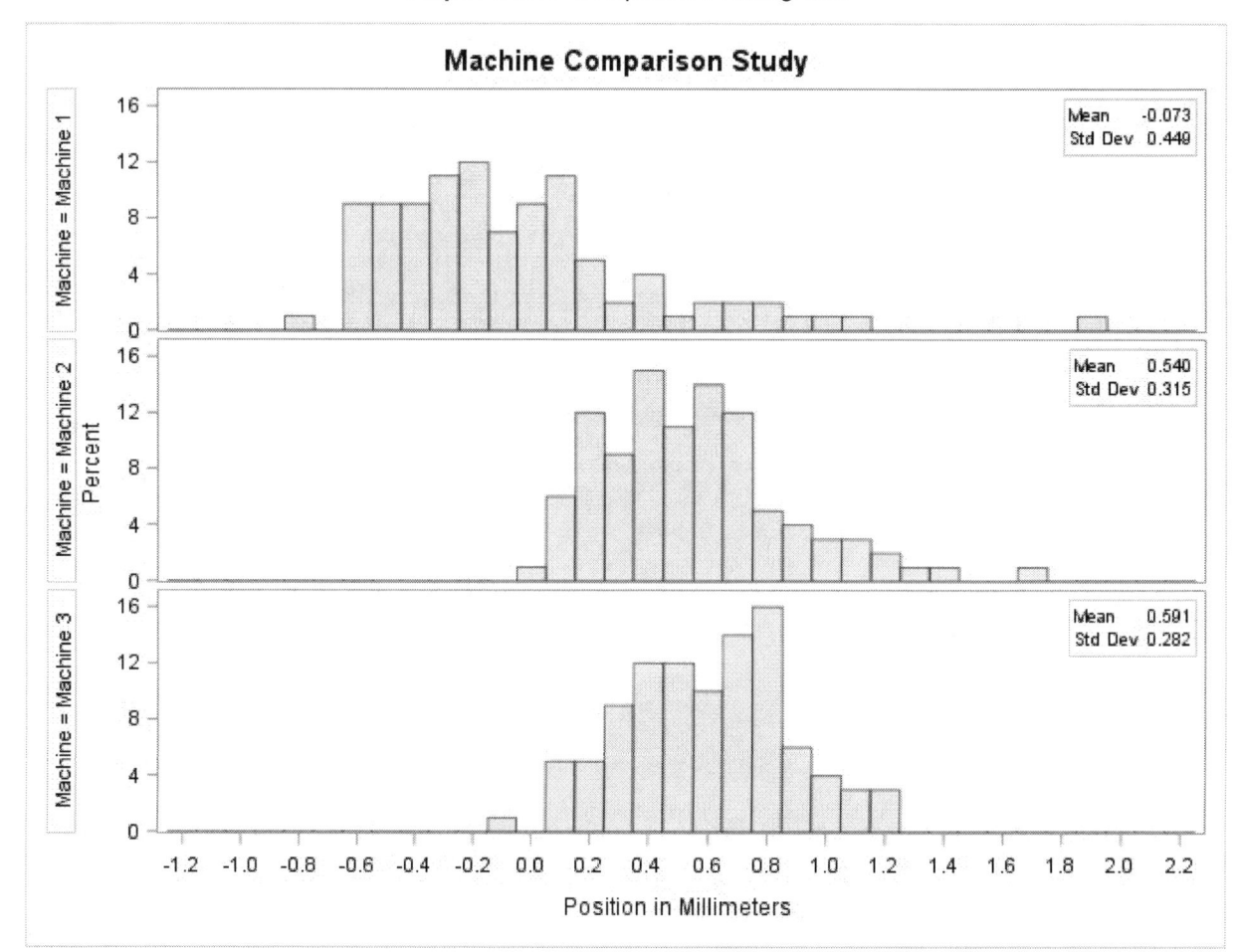

Output 4.17.1 Comparative Histograms

Output 4.17.1 shows that the average position for Machines 2 and 3 are similar and that the spread for Machine 1 is much larger than for Machines 2 and 3.

A sample program for this example, *uniex11.sas*, is available in the SAS Sample Library for Base SAS software.

Example 4.18: Binning a Histogram

This example, which is a continuation of Example 4.14, demonstrates various methods for binning a histogram. This example also illustrates how to save bin percentages in an OUTHISTOGRAM= data set.

The manufacturer from Example 4.14 now wants to enhance the histogram by using the ENDPOINTS= option to change the endpoints of the bins. The following statements create a histogram with bins that have end points 3.425 and 3.6 and width 0.025:

```
title 'Enhancing a Histogram';
ods select Histogram HistogramBins;
proc univariate data=Trans;
   histogram Thick / midpercents
                    endpoints = 3.425 to 3.6 by .025
                    odstitle  = title;
run;
```

The ODS SELECT statement restricts the output to the "HistogramBins" table and the "MyHist" histogram; see the section "ODS Table Names" on page 451. The ENDPOINTS= option specifies the endpoints for the histogram bins. By default, if the ENDPOINTS= option is not specified, the automatic binning algorithm computes values for the midpoints of the bins. The MIDPERCENTS option requests a table of the midpoints of each histogram bin and the percent of the observations that fall in each bin. This table is displayed in Output 4.18.1; the histogram is displayed in Output 4.18.2. The NAME= option specifies a name for the histogram that can be used in the ODS SELECT statement.

Output 4.18.1 Table of Bin Percentages Requested with MIDPERCENTS Option

Enhancing a Histogram

The UNIVARIATE Procedure

Histogram Bins for Thick	
Bin Minimum Point	Observed Percent
3.425	8.000
3.450	21.000
3.475	25.000
3.500	29.000
3.525	11.000
3.550	5.000
3.575	1.000

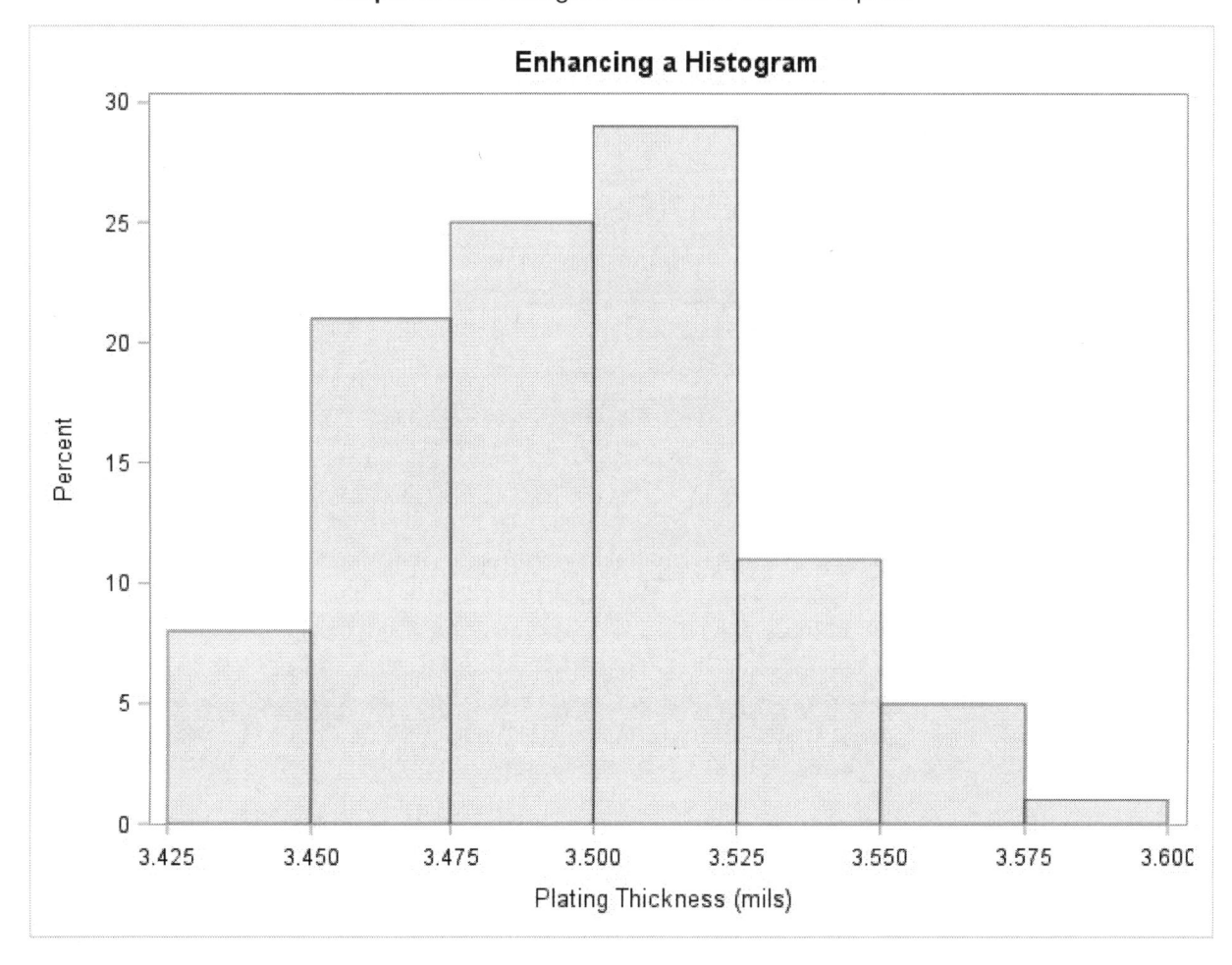

Output 4.18.2 Histogram with ENDPOINTS= Option

Enhancing a Histogram

The MIDPOINTS= option is an alternative to the ENDPOINTS= option for specifying histogram bins. The following statements create a histogram, shown in Output 4.18.3, which is similar to the one in Output 4.18.2:

```
title 'Enhancing a Histogram';
proc univariate data=Trans noprint;
   histogram Thick / midpoints    = 3.4375 to 3.5875 by .025
                    rtinclude
                    outhistogram = OutMdpts
                    odstitle     = title;
run;
```

Output 4.18.3 differs from Output 4.18.2 in two ways:

- The MIDPOINTS= option specifies the bins for the histogram by specifying the midpoints of the bins instead of specifying the endpoints. Note that the histogram displays midpoints instead of endpoints.

- The RTINCLUDE option requests that the right endpoint of each bin be included in the histogram interval instead of the default, which is to include the left endpoint in the interval. This changes the histogram slightly from Output 4.18.2. Six observations have a thickness equal to an endpoint of an interval. For instance, there is one observation with a thickness of 3.45 mils. In Output 4.18.3, this observation is included in the bin from 3.425 to 3.45.

Output 4.18.3 Histogram with MIDPOINTS= and RTINCLUDE Options

The OUTHISTOGRAM= option produces an output data set named OutMdpts, displayed in Output 4.18.4. This data set provides information about the bins of the histogram. For more information, see the section "OUTHISTOGRAM= Output Data Set" on page 446.

Output 4.18.4 The OUTHISTOGRAM= Data Set OutMdpts

Enhancing a Histogram

Obs	_VAR_	_MIDPT_	_OBSPCT_	_COUNT_
1	Thick	3.4375	9	9
2	Thick	3.4625	21	21
3	Thick	3.4875	26	26
4	Thick	3.5125	28	28
5	Thick	3.5375	11	11
6	Thick	3.5625	5	5

492 ✦ *Chapter 4: The UNIVARIATE Procedure*

A sample program for this example, *uniex08.sas*, is available in the SAS Sample Library for Base SAS software.

Example 4.19: Adding a Normal Curve to a Histogram

This example is a continuation of Example 4.14. The following statements fit a normal distribution to the thickness measurements in the Trans data set and superimpose the fitted density curve on the histogram:

```
title 'Analysis of Plating Thickness';
ods select Histogram ParameterEstimates GoodnessOfFit FitQuantiles Bins;
proc univariate data=Trans;
   histogram Thick / normal(percents=20 40 60 80 midpercents)
                    odstitle = title;
   inset n normal(ksdpval) / pos = ne format = 6.3;
run;
```

The ODS SELECT statement restricts the output to the "ParameterEstimates," "GoodnessOfFit," "FitQuantiles," and "Bins" tables; see the section "ODS Table Names" on page 451. The NORMAL option specifies that the normal curve be displayed on the histogram shown in Output 4.19.2. It also requests a summary of the fitted distribution, which is shown in Output 4.19.1. goodness-of-fit tests, parameter estimates, and quantiles of the fitted distribution. (If you specify the NORMALTEST option in the PROC UNIVARIATE statement, the Shapiro-Wilk test for normality is included in the tables of statistical output.)

Two secondary options are specified in parentheses after the NORMAL primary option. The PERCENTS= option specifies quantiles, which are to be displayed in the "FitQuantiles" table. The MIDPERCENTS option requests a table that lists the midpoints, the observed percentage of observations, and the estimated percentage of the population in each interval (estimated from the fitted normal distribution). See Table 4.6 for the secondary options that can be specified with after the NORMAL primary option.

Output 4.19.1 Summary of Fitted Normal Distribution

Analysis of Plating Thickness

The UNIVARIATE Procedure
Fitted Normal Distribution for Thick (Plating Thickness (mils))

Parameters for Normal Distribution		
Parameter	Symbol	Estimate
Mean	Mu	3.49533
Std Dev	Sigma	0.032117

Goodness-of-Fit Tests for Normal Distribution				
Test		Statistic	p Value	
Kolmogorov-Smirnov	D	0.05563823	Pr > D	>0.150
Cramer-von Mises	W-Sq	0.04307548	Pr > W-Sq	>0.250
Anderson-Darling	A-Sq	0.27840748	Pr > A-Sq	>0.250

Output 4.19.1 *continued*

Histogram Bin Percents for Normal Distribution

Bin Midpoint	Percent Observed	Percent Estimated
3.43	3.000	3.296
3.45	9.000	9.319
3.47	23.000	18.091
3.49	19.000	24.124
3.51	24.000	22.099
3.53	15.000	13.907
3.55	3.000	6.011
3.57	4.000	1.784

Quantiles for Normal Distribution

Percent	Quantile Observed	Quantile Estimated
20.0	3.46700	3.46830
40.0	3.48350	3.48719
60.0	3.50450	3.50347
80.0	3.52250	3.52236

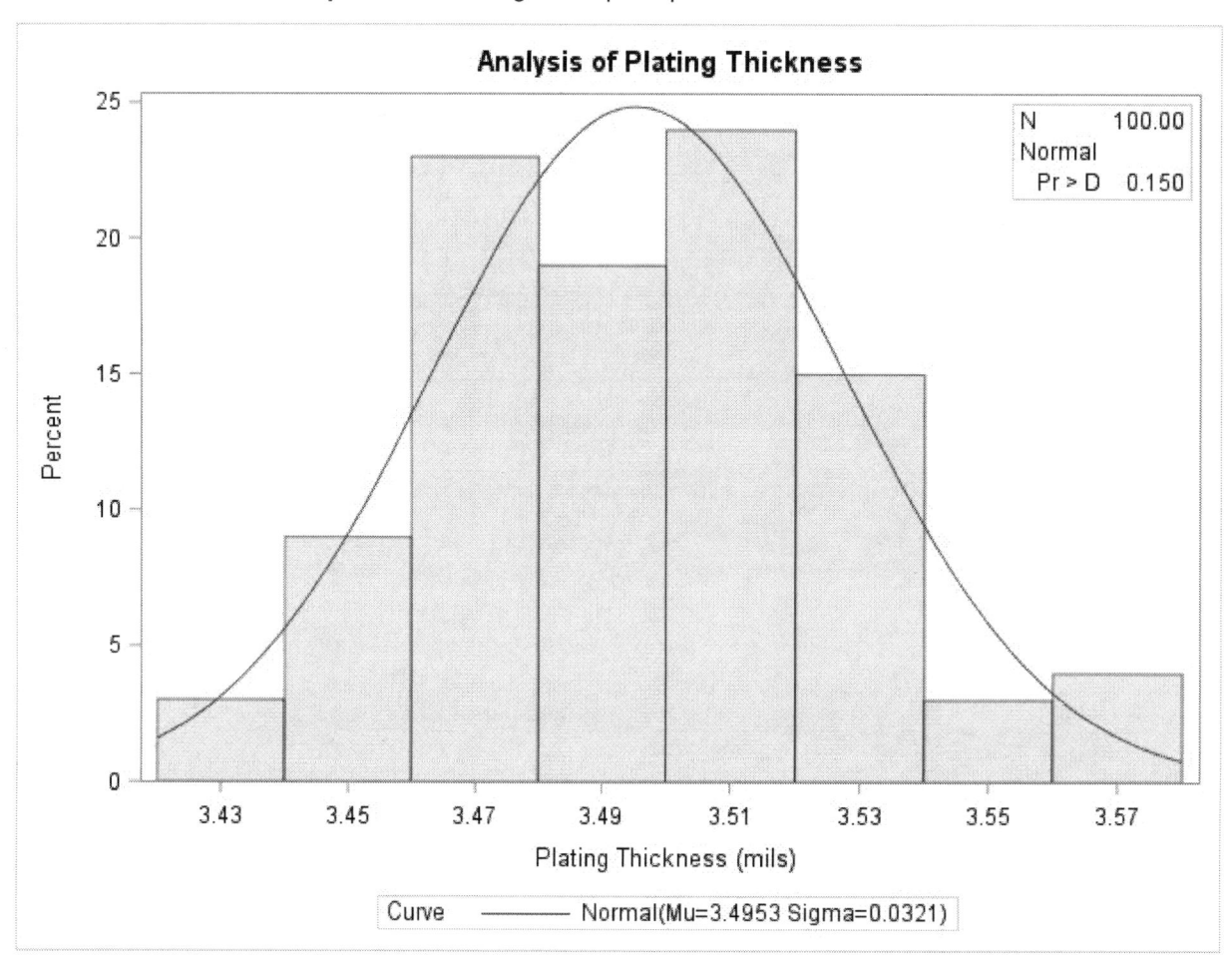

Output 4.19.2 Histogram Superimposed with Normal Curve

Analysis of Plating Thickness

N 100.00
Normal
Pr > D 0.150

Curve ——— Normal(Mu=3.4953 Sigma=0.0321)

The histogram of the variable Thick with a superimposed normal curve is shown in Output 4.19.2.

The estimated parameters for the normal curve ($\hat{\mu} = 3.50$ and $\hat{\sigma} = 0.03$) are shown in Output 4.19.1. By default, the parameters are estimated unless you specify values with the MU= and SIGMA= secondary options after the NORMAL primary option. The results of three goodness-of-fit tests based on the empirical distribution function (EDF) are displayed in Output 4.19.1. Because the *p*-values are all greater than 0.15, the hypothesis of normality is not rejected.

A sample program for this example, *uniex08.sas*, is available in the SAS Sample Library for Base SAS software.

Example 4.20: Adding Fitted Normal Curves to a Comparative Histogram

This example is a continuation of Example 4.15, which introduced the data set Channel. In Output 4.15.3, it appears that the channel lengths in each lot are normally distributed. The following statements use the NORMAL option to fit a normal distribution for each lot:

```
title 'Comparative Analysis of Lot Source';
proc univariate data=Channel noprint;
   class Lot;
   histogram Length / nrows     = 3
                      intertile = 1
                      odstitle  = title
                      cprop
                      normal(noprint);
   inset n = "N" / pos = nw;
run;
```

The NOPRINT option in the PROC UNIVARIATE statement suppresses the tables of statistical output produced by default; the NOPRINT option in parentheses after the NORMAL option suppresses the tables of statistical output related to the fit of the normal distribution. The normal parameters are estimated from the data for each lot, and the curves are superimposed on each component histogram. The INTERTILE= option specifies the space between the framed areas, which are referred to as tiles. The CPROP= option requests the shaded bars above each tile, which represent the relative frequencies of observations in each lot. The comparative histogram is displayed in Output 4.20.1.

A sample program for this example, *uniex09.sas*, is available in the SAS Sample Library for Base SAS software.

Output 4.20.1 Fitting Normal Curves to a Comparative Histogram

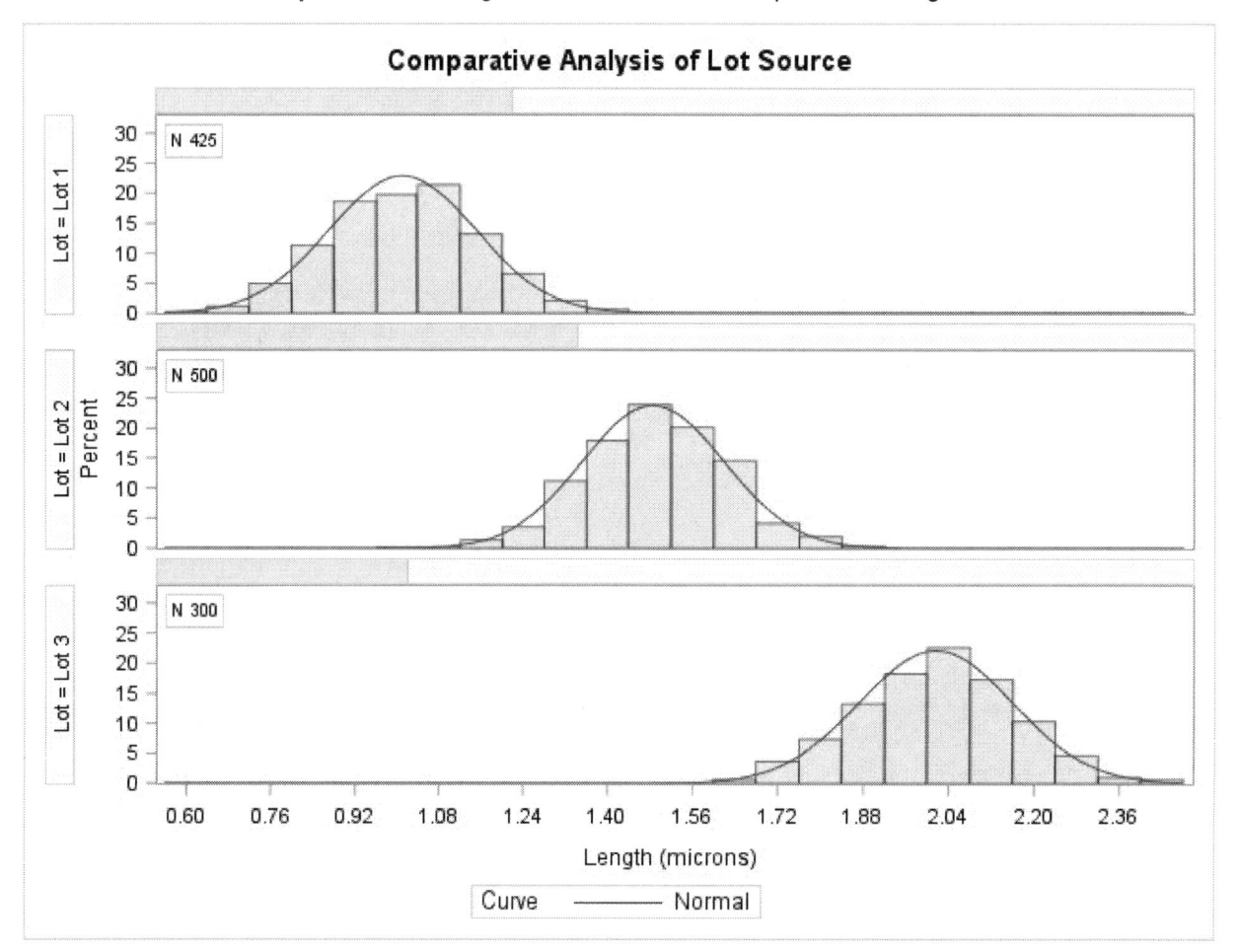

Example 4.21: Fitting a Beta Curve

You can use a beta distribution to model the distribution of a variable that is known to vary between lower and upper bounds. In this example, a manufacturing company uses a robotic arm to attach hinges on metal sheets. The attachment point should be offset 10.1 mm from the left edge of the sheet. The actual offset varies between 10.0 and 10.5 mm due to variation in the arm. The following statements save the offsets for 50 attachment points as the values of the variable Length in the data set Robots:

```
data Robots;
   input Length @@;
   label Length = 'Attachment Point Offset (in mm)';
   datalines;
10.147 10.070 10.032 10.042 10.102
10.034 10.143 10.278 10.114 10.127
10.122 10.018 10.271 10.293 10.136
10.240 10.205 10.186 10.186 10.080
10.158 10.114 10.018 10.201 10.065
10.061 10.133 10.153 10.201 10.109
10.122 10.139 10.090 10.136 10.066
10.074 10.175 10.052 10.059 10.077
10.211 10.122 10.031 10.322 10.187
10.094 10.067 10.094 10.051 10.174
;
```

The following statements create a histogram with a fitted beta density curve, shown in Output 4.21.1:

```
ods select ParameterEstimates FitQuantiles Histogram;
proc univariate data=Robots;
   histogram Length /
      beta(theta=10 scale=0.5 fill)
      href     = 10
      hreflabel = 'Lower Bound'
      odstitle = 'Fitted Beta Distribution of Offsets';
   inset n = 'Sample Size' /
      pos=ne  cfill=blank;
run;
```

The ODS SELECT statement restricts the output to the "ParameterEstimates" and "FitQuantiles" tables and the histogram; see the section "ODS Table Names" on page 451. The BETA primary option requests a fitted beta distribution. The THETA= secondary option specifies the lower threshold. The SCALE= secondary option specifies the range between the lower threshold and the upper threshold. Note that the default THETA= and SCALE= values are zero and one, respectively.

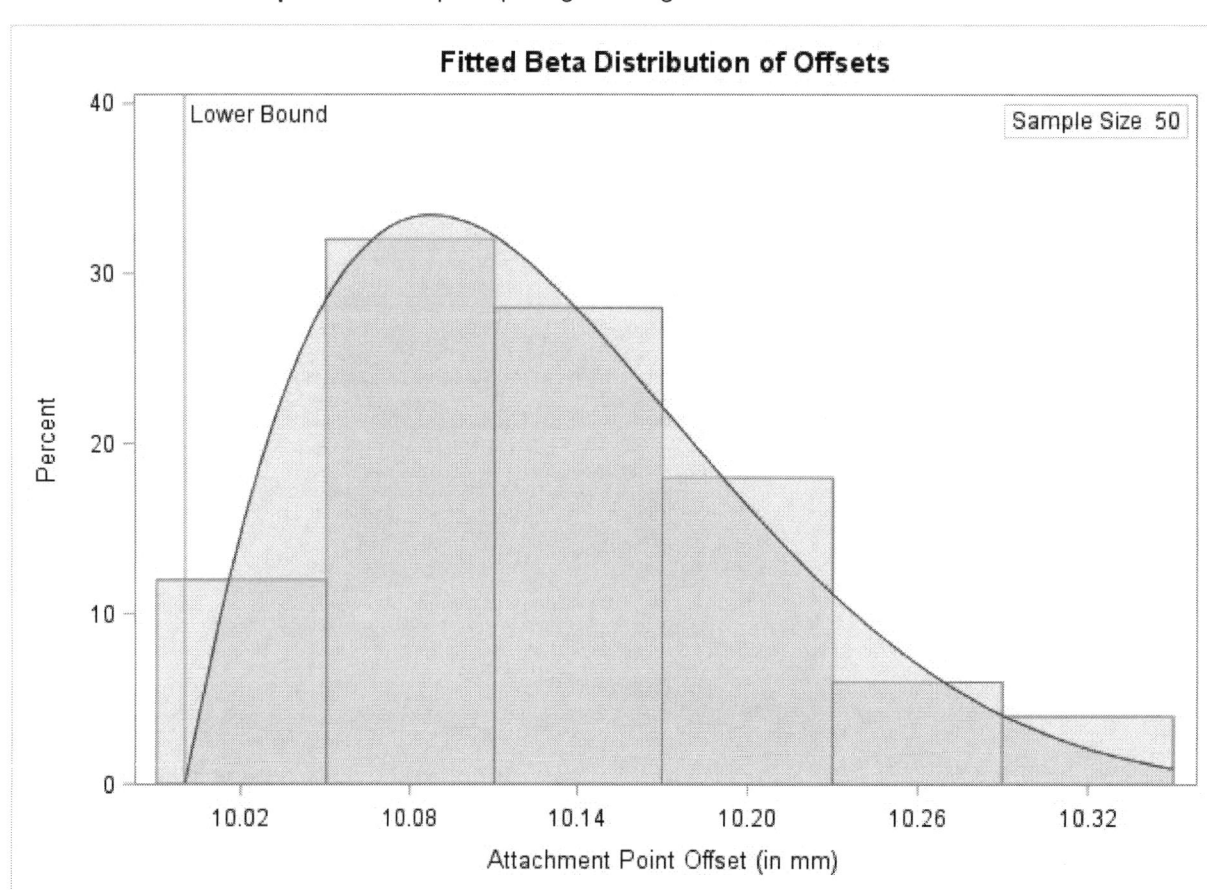

Output 4.21.1 Superimposing a Histogram with a Fitted Beta Curve

The FILL secondary option specifies that the area under the curve is to be filled. The HREF= option draws a reference line at the lower bound, and the HREFLABEL= option adds the label *Lower Bound*. The ODSTITLE= option adds a title to the histogram. The INSET statement adds an inset with the sample size positioned in the northeast corner of the plot.

In addition to displaying the beta curve, the BETA option requests a summary of the curve fit. This summary, which includes parameters for the curve and the observed and estimated quantiles, is shown in Output 4.21.2.

A sample program for this example, *uniex12.sas*, is available in the SAS Sample Library for Base SAS software.

Output 4.21.2 Summary of Fitted Beta Distribution

Comparative Analysis of Lot Source

The UNIVARIATE Procedure
Fitted Beta Distribution for Length (Attachment Point Offset (in mm))

Parameters for Beta Distribution		
Parameter	Symbol	Estimate
Threshold	Theta	10
Scale	Sigma	0.5
Shape	Alpha	2.06832
Shape	Beta	6.022479
Mean		10.12782
Std Dev		0.072339

Quantiles for Beta Distribution		
	Quantile	
Percent	Observed	Estimated
1.0	10.0180	10.0124
5.0	10.0310	10.0285
10.0	10.0380	10.0416
25.0	10.0670	10.0718
50.0	10.1220	10.1174
75.0	10.1750	10.1735
90.0	10.2255	10.2292
95.0	10.2780	10.2630
99.0	10.3220	10.3237

Example 4.22: Fitting Lognormal, Weibull, and Gamma Curves

To determine an appropriate model for a data distribution, you should consider curves from several distribution families. As shown in this example, you can use the HISTOGRAM statement to fit more than one distribution and display the density curves on a histogram.

The gap between two plates is measured (in cm) for each of 50 welded assemblies selected at random from the output of a welding process. The following statements save the measurements (Gap) in a data set named Plates:

```
data Plates;
   label Gap = 'Plate Gap in cm';
   input Gap @@;
   datalines;
0.746  0.357  0.376  0.327  0.485 1.741  0.241  0.777  0.768  0.409
0.252  0.512  0.534  1.656  0.742 0.378  0.714  1.121  0.597  0.231
0.541  0.805  0.682  0.418  0.506 0.501  0.247  0.922  0.880  0.344
0.519  1.302  0.275  0.601  0.388 0.450  0.845  0.319  0.486  0.529
1.547  0.690  0.676  0.314  0.736 0.643  0.483  0.352  0.636  1.080
;
```

The following statements fit three distributions (lognormal, Weibull, and gamma) and display their density curves on a single histogram:

```
title 'Distribution of Plate Gaps';
ods graphics on;
ods select Histogram ParameterEstimates GoodnessOfFit FitQuantiles;
proc univariate data=Plates;
   var Gap;
   histogram / midpoints=0.2 to 1.8 by 0.2
               lognormal
               weibull
               gamma
               odstitle = title;
   inset n mean(5.3) std='Std Dev'(5.3) skewness(5.3)
         / pos = ne   header = 'Summary Statistics';
run;
```

The ODS SELECT statement restricts the output to the "ParameterEstimates," "GoodnessOfFit," and "FitQuantiles" tables; see the section "ODS Table Names" on page 451. The LOGNORMAL, WEIBULL, and GAMMA primary options request superimposed fitted curves on the histogram in Output 4.22.1. Note that a threshold parameter $\theta = 0$ is assumed for each curve. In applications where the threshold is not zero, you can specify θ with the THETA= secondary option.

The LOGNORMAL, WEIBULL, and GAMMA options also produce the summaries for the fitted distributions shown in Output 4.22.2 through Output 4.22.4.

Output 4.22.2 provides three EDF goodness-of-fit tests for the lognormal distribution: the Anderson-Darling, the Cramér-von Mises, and the Kolmogorov-Smirnov tests. At the $\alpha = 0.10$ significance level, all tests support the conclusion that the two-parameter lognormal distribution with scale parameter $\hat{\zeta} = -0.58$ and shape parameter $\hat{\sigma} = 0.50$ provides a good model for the distribution of plate gaps.

Output 4.22.1 Superimposing a Histogram with Fitted Curves

Distribution of Plate Gaps

Summary Statistics
N   50
Mean   0.634
Std Dev   0.351
Skewness   1.573

Curves
— Lognormal(Theta=0 Sigma=0.5 Zeta=-.58)
— Weibull(Theta=0 C=1.96 Sigma=0.72)
— Gamma(Theta=0 Alpha=4.08 Sigma=0.16)

Output 4.22.2 Summary of Fitted Lognormal Distribution

Distribution of Plate Gaps

The UNIVARIATE Procedure
Fitted Lognormal Distribution for Gap (Plate Gap in cm)

Parameters for Lognormal Distribution		
Parameter	Symbol	Estimate
Threshold	Theta	0
Scale	Zeta	-0.58375
Shape	Sigma	0.499546
Mean		0.631932
Std Dev		0.336436

Output 4.22.2 *continued*

Goodness-of-Fit Tests for Lognormal Distribution				
Test		Statistic		p Value
Kolmogorov-Smirnov	D	0.06441431	Pr > D	>0.150
Cramer-von Mises	W-Sq	0.02823022	Pr > W-Sq	>0.500
Anderson-Darling	A-Sq	0.24308402	Pr > A-Sq	>0.500

Quantiles for Lognormal Distribution

	Quantile	
Percent	Observed	Estimated
1.0	0.23100	0.17449
5.0	0.24700	0.24526
10.0	0.29450	0.29407
25.0	0.37800	0.39825
50.0	0.53150	0.55780
75.0	0.74600	0.78129
90.0	1.10050	1.05807
95.0	1.54700	1.26862
99.0	1.74100	1.78313

Output 4.22.3 Summary of Fitted Weibull Distribution

Distribution of Plate Gaps

The UNIVARIATE Procedure
Fitted Weibull Distribution for Gap (Plate Gap in cm)

Parameters for Weibull Distribution		
Parameter	Symbol	Estimate
Threshold	Theta	0
Scale	Sigma	0.719208
Shape	C	1.961159
Mean		0.637641
Std Dev		0.339248

Goodness-of-Fit Tests for Weibull Distribution				
Test		Statistic		p Value
Cramer-von Mises	W-Sq	0.15937281	Pr > W-Sq	0.016
Anderson-Darling	A-Sq	1.15693542	Pr > A-Sq	<0.010

Example 4.22: Fitting Lognormal, Weibull, and Gamma Curves ✦ 503

Output 4.22.3 *continued*

	Quantiles for Weibull Distribution	
		Quantile
Percent	Observed	Estimated
1.0	0.23100	0.06889
5.0	0.24700	0.15817
10.0	0.29450	0.22831
25.0	0.37800	0.38102
50.0	0.53150	0.59661
75.0	0.74600	0.84955
90.0	1.10050	1.10040
95.0	1.54700	1.25842
99.0	1.74100	1.56691

Output 4.22.3 provides two EDF goodness-of-fit tests for the Weibull distribution: the Anderson-Darling and the Cramér–von Mises tests. The *p*-values for the EDF tests are all less than 0.10, indicating that the data do not support a Weibull model.

Output 4.22.4 Summary of Fitted Gamma Distribution

Distribution of Plate Gaps

The UNIVARIATE Procedure
Fitted Gamma Distribution for Gap (Plate Gap in cm)

Parameters for Gamma Distribution		
Parameter	Symbol	Estimate
Threshold	Theta	0
Scale	Sigma	0.155198
Shape	Alpha	4.082646
Mean		0.63362
Std Dev		0.313587

Goodness-of-Fit Tests for Gamma Distribution			
Test	Statistic		p Value
Kolmogorov-Smirnov	D 0.09695325	Pr > D	>0.250
Cramer-von Mises	W-Sq 0.07398467	Pr > W-Sq	>0.250
Anderson-Darling	A-Sq 0.58106613	Pr > A-Sq	0.137

Output 4.22.4 *continued*

Quantiles for Gamma Distribution		
	Quantile	
Percent	Observed	Estimated
1.0	0.23100	0.13326
5.0	0.24700	0.21951
10.0	0.29450	0.27938
25.0	0.37800	0.40404
50.0	0.53150	0.58271
75.0	0.74600	0.80804
90.0	1.10050	1.05392
95.0	1.54700	1.22160
99.0	1.74100	1.57939

Output 4.22.4 provides three EDF goodness-of-fit tests for the gamma distribution: the Anderson-Darling, the Cramér–von Mises, and the Kolmogorov-Smirnov tests. At the $\alpha = 0.10$ significance level, all tests support the conclusion that the gamma distribution with scale parameter $\sigma = 0.16$ and shape parameter $\alpha = 4.08$ provides a good model for the distribution of plate gaps.

Based on this analysis, the fitted lognormal distribution and the fitted gamma distribution are both good models for the distribution of plate gaps.

A sample program for this example, *uniex13.sas*, is available in the SAS Sample Library for Base SAS software.

Example 4.23: Computing Kernel Density Estimates

This example illustrates the use of kernel density estimates to visualize a nonnormal data distribution. This example uses the data set Channel, which is introduced in Example 4.15.

When you compute kernel density estimates, you should try several choices for the bandwidth parameter c because this determines the smoothness and closeness of the fit. You can specify a list of up to five C= values with the KERNEL option to request multiple density estimates, as shown in the following statements:

```
title 'FET Channel Length Analysis';
proc univariate data=Channel noprint;
   histogram Length / kernel(c = 0.25 0.50 0.75 1.00
                     l = 1 20 2 34
                     noprint)
                odstitle = title;
run;
```

The L= secondary option specifies distinct line types for the curves (the L= values are paired with the C= values in the order listed). Output 4.23.1 demonstrates the effect of c. In general, larger values of c yield smoother density estimates, and smaller values yield estimates that more closely fit the data distribution.

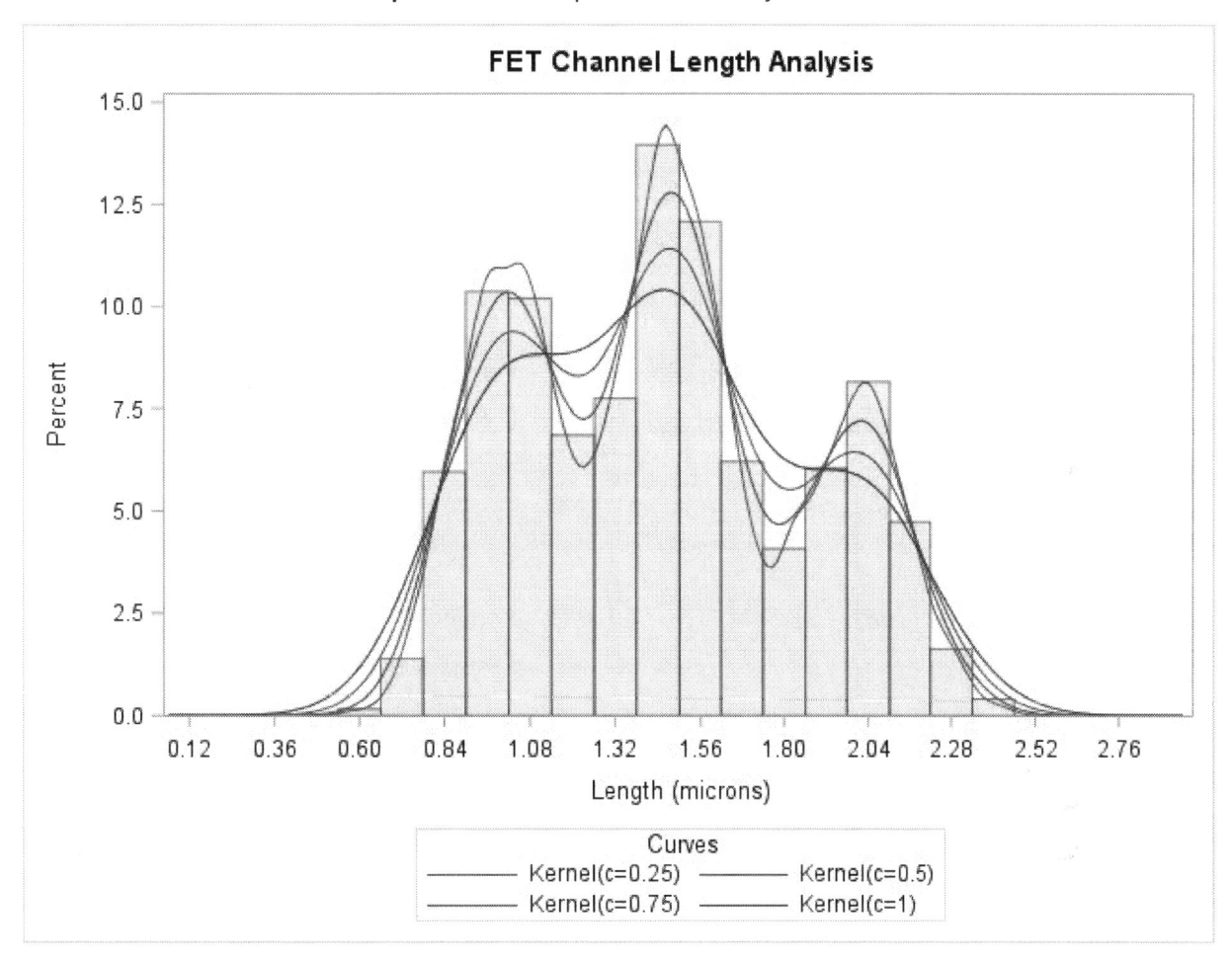

Output 4.23.1 Multiple Kernel Density Estimates

Output 4.23.1 reveals strong trimodality in the data, which is displayed with comparative histograms in Example 4.15.

A sample program for this example, *uniex09.sas*, is available in the SAS Sample Library for Base SAS software.

Example 4.24: Fitting a Three-Parameter Lognormal Curve

If you request a lognormal fit with the LOGNORMAL primary option, a two-parameter lognormal distribution is assumed. This means that the shape parameter σ and the scale parameter ζ are unknown (unless specified) and that the threshold θ is known (it is either specified with the THETA= option or assumed to be zero).

If it is necessary to estimate θ in addition to ζ and σ, the distribution is referred to as a three-parameter lognormal distribution. This example shows how you can request a three-parameter lognormal distribution.

A manufacturing process produces a plastic laminate whose strength must exceed a minimum of 25 pounds per square inch (PSI). Samples are tested, and a lognormal distribution is observed for the strengths. It is important to estimate θ to determine whether the process meets the strength requirement. The following statements save the strengths for 49 samples in the data set Plastic:

```
data Plastic;
   label Strength = 'Strength in psi';
   input Strength @@;
   datalines;
30.26 31.23 71.96 47.39 33.93 76.15 42.21
81.37 78.48 72.65 61.63 34.90 24.83 68.93
43.27 41.76 57.24 23.80 34.03 33.38 21.87
31.29 32.48 51.54 44.06 42.66 47.98 33.73
25.80 29.95 60.89 55.33 39.44 34.50 73.51
43.41 54.67 99.43 50.76 48.81 31.86 33.88
35.57 60.41 54.92 35.66 59.30 41.96 45.32
;
```

The following statements use the LOGNORMAL primary option in the HISTOGRAM statement to display the fitted three-parameter lognormal curve shown in Output 4.24.1:

```
title 'Three-Parameter Lognormal Fit';
ods graphics on;
proc univariate data=Plastic noprint;
   histogram Strength / lognormal(fill theta = est noprint)
                       odstitle = title;
   inset lognormal    / format=6.2 pos=ne;
run;
```

The NOPRINT option suppresses the tables of statistical output produced by default. Specifying THETA=EST requests a local maximum likelihood estimate (LMLE) for θ, as described by Cohen (1951). This estimate is then used to compute maximum likelihood estimates for σ and ζ.

NOTE: You can also specify THETA=EST with the WEIBULL primary option to fit a three-parameter Weibull distribution.

A sample program for this example, *uniex14.sas*, is available in the SAS Sample Library for Base SAS software.

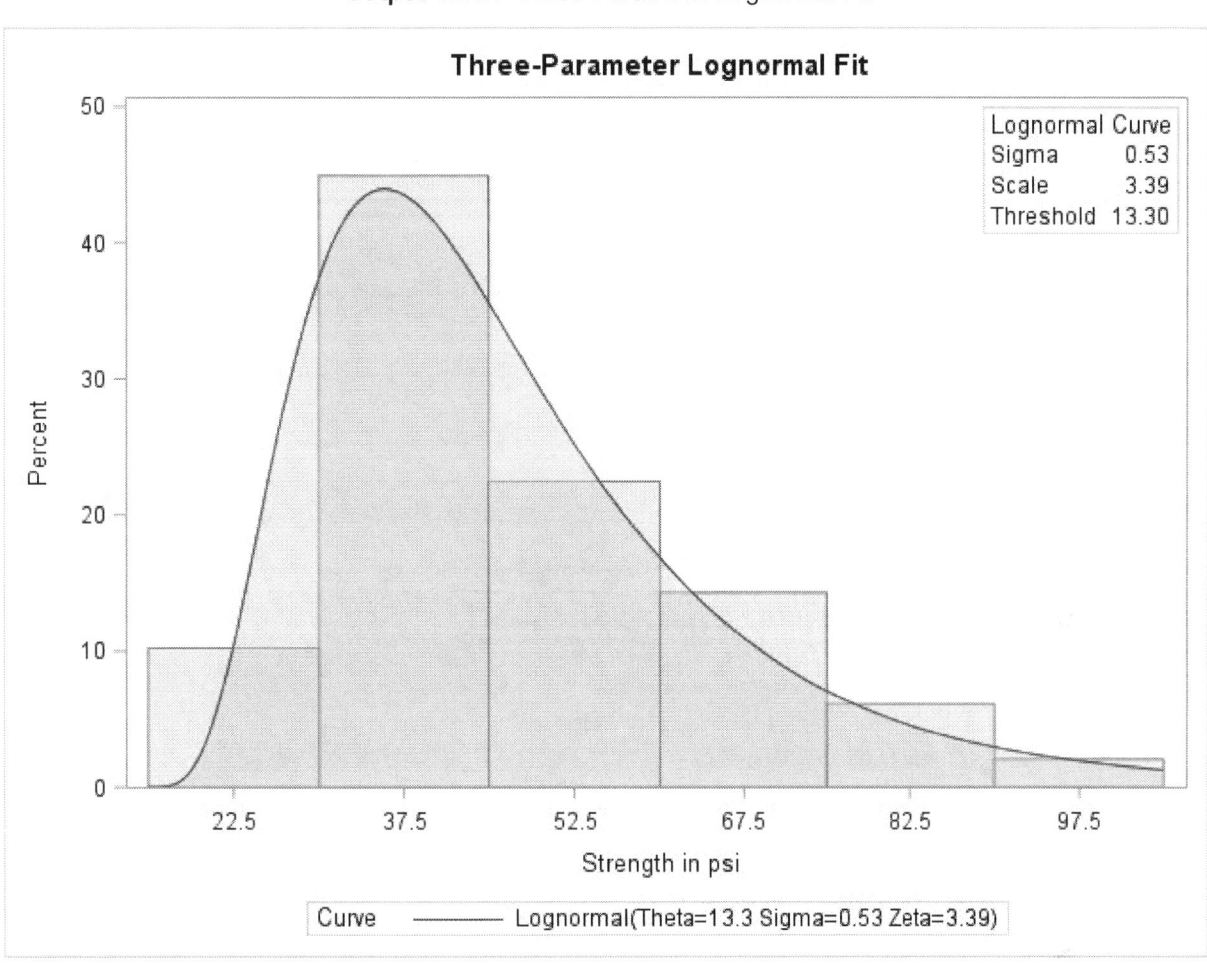

Output 4.24.1 Three-Parameter Lognormal Fit

Example 4.25: Annotating a Folded Normal Curve

This example shows how to display a fitted curve that is not supported by the HISTOGRAM statement. The offset of an attachment point is measured (in mm) for a number of manufactured assemblies, and the measurements (Offset) are saved in a data set named Assembly. The following statements create the data set Assembly:

```
data Assembly;
   label Offset = 'Offset (in mm)';
   input Offset @@;
   datalines;
11.11 13.07 11.42  3.92 11.08  5.40 11.22 14.69  6.27  9.76
 9.18  5.07  3.51 16.65 14.10  9.69 16.61  5.67  2.89  8.13
 9.97  3.28 13.03 13.78  3.13  9.53  4.58  7.94 13.51 11.43
11.98  3.90  7.67  4.32 12.69  6.17 11.48  2.82 20.42  1.01
 3.18  6.02  6.63  1.72  2.42 11.32 16.49  1.22  9.13  3.34
 1.29  1.70  0.65  2.62  2.04 11.08 18.85 11.94  8.34  2.07
 0.31  8.91 13.62 14.94  4.83 16.84  7.09  3.37  0.49 15.19
 5.16  4.14  1.92 12.70  1.97  2.10  9.38  3.18  4.18  7.22
```

```
15.84 10.85  2.35  1.93  9.19  1.39 11.40 12.20 16.07  9.23
 0.05  2.15  1.95  4.39  0.48 10.16  4.81  8.28  5.68 22.81
 0.23  0.38 12.71  0.06 10.11 18.38  5.53  9.36  9.32  3.63
12.93 10.39  2.05 15.49  8.12  9.52  7.77 10.70  6.37  1.91
 8.60 22.22  1.74  5.84 12.90 13.06  5.08  2.09  6.41  1.40
15.60  2.36  3.97  6.17  0.62  8.56  9.36 10.19  7.16  2.37
12.91  0.95  0.89  3.82  7.86  5.33 12.92  2.64  7.92 14.06
;
```

It is decided to fit a *folded normal distribution* to the offset measurements. A variable X has a folded normal distribution if $X = |Y|$, where Y is distributed as $N(\mu, \sigma)$. The fitted density is

$$h(x) = \frac{1}{\sqrt{2\pi}\sigma} \left[\exp\left(-\frac{(x-\mu)^2}{2\sigma^2}\right) + \exp\left(-\frac{(x+\mu)^2}{2\sigma^2}\right) \right]$$

where $x \geq 0$.

You can use SAS/IML to compute preliminary estimates of μ and σ based on a method of moments given by Elandt (1961). These estimates are computed by solving equation (19) Elandt (1961), which is given by

$$f(\theta) = \frac{\left(\frac{2}{\sqrt{2\pi}} e^{-\theta^2/2} - \theta\left[1 - 2\Phi(\theta)\right]\right)^2}{1 + \theta^2} = A$$

where $\Phi(\cdot)$ is the standard normal distribution function and

$$A = \frac{\bar{x}^2}{\frac{1}{n}\sum_{i=1}^{n} x_i^2}$$

Then the estimates of σ and μ are given by

$$\hat{\sigma}_0 = \sqrt{\frac{\frac{1}{n}\sum_{i=1}^{n} x_i^2}{1 + \hat{\theta}^2}}$$
$$\hat{\mu}_0 = \hat{\theta} \cdot \hat{\sigma}_0$$

Begin by using PROC MEANS to compute the first and second moments and by using the following DATA step to compute the constant A:

```
proc means data = Assembly noprint;
   var Offset;
   output out=stat mean=m1 var=var n=n min = min;
run;

* Compute constant A from equation (19) of Elandt (1961);
data stat;
   keep m2 a min;
   set stat;
   a  = (m1*m1);
   m2 = ((n-1)/n)*var + a;
   a  = a/m2;
run;
```

Next, use the SAS/IML subroutine NLPDD to solve equation (19) by minimizing $(f(\theta) - A)^2$, and compute $\hat{\mu}_0$ and $\hat{\sigma}_0$:

```
proc iml;
   use stat;
   read all var {m2}  into m2;
   read all var {a}   into a;
   read all var {min} into min;

   * f(t) is the function in equation (19) of Elandt (1961);
   start f(t) global(a);
     y = .39894*exp(-0.5*t*t);
     y = (2*y-(t*(1-2*probnorm(t))))**2/(1+t*t);
     y = (y-a)**2;
     return(y);
   finish;

   * Minimize (f(t)-A)**2 and estimate mu and sigma;
   if ( min < 0 ) then do;
      print "Warning: Observations are not all nonnegative.";
      print "         The folded normal is inappropriate.";
      stop;
      end;
   if ( a < 0.637 ) then do;
      print "Warning: the folded normal may be inappropriate";
      end;
   opt = { 0 0 };
   con = { 1e-6 };
   x0  = { 2.0 };
   tc  = { . . . . . 1e-8 . . . . . . . };
   call nlpdd(rc,etheta0,"f",x0,opt,con,tc);
   esig0 = sqrt(m2/(1+etheta0*etheta0));
   emu0  = etheta0*esig0;

   create prelim var {emu0 esig0 etheta0};
   append;
   close prelim;

   * Define the log likelihood of the folded normal;
   start g(p) global(x);
      y = 0.0;
      do i = 1 to nrow(x);
         z = exp( (-0.5/p[2])*(x[i]-p[1])*(x[i]-p[1]) );
         z = z + exp( (-0.5/p[2])*(x[i]+p[1])*(x[i]+p[1]) );
         y = y + log(z);
      end;
      y = y - nrow(x)*log( sqrt( p[2] ) );
      return(y);
   finish;
   * Maximize the log likelihood with subroutine NLPDD;
   use assembly;
   read all var {offset} into x;
   esig0sq = esig0*esig0;
   x0      = emu0||esig0sq;
   opt     = { 1 0 };
   con     = { . 0.0, . . };
```

```
     call nlpdd(rc,xr,"g",x0,opt,con);
     emu     = xr[1];
     esig    = sqrt(xr[2]);
     etheta  = emu/esig;
     create parmest var{emu esig etheta};
     append;
     close parmest;
quit;
```

The preliminary estimates are saved in the data set Prelim, as shown in Output 4.25.1.

Output 4.25.1 Preliminary Estimates of μ, σ, and θ

The Data Set Prelim

Obs	EMU0	ESIG0	ETHETA0
1	6.51735	6.54953	0.99509

Now, using $\hat{\mu}_0$ and $\hat{\sigma}_0$ as initial estimates, call the NLPDD subroutine to maximize the log likelihood, $l(\mu, \sigma)$, of the folded normal distribution, where, up to a constant,

$$l(\mu, \sigma) = -n \log \sigma + \sum_{i=1}^{n} \log \left[\exp\left(-\frac{(x_i - \mu)^2}{2\sigma^2}\right) + \exp\left(-\frac{(x_i + \mu)^2}{2\sigma^2}\right) \right]$$

```
* Define the log likelihood of the folded normal;
start g(p) global(x);
   y = 0.0;
   do i = 1 to nrow(x);
      z = exp( (-0.5/p[2])*(x[i]-p[1])*(x[i]-p[1]) );
      z = z + exp( (-0.5/p[2])*(x[i]+p[1])*(x[i]+p[1]) );
      y = y + log(z);
   end;
   y = y - nrow(x)*log( sqrt( p[2] ) );
   return(y);
finish;
* Maximize the log likelihood with subroutine NLPDD;
use assembly;
read all var {offset} into x;
esig0sq = esig0*esig0;
x0      = emu0||esig0sq;
opt     = { 1 0 };
con     = { . 0.0, . . };
call nlpdd(rc,xr,"g",x0,opt,con);
emu     = xr[1];
esig    = sqrt(xr[2]);
etheta  = emu/esig;
create parmest var{emu esig etheta};
append;
close parmest;
quit;
```

The data set ParmEst contains the maximum likelihood estimates $\hat{\mu}$ and $\hat{\sigma}$ (as well as $\hat{\mu}/\hat{\sigma}$), as shown in Output 4.25.2.

Output 4.25.2 Final Estimates of μ, σ, and θ

The Data Set ParmEst

Obs	EMU	ESIG	ETHETA
1	6.66761	6.39650	1.04239

To annotate the curve on a histogram, begin by computing the width and endpoints of the histogram intervals. The following statements save these values in a data set called OutCalc. Note that a plot is not produced at this point.

```
proc univariate data = Assembly noprint;
   histogram Offset / outhistogram = out normal(noprint) noplot;
run;

data OutCalc (drop = _MIDPT_);
   set out (keep = _MIDPT_) end = eof;
   retain _MIDPT1_ _WIDTH_;
   if _N_ = 1 then _MIDPT1_ = _MIDPT_;
   if eof then do;
      _MIDPTN_ = _MIDPT_;
      _WIDTH_ = (_MIDPTN_ - _MIDPT1_) / (_N_ - 1);
      output;
   end;
run;
```

Output 4.25.3 provides a listing of the data set OutCalc. The width of the histogram bars is saved as the value of the variable _WIDTH_; the midpoints of the first and last histogram bars are saved as the values of the variables _MIDPT1_ and _MIDPTN_.

Output 4.25.3 The Data Set OutCalc

The Data Set OutCalc

Obs	_MIDPT1_	_WIDTH_	_MIDPTN_
1	1.5	3	22.5

The following statements create an annotate data set named Anno, which contains the coordinates of the fitted curve:

```
data Anno;
   merge ParmEst OutCalc;
   length function color $ 8;
   function = 'point';
   color    = 'black';
   size     = 2;
   xsys     = '2';
   ysys     = '2';
   when     = 'a';
   constant = 39.894*_width_;;
   left     = _midpt1_ - .5*_width_;
   right    = _midptn_ + .5*_width_;
   inc      = (right-left)/100;
```

```
    do x = left to right by inc;
       z1 = (x-emu)/esig;
       z2 = (x+emu)/esig;
       y  = (constant/esig)*(exp(-0.5*z1*z1)+exp(-0.5*z2*z2));
       output;
       function = 'draw';
    end;
run;
```

The following statements read the ANNOTATE= data set and display the histogram and fitted curve:

```
title 'Folded Normal Distribution';
ods graphics off;
proc univariate data=assembly noprint;
   histogram Offset / annotate = anno;
run;
```

Output 4.25.4 displays the histogram and fitted curve.

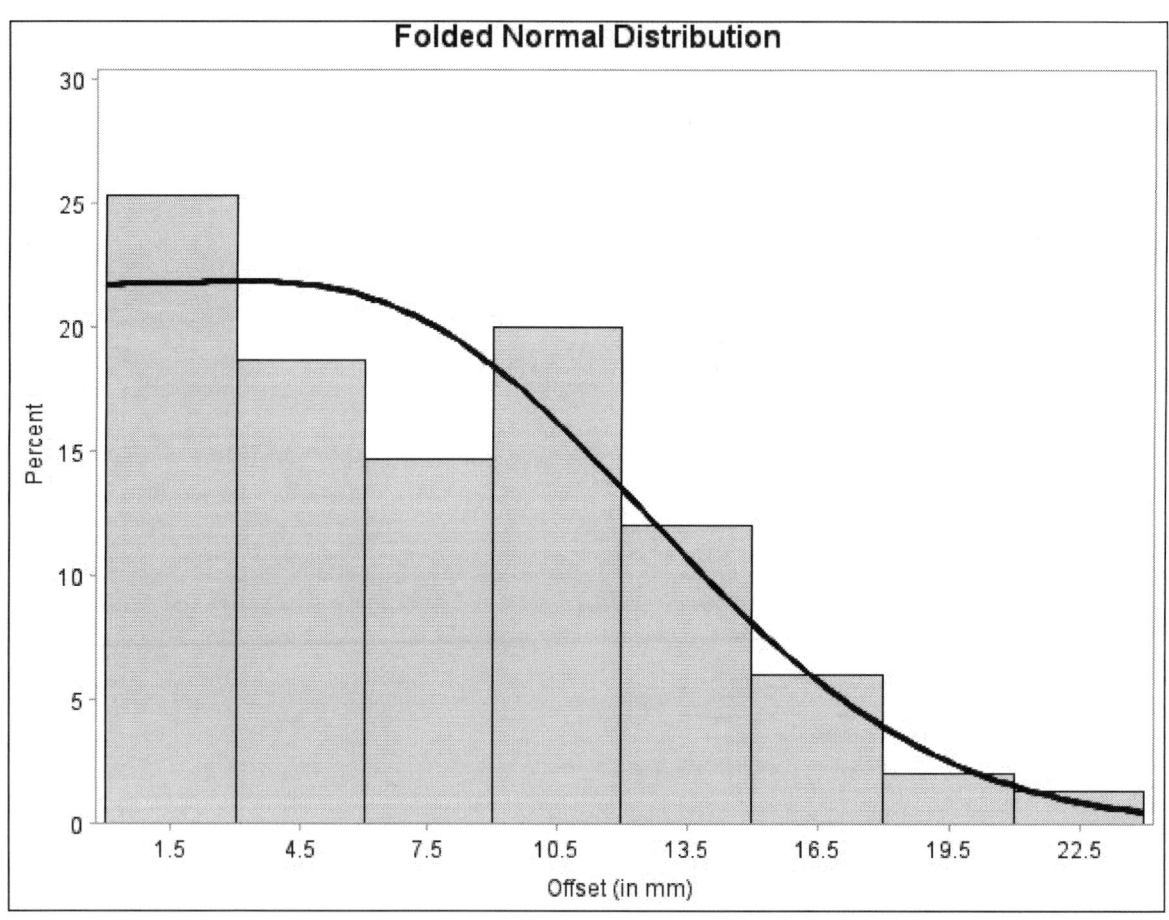

Output 4.25.4 Histogram with Annotated Folded Normal Curve

A sample program for this example, *uniex15.sas*, is available in the SAS Sample Library for Base SAS software.

Example 4.26: Creating Lognormal Probability Plots

This example is a continuation of the example explored in the section "Modeling a Data Distribution" on page 286.

In the normal probability plot shown in Output 4.6, the nonlinearity of the point pattern indicates a departure from normality in the distribution of Deviation. Because the point pattern is curved with slope increasing from left to right, a theoretical distribution that is skewed to the right, such as a lognormal distribution, should provide a better fit than the normal distribution. See the section "Interpretation of Quantile-Quantile and Probability Plots" on page 435.

You can explore the possibility of a lognormal fit with a lognormal probability plot. When you request such a plot, you must specify the shape parameter σ for the lognormal distribution. This value must be positive, and typical values of σ range from 0.1 to 1.0. You can specify values for σ with the SIGMA= secondary option in the LOGNORMAL primary option, or you can specify that σ is to be estimated from the data.

The following statements illustrate the first approach by creating a series of three lognormal probability plots for the variable Deviation introduced in the section "Modeling a Data Distribution" on page 286:

```
title 'Lognormal Probability Plot for Position Deviations';
ods graphics on;
proc univariate data=Aircraft noprint;
   probplot Deviation /
      lognormal(theta=est zeta=est sigma=0.7 0.9 1.1)
      odstitle = title
      href     = 95
      square;
run;
```

The LOGNORMAL primary option requests plots based on the lognormal family of distributions, and the SIGMA= secondary option requests plots for σ equal to 0.7, 0.9, and 1.1. These plots are displayed in Output 4.26.1, Output 4.26.2, and Output 4.26.3, respectively. Alternatively, you can specify σ to be estimated using the sample standard deviation by using the option SIGMA=EST.

The SQUARE option displays the probability plot in a square format, the HREF= option requests a reference line at the 95th percentile.

Output 4.26.1 Probability Plot Based on Lognormal Distribution with $\sigma = 0.7$

Example 4.26: Creating Lognormal Probability Plots ♦ 515

Output 4.26.2 Probability Plot Based on Lognormal Distribution with $\sigma = 0.9$

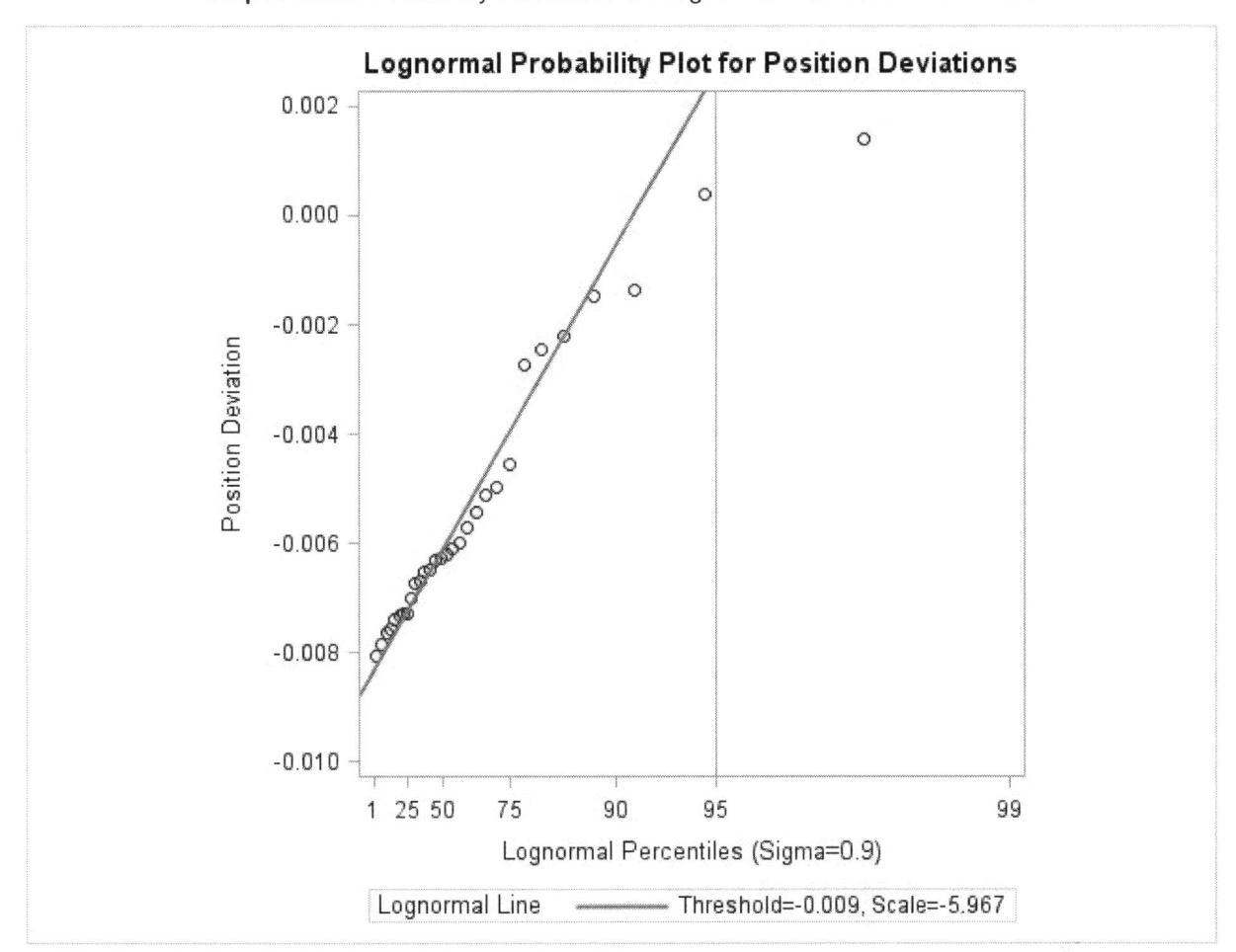

Output 4.26.3 Probability Plot Based on Lognormal Distribution with $\sigma = 1.1$

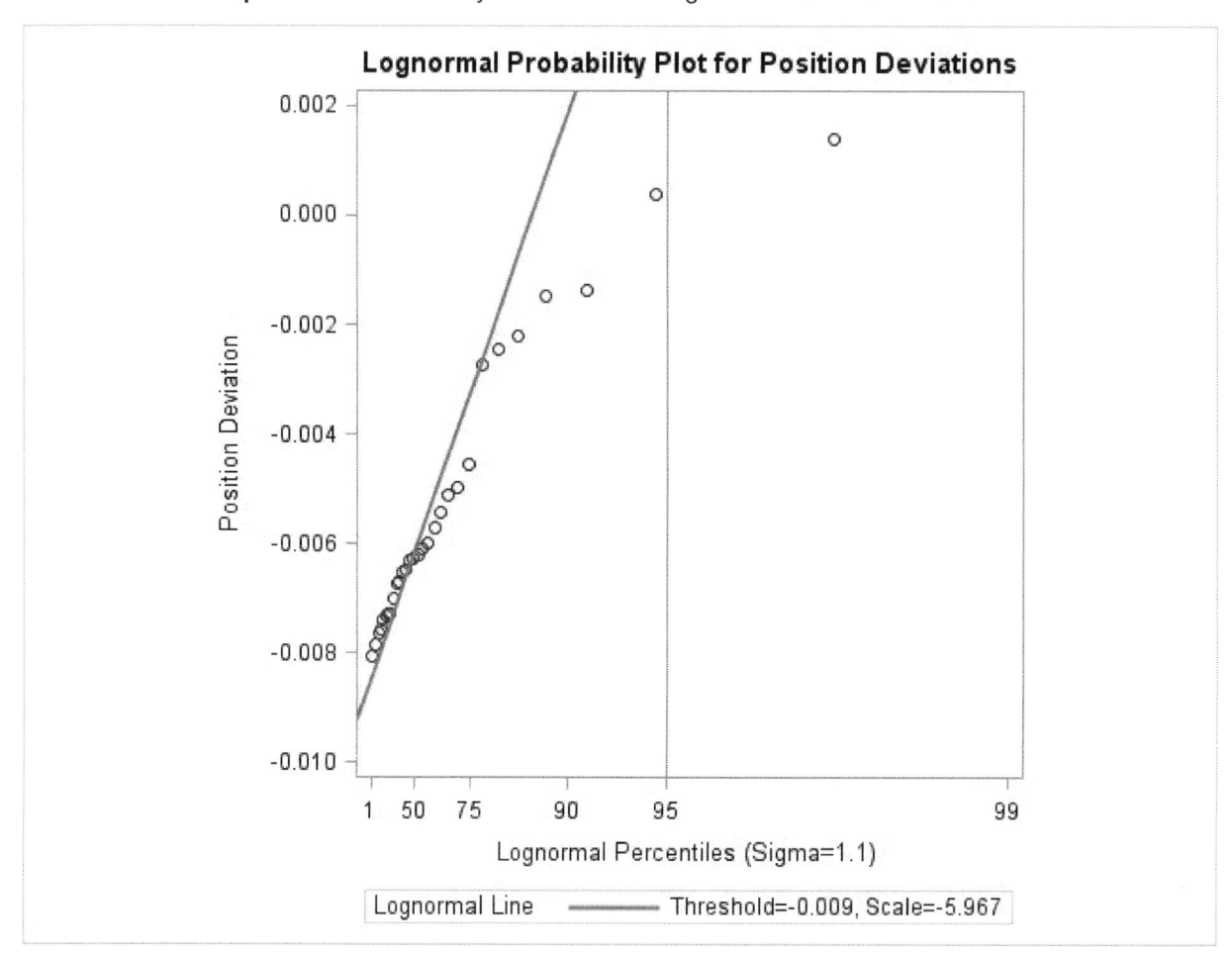

Lognormal Probability Plot for Position Deviations

[Probability plot with Position Deviation on y-axis ranging from -0.010 to 0.002, and Lognormal Percentiles (Sigma=1.1) on x-axis from 1 to 99. Legend shows: Lognormal Line, Threshold=-0.009, Scale=-5.967]

The value $\sigma = 0.9$ in Output 4.26.2 most nearly linearizes the point pattern. The 95th percentile of the position deviation distribution seen in Output 4.26.2 is approximately 0.001, because this is the value corresponding to the intersection of the point pattern with the reference line.

NOTE: After the σ that produces the most linear fit is found, you can then estimate the threshold parameter θ and the scale parameter ζ. See Example 4.31.

The following statements illustrate how you can create a lognormal probability plot for Deviation by using a local maximum likelihood estimate for σ.

```
title 'Lognormal Probability Plot for Position Deviations';
proc univariate data=Aircraft noprint;
   probplot Deviation / lognormal(theta=est zeta=est sigma=est)
                       href     = 95
                       odstitle = title
                       square;
run;
```

The plot is displayed in Output 4.26.4. Note that the maximum likelihood estimate of σ (in this case, 0.882) does not necessarily produce the most linear point pattern.

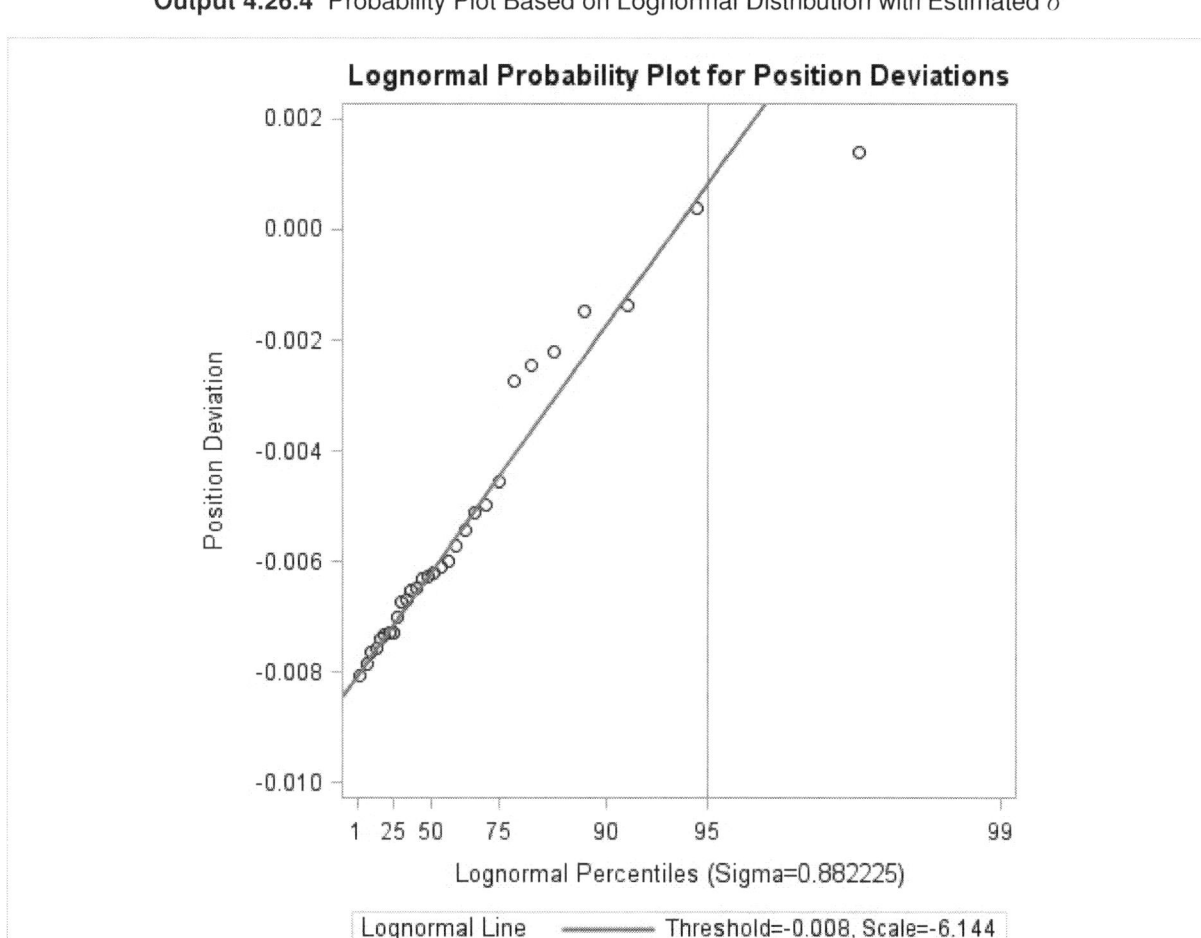

Output 4.26.4 Probability Plot Based on Lognormal Distribution with Estimated σ

A sample program for this example, *uniex16.sas*, is available in the SAS Sample Library for Base SAS software.

Example 4.27: Creating a Histogram to Display Lognormal Fit

This example uses the data set Aircraft from Example 4.26 to illustrate how to display a lognormal fit with a histogram. To determine whether the lognormal distribution is an appropriate model for a distribution, you should consider the graphical fit as well as conduct goodness-of-fit tests. The following statements fit a lognormal distribution and display the density curve on a histogram:

```
title 'Distribution of Position Deviations';
ods select Histogram Lognormal.ParameterEstimates Lognormal.GoodnessOfFit;
proc univariate data=Aircraft;
   var Deviation;
   histogram / lognormal(w=3 theta=est)
               odstitle = title;
   inset n mean (5.3) std='Std Dev' (5.3) skewness (5.3) /
      pos    = ne
      header = 'Summary Statistics';
run;
```

518 ✦ *Chapter 4: The UNIVARIATE Procedure*

The ODS SELECT statement restricts the output to the "ParameterEstimates" and "GoodnessOfFit" tables; see the section "ODS Table Names" on page 451. The LOGNORMAL primary option superimposes a fitted curve on the histogram in Output 4.27.1. The W= option specifies the line width for the curve. The INSET statement specifies that the mean, standard deviation, and skewness be displayed in an inset in the northeast corner of the plot. Note that the default value of the threshold parameter θ is zero. In applications where the threshold is not zero, you can specify θ with the THETA= option. The variable Deviation includes values that are less than the default threshold; therefore, the option THETA= EST is used.

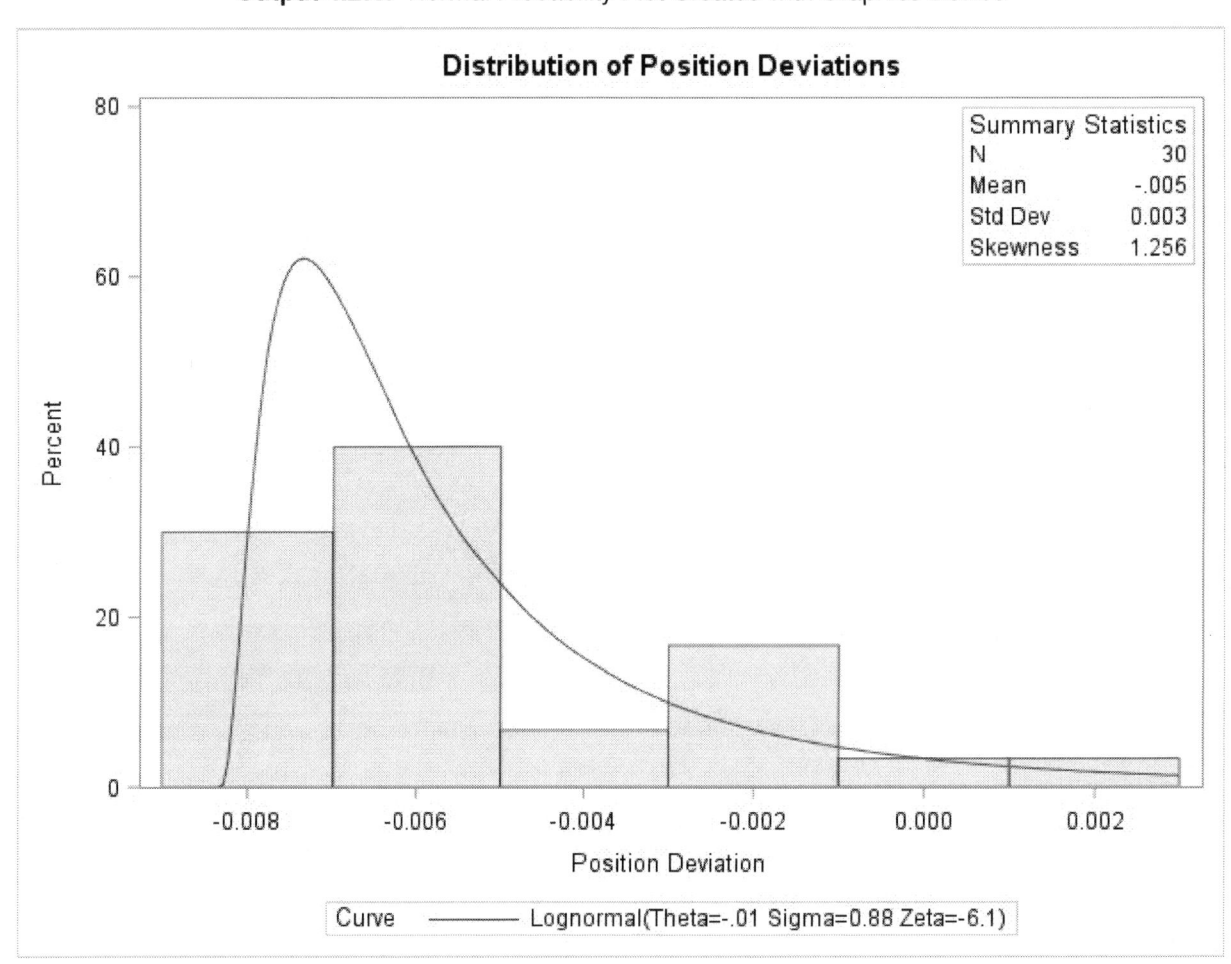

Output 4.27.1 Normal Probability Plot Created with Graphics Device

Distribution of Position Deviations

Summary Statistics	
N	30
Mean	-.005
Std Dev	0.003
Skewness	1.256

Curve —— Lognormal(Theta=-.01 Sigma=0.88 Zeta=-6.1)

Output 4.27.2 provides three EDF goodness-of-fit tests for the lognormal distribution: the Anderson-Darling, the Cramér–von Mises, and the Kolmogorov-Smirnov tests. The null hypothesis for the three tests is that a lognormal distribution holds for the sample data.

Output 4.27.2 Summary of Fitted Lognormal Distribution

Distribution of Position Deviations

The UNIVARIATE Procedure
Fitted Lognormal Distribution for Deviation (Position Deviation)

Parameters for Lognormal Distribution		
Parameter	Symbol	Estimate
Threshold	Theta	-0.00834
Scale	Zeta	-6.14382
Shape	Sigma	0.882225
Mean		-0.00517
Std Dev		0.003438

Goodness-of-Fit Tests for Lognormal Distribution				
Test		Statistic		p Value
Kolmogorov-Smirnov	D	0.09419634	Pr > D	>0.500
Cramer-von Mises	W-Sq	0.02919815	Pr > W-Sq	>0.500
Anderson-Darling	A-Sq	0.21606642	Pr > A-Sq	>0.500

The *p*-values for all three tests are greater than 0.5, so the null hypothesis is not rejected. The tests support the conclusion that the two-parameter lognormal distribution with scale parameter $\hat{\zeta} = -6.14$ and shape parameter $\hat{\sigma} = 0.88$ provides a good model for the distribution of position deviations. For further discussion of goodness-of-fit interpretation, see the section "Goodness-of-Fit Tests" on page 428.

A sample program for this example, *uniex16.sas*, is available in the SAS Sample Library for Base SAS software.

Example 4.28: Creating a Normal Quantile Plot

This example illustrates how to create a normal quantile plot. An engineer is analyzing the distribution of distances between holes cut in steel sheets. The following statements save measurements of the distance between two holes cut into 50 steel sheets as values of the variable Distance in the data set Sheets:

```
data Sheets;
   input Distance @@;
   label Distance = 'Hole Distance (cm)';
   datalines;
 9.80 10.20 10.27  9.70  9.76
10.11 10.24 10.20 10.24  9.63
 9.99  9.78 10.10 10.21 10.00
 9.96  9.79 10.08  9.79 10.06
10.10  9.95  9.84 10.11  9.93
10.56 10.47  9.42 10.44 10.16
10.11 10.36  9.94  9.77  9.36
 9.89  9.62 10.05  9.72  9.82
 9.99 10.16 10.58 10.70  9.54
10.31 10.07 10.33  9.98 10.15
;
```

The engineer decides to check whether the distribution of distances is normal. The following statements create a Q-Q plot for Distance, shown in Output 4.28.1:

```
title 'Normal Quantile-Quantile Plot for Hole Distance';
ods graphics on;
proc univariate data=Sheets noprint;
   qqplot Distance / odstitle = title;
run;
```

The plot compares the ordered values of Distance with quantiles of the normal distribution. The linearity of the point pattern indicates that the measurements are normally distributed. Note that a normal Q-Q plot is created by default.

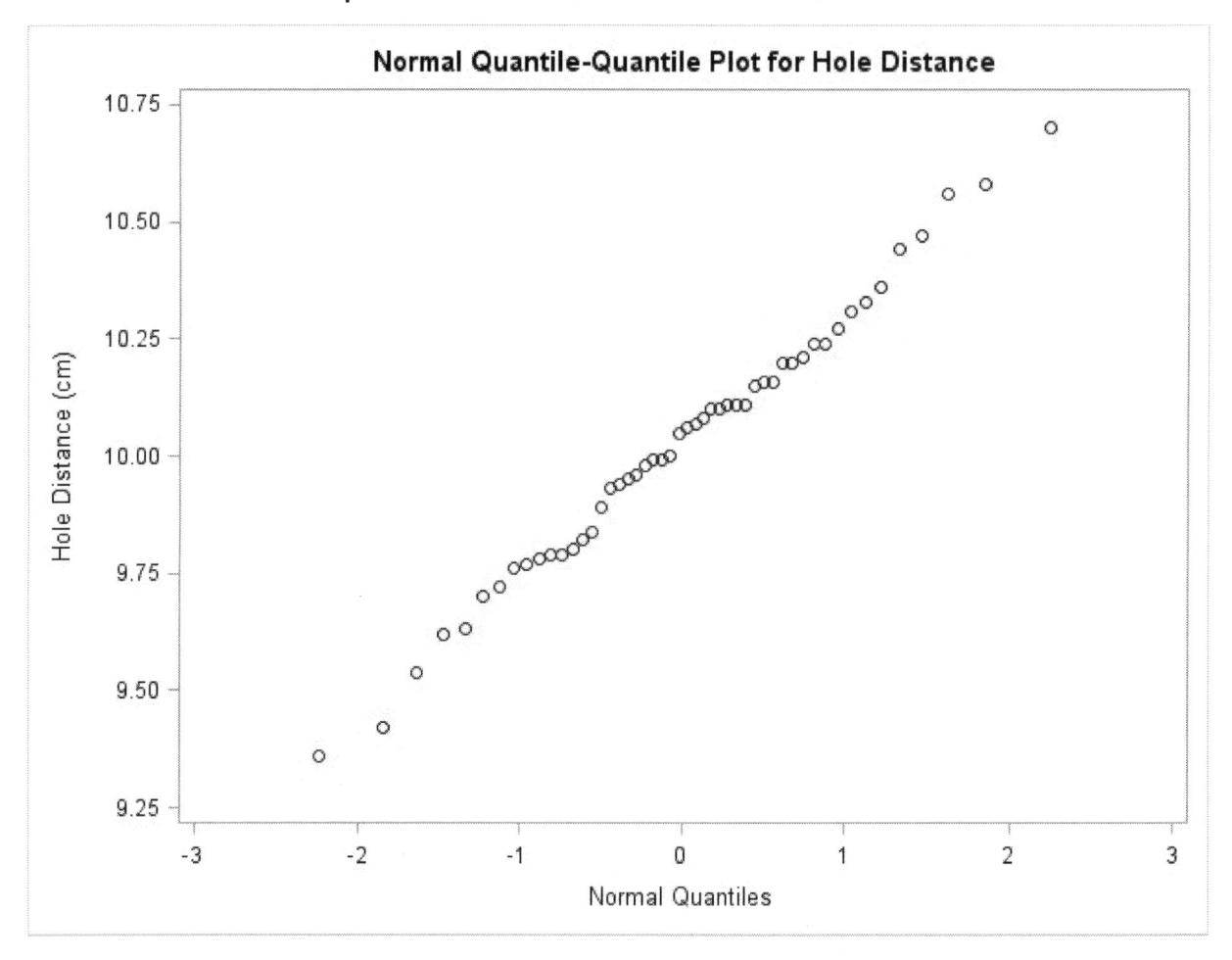

Output 4.28.1 Normal Quantile-Quantile Plot for Distance

A sample program for this example, *uniex17.sas*, is available in the SAS Sample Library for Base SAS software.

Example 4.29: Adding a Distribution Reference Line

This example, which is a continuation of Example 4.28, illustrates how to add a reference line to a normal Q-Q plot, which represents the normal distribution with mean μ_0 and standard deviation σ_0. The following statements reproduce the Q-Q plot in Output 4.28.1 and add the reference line:

```
title 'Normal Quantile-Quantile Plot for Hole Distance';
proc univariate data=Sheets noprint;
   qqplot Distance / normal(mu=est sigma=est)
                   odstitle = title
                   square;
run;
```

The plot is displayed in Output 4.29.1.

Specifying MU=EST and SIGMA=EST with the NORMAL primary option requests the reference line for which μ_0 and σ_0 are estimated by the sample mean and standard deviation. Alternatively, you can specify numeric values for μ_0 and σ_0 with the MU= and SIGMA= secondary options. The COLOR= and L= options specify the color and type of the line, and the SQUARE option displays the plot in a square format. The NOPRINT options in the PROC UNIVARIATE statement and after the NORMAL option suppress all the tables of statistical output produced by default.

Output 4.29.1 Adding a Distribution Reference Line to a Q-Q Plot

The data clearly follow the line, which indicates that the distribution of the distances is normal.

A sample program for this example, *uniex17.sas*, is available in the SAS Sample Library for Base SAS software.

Example 4.30: Interpreting a Normal Quantile Plot

This example illustrates how to interpret a normal quantile plot when the data are from a non-normal distribution. The following statements create the data set Measures, which contains the measurements of the diameters of 50 steel rods in the variable Diameter:

```
data Measures;
   input Diameter @@;
   label Diameter = 'Diameter (mm)';
   datalines;
5.501  5.251  5.404  5.366  5.445  5.576  5.607
5.200  5.977  5.177  5.332  5.399  5.661  5.512
5.252  5.404  5.739  5.525  5.160  5.410  5.823
5.376  5.202  5.470  5.410  5.394  5.146  5.244
5.309  5.480  5.388  5.399  5.360  5.368  5.394
5.248  5.409  5.304  6.239  5.781  5.247  5.907
5.208  5.143  5.304  5.603  5.164  5.209  5.475
5.223
;
```

The following statements request the normal Q-Q plot in Output 4.30.1:

```
title 'Normal Q-Q Plot for Diameters';
ods graphics on;
proc univariate data=Measures noprint;
   qqplot Diameter / normal
                     square
                     odstitle = title;
run;
```

The nonlinearity of the points in Output 4.30.1 indicates a departure from normality. Because the point pattern is curved with slope increasing from left to right, a theoretical distribution that is skewed to the right, such as a lognormal distribution, should provide a better fit than the normal distribution. The mild curvature suggests that you should examine the data with a series of lognormal Q-Q plots for small values of the shape parameter σ, as illustrated in Example 4.31. For details on interpreting a Q-Q plot, see the section "Interpretation of Quantile-Quantile and Probability Plots" on page 435.

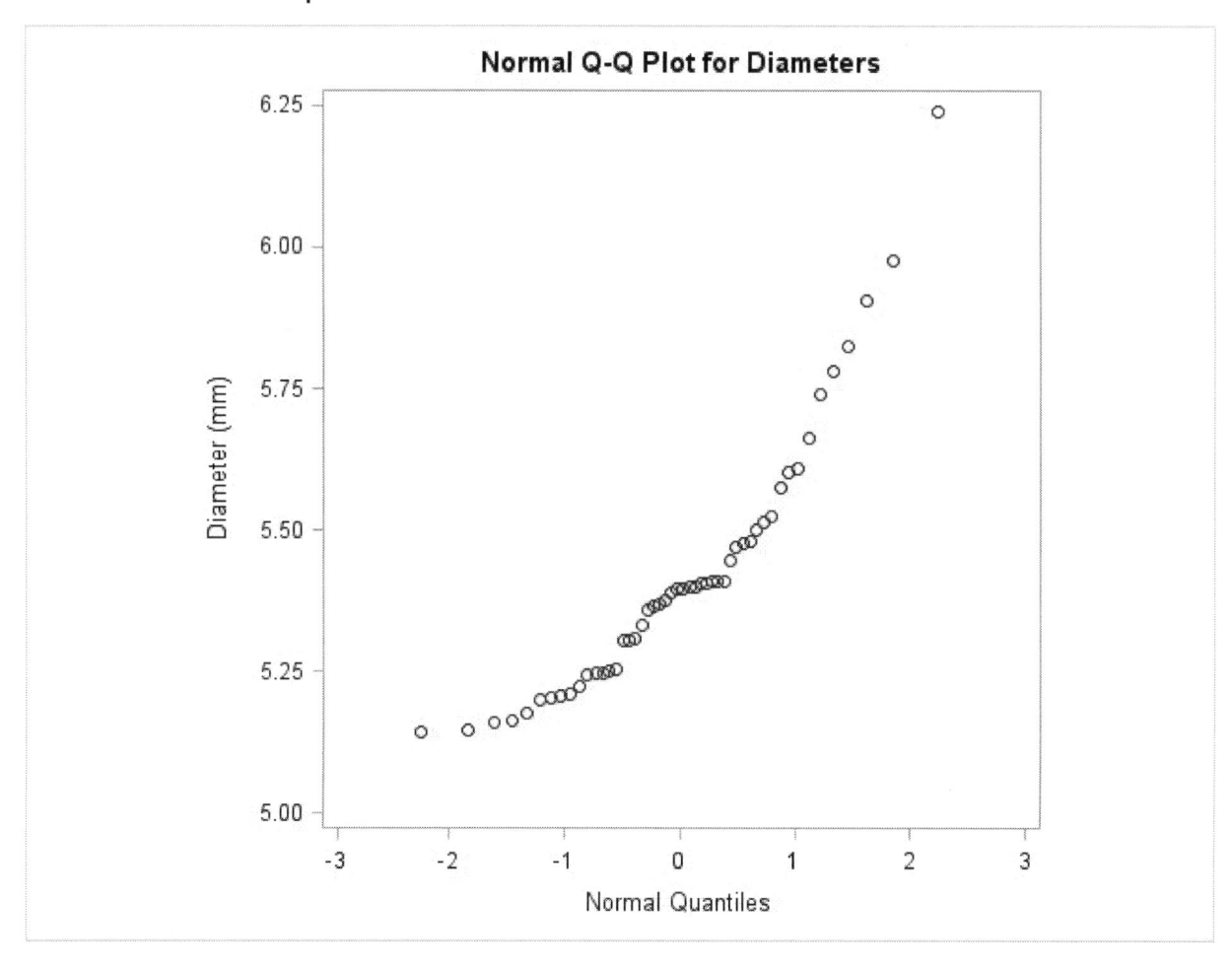

Output 4.30.1 Normal Quantile-Quantile Plot of Nonnormal Data

A sample program for this example, *uniex18.sas*, is available in the SAS Sample Library for Base SAS software.

Example 4.31: Estimating Three Parameters from Lognormal Quantile Plots

This example, which is a continuation of Example 4.30, demonstrates techniques for estimating the shape, location, and scale parameters, and the theoretical percentiles for a three-parameter lognormal distribution.

The three-parameter lognormal distribution depends on a threshold parameter θ, a scale parameter ζ, and a shape parameter σ. You can estimate σ from a series of lognormal Q-Q plots which use the SIGMA= secondary option to specify different values of σ; the estimate of σ is the value that linearizes the point pattern. You can then estimate the threshold and scale parameters from the intercept and slope of the point pattern. The following statements create the series of plots in Output 4.31.1, Output 4.31.2, and Output 4.31.3 for σ values of 0.2, 0.5, and 0.8, respectively:

Example 4.31: Estimating Three Parameters from Lognormal Quantile Plots ✦ 525

```
title 'Lognormal Q-Q Plot for Diameters';
proc univariate data=Measures noprint;
   qqplot Diameter / lognormal(sigma=0.2 0.5 0.8)
                    square
                    odstitle = title;
run;
```

NOTE: You must specify a value for the shape parameter σ for a lognormal Q-Q plot with the SIGMA= option or its alias, the SHAPE= option.

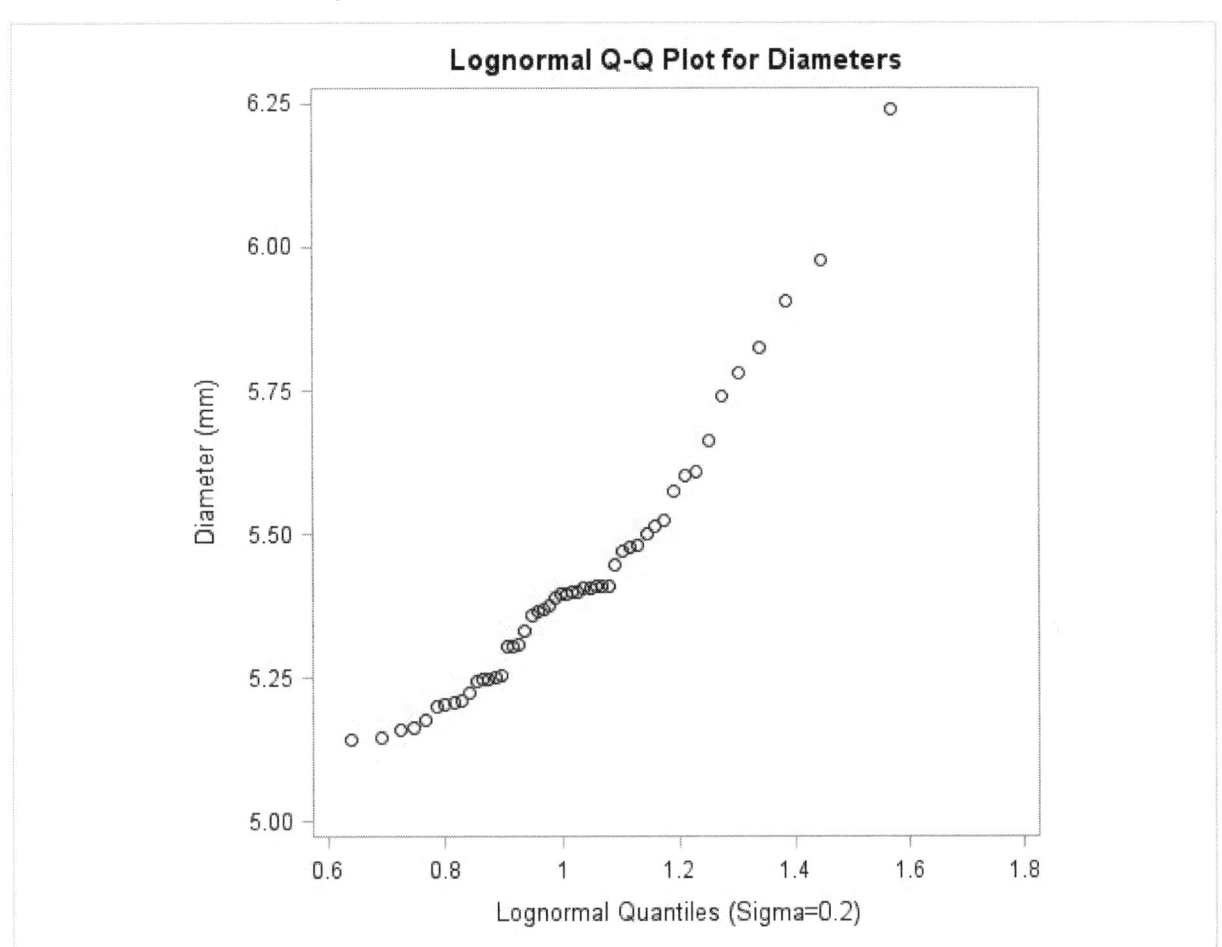

Output 4.31.1 Lognormal Quantile-Quantile Plot (σ =0.2)

Output 4.31.2 Lognormal Quantile-Quantile Plot ($\sigma = 0.5$)

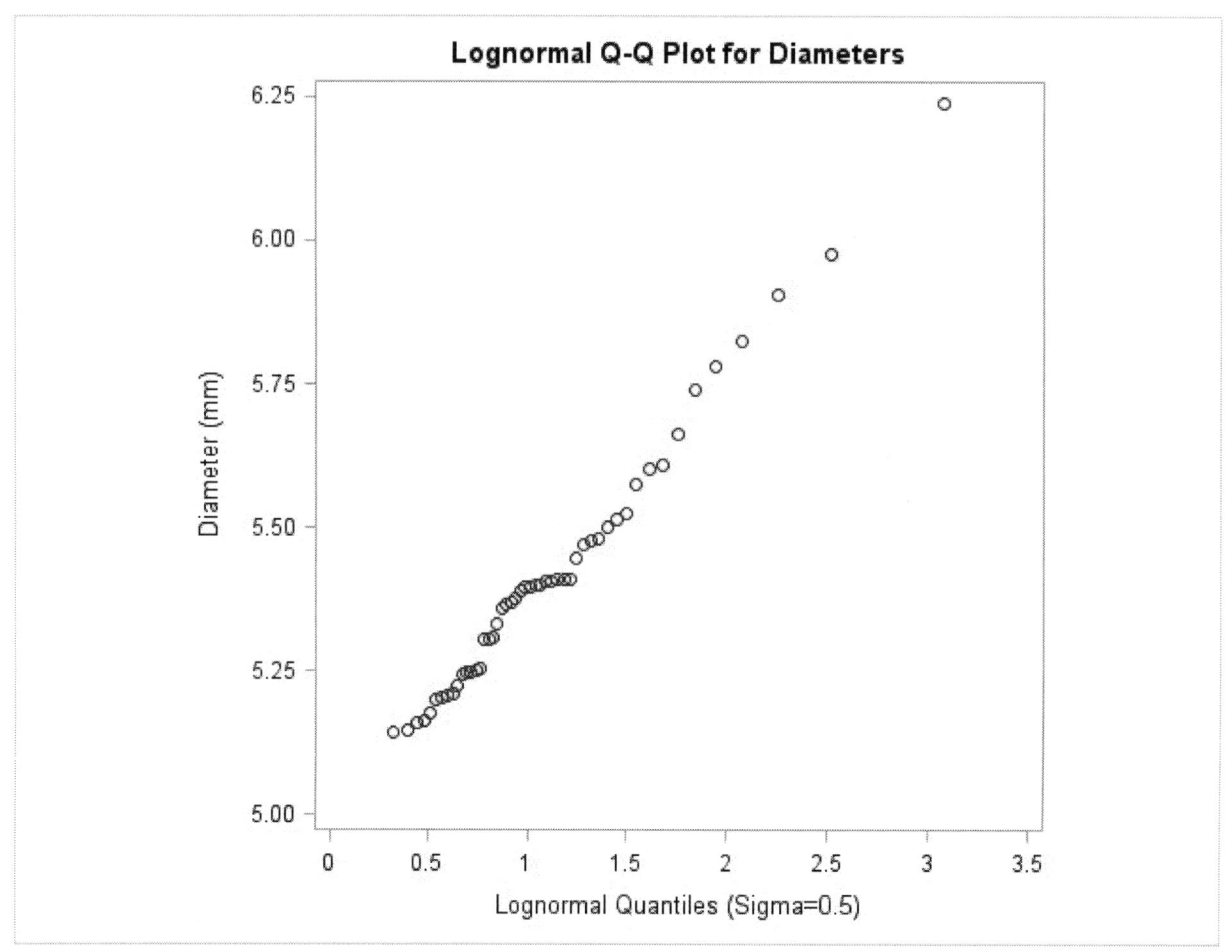

Example 4.31: Estimating Three Parameters from Lognormal Quantile Plots ♦ 527

Output 4.31.3 Lognormal Quantile-Quantile Plot ($\sigma = 0.8$)

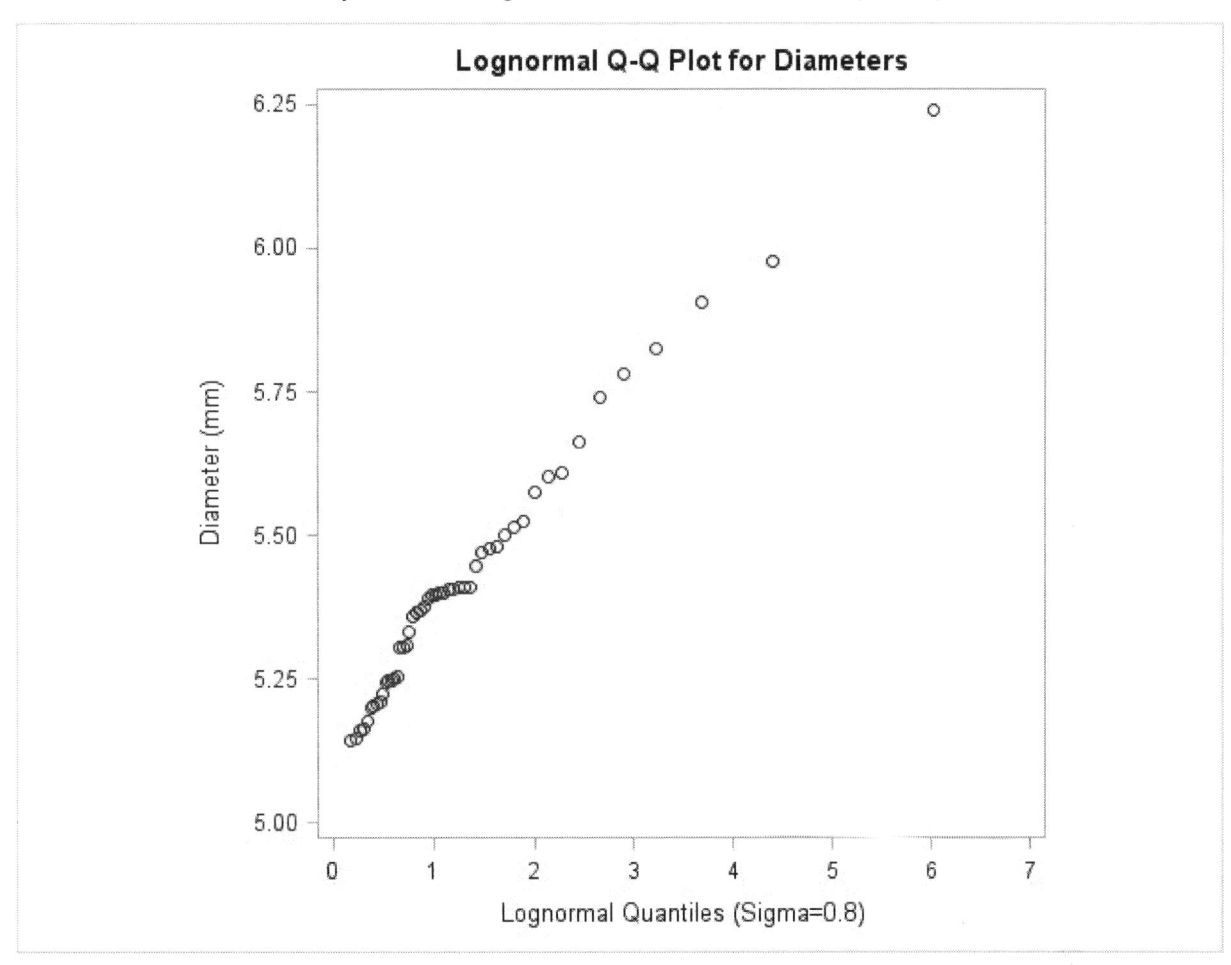

Lognormal Q-Q Plot for Diameters

The plot in Output 4.31.2 displays the most linear point pattern, indicating that the lognormal distribution with $\sigma = 0.5$ provides a reasonable fit for the data distribution.

Data with this particular lognormal distribution have the following density function:

$$p(x) = \begin{cases} \frac{\sqrt{2}}{\sqrt{\pi}(x-\theta)} \exp\left(-2(\log(x-\theta)-\zeta)^2\right) & \text{for } x > \theta \\ 0 & \text{for } x \leq \theta \end{cases}$$

The points in the plot fall on or near the line with intercept θ and slope $\exp(\zeta)$. Based on Output 4.31.2, $\theta \approx 5$ and $\exp(\zeta) \approx \frac{1.2}{3} = 0.4$, giving $\zeta \approx \log(0.4) \approx -0.92$.

You can also request a reference line by using the SIGMA=, THETA=, and ZETA= options together. The following statements produce the lognormal Q-Q plot in Output 4.31.4:

```
title 'Lognormal Q-Q Plot for Diameters';
proc univariate data=Measures noprint;
   qqplot Diameter / lognormal(theta=5 zeta=est sigma=est)
                  square
                  odstitle = title;
run;
```

Output 4.31.1 through Output 4.31.3 show that the threshold parameter θ is not equal to zero. Specifying THETA=5 overrides the default value of zero. The SIGMA=EST and ZETA=EST secondary options request estimates for σ and $\exp(\zeta)$ that use the sample mean and standard deviation.

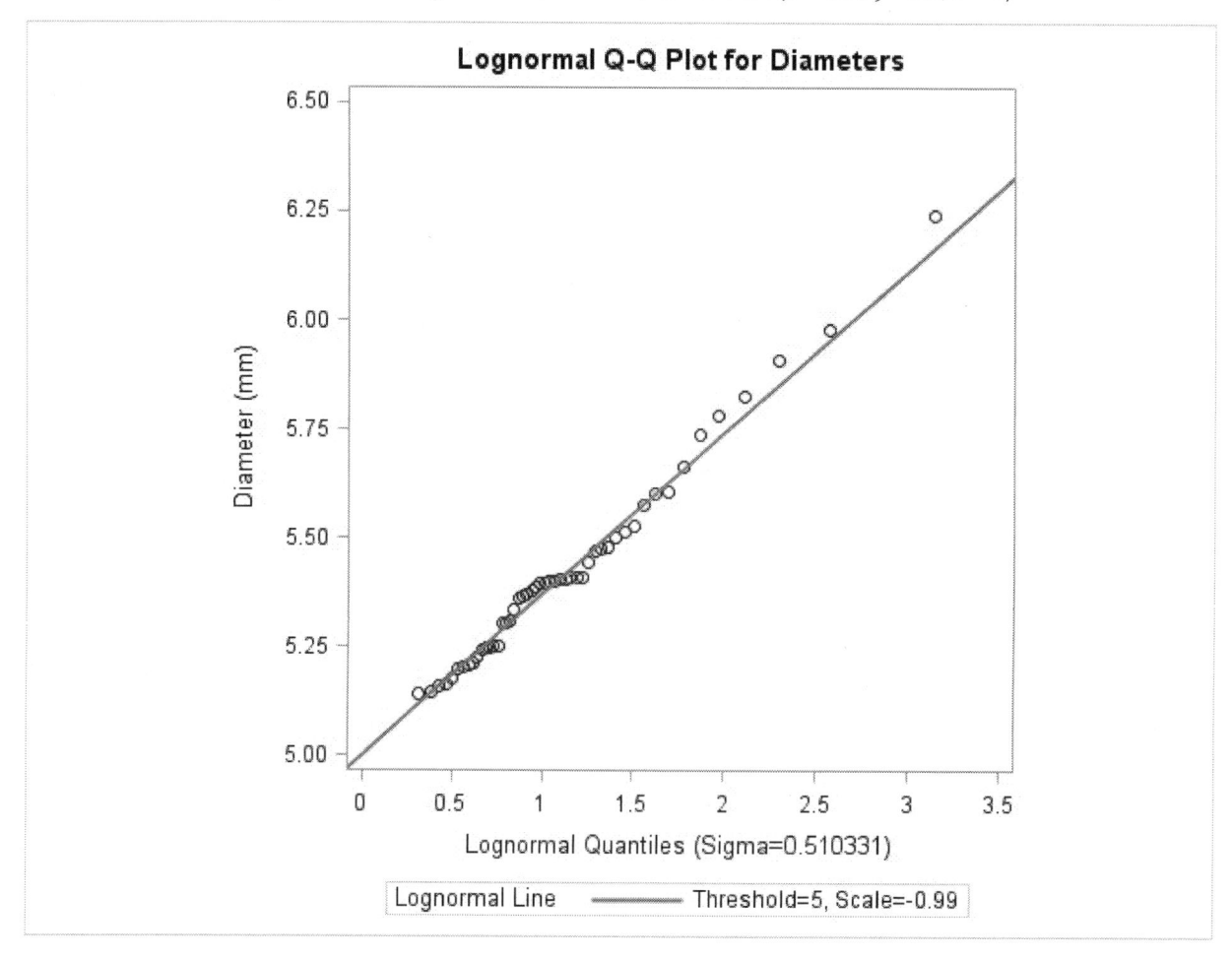

Output 4.31.4 Lognormal Quantile-Quantile Plot (σ =est, ζ =est, θ =5)

From the plot in Output 4.31.2, σ can be estimated as 0.51, which is consistent with the estimate of 0.5 derived from the plot in Output 4.31.2. Example 4.32 illustrates how to estimate percentiles by using lognormal Q-Q plots.

A sample program for this example, *uniex18.sas*, is available in the SAS Sample Library for Base SAS software.

Example 4.32: Estimating Percentiles from Lognormal Quantile Plots

This example, which is a continuation of Example 4.31, shows how to use a Q-Q plot to estimate percentiles such as the 95th percentile of the lognormal distribution. A probability plot can also be used for this purpose, as illustrated in Example 4.26.

The point pattern in Output 4.31.4 has a slope of approximately 0.39 and an intercept of 5. The following statements reproduce this plot, adding a lognormal reference line with this slope and intercept:

```
title 'Lognormal Q-Q Plot for Diameters';
proc univariate data=Measures noprint;
   qqplot Diameter / lognormal(sigma=0.5 theta=5 slope=0.39)
                     pctlaxis(grid)
                     vref     = 5.8 5.9 6.0
                     odstitle = title
                     square;
run;
```

The result is shown in Output 4.32.1.

Output 4.32.1 Lognormal Q-Q Plot Identifying Percentiles

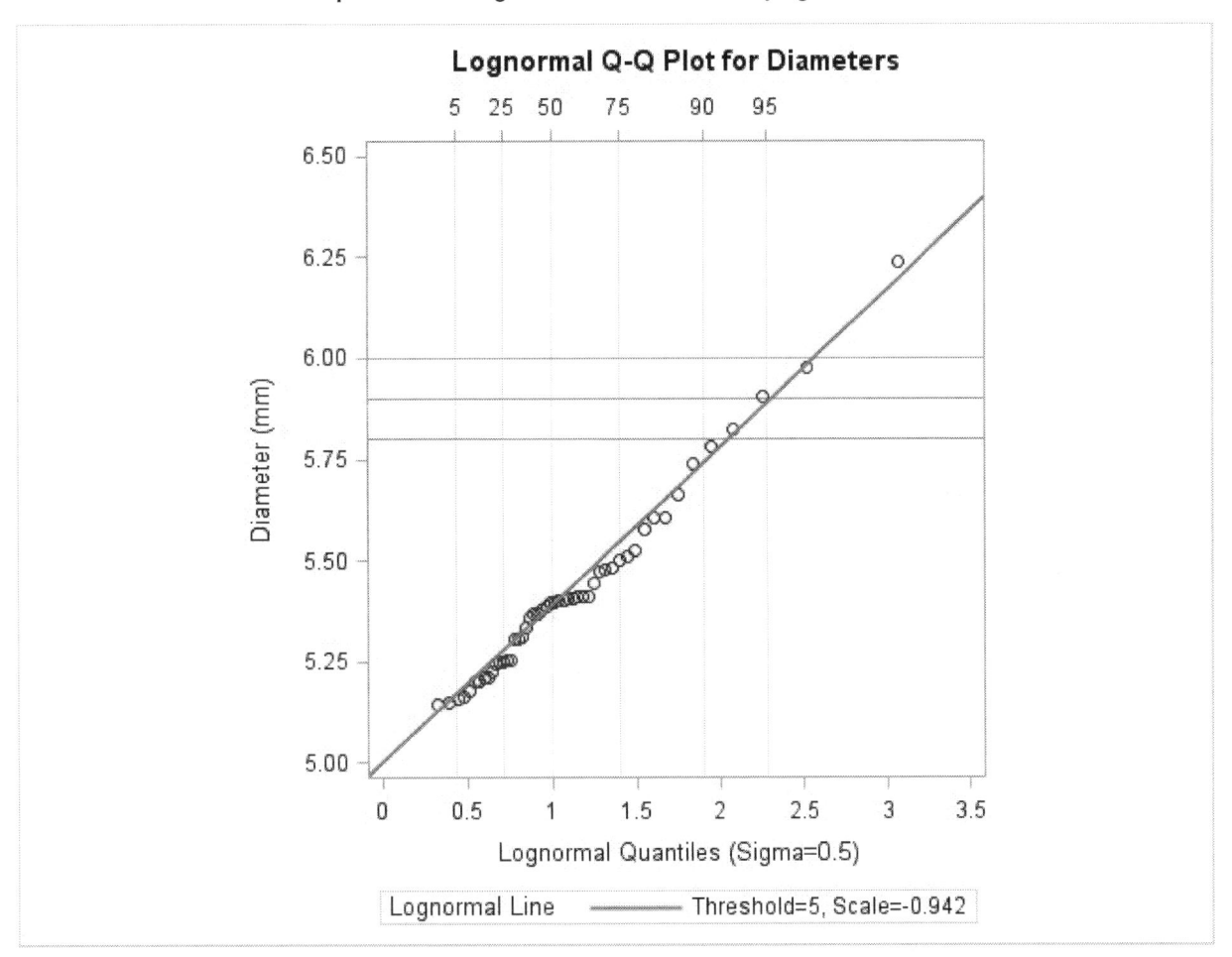

The PCTLAXIS option labels the major percentiles, and the GRID option draws percentile axis reference lines. The 95th percentile is 5.9, because the intersection of the distribution reference line and the 95th reference line occurs at this value on the vertical axis.

Alternatively, you can compute this percentile from the estimated lognormal parameters. The αth percentile of the lognormal distribution is

$$P_\alpha = \exp(\sigma \Phi^{-1}(\alpha) + \zeta) + \theta$$

where $\Phi^{-1}(\cdot)$ is the inverse cumulative standard normal distribution. Consequently,

$$\hat{P}_{0.95} = \exp\left(\tfrac{1}{2}\Phi^{-1}(0.95) + \log(0.39)\right) + 5 = 5.89$$

A sample program for this example, *uniex18.sas*, is available in the SAS Sample Library for Base SAS software.

Example 4.33: Estimating Parameters from Lognormal Quantile Plots

This example, which is a continuation of Example 4.31, demonstrates techniques for estimating the shape, location, and scale parameters, and the theoretical percentiles for a two-parameter lognormal distribution.

If the threshold parameter is known, you can construct a two-parameter lognormal Q-Q plot by subtracting the threshold from the data values and making a normal Q-Q plot of the log-transformed differences, as illustrated in the following statements:

```
data ModifiedMeasures;
   set Measures;
   LogDiameter = log(Diameter-5);
   label LogDiameter = 'log(Diameter-5)';
run;

title 'Two-Parameter Lognormal Q-Q Plot for Diameters';
proc univariate data=ModifiedMeasures noprint;
   qqplot LogDiameter / normal(mu=est sigma=est)
                        square
                        odstitle = title;
   inset n mean (5.3) std (5.3) /
       pos = nw header = 'Summary Statistics';
run;
```

Output 4.33.1 Two-Parameter Lognormal Q-Q Plot for Diameters

Because the point pattern in Output 4.33.1 is linear, you can estimate the lognormal parameters ζ and σ as the normal plot estimates of μ and σ, which are –0.99 and 0.51. These values correspond to the previous estimates of –0.92 for ζ and 0.5 for σ from Example 4.31. A sample program for this example, *uniex18.sas*, is available in the SAS Sample Library for Base SAS software.

Example 4.34: Comparing Weibull Quantile Plots

This example compares the use of three-parameter and two-parameter Weibull Q-Q plots for the failure times in months for 48 integrated circuits. The times are assumed to follow a Weibull distribution. The following statements save the failure times as the values of the variable Time in the data set Failures:

```
data Failures;
   input Time @@;
   label Time = 'Time in Months';
   datalines;
29.42 32.14 30.58 27.50 26.08 29.06 25.10 31.34
29.14 33.96 30.64 27.32 29.86 26.28 29.68 33.76
29.32 30.82 27.26 27.92 30.92 24.64 32.90 35.46
30.28 28.36 25.86 31.36 25.26 36.32 28.58 28.88
26.72 27.42 29.02 27.54 31.60 33.46 26.78 27.82
29.18 27.94 27.66 26.42 31.00 26.64 31.44 32.52
;
```

If no assumption is made about the parameters of this distribution, you can use the WEIBULL option to request a three-parameter Weibull plot. As in the previous example, you can visually estimate the shape parameter c by requesting plots for different values of c and choosing the value of c that linearizes the point pattern. Alternatively, you can request a maximum likelihood estimate for c, as illustrated in the following statements:

```
title 'Three-Parameter Weibull Q-Q Plot for Failure Times';
ods graphics on;
proc univariate data=Failures noprint;
   qqplot Time / weibull(c=est theta=est sigma=est)
              square
              href     = 0.5 1 1.5 2
              vref     = 25 27.5 30 32.5 35
              odstitle = title;
run;
```

NOTE: When using the WEIBULL option, you must either specify a list of values for the Weibull shape parameter c with the C= option or specify C=EST.

Output 4.34.1 displays the plot for the estimated value $\hat{c} = 1.99$. The reference line corresponds to the estimated values for the threshold and scale parameters of $\hat{\theta} = 24.19$ and $\hat{\sigma}_0 = 5.83$, respectively.

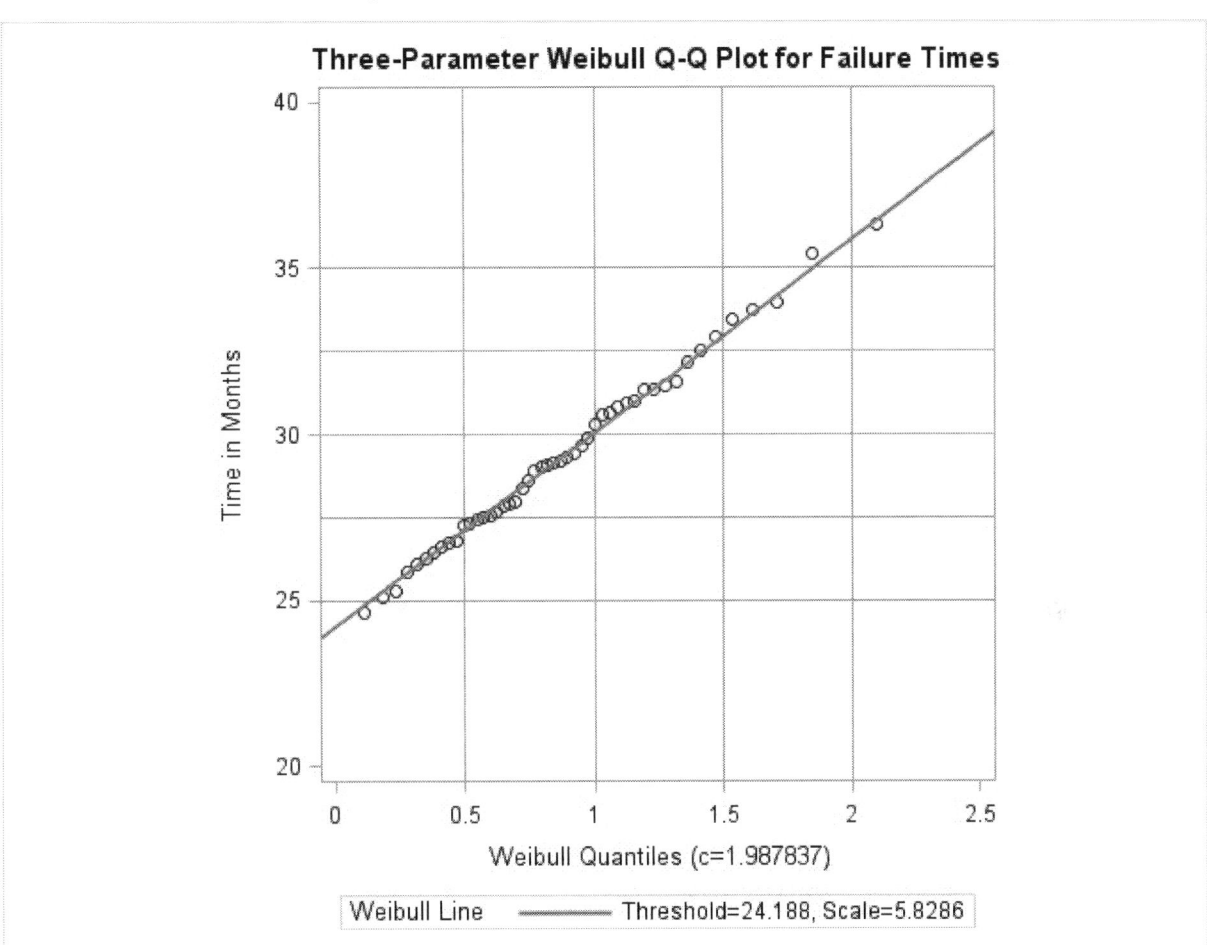

Output 4.34.1 Three-Parameter Weibull Q-Q Plot

[Figure: Three-Parameter Weibull Q-Q Plot for Failure Times. X-axis: Weibull Quantiles (c=1.987837), 0 to 2.5. Y-axis: Time in Months, 20 to 40. Legend: Weibull Line — Threshold=24.188, Scale=5.8286.]

Now, suppose it is known that the circuit lifetime is at least 24 months. The following statements use the known threshold value $\theta_0 = 24$ to produce the two-parameter Weibull Q-Q plot shown in Output 4.31.4:

```
title 'Two-Parameter Weibull Q-Q Plot for Failure Times';
proc univariate data=Failures noprint;
   qqplot Time / weibull(theta=24 c=est sigma=est)
             square
             vref    = 25 to 35 by 2.5
             href    = 0.5 to 2.0 by 0.5
             odstitle = title;
run;
```

The reference line is based on maximum likelihood estimates $\hat{c} = 2.08$ and $\hat{\sigma} = 6.05$.

Output 4.34.2 Two-Parameter Weibull Q-Q Plot for $\theta_0 = 24$

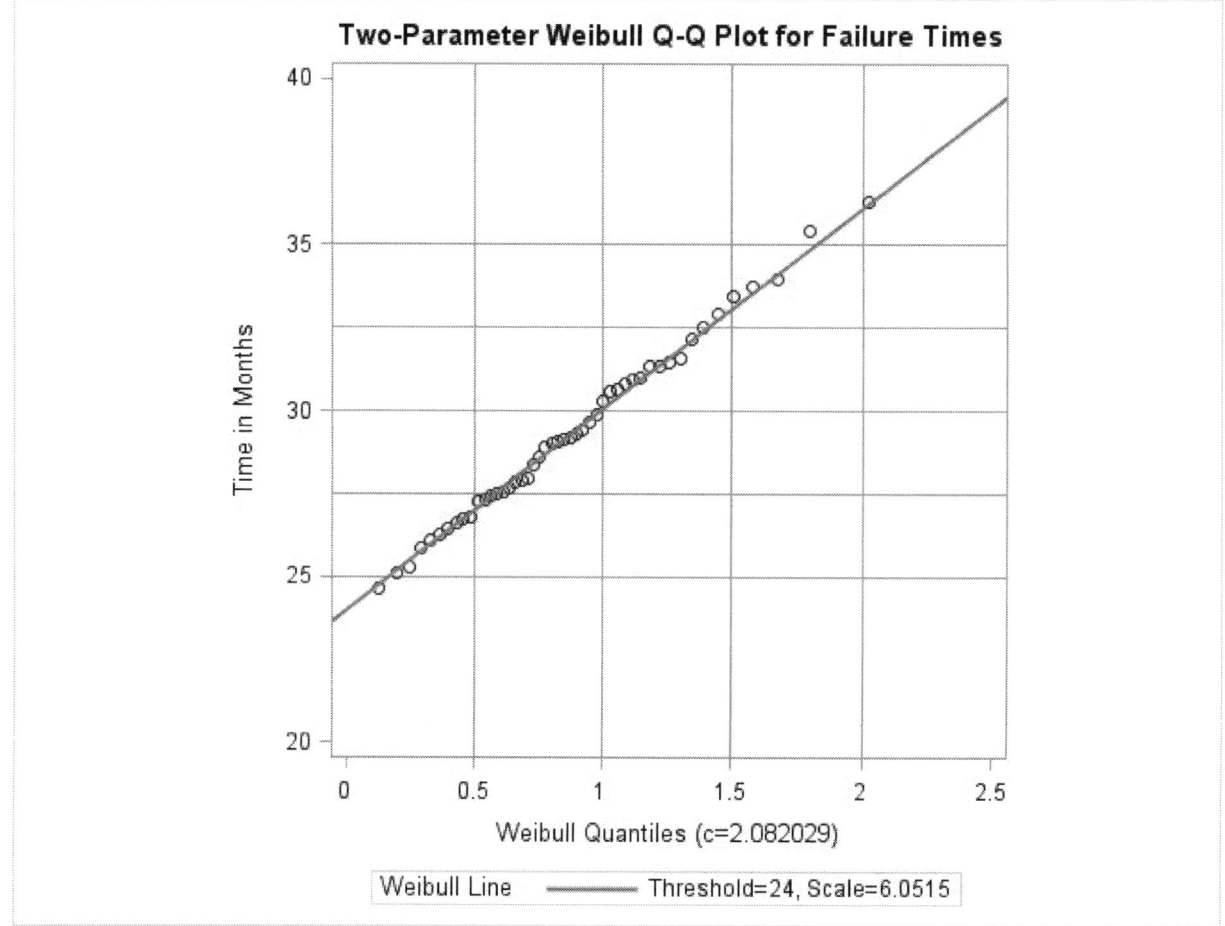

A sample program for this example, *uniex19.sas*, is available in the SAS Sample Library for Base SAS software.

Example 4.35: Creating a Cumulative Distribution Plot

A company that produces fiber-optic cord is interested in the breaking strength of the cord. The following statements create a data set named Cord, which contains 50 breaking strengths measured in pounds per square inch (PSI):

```
data Cord;
   label Strength="Breaking Strength (psi)";
   input Strength @@;
datalines;
6.94 6.97 7.11 6.95 7.12 6.70 7.13 7.34 6.90 6.83
7.06 6.89 7.28 6.93 7.05 7.00 7.04 7.21 7.08 7.01
7.05 7.11 7.03 6.98 7.04 7.08 6.87 6.81 7.11 6.74
6.95 7.05 6.98 6.94 7.06 7.12 7.19 7.12 7.01 6.84
6.91 6.89 7.23 6.98 6.93 6.83 6.99 7.00 6.97 7.01
;
```

You can use the CDFPLOT statement to fit any of six theoretical distributions (beta, exponential, gamma, lognormal, normal, and Weibull) and superimpose them on the cdf plot. The following statements use the NORMAL option to display a fitted normal distribution function on a cdf plot of breaking strengths:

```
title 'Cumulative Distribution Function of Breaking Strength';
ods graphics on;
proc univariate data=Cord noprint;
   cdf Strength / normal odstitle = title;
   inset normal(mu sigma);
run;
```

The NORMAL option requests the fitted curve. The INSET statement requests an inset containing the parameters of the fitted curve, which are the sample mean and standard deviation. For more information about the INSET statement, see "INSET Statement" on page 330. The resulting plot is shown in Output 4.35.1.

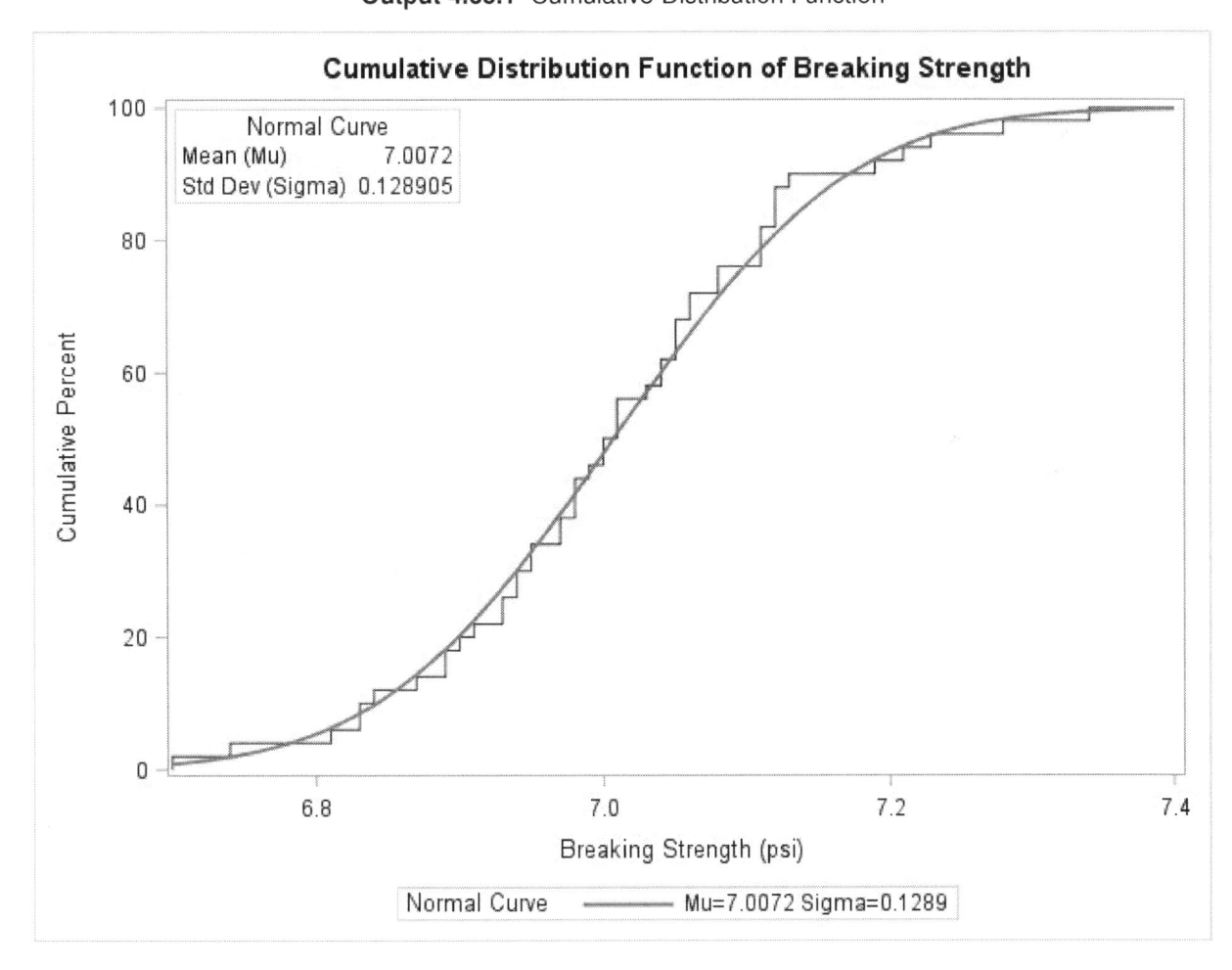

Output 4.35.1 Cumulative Distribution Function

The plot shows a symmetric distribution with observations concentrated 6.9 and 7.1. The agreement between the empirical and the normal distribution functions in Output 4.35.1 is evidence that the normal distribution is an appropriate model for the distribution of breaking strengths.

Example 4.36: Creating a P-P Plot

The distances between two holes cut into 50 steel sheets are measured and saved as values of the variable Distance in the following data set:

```
data Sheets;
   input Distance @@;
   label Distance='Hole Distance in cm';
   datalines;
 9.80 10.20 10.27  9.70  9.76
10.11 10.24 10.20 10.24  9.63
 9.99  9.78 10.10 10.21 10.00
 9.96  9.79 10.08  9.79 10.06
10.10  9.95  9.84 10.11  9.93
10.56 10.47  9.42 10.44 10.16
10.11 10.36  9.94  9.77  9.36
 9.89  9.62 10.05  9.72  9.82
 9.99 10.16 10.58 10.70  9.54
10.31 10.07 10.33  9.98 10.15
;
```

It is decided to check whether the distances are normally distributed. The following statements create a P-P plot, shown in Output 4.36.1, which is based on the normal distribution with mean $\mu = 10$ and standard deviation $\sigma = 0.3$:

```
title 'Normal Probability-Probability Plot for Hole Distance';
ods graphics on;
proc univariate data=Sheets noprint;
   ppplot Distance / normal(mu=10 sigma=0.3)
                     square
                     odstitle = title;
run;
```

The NORMAL option in the PPPLOT statement requests a P-P plot based on the normal cumulative distribution function, and the MU= and SIGMA= *normal-options* specify μ and σ. Note that a P-P plot is always based on a *completely specified* distribution—in other words, a distribution with specific parameters. In this example, if you did not specify the MU= and SIGMA= *normal-options*, the sample mean and sample standard deviation would be used for μ and σ.

Output 4.36.1 Normal P-P Plot with Diagonal Reference Line

The linearity of the pattern in Output 4.36.1 is evidence that the measurements are normally distributed with mean 10 and standard deviation 0.3. The SQUARE option displays the plot in a square format.

References

Blom, G. (1958). *Statistical Estimates and Transformed Beta Variables*. New York: John Wiley & Sons.

Bowman, K. O., and Shenton, L. R. (1983). "Johnson's System of Distributions." In *Encyclopedia of Statistical Sciences*, vol. 4, edited by S. Kotz, N. L. Johnson, and C. B. Read. New York: John Wiley & Sons.

Chambers, J. M., Cleveland, W. S., Kleiner, B., and Tukey, P. A. (1983). *Graphical Methods for Data Analysis*. Belmont, CA: Wadsworth International Group.

Cohen, A. C. (1951). "Estimating Parameters of Logarithmic-Normal Distributions by Maximum Likelihood." *Journal of the American Statistical Association* 46:206–212.

Conover, W. J. (1980). *Practical Nonparametric Statistics*. 2nd ed. New York: John Wiley & Sons.

Croux, C., and Rousseeuw, P. J. (1992). "Time-Efficient Algorithms for Two Highly Robust Estimators of Scale." *Computational Statistics* 1:411–428.

D'Agostino, R. B., and Stephens, M., eds. (1986). *Goodness-of-Fit Techniques*. New York: Marcel Dekker.

Dixon, W. J., and Tukey, J. W. (1968). "Approximate Behavior of the Distribution of Winsorized *t* (Trimming/Winsorization 2)." *Technometrics* 10:83–98.

Elandt, R. C. (1961). "The Folded Normal Distribution: Two Methods of Estimating Parameters from Moments." *Technometrics* 3:551–562.

Fisher, R. A. (1973). *Statistical Methods for Research Workers*. 14th ed. New York: Hafner Publishing.

Fowlkes, E. B. (1987). *A Folio of Distributions: A Collection of Theoretical Quantile-Quantile Plots*. New York: Marcel Dekker.

Hahn, G. J., and Meeker, W. Q. (1991). *Statistical Intervals: A Guide for Practitioners*. New York: John Wiley & Sons.

Hampel, F. R. (1974). "The Influence Curve and Its Role in Robust Estimation." *Journal of the American Statistical Association* 69:383–393.

Iman, R. L. (1974). "Use of a *t*-Statistic as an Approximation to the Exact Distribution of the Wilcoxon Signed Rank Statistic." *Communications in Statistics* 3:795–806.

Johnson, N. L., Kotz, S., and Balakrishnan, N. (1994). *Continuous Univariate Distributions*. 2nd ed. Vol. 1. New York: John Wiley & Sons.

Johnson, N. L., Kotz, S., and Balakrishnan, N. (1995). *Continuous Univariate Distributions*. 2nd ed. Vol. 2. New York: John Wiley & Sons.

Jones, M. C., Marron, J. S., and Sheather, S. J. (1996). "A Brief Survey of Bandwidth Selection for Density Estimation." *Journal of the American Statistical Association* 91:401–407.

Lehmann, E. L., and D'Abrera, H. J. M. (1975). *Nonparametrics: Statistical Methods Based on Ranks*. San Francisco: Holden-Day.

Odeh, R. E., and Owen, D. B. (1980). *Tables for Normal Tolerance Limits, Sampling Plans, and Screening.* New York: Marcel Dekker.

Owen, D. B., and Hua, T. A. (1977). "Tables of Confidence Limits on the Tail Area of the Normal Distribution." *Communications in Statistics—Simulation and Computation* 6:285–311.

Rousseeuw, P. J., and Croux, C. (1993). "Alternatives to the Median Absolute Deviation." *Journal of the American Statistical Association* 88:1273–1283.

Royston, J. P. (1992). "Approximating the Shapiro-Wilk W Test for Nonnormality." *Statistics and Computing* 2:117–119.

Shapiro, S. S., and Wilk, M. B. (1965). "An Analysis of Variance Test for Normality (Complete Samples)." *Biometrika* 52:591–611.

Silverman, B. W. (1986). *Density Estimation for Statistics and Data Analysis.* New York: Chapman & Hall.

Slifker, J. F., and Shapiro, S. S. (1980). "The Johnson System: Selection and Parameter Estimation." *Technometrics* 22:239–246.

Terrell, G. R., and Scott, D. W. (1985). "Oversmoothed Nonparametric Density Estimates." *Journal of the American Statistical Association* 80:209–214.

Tukey, J. W. (1977). *Exploratory Data Analysis.* Reading, MA: Addison-Wesley.

Tukey, J. W., and McLaughlin, D. H. (1963). "Less Vulnerable Confidence and Significance Procedures for Location Based on a Single Sample: Trimming/Winsorization 1." *Sankhyā, Series A* 25:331–352.

Velleman, P. F., and Hoaglin, D. C. (1981). *Applications, Basics, and Computing of Exploratory Data Analysis.* Boston: Duxbury Press.

Wainer, H. (1974). "The Suspended Rootogram and Other Visual Displays: An Empirical Validation." *American Statistician* 28:143–145.

Subject Index

AC1 agreement coefficient
    FREQ procedure, 211
adjusted odds ratio
    FREQ procedure, 216
adjusted relative risks
    FREQ procedure, 217
agreement plots
    FREQ procedure, 74
agreement, measures of
    FREQ procedure, 205
Agresti-Caffo confidence limits
    risk difference (FREQ), 180
Agresti-Coull confidence limits
    proportions (FREQ), 171
Anderson-Darling statistic, 430
Anderson-Darling test, 292
annotating
    histograms, 385
ANOVA (row mean scores) statistic
    Mantel-Haenszel (FREQ), 214
association, measures of
    FREQ procedure, 160

bar charts
    FREQ procedure, 134
Barnard's test
    FREQ procedure, 187
beta distribution, 416, 437
    cdf plots, 302
    deviation from theoretical distribution, 429
    EDF goodness-of-fit test, 429
    estimation of parameters, 319
    fitting, 319, 416
    formulas for, 416
    P-P plots, 351
    probability plots, 365, 437
    quantile plots, 376, 437
binomial proportions
    Clopper-Pearson test (FREQ), 174
    confidence limits (FREQ), 170
    equivalence tests (FREQ), 176
    exact test (FREQ), 174
    FREQ procedure, 170
    noninferiority tests (FREQ), 174
    superiority tests (FREQ), 176
    tests (FREQ), 173
    TOST (FREQ), 176
Blaker confidence limits
    proportions (FREQ), 171
Bowker's symmetry test
    FREQ procedure, 205, 206
box plots, line printer, 293, 406
    side-by-side, 292, 407
Breslow-Day test
    FREQ procedure, 218
    Tarone's adjustment (FREQ), 218

case-control studies
    odds ratio (FREQ), 192
categorical data analysis
    FREQ procedure, 64
cdf plots, 297
    axes, specifying, 308
    beta distribution, 303
    creating, 534
    example, 534
    exponential distribution, 303
    gamma distribution, 304
    gumbel distribution, 304
    igauss distribution, 305
    lognormal distribution, 305
    normal distribution, 306
    normal distribution, example, 534
    pareto distribution, 306
    power function distribution, 307
    rayleigh distribution, 307
    suppressing empirical cdf, 306
    suppressing legend, 306
    Weibull distribution, 308
cell count data
    example (FREQ), 244
    FREQ procedure, 148
chi-square goodness-of-fit test
    FREQ procedure, 155
chi-square tests
    FREQ procedure, 155
Cicchetti-Allison weights
    kappa coefficient (FREQ), 210
Clopper-Pearson confidence limits
    proportions (FREQ), 170
Cochran's Q test
    FREQ procedure, 205, 212
Cochran-Armitage test for trend
    FREQ procedure, 202
Cochran-Mantel-Haenszel statistics
    FREQ procedure, 212

cohort studies
    relative risks (FREQ), 195
common odds ratio
    exact confidence limits (FREQ), 219
    exact test (FREQ), 219
    logit (FREQ), 216
    Mantel-Haenszel (FREQ), 216
common relative risks
    logit (FREQ), 217
    Mantel-Haenszel (FREQ), 217
comparative plots, 309, 310, 411
    histograms, 325, 389, 482, 485, 495
concordant observations
    FREQ procedure, 160
confidence ellipse, 31
confidence limits
    exact (FREQ), 78
    for percentiles, 398
    means, for, 401
    measures of association (FREQ), 160
    parameters of normal distribution, for, 401
    proportions (FREQ), 170
    standard deviations, for, 401
    variances, for, 402
confidence limits for the correlation
    Fisher's z transformation, 25
contingency coefficient
    FREQ procedure, 160
contingency tables
    FREQ procedure, 64, 99
continuity-adjusted chi-square test
    FREQ procedure, 158
CORR procedure
    concepts, 18
    details, 18
    examples, 37
    missing values, 32
    ODS graph names, 36
    ODS table names, 35
    output, 33
    output data sets, 34
    overview, 4
    results, 32
    syntax, 8
    task tables, 8
corrected sums of squares and crossproducts, 9
correlation coefficients, 4
    limited combinations of, 18
    printing, for each variable, 9
    suppressing probabilities, 9
correlation statistic
    Mantel-Haenszel (FREQ), 214
covariances, 9
Cramér's V statistic
    FREQ procedure, 160
Cramér–von Mises statistic, 430
Cramér–von Mises test, 292
Cronbach's coefficient alpha, 29
    calculating and printing, 9
    example, 51
    for estimating reliability, 4
crosstabulation tables
    FREQ procedure, 64, 99, 232

data summarization tools, 280
density estimation, see kernel density estimation
descriptive statistics
    computing, 394
discordant observations
    FREQ procedure, 160
distribution of variables, 280
dot plots
    FREQ procedure, 134, 247

EDF, see empirical distribution function
EDF goodness-of-fit tests, 429
    probability values of, 431
empirical distribution function
    definition of, 429
    EDF test statistics, 429
equivalence tests
    binomial proportions, 176
    relative risk (FREQ), 200
    risk difference (FREQ), 187
exact confidence limits
    odds ratio (FREQ), 194
    proportion difference (FREQ), 182
    proportions (FREQ), 170
    ratio of proportions (FREQ), 198
    relative risks (FREQ), 198
    risk difference (FREQ), 182
exact p-values
    FREQ procedure, 223
exact tests
    computational algorithms (FREQ), 222
    computational resources (FREQ), 224
    FREQ procedure, 78, 221, 262
    Monte Carlo estimation (FREQ), 86
    network algorithm (FREQ), 222
exponential distribution, 417, 438
    cdf plots, 303
    deviation from theoretical distribution, 429
    EDF goodness-of-fit test, 429
    estimation of parameters, 322
    fitting, 417
    formulas for, 417
    P-P plots, 352, 353
    probability plots, 366, 438

quantile plots, 377, 438
extreme observations, 330, 457
extreme values, 457

Farrington-Manning test
    risk difference (FREQ), 185
Fisher's exact test
    FREQ procedure, 158
Fisher's z transformation, 9
Fisher's z transformation, 24
    applications, 26
    confidence limits for the correlation, 25
fitted parametric distributions, 416
    beta distribution, 416
    exponential distribution, 417
    folded normal distribution, 507
    gamma distribution, 418
    gumbel distribution, 419
    inverse Gaussian distribution, 419
    Johnson S_B distribution, 424
    Johnson S_U distribution, 426
    lognormal distribution, 420
    normal distribution, 421
    pareto distribution, 421
    power function distribution, 422
    rayleigh distribution, 423
    Weibull distribution, 427
Fleiss-Cohen weights
    kappa coefficient (FREQ), 210
folded normal distribution, 507
Freeman-Halton test
    FREQ procedure, 159
FREQ procedure
    AC1 agreement coefficient, 211
    adjusted odds ratio (Mantel-Haenszel), 216
    adjusted relative risks (Mantel-Haenszel), 217
    Agresti-Caffo confidence limits, 180
    Agresti-Coull confidence limits, 171
    ANOVA (row mean scores) statistic, 214
    bar charts, 134
    Barnard's test, 187
    binomial proportions, 170
    Blaker confidence limits, 171
    Bowker's symmetry test, 205, 206
    Breslow-Day test, 218
    cell count data, 148
    chi-square goodness-of-fit test, 155
    chi-square tests, 155
    Clopper-Pearson confidence limits, 170
    Cochran's Q test, 205, 212
    Cochran-Armitage test for trend, 202
    common odds ratio, 219
    computational resources, 226
    computational resources (exact tests), 224

contingency coefficient, 160
continuity-adjusted chi-square test, 158
correlation statistic, 214
Cramér's V statistic, 160
crosstabulation tables, 232
default tables, 99
displayed output, 229
dot plots, 134, 247
equivalence tests, 176
equivalence tests (relative risk), 200
equivalence tests (risk difference), 187
exact confidence limits, 78
exact p-values, 223
exact tests, 78, 221, 262
exact unconditional confidence limits, 182
Farrington-Manning test, 185
Fisher's exact test, 158
Freeman-Halton test, 159
Friedman's chi-square test, 266
Gail-Simon test, 221
gamma statistic, 160, 162
general association statistic, 215
grouping with formats, 149
Hauck-Anderson confidence limits, 180
in-database computation, 152
input data sets, 76, 148
introductory examples, 65
Jeffreys confidence limits, 172
Jonckheere-Terpstra test, 203
kappa coefficient, 205, 206
Kendall's tau-b statistic, 160, 162
lambda asymmetric, 160, 167
lambda symmetric, 160, 168
likelihood ratio chi-square test, 157
Likelihood ratio confidence limits, 172
Logit confidence limits, 172
Mantel-Fleiss criterion, 215
Mantel-Haenszel chi-square test, 158
Mantel-Haenszel statistics, 212
maximum time (exact tests), 86
McNemar's test, 205
measures of agreement, 205
measures of association, 160
Mid-p confidence limits, 172, 195
Miettinen-Nurminen confidence limits, 180
missing values, 150
Monte Carlo estimation (exact tests), 78, 86, 225
mosaic plots, 127
multiway tables, 231
network algorithm, 222
Newcombe confidence limits, 181, 186
noninferiority tests, 174
noninferiority tests (relative risk), 199
noninferiority tests (risk difference), 184

odds ratio, 192
ODS graph names, 243
ODS table names, 239
one-way frequency tables, 230
ordering of levels, 77
output data sets, 87, 226
overall kappa coefficient, 211
Pearson chi-square test, 156
Pearson correlation coefficient, 160, 164
phi coefficient, 159
polychoric correlation coefficient, 160, 166
prevalence-adjusted bias-adjusted kappa, 210
relative risks, 195
risk difference, 178
score confidence limits, 180, 197
scores, 154
simple kappa coefficient, 206
Somers' D statistics, 160, 163
Spearman rank correlation coefficient, 160, 165
standardized residuals, 156
Stuart's tau-c statistic, 160, 163
superiority tests, 176
superiority tests (relative risk), 200
superiority tests (risk difference), 186
tetrachoric correlation coefficient, 166
uncertainty coefficients, 160, 169
Wald confidence limits (risk difference), 182
weighted kappa coefficient, 205, 208
Wilson confidence limits, 173
Yule's Q statistic, 162
Zelen's exact test, 218
frequency plots
    FREQ procedure, 68
frequency tables
    creating (UNIVARIATE), 460
    FREQ procedure, 64, 99
    one-way (FREQ), 230
Friedman's chi-square test
    FREQ procedure, 266

Gail-Simon test
    FREQ procedure, 221
gamma distribution, 418, 438
    cdf plots, 304
    deviation from theoretical distribution, 429
    EDF goodness-of-fit test, 429
    estimation of parameters, 322
    fitting, 322, 418
    formulas for, 418
    P-P plots, 353, 354
    probability plots, 366, 438
    quantile plots, 377, 438
gamma statistic
    FREQ procedure, 160, 162

general association statistic
    Mantel-Haenszel (FREQ), 215
Gini's mean difference, 404
goodness-of-fit tests, 292, 428, *see* empirical
        distribution function, 500
    Anderson-Darling, 430
    Cramér–von Mises, 430
    EDF, 429, 431
    Kolmogorov D, 430
    Shapiro-Wilk, 428
graphics, 280
    annotating, 290
    descriptions, 386
    high-resolution, 280, 407
    insets, 330, 412–414
    line printer, 405
    naming, 388
    probability plots, 360
    quantile plots, 371
    saving, 291
Gumbel distribution, 438
    P-P plots, 354
    probability plots, 438
    quantile plots, 438
gumbel distribution, 419
    cdf plots, 304
    estimation of parameters, 323
    fitting, 323, 419
    formulas for, 419
    probability plots, 366
    quantile plots, 378

Hauck-Anderson confidence limits
    risk difference (FREQ), 180
high-resolution graphics, 280, 407
histograms, 311, 452
    adding a grid, 322
    annotating, 385
    appearance, 319, 320, 324, 326, 328–330,
        385–389, 391, 392
    axis color, 385
    axis scaling, 329
    bar labels, 319
    bar width, 319, 325
    bars, suppressing, 326
    beta curve, superimposed, 319
    binning, 489
    color, options, 320, 385, 386
    comparative, 325, 389, 482, 485, 495
    creating, 480
    endpoints of intervals, 328
    exponential curve, superimposed, 322
    extreme observations, 457
    filling area under density curve, 322

gamma curve, superimposed, 322
gumbel curve, superimposed, 323
hanging, 323
insets, 487
intervals, 327, 452
inverse Gaussian curve, superimposed, 324
Johnson S_B curve, superimposed, 328
Johnson S_U curve, superimposed, 329
kernel density estimation, options, 320, 324, 325, 329
kernel density estimation, superimposed, 432, 504
line type, 388
lognormal curve, superimposed, 324
lognormal distribution, 517
midpoints, 325
multiple distributions, example, 500
normal curve, superimposed, 326
normal distribution, 326
output data sets, 446, 447
parameters for fitted density curves, 319, 320, 326, 328–330
pareto curve, superimposed, 327
plots, suppressing, 326
power curve, superimposed, 328
quantiles, 452
rayleigh curve, superimposed, 328
reference lines, options, 322, 385–388, 392
saving histogram intervals, 327
suppressing legend, 326
tables of statistical output, 452
tables of statistical output, suppressing, 326
three-parameter lognormal distribution, superimposed, 506
three-parameter Weibull distribution, superimposed, 506
tick marks on horizontal axis, 387
tiles for comparative plots, 388
Weibull curve, superimposed, 330
Hoeffding's measure of dependence, 4, 22
calculating and printing, 9
example, 37
output data set with, 9
probability values, 22
hypothesis tests
exact (FREQ), 78

igauss distribution
cdf plots, 305
in-database computation
FREQ procedure, 152
insets, 330, 487
appearance, 339
appearance, color, 338, 339
positioning, 339, 340, 412

positioning in margins, 413
positioning with compass point, 412
positioning with coordinates, 414
statistics associated with distributions, 335
insets for descriptive statistics, *see* insets
interquartile range, 404
inverse Gaussian distribution, 419
estimation of parameters, 324
fitting, 324, 419
formulas for, 419
P-P plots, 354, 355

Jeffreys confidence limits
proportions (FREQ), 172
Johnson S_B distribution, 424
estimation of parameters, 328
fitting, 328, 424
formulas for, 424
Johnson S_U distribution, 426
estimation of parameters, 329
fitting, 329, 426
formulas for, 426
Jonckheere-Terpstra test
FREQ procedure, 203

kappa coefficient
FREQ procedure, 205, 206
weights (FREQ), 210
Kendall correlation statistics, 9
Kendall's partial tau-b, 4, 17
Kendall's tau-b, 4, 21
probability values, 21
Kendall's tau-*b* statistic
FREQ procedure, 160, 162
kernel density estimation, 432, 504
adding density curve to histogram, 324
bandwidth parameter, specifying, 320
kernel function, specifying type of, 324
line type for density curve, 388
lower bound, specifying, 325
upper bound, specifying, 329
kernel function, *see* kernel density estimation
key cell for comparative plots, 310
Kolmogorov D statistic, 430
Kolmogorov-Smirnov test, 292

lambda asymmetric
FREQ procedure, 160, 167
lambda symmetric
FREQ procedure, 160, 168
likelihood ratio chi-square test
FREQ procedure, 157
Likelihood ratio confidence limits
proportions (FREQ), 172
line printer plots, 405

box plots, 406, 407
normal probability plots, 406
stem-and-leaf plots, 405
listwise deletion, 32
location estimates
robust, 295, 296
location parameters, 442
probability plots, estimation with, 442
quantile plots, estimation with, 442
location, tests for
UNIVARIATE procedure, 478
Logit confidence limits
proportions (FREQ), 172
lognormal distribution, 420, 439
cdf plots, 305
deviation from theoretical distribution, 429
EDF goodness-of-fit test, 429
estimation of parameters, 324
fitting, 324, 420
formulas for, 420
histograms, 506, 517
P-P plots, 355, 356
probability plots, 367, 439, 513
quantile plots, 378, 439, 530

Mantel-Fleiss criterion
FREQ procedure, 215
Mantel-Haenszel chi-square test
FREQ procedure, 158
Mantel-Haenszel statistics
ANOVA (row mean scores) statistic (FREQ), 214
correlation statistic (FREQ), 214
FREQ procedure, 212
general association statistic (FREQ), 215
Mantel-Fleiss criterion (FREQ), 215
McNemar's test
FREQ procedure, 205
measures of association, 37
exact tests (FREQ), 161
nonparametric, 4
tests (FREQ), 161
measures of location
means, 394
modes, 397, 456
trimmed means, 403
Winsorized means, 402
median absolute deviation (MAD), 404
Mehta-Patel network algorithm
exact tests (FREQ), 222
Mid-p confidence limits
odds ratio (FREQ), 195
Mid-p confidence limits
proportions (FREQ), 172
Miettinen-Nurminen confidence limits

risk difference (FREQ), 180
missing values
FREQ procedure, 150
UNIVARIATE procedure, 392
mode calculation, 397
modified ridit scores
FREQ procedure, 154
Monte Carlo estimation
exact tests (FREQ), 78, 86, 225
mosaic plots
FREQ procedure, 127
multiway tables
FREQ procedure, 64, 99, 231

network algorithm
exact tests (FREQ), 222
Newcombe confidence limits
risk difference (FREQ), 181, 186
Newton-Raphson approximation
gamma shape parameter, 384
noninferiority tests
binomial proportions, 174
relative risk (FREQ), 199
risk difference (FREQ), 184
nonparametric density estimation, *see* kernel density estimation
nonparametric measures of association, 4
normal distribution, 421, 439
cdf plots, 306
cdf plots, example, 534
deviation from theoretical distribution, 429
EDF goodness-of-fit test, 429
estimation of parameters, 326
fitting, 326, 421
formulas for, 421
histograms, 326
P-P plots, 356
probability plots, 360, 367, 439
quantile plots, 379, 439, 519
normal probability plots, *see* probability plots
line printer, 293, 406

odds ratio
Breslow-Day test (FREQ), 218
case-control studies (FREQ), 192
confidence limits (FREQ), 193, 196
exact confidence limits (FREQ), 194
FREQ procedure, 192
likelihood ratio confidence limits (FREQ), 194
logit adjusted (FREQ), 216
Mantel-Haenszel adjusted (FREQ), 216
mid-p confidence limits (FREQ), 195
score confidence limits (FREQ), 193
Wald (log) confidence limits (FREQ), 193

Zelen's exact test (FREQ), 218
ODS (Output Delivery System)
    CORR procedure and, 35
    UNIVARIATE procedure table names, 451
ODS graph names
    CORR procedure, 36
    FREQ procedure, 243
output data sets
    saving correlations in, 53

P-P plots, 346
    beta distribution, 351, 352
    distribution options, 348, 350
    distribution reference line, 348, 537
    exponential distribution, 352
    gamma distribution, 353
    Gumbel distribution, 354
    gumbel distribution, 354
    inverse Gaussian distribution, 354, 355
    lognormal distribution, 355, 356
    normal distribution, 356
    options summarized by function, 350, 351
    Pareto distribution, 357
    plot layout, 350
    Power distribution, 357
    Rayleigh distribution, 358
    Weibull distribution, 359
paired data, 400, 479
pairwise deletion, 32
parameters for fitted density curves, 319, 320, 326, 328–330
Pareto distribution, 439
    P-P plots, 357
    probability plots, 439
    quantile plots, 439
pareto distribution, 421
    cdf plots, 306
    estimation of parameters, 327
    fitting, 327, 421
    formulas for, 421
    probability plots, 367
    quantile plots, 379
partial correlations, 22
    probability values, 24
Pearson chi-square test
    FREQ procedure, 156
Pearson correlation coefficient
    FREQ procedure, 160, 164
Pearson correlation statistics, 4
    example, 37
    in output data set, 9
    Pearson partial correlation, 4, 17
    Pearson product-moment correlation, 4, 9, 18, 37

Pearson weighted product-moment correlation, 4, 17
    probability values, 20
    suppressing, 9
percent plots, see P-P plots
percentiles
    axes, quantile plots, 379, 380, 443
    calculating, 397
    confidence limits for, 398, 475
    defining, 293, 397
    empirical distribution function, 397
    options, 344–346
    probability plots and, 360
    quantile plots and, 371
    saving to an output data set, 472
    visual estimates, probability plots, 443
    visual estimates, quantile plots, 443
    weighted, 398
    weighted average, 397
phi coefficient
    FREQ procedure, 159
plot statements, UNIVARIATE procedure, 279
plots
    box plots, 292, 293, 406, 407
    comparative, 309, 310, 411
    comparative histograms, 325, 389, 482, 485, 495
    line printer plots, 405
    normal probability plots, 293, 406
    probability plots, 360, 436
    quantile plots, 371, 436
    reference lines, options, 391
    size of, 294
    stem-and-leaf, 293, 405
    suppressing, 326
polychoric correlation, 27
polychoric correlation coefficient
    FREQ procedure, 160, 166
polyserial correlation, 27
polyserial correlation statistics, 4, 9
Power distribution
    P-P plots, 357
power function distribution, 422, 440
    cdf plots, 307
    estimation of parameters, 328
    fitting, 328, 422
    formulas for, 422
    probability plots, 368, 440
    quantile plots, 380, 440
prediction ellipse, 31
prevalence-adjusted bias-adjusted kappa
    FREQ procedure, 210
probability plots, 360
    appearance, 366, 368
    beta distribution, 365, 437

distribution reference lines, 369
distributions for, 436
exponential distribution, 366, 438
gamma distribution, 366, 438
generalized Pareto distribution, 439, 440
Gumbel distribution, 438
gumbel distribution, 366
location parameters, estimation of, 442
lognormal distribution, 367, 439, 513, 517
normal distribution, 360, 367, 439
overview, 360
parameters for distributions, 365–370
Pareto distribution, 439
pareto distribution, 367
percentile axis, 368
percentiles, estimates of, 443
power function distribution, 368, 440
Rayleigh distribution, 440
rayleigh distribution, 368
reference lines, 366
reference lines, options, 366
scale parameters, estimation of, 442
shape parameters, 441
suppressing legend, 367
three-parameter Weibull distribution, 440
threshold parameter, 370
threshold parameters, estimation of, 442
two-parameter Weibull distribution, 441
Weibull distribution, 370
proportion difference
FREQ procedure, 178
proportions, *see* binomial proportions

Q-Q plots, *see* quantile plots
Q_n, 404
quantile plots, 371
appearance, 377, 378, 381
axes, percentile scale, 379, 380, 443
beta distribution, 376, 437
creating, 434
diagnostics, 435
distribution reference lines, 381, 521
distributions for, 436
exponential distribution, 377, 438
gamma distribution, 377, 438
Gumbel distribution, 438
gumbel distribution, 378
interpreting, 435
legends, suppressing (UNIVARIATE), 521
location parameters, estimation of, 442
lognormal distribution, 378, 439, 524, 530
lognormal distribution, percentiles, 528
nonnormal data, 523
normal distribution, 379, 439, 519

overview, 371
parameters for distributions, 376–378, 380–382
Pareto distribution, 439
pareto distribution, 379
percentiles, estimates of, 443
power function distribution, 380, 440
Rayleigh distribution, 440
rayleigh distribution, 380
reference lines, 377, 379
reference lines, options, 378
scale parameters, estimation of, 442
shape parameters, 441
three-parameter Weibull distribution, 440
threshold parameter, 382
threshold parameters, estimation of, 442
two-parameter Weibull distribution, 441
Weibull distribution, 382, 532
quantile-quantile plots, *see* quantile plots
quantiles
defining, 397
empirical distribution function, 397
histograms and, 452
weighted average, 397
quantiles plots
suppressing legend, 378

rank scores
FREQ procedure, 154
Rayleigh distribution, 440
P-P plots, 358
probability plots, 440
quantile plots, 440
rayleigh distribution, 423
cdf plots, 307
estimation of parameters, 328
fitting, 328, 423
formulas for, 423
probability plots, 368
quantile plots, 380
relative risk
equivalence tests (FREQ), 200
likelihood ratio confidence limits (FREQ), 197
noninferiority tests (FREQ), 199
superiority tests (FREQ), 200
tests (FREQ), 199
Wald confidence limits (FREQ), 196
relative risks
cohort studies (FREQ), 195
exact confidence limits (FREQ), 198
FREQ procedure, 195
logit adjusted (FREQ), 217
Mantel-Haenszel adjusted (FREQ), 217
reliability estimation, 4
ridit scores

FREQ procedure, 154
risk difference
    confidence limits (FREQ), 179
    equivalence tests (FREQ), 187
    exact confidence limits (FREQ), 182
    FREQ procedure, 178
    noninferiority tests (FREQ), 184
    superiority tests (FREQ), 186
    tests (FREQ), 183
    TOST (FREQ), 187, 201
risks, *see also* binomial proportions
    FREQ procedure, 178
robust estimates, 294–296
robust estimators, 402, 476
    Gini's mean difference, 404
    interquartile range, 404
    median absolute deviation (MAD), 404
    Q_n, 404
    S_n, 404
    trimmed means, 403
    Winsorized means, 402
robust measures of scale, 404
rounding, 294, 393
    UNIVARIATE procedure, 393
row mean scores statistic
    Mantel-Haenszel (FREQ), 214

saving correlations
    example, 53
scale estimates
    robust, 294
scale parameters, 442
    probability plots, 442
    quantile plots, 442
score confidence limits
    odds ratio (FREQ), 193
    relative risk (FREQ), 197
    risk difference (FREQ), 180
shape parameters, 441
Shapiro-Wilk statistic, 428
Shapiro-Wilk test, 292
sign test, 399, 400
    paired data and, 479
signed rank statistic, computing, 401
singularity of variables, 9
smoothing data distribution, *see* kernel density
        estimation
S_n, 404
Somers' *D* statistics
    FREQ procedure, 160, 163
Spearman correlation statistics, 9
    probability values, 20
    Spearman partial correlation, 4, 17
    Spearman rank-order correlation, 4, 20, 37

Spearman rank correlation coefficient
    FREQ procedure, 160, 165
standard deviation, 9
    specifying, 306
standardized residuals
    FREQ procedure, 156
stem-and-leaf plots, 293, 405
stratified analysis
    FREQ procedure, 64, 99
Stuart's tau-*c* statistic
    FREQ procedure, 160, 163
Student's *t* test, 399, 400
summary statistics
    insets of, 330
    saving, 293, 448
sums of squares and crossproducts, 9
superiority tests
    binomial proportions, 176
    relative risk (FREQ), 200
    risk difference (FREQ), 186

t test
    Student's, 399, 400
table scores
    FREQ procedure, 154
tables
    contingency (FREQ), 64
    crosstabulation (FREQ), 64, 232
    multiway (FREQ), 64, 231
    one-way frequency (FREQ), 64, 230
Tarone's adjustment
    Breslow-Day test (FREQ), 218
tests for location, 399, 478
    paired data, 400, 479
    sign test, 400
    Student's *t* test, 400
    Wilcoxon signed rank test, 401
tetrachoric correlation, 27
tetrachoric correlation coefficient
    FREQ procedure, 166
theoretical distributions, 436
three-parameter Weibull distribution, 440
    probability plots, 440
    quantile plots, 440
threshold parameter
    probability plots, 370
    quantile plots, 382
threshold parameters
    probability plots, 442
    quantile plots, 442
tiles for comparative plots
    histograms, 388
TOST
    equivalence tests (FREQ), 176, 187, 201

trend test
    FREQ procedure, 202
trimmed means, 295, 403
two-parameter Weibull distribution, 441
    probability plots, 441
    quantile plots, 441

uncertainty coefficients
    FREQ procedure, 160, 169
univariate analysis
    for multiple variables, 454
UNIVARIATE procedure
    basic summary plots, 461
    calculating modes, 456
    classification levels, 309
    comparative plots, 309, 310, 411
    computational resources, 453
    concepts, 392
    confidence limits, 290, 401, 473
    descriptive statistics, 394, 454
    examples, 454
    extreme observations, 330, 457
    extreme values, 457
    fitted continuous distributions, 416
    frequency variables, 469
    goodness-of-fit tests, 428
    high-resolution graphics, 407
    histograms, 452, 457
    insets for descriptive statistics, 330
    keywords for insets, 330
    keywords for output data sets, 340
    line printer plots, 405
    missing values, 309, 392
    mode calculation, 397
    normal probability plots, 406
    ODS graph names, 452
    ODS table names, 451
    output data sets, 340, 444, 470
    overview, 280
    percentiles, 360, 371, 397
    percentiles, confidence limits, 290, 475
    plot statements, 279
    probability plots, 360, 436
    quantile plots, 371, 436
    quantiles, confidence limits, 290
    results, 450
    robust estimates, 476
    robust estimators, 402
    robust location estimates, 295, 296
    robust scale estimates, 294
    rounding, 393
    sign test, 400, 479
    specifying analysis variables, 383
    task tables, 371
    testing for location, 478
    tests for location, 399
    weight variables, 383
UNIVARIATE procedure, OUTPUT statement
    output data set, 444

variances, 9

Wald confidence limits
    risk difference (FREQ), 182
Weibull distribution, 427
    cdf plots, 308
    deviation from theoretical distribution, 429
    EDF goodness-of-fit test, 429
    estimation of parameters, 330
    fitting, 330, 427
    formulas for, 427
    histograms, 506
    P-P plots, 359
    probability plots, 370
    quantile plots, 382, 532
    three-parameter, 440
    two-parameter, 441
weight values, 291
weighted kappa coefficient
    FREQ procedure, 205, 208
weighted percentiles, 398
Wilcoxon signed rank test, 399, 401
Wilson confidence limits
    proportions (FREQ), 173
Winsorized means, 296, 402

Yule's Q statistic
    FREQ procedure, 162

Zelen's test
    equal odds ratios (FREQ), 218
zeros, structural and random
    agreement statistics (FREQ), 212

Syntax Index

ADJUST option (PLCORR)
    TABLES statement (FREQ), 122
AGREE option
    EXACT statement (FREQ), 80
    OUTPUT statement (FREQ), 91
    TABLES statement (FREQ), 102
    TEST statement (FREQ), 145
AJCHI option
    OUTPUT statement (FREQ), 91
ALL option
    OUTPUT statement (FREQ), 91
    PLOTS= option (CORR), 13
    PROC UNIVARIATE statement, 289
    TABLES statement (FREQ), 104
ALPHA option
    PROC CORR statement, 9
ALPHA= option
    CDFPLOT statement (UNIVARIATE), 302
    EXACT statement (FREQ), 86
    HISTOGRAM statement (UNIVARIATE), 319, 417, 423
    PLOTS=SCATTER option (CORR), 13
    PPPLOT statement (UNIVARIATE), 351
    PROBPLOT statement (UNIVARIATE), 365
    PROC UNIVARIATE statement, 289
    QQPLOT statement (UNIVARIATE), 376
    TABLES statement (FREQ), 105
ALPHADELTA= option
    plot statements (UNIVARIATE), 384
ALPHAINITIAL= option
    plot statements (UNIVARIATE), 384
ANNOKEY option
    plot statements (UNIVARIATE), 384
ANNOTATE= option
    HISTOGRAM statement (UNIVARIATE), 507
    plot statements (UNIVARIATE), 384
    PROC UNIVARIATE statement, 290, 444

BARLABEL= option
    HISTOGRAM statement (UNIVARIATE), 319
BARNARD option
    EXACT statement (FREQ), 80
BARWIDTH= option
    HISTOGRAM statement (UNIVARIATE), 319
BDCHI option
    OUTPUT statement (FREQ), 91
BDT option (CMH)
    TABLES statement (FREQ), 113

BEST= option
    PROC CORR statement, 10
BETA option
    CDFPLOT statement (UNIVARIATE), 302
    HISTOGRAM statement (UNIVARIATE), 319, 416, 497
    PPPLOT statement (UNIVARIATE), 351
    PROBPLOT statement (UNIVARIATE), 365, 437
    QQPLOT statement (UNIVARIATE), 376, 437
BETA= option
    CDFPLOT statement (UNIVARIATE), 303
    HISTOGRAM statement (UNIVARIATE), 320, 417
    PPPLOT statement (UNIVARIATE), 352
    PROBPLOT statement (UNIVARIATE), 365
    QQPLOT statement (UNIVARIATE), 377
BINOMIAL option
    EXACT statement (FREQ), 80
    OUTPUT statement (FREQ), 91
    TABLES statement (FREQ), 105
BOWKER option
    OUTPUT statement (FREQ), 98
BY statement
    CORR procedure, 16
    FREQ procedure, 77
    UNIVARIATE procedure, 296

C= option
    CDFPLOT statement (UNIVARIATE), 303
    HISTOGRAM statement (UNIVARIATE), 320, 432, 433, 504
    PPPLOT statement (UNIVARIATE), 352
    PROBPLOT statement (UNIVARIATE), 366
    QQPLOT statement (UNIVARIATE), 377, 532
CAXIS= option
    plot statements (UNIVARIATE), 385
CBARLINE= option
    HISTOGRAM statement (UNIVARIATE), 320
CDFPLOT statement
    examples, 534
    UNIVARIATE procedure, 297
CELLCHI2 option
    TABLES statement (FREQ), 109
CFILL= option
    HISTOGRAM statement (UNIVARIATE), 320
    INSET statement (UNIVARIATE), 338
CFILLH= option
    INSET statement (UNIVARIATE), 339

CFRAME= option
    INSET statement (UNIVARIATE), 339
    plot statements (UNIVARIATE), 385
CFRAMESIDE= option
    plot statements (UNIVARIATE), 385
CFRAMETOP= option
    plot statements (UNIVARIATE), 385
CGRID= option
    HISTOGRAM statement (UNIVARIATE), 320
    PROBPLOT statement (UNIVARIATE), 366
    QQPLOT statement (UNIVARIATE), 377
CHEADER= option
    INSET statement (UNIVARIATE), 339
CHISQ option
    EXACT statement (FREQ), 81, 256
    OUTPUT statement (FREQ), 92
    TABLES statement (FREQ), 110, 256
CHREF= option
    plot statements (UNIVARIATE), 385
CIBASIC option
    PROC UNIVARIATE statement, 290, 473
CIPCTLDF option
    PROC UNIVARIATE statement, 290
CIPCTLNORMAL option
    PROC UNIVARIATE statement, 290
CIQUANTDF option
    PROC UNIVARIATE statement, 475
CIQUANTNORMAL option
    PROC UNIVARIATE statement, 290, 475
CL option
    TABLES statement (FREQ), 112
CL= option (BINOMIAL)
    TABLES statement (FREQ), 106
CL= option (COMMONRISKDIFF)
    TABLES statement (FREQ), 114
CL= option (RELRISK)
    TABLES statement (FREQ), 135
CL= option (RISKDIFF)
    TABLES statement (FREQ), 139
CL=AGRESTICAFFO option (RISKDIFF)
    TABLES statement (FREQ), 139
CL=AGRESTICOULL option (BINOMIAL)
    TABLES statement (FREQ), 106
CL=BLAKER option (BINOMIAL)
    TABLES statement (FREQ), 107
CL=CLOPPERPEARSON option (BINOMIAL)
    TABLES statement (FREQ), 107
CL=EXACT option (BINOMIAL)
    TABLES statement (FREQ), 107
CL=EXACT option (RISKDIFF)
    TABLES statement (FREQ), 140
CL=HA option (RISKDIFF)
    TABLES statement (FREQ), 140
CL=JEFFREYS option (BINOMIAL)
    TABLES statement (FREQ), 107
CL=LIKELIHOODRATIO option (BINOMIAL)
    TABLES statement (FREQ), 107
CL=LOGIT option (BINOMIAL)
    TABLES statement (FREQ), 107
CL=MH option (COMMONRISKDIFF)
    TABLES statement (FREQ), 114
CL=MIDP option (BINOMIAL)
    TABLES statement (FREQ), 107
CL=MR option (COMMONRISKDIFF)
    TABLES statement (FREQ), 114
CL=NEWCOMBE option (COMMONRISKDIFF)
    TABLES statement (FREQ), 114
CL=NEWCOMBE option (RISKDIFF)
    TABLES statement (FREQ), 140
CL=NEWCOMBEMR option (COMMONRISKDIFF)
    TABLES statement (FREQ), 114
CL=SCORE option (COMMONRISKDIFF)
    TABLES statement (FREQ), 114
CL=WALD option (BINOMIAL)
    TABLES statement (FREQ), 107
CL=WALD option (RISKDIFF)
    TABLES statement (FREQ), 140
CL=WILSON option (BINOMIAL)
    TABLES statement (FREQ), 107
CLASS statement
    UNIVARIATE procedure, 309
CLIPCURVES option
    HISTOGRAM statement (UNIVARIATE), 320
CLIPREF option
    HISTOGRAM statement (UNIVARIATE), 320
CMH option
    OUTPUT statement (FREQ), 92
    TABLES statement (FREQ), 112
CMH1 option
    OUTPUT statement (FREQ), 92
    TABLES statement (FREQ), 113
CMH2 option
    OUTPUT statement (FREQ), 92
    TABLES statement (FREQ), 113
CMHCOR option
    OUTPUT statement (FREQ), 92
CMHGA option
    OUTPUT statement (FREQ), 92
CMHRMS option
    OUTPUT statement (FREQ), 92
COCHQ option
    OUTPUT statement (FREQ), 93
COLOR= option
    plot statements (UNIVARIATE), 385
COLUMN= option (COMMONRISKDIFF)
    TABLES statement (FREQ), 115
COLUMN= option (RELRISK)
    EXACT statement (FREQ), 83

TABLES statement (FREQ), 136
COLUMN= option (RISKDIFF)
    EXACT statement (FREQ), 84
    TABLES statement (FREQ), 140
COMMON option (RISKDIFF)
    TABLES statement (FREQ), 141
COMMONRISKDIFF option
    TABLES statement (FREQ), 114
COMOR option
    EXACT statement (FREQ), 81
    OUTPUT statement (FREQ), 95
COMPRESS option
    PROC FREQ statement, 76
CONTENTS= option
    HISTOGRAM statement (UNIVARIATE), 321
    plot statements (UNIVARIATE), 385
    TABLES statement (FREQ), 115
CONTGY option
    OUTPUT statement (FREQ), 93
CONVERGE option
    POLYCHORIC option (CORR), 14
    POLYSERIAL option (CORR), 15
CONVERGE= option (PLCORR)
    TABLES statement (FREQ), 122
CORR procedure
    syntax, 8
CORR procedure, BY statement, 16
CORR procedure, FREQ statement, 16
CORR procedure, ID statement, 17
CORR procedure, PARTIAL statement, 17
CORR procedure, PLOTS= option
    ALL option, 13
    HISTOGRAM option, 13
    MATRIX option, 13
    MAXPOINTS= option, 12
    NONE option, 13
    NVAR= option, 13, 14
    NWITH= option, 13, 14
    SCATTER option, 13
CORR procedure, PLOTS=SCATTER option
    ALPHA=, 13
    ELLIPSE=, 13
    NOINSET, 14
CORR procedure, POLYCHORIC
    CONVERGE option, 14
    MAXITER option, 14
    NGROUPS option, 14
CORR procedure, POLYCHORIC option, 14
CORR procedure, POLYSERIAL
    CONVERGE option, 15
    MAXITER option, 15
    NGROUPS option, 15
    ORDINAL option, 15
CORR procedure, POLYSERIAL option, 14

CORR procedure, PROC CORR statement, 8
    ALPHA option, 9
    BEST= option, 10
    COV option, 10
    CSSCP option, 10
    DATA= option, 10
    EXCLNPWGT option, 10
    FISHER option, 10
    HOEFFDING option, 11
    KENDALL option, 11
    NOCORR option, 11
    NOMISS option, 11
    NOPRINT option, 11
    NOPROB option, 11
    NOSIMPLE option, 11
    OUT= option, 12
    OUTH= option, 11
    OUTK= option, 11
    OUTP= option, 12
    OUTPLC= option, 12
    OUTPLS= option, 12
    OUTS= option, 12
    PEARSON option, 12
    RANK option, 15
    SINGULAR= option, 15
    SPEARMAN option, 15
    SSCP option, 15
    VARDEF= option, 15
CORR procedure, VAR statement, 17
CORR procedure, WEIGHT statement, 17
CORR procedure, WITH statement, 18
CORRECT option (BINOMIAL)
    TABLES statement (FREQ), 108
CORRECT option (RISKDIFF)
    TABLES statement (FREQ), 141
CORRECT=NO option (COMMONRISKDIFF)
    TABLES statement (FREQ), 115
COV option
    PROC CORR statement, 10
CPROP= option
    HISTOGRAM statement (UNIVARIATE), 495
    plot statements (UNIVARIATE), 386
CRAMV option
    OUTPUT statement (FREQ), 93
CROSSLIST option
    TABLES statement (FREQ), 116
CSHADOW= option
    INSET statement (UNIVARIATE), 339
CSSCP option
    PROC CORR statement, 10
CSTATREF= option
    plot statements (UNIVARIATE), 386
CTEXT= option
    INSET statement (UNIVARIATE), 339

plot statements (UNIVARIATE), 386
CTEXTSIDE= option
    plot statements (UNIVARIATE), 386
CTEXTTOP= option
    plot statements (UNIVARIATE), 386
CUMCOL option
    TABLES statement (FREQ), 117
CVREF= option
    plot statements (UNIVARIATE), 386

DATA option
    INSET statement (UNIVARIATE), 339
DATA= option
    INSET statement (UNIVARIATE), 334
    PROC CORR statement, 10
    PROC FREQ statement, 76
    PROC UNIVARIATE statement, 291, 444
DELTA= option
    HISTOGRAM statement (UNIVARIATE), 321
DESCRIPTION= option
    plot statements (UNIVARIATE), 386
DEVIATION option
    TABLES statement (FREQ), 117
DF= option (CHISQ)
    TABLES statement (FREQ), 110

EDFNSAMPLES= option
    HISTOGRAM statement (UNIVARIATE), 321
EDFSEED= option
    HISTOGRAM statement (UNIVARIATE), 321
ELLIPSE= option
    PLOTS=SCATTER option (CORR), 13
ENDPOINTS= option
    HISTOGRAM statement (UNIVARIATE), 321, 489
EQKAP option
    OUTPUT statement (FREQ), 93
EQOR option
    EXACT statement (FREQ), 81
    OUTPUT statement (FREQ), 93
EQUAL option (RELRISK)
    TABLES statement (FREQ), 136
EQUAL option (RISKDIFF)
    TABLES statement (FREQ), 141
EQUIVALENCE option (BINOMIAL)
    TABLES statement (FREQ), 108
EQUIVALENCE option (RELRISK)
    TABLES statement (FREQ), 137
EQUIVALENCE option (RISKDIFF)
    TABLES statement (FREQ), 141
EQWKP option
    OUTPUT statement (FREQ), 93
EXACT option
    OUTPUT statement (FREQ), 93
    TABLES statement (FREQ), 117
EXACT statement
    FREQ procedure, 78
EXCLNPWGT option
    PROC CORR statement, 10
    PROC UNIVARIATE statement, 291
EXPECTED option
    TABLES statement (FREQ), 117
EXPONENTIAL option
    CDFPLOT statement (UNIVARIATE), 303
    HISTOGRAM statement (UNIVARIATE), 322, 417
    PPPLOT statement (UNIVARIATE), 352
    PROBPLOT statement (UNIVARIATE), 366, 438
    QQPLOT statement (UNIVARIATE), 377, 438

FILL option
    HISTOGRAM statement (UNIVARIATE), 322
FISHER option
    EXACT statement (FREQ), 81
    OUTPUT statement (FREQ), 93
    PROC CORR statement, 10
    TABLES statement (FREQ), 117
FITINTERVAL= option
    plot statements (UNIVARIATE), 386
FITMETHOD= option
    plot statements (UNIVARIATE), 386
FITTOLERANCE= option
    plot statements (UNIVARIATE), 387
FONT= option
    INSET statement (UNIVARIATE), 339
    plot statements (UNIVARIATE), 387
FORCEQN option
    PROC UNIVARIATE statement, 291
FORCESN option
    PROC UNIVARIATE statement, 291
FORMAT= option
    INSET statement (UNIVARIATE), 339
    TABLES statement (FREQ), 117
FORMCHAR= option
    PROC FREQ statement, 76
FREQ option
    PROC UNIVARIATE statement, 291, 460
FREQ procedure
    syntax, 75
FREQ procedure, BY statement, 77
FREQ procedure, EXACT statement, 78
    AGREE option, 80
    ALPHA= option, 86
    BARNARD option, 80
    BINOMIAL option, 80
    CHISQ option, 81, 256
    COLUMN= option (RELRISK), 83
    COLUMN= option (RISKDIFF), 84

Syntax Index

COMOR option, 81
EQOR option, 81
FISHER option, 81
JT option, 81
KAPPA option, 81
KENTB option, 82
LRCHI option, 82
MAXTIME= option, 86
MC option, 86
MCNEM option, 82
MEASURES option, 82
METHOD= option (RELRISK), 83, 84
METHOD= option (RISKDIFF), 84
MHCHI option, 82
MIDP option, 86
N= option, 87
OR option, 82, 256
PCHI option, 83
PCORR option, 83
PFORMAT= option, 87
POINT option, 87
RELRISK option, 83
RISKDIFF option, 84
SCORR option, 85
SEED= option, 87
SMDCR option, 85
SMDRC option, 85
STUTC option, 85
SYMMETRY option, 85
TREND option, 85, 262
WTKAPPA option, 85
ZELEN option, 81
FREQ procedure, OUTPUT statement, 87
AGREE option, 91
AJCHI option, 91
ALL option, 91
BDCHI option, 91
BINOMIAL option, 91
BOWKER option, 98
CHISQ option, 92
CMH option, 92
CMH1 option, 92
CMH2 option, 92
CMHCOR option, 92
CMHGA option, 92
CMHRMS option, 92
COCHQ option, 93
COMOR option, 95
CONTGY option, 93
CRAMV option, 93
EQKAP option, 93
EQOR option, 93
EQWKP option, 93
EXACT option, 93

FISHER option, 93
GAILSIMON option, 93
GAMMA option, 93
JT option, 94
KAPPA option, 94
KENTB option, 94
LAMCR option, 94
LAMDAS option, 94
LAMRC option, 94
LGOR option, 94
LGRRC1 option, 94
LGRRC2 option, 94
LRCHI option, 95
MCNEM option, 95
MEASURES option, 95
MHCHI option, 95
MHOR option, 95
MHRRC1 option, 95
MHRRC2 option, 95
N option, 95
NMISS option, 95
OR option, 96
OUT= option, 88
output-options, 88
PCHI option, 96
PCORR option, 96
PLCORR option, 96
RDIF1 option, 96
RDIF2 option, 96
RELRISK option, 96
RISK1 option, 97
RISK11 option, 97
RISK12 option, 97
RISK2 option, 98
RISK21 option, 98
RISK22 option, 98
RISKDIFF option, 97
RISKDIFF1 option, 97
RISKDIFF2 option, 97
RRC1 option, 97
RRC2 option, 97
SCORR option, 98
SMDCR option, 98
SMDRC option, 98
STUTC option, 98
TAUB option, 94
TAUC option, 98
TREND option, 98
TSYMM option, 98
U option, 99
UCR option, 99
URC option, 99
WTKAPPA option, 99
ZELEN option, 93

FREQ procedure, PROC FREQ statement, 75
    COMPRESS option, 76
    DATA= option, 76
    FORMCHAR= option, 76
    NLEVELS option, 77
    NOPRINT option, 77
    ORDER= option, 77
    PAGE option, 77
FREQ procedure, TABLES statement, 99
    ADJUST option (PLCORR), 122
    AGREE option, 102
    ALL option, 104
    ALPHA= option, 105
    BDT option (CMH), 113
    BINOMIAL option, 105
    CELLCHI2 option, 109
    CHISQ option, 110, 256
    CL option, 112
    CL= option (BINOMIAL), 106
    CL= option (COMMONRISKDIFF), 114
    CL= option (RELRISK), 135
    CL= option (RISKDIFF), 139
    CL=AGRESTICAFFO option (RISKDIFF), 139
    CL=AGRESTICOULL option (BINOMIAL), 106
    CL=BLAKER option (BINOMIAL), 107
    CL=CLOPPERPEARSON option (BINOMIAL), 107
    CL=EXACT option (BINOMIAL), 107
    CL=EXACT option (RISKDIFF), 140
    CL=HA option (RISKDIFF), 140
    CL=JEFFREYS option (BINOMIAL), 107
    CL=LIKELIHOODRATIO option (BINOMIAL), 107
    CL=LOGIT option (BINOMIAL), 107
    CL=MH option (COMMONRISKDIFF), 114
    CL=MIDP option (BINOMIAL), 107
    CL=MR option (COMMONRISKDIFF), 114
    CL=NEWCOMBE option (COMMONRISKDIFF), 114
    CL=NEWCOMBE option (RISKDIFF), 140
    CL=NEWCOMBEMR option (COMMONRISKDIFF), 114
    CL=SCORE option (COMMONRISKDIFF), 114
    CL=WALD option (BINOMIAL), 107
    CL=WALD option (RISKDIFF), 140
    CL=WILSON option (BINOMIAL), 107
    CMH option, 112
    CMH1 option, 113
    CMH2 option, 113
    COLUMN= option (COMMONRISKDIFF), 115
    COLUMN= option (RELRISK), 136
    COLUMN= option (RISKDIFF), 140
    COMMON option (RISKDIFF), 141
    COMMONRISKDIFF option, 114

CONTENTS= option, 115
CONVERGE= option (PLCORR), 122
CORRECT option (BINOMIAL), 108
CORRECT option (RISKDIFF), 141
CORRECT=NO option (COMMONRISKDIFF), 115
CROSSLIST option, 116
CUMCOL option, 117
DEVIATION option, 117
DF= option (CHISQ), 110
EQUAL option (RELRISK), 136
EQUAL option (RISKDIFF), 141
EQUIVALENCE option (BINOMIAL), 108
EQUIVALENCE option (RELRISK), 137
EQUIVALENCE option (RISKDIFF), 141
EXACT option, 117
EXPECTED option, 117
FISHER option, 117
FORMAT= option, 117
GAILSIMON option, 118
JT option, 118
LEVEL= option (BINOMIAL), 108
LIST option, 118
LRCHI option (CHISQ), 110
MANTELFLEISS option (CMH), 113
MARGIN= option (BINOMIAL), 108
MARGIN= option (RELRISK), 137
MARGIN= option (RISKDIFF), 141
MAXITER= option (PLCORR), 122
MAXLEVELS= option, 118
MEASURES option, 118
METHOD= option (RELRISK), 137
METHOD= option (RISKDIFF), 142
METHOD=FM option (RELRISK), 137
METHOD=FM option (RISKDIFF), 142
METHOD=HA option (RISKDIFF), 142
METHOD=LR option (RELRISK), 137
METHOD=NEWCOMBE option (RISKDIFF), 142
METHOD=WALD option (RELRISK), 137
METHOD=WALD option (RISKDIFF), 142
METHOD=WALDMODIFIED option (RELRISK), 137
MISSING option, 119
MISSPRINT option, 119
NOCOL option, 119
NOCUM option, 119
NOFREQ option, 119
NONINFERIORITY option (BINOMIAL), 109
NONINFERIORITY option (RELRISK), 138
NONINFERIORITY option (RISKDIFF), 142
NOPERCENT option, 119
NOPRINT option, 120
NORISKS option (RISKDIFF), 142

NOROW option, 120
NOSPARSE option, 120
NOWARN option, 120
OR option, 120
OUT= option, 121
OUTCUM option, 121
OUTEXPECT option, 122, 245
OUTLEVEL option (BINOMIAL), 109
OUTPCT option, 122
P= option (BINOMIAL), 109
PEARSONRES option (CROSSLIST), 116
PLCORR option, 122
PLOTS= option, 123
PLOTS=AGREEPLOT option, 124
PLOTS=CUMFREQPLOT option, 125
PLOTS=DEVIATIONPLOT option, 125
PLOTS=FREQPLOT option, 125
PLOTS=KAPPAPLOT option, 126
PLOTS=MOSAICPLOT option, 127
PLOTS=NONE option, 128
PLOTS=ODDSRATIOPLOT option, 128
PLOTS=RELRISKPLOT option, 128
PLOTS=RISKDIFFPLOT option, 129
PLOTS=WTKAPPAPLOT, 129
POLYCHORIC option, 122
PRINTALL option (RELRISK), 138
PRINTKWTS option, 104, 135
PRINTWTS option (COMMONRISKDIFF), 115
RELRISK option, 135, 256
RISKDIFF option, 138
SCORES= option, 143, 266
SCOROUT option, 143
SPARSE option, 143, 245
STDRES option (CROSSLIST), 116
SUPERIORITY option (BINOMIAL), 109
SUPERIORITY option (RELRISK), 138
SUPERIORITY option (RISKDIFF), 143
TEST option (COMMONRISKDIFF), 115
TEST=MH option (COMMONRISKDIFF), 115
TEST=MR option (COMMONRISKDIFF), 115
TEST=SCORE option (COMMONRISKDIFF), 115
TESTF= option, 155
TESTF= option (CHISQ), 111
TESTP= option, 156, 251
TESTP= option (CHISQ), 111
TOTPCT option, 144
TREND option, 144, 262
VAR= option (BINOMIAL), 109
VAR= option (RISKDIFF), 143
WARN= option (CHISQ), 112
FREQ procedure, TEST statement, 144
    AGREE option, 145
    GAMMA option, 145
    KAPPA option, 145
    KENTB option, 146
    MEASURES option, 146
    PCORR option, 146
    PLCORR option, 146
    SCORR option, 146
    SMDCR option, 146, 262
    SMDRC option, 147
    STUTC option, 147
    TAUB option, 146
    TAUC option, 147
    WTKAPPA option, 147
FREQ procedure, WEIGHT statement, 147
    ZEROS option, 148
FREQ statement
    CORR procedure, 16
    UNIVARIATE procedure, 311
FRONTREF option
    HISTOGRAM statement (UNIVARIATE), 322

GAILSIMON option
    OUTPUT statement (FREQ), 93
    TABLES statement (FREQ), 118
GAMMA option
    CDFPLOT statement (UNIVARIATE), 304
    HISTOGRAM statement (UNIVARIATE), 322, 418, 500
    OUTPUT statement (FREQ), 93
    PPPLOT statement (UNIVARIATE), 353
    PROBPLOT statement (UNIVARIATE), 366, 438
    QQPLOT statement (UNIVARIATE), 377, 438
    TEST statement (FREQ), 145
GAMMA= option
    HISTOGRAM statement (UNIVARIATE), 322
GOUT= option
    PROC UNIVARIATE statement, 291
GRID option
    HISTOGRAM statement (UNIVARIATE), 322
    PROBPLOT statement (UNIVARIATE), 366
    QQPLOT statement (UNIVARIATE), 377, 379, 528
GUMBEL option
    CDFPLOT statement (UNIVARIATE), 304
    HISTOGRAM statement (UNIVARIATE), 323, 419
    PPPLOT statement (UNIVARIATE), 354
    PROBPLOT statement (UNIVARIATE), 366, 438
    QQPLOT statement (UNIVARIATE), 378, 438
GUTTER= option
    INSET statement (UNIVARIATE), 339

HANGING option
    HISTOGRAM statement (UNIVARIATE), 323
HAXIS= option

plot statements (UNIVARIATE), 387
HEADER= option
    INSET statement (UNIVARIATE), 339
HEIGHT= option
    INSET statement (UNIVARIATE), 339
    plot statements (UNIVARIATE), 387
HISTOGRAM
    PLOTS= option (CORR), 13
HISTOGRAM statement
    UNIVARIATE procedure, 311
HMINOR= option
    plot statements (UNIVARIATE), 387
HOEFFDING option
    PROC CORR statement, 11
HOFFSET= option
    HISTOGRAM statement (UNIVARIATE), 324
HREF= option
    plot statements (UNIVARIATE), 387
HREFLABELS= option
    plot statements (UNIVARIATE), 387
HREFLABPOS= option
    plot statements (UNIVARIATE), 387

ID statement
    CORR procedure, 17
    UNIVARIATE procedure, 330
IDOUT option
    PROC UNIVARIATE statement, 291
IGAUSS option
    CDFPLOT statement (UNIVARIATE), 305
    HISTOGRAM statement (UNIVARIATE), 324, 419
    PPPLOT statement (UNIVARIATE), 354
INFONT= option
    plot statements (UNIVARIATE), 387
INHEIGHT= option
    plot statements (UNIVARIATE), 388
INSET statement
    UNIVARIATE procedure, 330
INTERBAR= option
    HISTOGRAM statement (UNIVARIATE), 324
INTERTILE= option
    HISTOGRAM statement (UNIVARIATE), 495
    plot statements (UNIVARIATE), 388
ITPRINT option
    plot statements (UNIVARIATE), 388

JT option
    EXACT statement (FREQ), 81
    OUTPUT statement (FREQ), 94
    TABLES statement (FREQ), 118

K= option
    HISTOGRAM statement (UNIVARIATE), 324, 432, 433

KAPPA option
    EXACT statement (FREQ), 81
    OUTPUT statement (FREQ), 94
    TEST statement (FREQ), 145
KENDALL option
    PROC CORR statement, 11
KENTB option
    EXACT statement (FREQ), 82
    OUTPUT statement (FREQ), 94
    TEST statement (FREQ), 146
KERNEL option
    HISTOGRAM statement (UNIVARIATE), 324, 432, 433, 504
KEYLEVEL= option
    CLASS statement (UNIVARIATE), 310
    PROC UNIVARIATE statement, 485

L= option
    plot statements (UNIVARIATE), 388
LABEL= option
    QQPLOT statement (UNIVARIATE), 379
LAMBDA= option
    CDFPLOT statement (UNIVARIATE), 305
    HISTOGRAM statement (UNIVARIATE), 324
    PPPLOT statement (UNIVARIATE), 355
LAMCR option
    OUTPUT statement (FREQ), 94
LAMDAS option
    OUTPUT statement (FREQ), 94
LAMRC option
    OUTPUT statement (FREQ), 94
LEVEL= option (BINOMIAL)
    TABLES statement (FREQ), 108
LGOR option
    OUTPUT statement (FREQ), 94
LGRID= option
    HISTOGRAM statement (UNIVARIATE), 324
    PROBPLOT statement (UNIVARIATE), 366
    QQPLOT statement (UNIVARIATE), 378, 379
LGRRC1 option
    OUTPUT statement (FREQ), 94
LGRRC2 option
    OUTPUT statement (FREQ), 94
LHREF= option
    plot statements (UNIVARIATE), 388
LIST option
    TABLES statement (FREQ), 118
LOCCOUNT option
    PROC UNIVARIATE statement, 291, 478
LOGNORMAL option
    CDFPLOT statement (UNIVARIATE), 305
    HISTOGRAM statement (UNIVARIATE), 324, 420, 500, 506, 517
    PPPLOT statement (UNIVARIATE), 355

PROBPLOT statement (UNIVARIATE), 367, 439, 513
QQPLOT statement (UNIVARIATE), 378, 439
LOWER= option
    HISTOGRAM statement (UNIVARIATE), 325
LRCHI option
    EXACT statement (FREQ), 82
    OUTPUT statement (FREQ), 95
LRCHI option (CHISQ)
    TABLES statement (FREQ), 110
LSTATREF= option
    plot statements (UNIVARIATE), 388
LVREF= option
    plot statements (UNIVARIATE), 388

MANTELFLEISS option (CMH)
    TABLES statement (FREQ), 113
MARGIN= option (BINOMIAL)
    TABLES statement (FREQ), 108
MARGIN= option (RELRISK)
    TABLES statement (FREQ), 137
MARGIN= option (RISKDIFF)
    TABLES statement (FREQ), 141
MATRIX option
    PLOTS= option (CORR), 13
MAXITER option
    POLYCHORIC option (CORR), 14
    POLYSERIAL option (CORR), 15
MAXITER= option
    plot statements (UNIVARIATE), 388
MAXITER= option (PLCORR)
    TABLES statement (FREQ), 122
MAXLEVELS= option
    TABLES statement (FREQ), 118
MAXNBIN= option
    HISTOGRAM statement (UNIVARIATE), 325
MAXPOINTS= option
    PLOTS= option (CORR), 12
MAXSIGMAS= option
    HISTOGRAM statement (UNIVARIATE), 325
MAXTIME= option
    EXACT statement (FREQ), 86
MC option
    EXACT statement (FREQ), 86
MCNEM option
    EXACT statement (FREQ), 82
    OUTPUT statement (FREQ), 95
MEASURES option
    EXACT statement (FREQ), 82
    OUTPUT statement (FREQ), 95
    TABLES statement (FREQ), 118
    TEST statement (FREQ), 146
METHOD= option (RELRISK)
    EXACT statement (FREQ), 83, 84

    TABLES statement (FREQ), 137
METHOD= option (RISKDIFF)
    EXACT statement (FREQ), 84
    TABLES statement (FREQ), 142
METHOD=FM option (RELRISK)
    TABLES statement (FREQ), 137
METHOD=FM option (RISKDIFF)
    TABLES statement (FREQ), 142
METHOD=HA option (RISKDIFF)
    TABLES statement (FREQ), 142
METHOD=LR option (RELRISK)
    TABLES statement (FREQ), 137
METHOD=NEWCOMBE option (RISKDIFF)
    TABLES statement (FREQ), 142
METHOD=WALD option (RELRISK)
    TABLES statement (FREQ), 137
METHOD=WALD option (RISKDIFF)
    TABLES statement (FREQ), 142
METHOD=WALDMODIFIED option (RELRISK)
    TABLES statement (FREQ), 137
MHCHI option
    EXACT statement (FREQ), 82
    OUTPUT statement (FREQ), 95
MHOR option
    OUTPUT statement (FREQ), 95
MHRRC1 option
    OUTPUT statement (FREQ), 95
MHRRC2 option
    OUTPUT statement (FREQ), 95
MIDP option
    EXACT statement (NPAR1WAY), 86
MIDPERCENTS option
    HISTOGRAM statement (UNIVARIATE), 325, 492
MIDPOINTS= option
    HISTOGRAM statement (UNIVARIATE), 325, 487, 489
MISSING option
    CLASS statement (UNIVARIATE), 309
    TABLES statement (FREQ), 119
MISSPRINT option
    TABLES statement (FREQ), 119
MODES option
    PROC UNIVARIATE statement, 291, 456
MU0= option
    PROC UNIVARIATE statement, 292
MU= option
    CDFPLOT statement (UNIVARIATE), 305
    HISTOGRAM statement (UNIVARIATE), 326, 492
    PPPLOT statement (UNIVARIATE), 356
    PROBPLOT statement (UNIVARIATE), 367
    QQPLOT statement (UNIVARIATE), 378, 521

N option
    OUTPUT statement (FREQ), 95
N= option
    EXACT statement (FREQ), 87
NADJ= option
    PROBPLOT statement (UNIVARIATE), 367
    QQPLOT statement (UNIVARIATE), 378, 434
NAME= option
    plot statements (UNIVARIATE), 388
NCOLS= option
    INSET statement (UNIVARIATE), 340
    plot statements (UNIVARIATE), 389
NENDPOINTS= option
    HISTOGRAM statement (UNIVARIATE), 326
NEXTROBS= option
    PROC UNIVARIATE statement, 292, 457
NEXTRVAL= option
    PROC UNIVARIATE statement, 292, 457
NGROUPS option
    POLYCHORIC option (CORR), 14
    POLYSERIAL option (CORR), 15
NLEVELS option
    PROC FREQ statement, 77
NMIDPOINTS= option
    HISTOGRAM statement (UNIVARIATE), 326
NMISS option
    OUTPUT statement (FREQ), 95
NOBARS option
    HISTOGRAM statement (UNIVARIATE), 326
NOBYPLOT option
    PROC UNIVARIATE statement, 292
NOCDFLEGEND option
    CDFPLOT statement (UNIVARIATE), 306
NOCOL option
    TABLES statement (FREQ), 119
NOCORR option
    PROC CORR statement, 11
NOCUM option
    TABLES statement (FREQ), 119
NOCURVELEGEND option
    HISTOGRAM statement (UNIVARIATE), 326
NOECDF option
    CDFPLOT statement (UNIVARIATE), 306
NOFRAME option
    INSET statement (UNIVARIATE), 340
    plot statements (UNIVARIATE), 389
NOFREQ option
    TABLES statement (FREQ), 119
NOHLABEL option
    plot statements (UNIVARIATE), 389
NOINSET option
    PLOTS=SCATTER option (CORR), 14
NOKEYMOVE option
    CLASS statement (UNIVARIATE), 311

NOLINE option
    PPPLOT statement (UNIVARIATE), 356
NOLINELEGEND option
    PROBPLOT statement (UNIVARIATE), 367
    QQPLOT statement (UNIVARIATE), 378
NOMISS option
    PROC CORR statement, 11
NONE option
    PLOTS= option (CORR), 13
NONINFERIORITY option (BINOMIAL)
    TABLES statement (FREQ), 109
NONINFERIORITY option (RELRISK)
    TABLES statement (FREQ), 138
NONINFERIORITY option (RISKDIFF)
    TABLES statement (FREQ), 142
NOPERCENT option
    TABLES statement (FREQ), 119
NOPLOT option
    HISTOGRAM statement (UNIVARIATE), 326
NOPRINT option
    HISTOGRAM statement (UNIVARIATE), 326
    PROC CORR statement, 11
    PROC FREQ statement, 77
    PROC UNIVARIATE statement, 292
    TABLES statement (FREQ), 120
NOPROB option
    PROC CORR statement, 11
NORISKS option (RISKDIFF)
    TABLES statement (FREQ), 142
NORMAL option
    CDFPLOT statement (UNIVARIATE), 306
    HISTOGRAM statement (UNIVARIATE), 326, 421, 492
    PPPLOT statement (UNIVARIATE), 356
    PROBPLOT statement (UNIVARIATE), 367, 439
    PROC UNIVARIATE statement, 292
    QQPLOT statement (UNIVARIATE), 379, 439
NORMALTEST option
    PROC UNIVARIATE statement, 292
NOROW option
    TABLES statement (FREQ), 120
NOSIMPLE option
    PROC CORR statement, 11
NOSPARSE option
    TABLES statement (FREQ), 120
NOTABCONTENTS option
    HISTOGRAM statement (UNIVARIATE), 327
    PROC UNIVARIATE statement, 292
NOVARCONTENTS option
    PROC UNIVARIATE statement, 292
NOVLABEL option
    plot statements (UNIVARIATE), 389
NOVTICK option
    plot statements (UNIVARIATE), 389

NOWARN option
: TABLES statement (FREQ), 120
NROWS= option
: HISTOGRAM statement (UNIVARIATE), 482
plot statements (UNIVARIATE), 389
NVAR= option
: PLOTS= option (CORR), 13, 14
NWITH= option
: PLOTS= option (CORR), 13, 14

ODSFOOTNOTE2= option
: plot statements (UNIVARIATE), 389
ODSFOOTNOTE= option
: plot statements (UNIVARIATE), 389
ODSTITLE2= option
: plot statements (UNIVARIATE), 390
ODSTITLE= option
: plot statements (UNIVARIATE), 389
OR option
: EXACT statement (FREQ), 82, 256
OUTPUT statement (FREQ), 96
TABLES statement (FREQ), 120
ORDER= option
: CLASS statement (UNIVARIATE), 309
PROC FREQ statement, 77
ORDINAL option
: POLYSERIAL option (CORR), 15
OUT= option
: OUTPUT statement (FREQ), 88
OUTPUT statement (UNIVARIATE), 340
PROC CORR statement, 12
TABLES statement (FREQ), 121
OUTCUM option
: TABLES statement (FREQ), 121
OUTEXPECT option
: TABLES statement (FREQ), 122, 245
OUTH= option
: PROC CORR statement, 11
OUTHISTOGRAM= option
: HISTOGRAM statement (UNIVARIATE), 327, 446, 489
OUTK= option
: PROC CORR statement, 11
OUTKERNEL= option
: HISTOGRAM statement (UNIVARIATE), 327, 447
OUTLEVEL option (BINOMIAL)
: TABLES statement (FREQ), 109
OUTP= option
: PROC CORR statement, 12
OUTPCT option
: TABLES statement (FREQ), 122
OUTPLC= option
: PROC CORR statement, 12

OUTPLS= option
: PROC CORR statement, 12
OUTPUT statement
: FREQ procedure, 87
UNIVARIATE procedure, 340, 383
OUTS= option
: PROC CORR statement, 12
OUTTABLE= option
: PROC UNIVARIATE statement, 293, 448
OVERLAY option
: plot statements (UNIVARIATE), 390

P= option (BINOMIAL)
: TABLES statement (FREQ), 109
PAGE option
: PROC FREQ statement, 77
PARETO option
: CDFPLOT statement (UNIVARIATE), 306
HISTOGRAM statement (UNIVARIATE), 327, 421
PPPLOT statement (UNIVARIATE), 357
PROBPLOT statement (UNIVARIATE), 367, 439
QQPLOT statement (UNIVARIATE), 379, 439
PARTIAL statement
: CORR procedure, 17
PCHI option
: EXACT statement (FREQ), 83
OUTPUT statement (FREQ), 96
PCORR option
: EXACT statement (FREQ), 83
OUTPUT statement (FREQ), 96
TEST statement (FREQ), 146
PCTLAXIS option
: QQPLOT statement (UNIVARIATE), 379, 443, 528
PCTLDEF= option
: PROC UNIVARIATE statement, 293, 397
PCTLGROUP= option
: OUTPUT statement (UNIVARIATE), 344
PCTLMINOR option
: PROBPLOT statement (UNIVARIATE), 368
QQPLOT statement (UNIVARIATE), 379
PCTLNAME= option
: OUTPUT statement (UNIVARIATE), 344
PCTLNDEC= option
: OUTPUT statement (UNIVARIATE), 345
PCTLORDER= option
: PROBPLOT statement (UNIVARIATE), 368
PCTLPRE= option
: OUTPUT statement (UNIVARIATE), 345
PCTLPTS= option
: OUTPUT statement (UNIVARIATE), 346
PCTLSCALE option
: QQPLOT statement (UNIVARIATE), 380, 443

PEARSON option
    PROC CORR statement, 12
PEARSONRES option (CROSSLIST)
    TABLES statement (FREQ), 116
PERCENTS= option
    HISTOGRAM statement (UNIVARIATE), 328
PFILL= option
    HISTOGRAM statement (UNIVARIATE), 328
PFORMAT= option
    EXACT statement (FREQ), 87
PLCORR option
    OUTPUT statement (FREQ), 96
    TABLES statement (FREQ), 122
    TEST statement (FREQ), 146
plot statements
    UNIVARIATE procedure, 384
PLOTS option
    PROC UNIVARIATE statement, 293, 461
PLOTS= option
    TABLES statement (FREQ), 123
PLOTS=AGREEPLOT option
    TABLES statement (FREQ), 124
PLOTS=CUMFREQPLOT option
    TABLES statement (FREQ), 125
PLOTS=DEVIATIONPLOT option
    TABLES statement (FREQ), 125
PLOTS=FREQPLOT option
    TABLES statement (FREQ), 125
PLOTS=KAPPAPLOT option
    TABLES statement (FREQ), 126
PLOTS=MOSAICPLOT option
    TABLES statement (FREQ), 127
PLOTS=NONE option
    TABLES statement (FREQ), 128
PLOTS=ODDSRATIOPLOT option
    TABLES statement (FREQ), 128
PLOTS=RELRISKPLOT option
    TABLES statement (FREQ), 128
PLOTS=RISKDIFFPLOT option
    TABLES statement (FREQ), 129
PLOTS=WTKAPPAPLOT option
    TABLES statement (FREQ), 129
PLOTSIZE= option
    PROC UNIVARIATE statement, 294
POINT option
    EXACT statement (FREQ), 87
POLYCHORIC option
    TABLES statement (FREQ), 122
POSITION= option
    INSET statement (UNIVARIATE), 340
POWER option
    CDFPLOT statement (UNIVARIATE), 307
    HISTOGRAM statement (UNIVARIATE), 328, 422

PPPLOT statement (UNIVARIATE), 357
PROBPLOT statement (UNIVARIATE), 368, 440
QQPLOT statement (UNIVARIATE), 380, 440
PPPLOT statement
    options dictionary, 351
    options summarized by function, 350, 351
    UNIVARIATE procedure, 346
PRINTALL option (RELRISK)
    TABLES statement (FREQ), 138
PRINTKWTS option
    TABLES statement (FREQ), 104, 135
PRINTWTS option (COMMONRISKDIFF)
    TABLES statement (FREQ), 115
PROBPLOT statement
    UNIVARIATE procedure, 360
PROC CORR statement, 8, *see* CORR procedure
    CORR procedure, 8
PROC FREQ statement, *see* FREQ procedure
PROC UNIVARIATE statement, 288, *see*
    UNIVARIATE procedure

QQPLOT statement
    UNIVARIATE procedure, 371

RANK option
    PROC CORR statement, 15
RANKADJ= option
    PROBPLOT statement (UNIVARIATE), 368
    QQPLOT statement (UNIVARIATE), 380, 434
RAYLEIGH option
    CDFPLOT statement (UNIVARIATE), 307
    HISTOGRAM statement (UNIVARIATE), 328, 423
    PPPLOT statement (UNIVARIATE), 358
    PROBPLOT statement (UNIVARIATE), 368, 440
    QQPLOT statement (UNIVARIATE), 380, 440
RDIF1 option
    OUTPUT statement (FREQ), 96
RDIF2 option
    OUTPUT statement (FREQ), 96
REFPOINT= option
    INSET statement (UNIVARIATE), 340
RELRISK option
    EXACT statement (FREQ), 83
    OUTPUT statement (FREQ), 96
    TABLES statement (FREQ), 135, 256
RISK1 option
    OUTPUT statement (FREQ), 97
RISK11 option
    OUTPUT statement (FREQ), 97
RISK12 option
    OUTPUT statement (FREQ), 97
RISK2 option
    OUTPUT statement (FREQ), 98

RISK21 option
    OUTPUT statement (FREQ), 98
RISK22 option
    OUTPUT statement (FREQ), 98
RISKDIFF option
    EXACT statement (FREQ), 84
    OUTPUT statement (FREQ), 97
    TABLES statement (FREQ), 138
RISKDIFF1 option
    OUTPUT statement (FREQ), 97
RISKDIFF2 option
    OUTPUT statement (FREQ), 97
ROBUSTSCALE option
    PROC UNIVARIATE statement, 294, 476
ROTATE option
    PROBPLOT statement (UNIVARIATE), 368
    QQPLOT statement (UNIVARIATE), 380
ROUND= option
    PROC UNIVARIATE statement, 294
RRC1 option
    OUTPUT statement (FREQ), 97
RRC2 option
    OUTPUT statement (FREQ), 97
RTINCLUDE option
    HISTOGRAM statement (UNIVARIATE), 328, 489

SB option
    HISTOGRAM statement (UNIVARIATE), 328, 424
SCALE= option
    HISTOGRAM statement (UNIVARIATE), 417, 418, 420, 497
    plot statements (UNIVARIATE), 390
SCATTER option
    PLOTS= option (CORR), 13
SCORES= option
    TABLES statement (FREQ), 143, 266
SCOROUT option
    TABLES statement (FREQ), 143
SCORR option
    EXACT statement (FREQ), 85
    OUTPUT statement (FREQ), 98
    TEST statement (FREQ), 146
SEED= option
    EXACT statement (FREQ), 87
SHAPE= option
    plot statements (UNIVARIATE), 390
SIGMA= option
    CDFPLOT statement (UNIVARIATE), 307
    HISTOGRAM statement (UNIVARIATE), 328, 417, 423, 424, 492
    PPPLOT statement (UNIVARIATE), 358
    PROBPLOT statement (UNIVARIATE), 368, 513

QQPLOT statement (UNIVARIATE), 380, 521, 524
SINGULAR= option
    PROC CORR statement, 15
SLOPE= option
    PROBPLOT statement (UNIVARIATE), 369
    QQPLOT statement (UNIVARIATE), 381
SMDCR option
    EXACT statement (FREQ), 85
    OUTPUT statement (FREQ), 98
    TEST statement (FREQ), 146, 262
SMDRC option
    EXACT statement (FREQ), 85
    OUTPUT statement (FREQ), 98
    TEST statement (FREQ), 147
SPARSE option
    TABLES statement (FREQ), 143, 245
SPEARMAN option
    PROC CORR statement, 15
SQUARE option
    PPPLOT statement (UNIVARIATE), 358, 537
    PROBPLOT statement (UNIVARIATE), 369, 513
    QQPLOT statement, 521
    QQPLOT statement (UNIVARIATE), 381
SSCP option
    PROC CORR statement, 15
STATREF= option
    plot statements (UNIVARIATE), 390
STATREFLABELS= option
    UNIVARIATE procedure, 391
STATREFSUBCHAR= option
    plot statements (UNIVARIATE), 391
STDRES option (CROSSLIST)
    TABLES statement (FREQ), 116
STUTC option
    EXACT statement (FREQ), 85
    OUTPUT statement (FREQ), 98
    TEST statement (FREQ), 147
SU option
    HISTOGRAM statement (UNIVARIATE), 329, 426
SUMMARYCONTENTS= option
    PROC UNIVARIATE statement, 295
SUPERIORITY option (BINOMIAL)
    TABLES statement (FREQ), 109
SUPERIORITY option (RELRISK)
    TABLES statement (FREQ), 138
SUPERIORITY option (RISKDIFF)
    TABLES statement (FREQ), 143
SYMMETRY option
    EXACT statement (FREQ), 85

TABLES statement
    FREQ procedure, 99

TAUB option
    OUTPUT statement (FREQ), 94
    TEST statement (FREQ), 146
TAUC option
    OUTPUT statement (FREQ), 98
    TEST statement (FREQ), 147
TEST statement
    FREQ procedure, 144
TEST= option (COMMONRISKDIFF)
    TABLES statement (FREQ), 115
TEST=MH option (COMMONRISKDIFF)
    TABLES statement (FREQ), 115
TEST=MR option (COMMONRISKDIFF)
    TABLES statement (FREQ), 115
TEST=SCORE option (COMMONRISKDIFF)
    TABLES statement (FREQ), 115
TESTF= option
    TABLES statement (FREQ), 155
TESTF= option (CHISQ)
    TABLES statement (FREQ), 111
TESTP= option
    TABLES statement (FREQ), 156, 251
TESTP= option (CHISQ)
    TABLES statement (FREQ), 111
THETA= option
    CDFPLOT statement (UNIVARIATE), 308
    HISTOGRAM statement (UNIVARIATE), 329,
        417, 423, 424, 497, 506, 517
    PPPLOT statement (UNIVARIATE), 359
    PROBPLOT statement (UNIVARIATE), 370
    QQPLOT statement (UNIVARIATE), 381
THRESHOLD= option
    HISTOGRAM statement (UNIVARIATE), 329,
        418
    PROBPLOT statement (UNIVARIATE), 370
    QQPLOT statement (UNIVARIATE), 382
TOTPCT option
    TABLES statement (FREQ), 144
TREND option
    EXACT statement (FREQ), 85, 262
    OUTPUT statement (FREQ), 98
    TABLES statement (FREQ), 144, 262
TRIMMED= option
    PROC UNIVARIATE statement, 295, 476
TSYMM option
    OUTPUT statement (FREQ), 98
TURNVLABELS option
    plot statements (UNIVARIATE), 391

U option
    OUTPUT statement (FREQ), 99
UCR option
    OUTPUT statement (FREQ), 99
UNIVARIATE procedure

syntax, 288
UNIVARIATE procedure, BY statement, 296
UNIVARIATE procedure, CDFPLOT statement, 297
    ALPHA= option, 302
    BETA option, 302
    BETA= option, 303
    C= option, 303
    EXPONENTIAL option, 303
    GAMMA option, 304
    GAMMA= option, 322
    GUMBEL option, 304
    IGAUSS option, 305
    LAMBDA= option, 305
    LOGNORMAL option, 305
    MU= option, 305
    NOCDFLEGEND option, 306
    NOECDF option, 306
    NORMAL option, 306
    PARETO option, 306
    POWER option, 307
    RAYLEIGH option, 307
    SIGMA= option, 307
    THETA= option, 308
    THRESHOLD= option, 308
    VSCALE= option, 308
    WEIBULL option, 308
    ZETA= option, 309
UNIVARIATE procedure, CLASS statement, 309
    KEYLEVEL= option, 310
    MISSING option, 309
    NOKEYMOVE option, 311
    ORDER= option, 309
UNIVARIATE procedure, FREQ statement, 311
UNIVARIATE procedure, HISTOGRAM statement,
    311
    ALPHA= option, 319, 417, 423
    ANNOTATE= option, 507
    BARLABEL= option, 319
    BARWIDTH= option, 319
    BETA option, 319, 416, 497
    BETA= option, 320, 417
    C= option, 320, 432, 433, 504
    CBARLINE= option, 320
    CFILL= option, 320
    CGRID= option, 320
    CLIPCURVES option, 320
    CLIPREF option, 320
    CONTENTS= option, 321
    CPROP= option, 495
    DELTA= option, 321
    EDFNSAMPLES= option, 321
    EDFSEED= option, 321
    ENDPOINTS= option, 321, 489
    EXPONENTIAL option, 322, 417

FILL option, 322
FRONTREF option, 322
GAMMA option, 322, 418, 500
GRID option, 322
GUMBEL option, 323, 419
HANGING option, 323
HOFFSET= option, 324
IGAUSS option, 324, 419
INTERBAR= option, 324
INTERTILE= option, 495
K= option, 324, 432, 433
KERNEL option, 324, 432, 433, 504
LAMBDA= option, 324, 355
LGRID= option, 324
LOGNORMAL option, 324, 420, 500, 506, 517
LOWER= option, 325
MAXNBIN= option, 325
MAXSIGMAS= option, 325
MIDPERCENTS option, 325, 492
MIDPOINTS= option, 325, 487, 489
MU= option, 326, 492
NENDPOINTS= option, 326
NMIDPOINTS= option, 326
NOBARS option, 326
NOCURVELEGEND option, 326
NOPLOT option, 326
NOPRINT option, 326
NORMAL option, 326, 421, 492
NOTABCONTENTS option, 327
NROWS= option, 482
OUTHISTOGRAM= option, 327, 446, 489
OUTKERNEL= option, 327, 447
PARETO option, 327, 421
PERCENTS= option, 328
PFILL= option, 328
POWER option, 328, 422
RAYLEIGH option, 328, 423
RTINCLUDE option, 328, 489
SB option, 328, 424
SCALE= option, 417, 418, 420, 497
SIGMA= option, 328, 417, 423, 424, 492
SU option, 329, 426
THETA= option, 329, 417, 423, 424, 497, 506, 517
THRESHOLD= option, 329, 418
UPPER= option, 329
VOFFSET= option, 329
VSCALE= option, 329
WBARLINE= option, 330
WEIBULL option, 330, 427, 500
WGRID= option, 330
ZETA= option, 330
UNIVARIATE procedure, ID statement, 330
UNIVARIATE procedure, INSET statement, 330

CFILL= option, 338
CFILLH= option, 339
CFRAME= option, 339
CHEADER= option, 339
CSHADOW= option, 339
CTEXT= option, 339
DATA option, 339
DATA= option, 334
FONT= option, 339
FORMAT= option, 339
GUTTER= option, 339
HEADER= option, 339
HEIGHT= option, 339
NCOLS= option, 340
NOFRAME option, 340
POSITION= option, 340
REFPOINT= option, 340
UNIVARIATE procedure, OUTPUT statement, 340, 383
OUT= option, 340
PCTLGROUP= option, 344
PCTLNAME= option, 344
PCTLNDEC= option, 345
PCTLPRE= option, 345
PCTLPTS= option, 346
UNIVARIATE procedure, plot statements, 384
ALPHADELTA= option, 384
ALPHAINITIAL= option, 384
ANNOKEY option, 384
ANNOTATE= option, 384
CAXIS= option, 385
CFRAME= option, 385
CFRAMESIDE= option, 385
CFRAMETOP= option, 385
CHREF= option, 385
COLOR= option, 385
CONTENTS= option, 385
CPROP= option, 386
CSTATREF= option, 386
CTEXT= option, 386
CTEXTSIDE= option, 386
CTEXTTOP= option, 386
CVREF= option, 386
DESCRIPTION= option, 386
FITINTERVAL= option, 386
FITMETHOD= option, 386
FITTOLERANCE= option, 387
FONT= option, 387
HAXIS= option, 387
HEIGHT= option, 387
HMINOR= option, 387
HREF= option, 387
HREFLABELS= option, 387
HREFLABPOS= option, 387

INFONT= option, 387
INHEIGHT= option, 388
INTERTILE= option, 388
ITPRINT option, 388
L= option, 388
LHREF= option, 388
LSTATREF= option, 388
LVREF= option, 388
MAXITER= option, 388
NAME= option, 388
NCOLS= option, 389
NOFRAME option, 389
NOHLABEL option, 389
NOVLABEL option, 389
NOVTICK option, 389
NROWS= option, 389
ODSFOOTNOTE2= option, 389
ODSFOOTNOTE= option, 389
ODSTITLE2= option, 390
ODSTITLE= option, 389
OVERLAY option, 390
SCALE= option, 390
SHAPE= option, 390
STATREF= option, 390
STATREFLABELS= option, 391
STATREFSUBCHAR= option, 391
TURNVLABELS option, 391
VAXIS= option, 391
VAXISLABEL= option, 391
VMINOR= option, 392
VREF= option, 392
VREFLABELS= option, 392
VREFLABPOS= option, 392
W= option, 392
WAXIS= option, 392
UNIVARIATE procedure, PPPLOT statement, 346
ALPHA= option, 351, 354, 355
BETA option, 351
BETA= option, 352
C= option, 352, 359
EXPONENTIAL option, 352
GAMMA option, 353
GUMBEL option, 354
IGAUSS option, 354
LOGNORMAL option, 355
MU= option, 348, 355–357
NOLINE option, 356
NORMAL option, 356
PARETO option, 357
POWER option, 357
RAYLEIGH option, 358
SCALE= option, 354, 356
SHAPE= option, 354, 356
SIGMA= option, 348, 354, 356–359

SQUARE option, 358, 537
THETA= option, 354, 356, 359
THRESHOLD= option, 354, 356, 359
WEIBULL option, 359
ZETA= option, 356, 359
UNIVARIATE procedure, PROBPLOT statement, 360
ALPHA= option, 365
BETA option, 365, 437
BETA= option, 365
C= option, 366
CGRID= option, 366
EXPONENTIAL option, 366, 438
GAMMA option, 366, 438
GRID option, 366
GUMBEL option, 366, 438
LGRID= option, 366
LOGNORMAL option, 367, 439, 513
MU= option, 367
NADJ= option, 367
NOLINELEGEND option, 367
NORMAL option, 367, 439
PARETO option, 367, 439
PCTLMINOR option, 368
PCTORDER= option, 368
POWER option, 368, 440
RANKADJ= option, 368
RAYLEIGH option, 368, 440
ROTATE option, 368
SIGMA= option, 368, 513
SLOPE= option, 369
SQUARE option, 369, 513
THETA= option, 370
THRESHOLD= option, 370
WEIBULL option, 370, 440
WEIBULL2 option, 370, 441
WGRID= option, 370
ZETA= option, 370
UNIVARIATE procedure, PROC UNIVARIATE statement, 288
ALL option, 289
ALPHA= option, 289
ANNOTATE= option, 290, 444
CIBASIC option, 290, 473
CIPCTLDF option, 290
CIPCTLNORMAL option, 290
CIQUANTDF option, 475
CIQUANTNORMAL option, 290, 475
DATA= option, 291, 444
EXCLNPWGT option, 291
FORCEQN option, 291
FORCESN option, 291
FREQ option, 291, 460
GOUT= option, 291
IDOUT option, 291

KEYLEVEL= option, 485
LOCCOUNT option, 291, 478
MODES option, 291, 456
MU0= option, 292
NEXTROBS= option, 292, 457
NEXTRVAL= option, 292, 457
NOBYPLOT option, 292
NOPRINT option, 292
NORMAL option, 292
NORMALTEST option, 292
NOTABCONTENTS option, 292
NOVARCONTENTS option, 292
OUTTABLE= option, 293, 448
PCTLDEF= option, 293, 397
PLOTS option, 293, 461
PLOTSIZE= option, 294
ROBUSTSCALE option, 294, 476
ROUND= option, 294
SUMMARYCONTENTS= option, 295
TRIMMED= option, 295, 476
VARDEF= option, 295
WINSORIZED= option, 296, 476
UNIVARIATE procedure, QQPLOT statement, 371
ALPHA= option, 376
BETA option, 376, 437
BETA= option, 377
C= option, 377, 532
CGRID= option, 377
EXPONENTIAL option, 377, 438
GAMMA option, 377, 438
GRID option, 377, 379, 528
GUMBEL option, 378, 438
LABEL= option, 379
LGRID= option, 378, 379
LOGNORMAL option, 378, 439
MU= option, 378, 521
NADJ= option, 378, 434
NOLINELEGEND option, 378
NORMAL option, 379, 439
PARETO option, 379, 439
PCTLAXIS option, 379, 443, 528
PCTLMINOR option, 379
PCTLSCALE option, 380, 443
POWER option, 380, 440
RANKADJ= option, 380, 434
RAYLEIGH option, 380, 440
ROTATE option, 380
SIGMA= option, 380, 521, 524
SLOPE= option, 381
SQUARE option, 381, 521
THETA= option, 381
THRESHOLD= option, 382
WEIBULL option, 382, 440, 532
WEIBULL2 option, 382, 441

WGRID= option, 382
ZETA= option, 382, 524
UNIVARIATE procedure, VAR statement, 383
UNIVARIATE procedure, WEIGHT statement, 383
UPPER= option
    HISTOGRAM statement (UNIVARIATE), 329
URC option
    OUTPUT statement (FREQ), 99

VAR statement
    CORR procedure, 17
    UNIVARIATE procedure, 383
VAR= option (BINOMIAL)
    TABLES statement (FREQ), 109
VAR= option (RISKDIFF)
    TABLES statement (FREQ), 143
VARDEF= option
    PROC CORR statement, 15
    PROC UNIVARIATE statement, 295
VAXIS= option
    plot statements (UNIVARIATE), 391
VAXISLABEL= option
    plot statements (UNIVARIATE), 391
VMINOR= option
    plot statements (UNIVARIATE), 392
VOFFSET= option
    HISTOGRAM statement (UNIVARIATE), 329
VREF= option
    plot statements (UNIVARIATE), 392
VREFLABELS= option
    plot statements (UNIVARIATE), 392
VREFLABPOS= option
    plot statements (UNIVARIATE), 392
VSCALE= option
    CDFPLOT statement (UNIVARIATE), 308
    HISTOGRAM statement (UNIVARIATE), 329

W= option
    plot statements (UNIVARIATE), 392
WARN= option (CHISQ)
    TABLES statement (FREQ), 112
WAXIS= option
    plot statements (UNIVARIATE), 392
WBARLINE= option
    HISTOGRAM statement (UNIVARIATE), 330
WEIBULL option
    CDFPLOT statement (UNIVARIATE), 308
    HISTOGRAM statement (UNIVARIATE), 330, 427, 500
    PPPLOT statement (UNIVARIATE), 359
    PROBPLOT statement (UNIVARIATE), 370, 440
    QQPLOT statement (UNIVARIATE), 382, 440, 532
WEIBULL2 option

PROBPLOT statement (UNIVARIATE), 370, 441
QQPLOT statement (UNIVARIATE), 382, 441
WEIGHT statement
    CORR procedure, 17
    FREQ procedure, 147
    UNIVARIATE procedure, 383
WGRID= option
    HISTOGRAM statement (UNIVARIATE), 330
    PROBPLOT statement (UNIVARIATE), 370
    QQPLOT statement (UNIVARIATE), 382
WINSORIZED= option
    PROC UNIVARIATE statement, 296, 476
WITH statement
    CORR procedure, 18
WTKAPPA option
    EXACT statement (FREQ), 85
    OUTPUT statement (FREQ), 99
    TEST statement (FREQ), 147

ZELEN option
    EXACT statement (FREQ), 81
    OUTPUT statement (FREQ), 93
ZEROS option
    WEIGHT statement (FREQ), 148
ZETA= option
    CDFPLOT statement (UNIVARIATE), 309
    HISTOGRAM statement (UNIVARIATE), 330
    PPPLOT statement (UNIVARIATE), 359
    PROBPLOT statement (UNIVARIATE), 370
    QQPLOT statement (UNIVARIATE), 382, 524

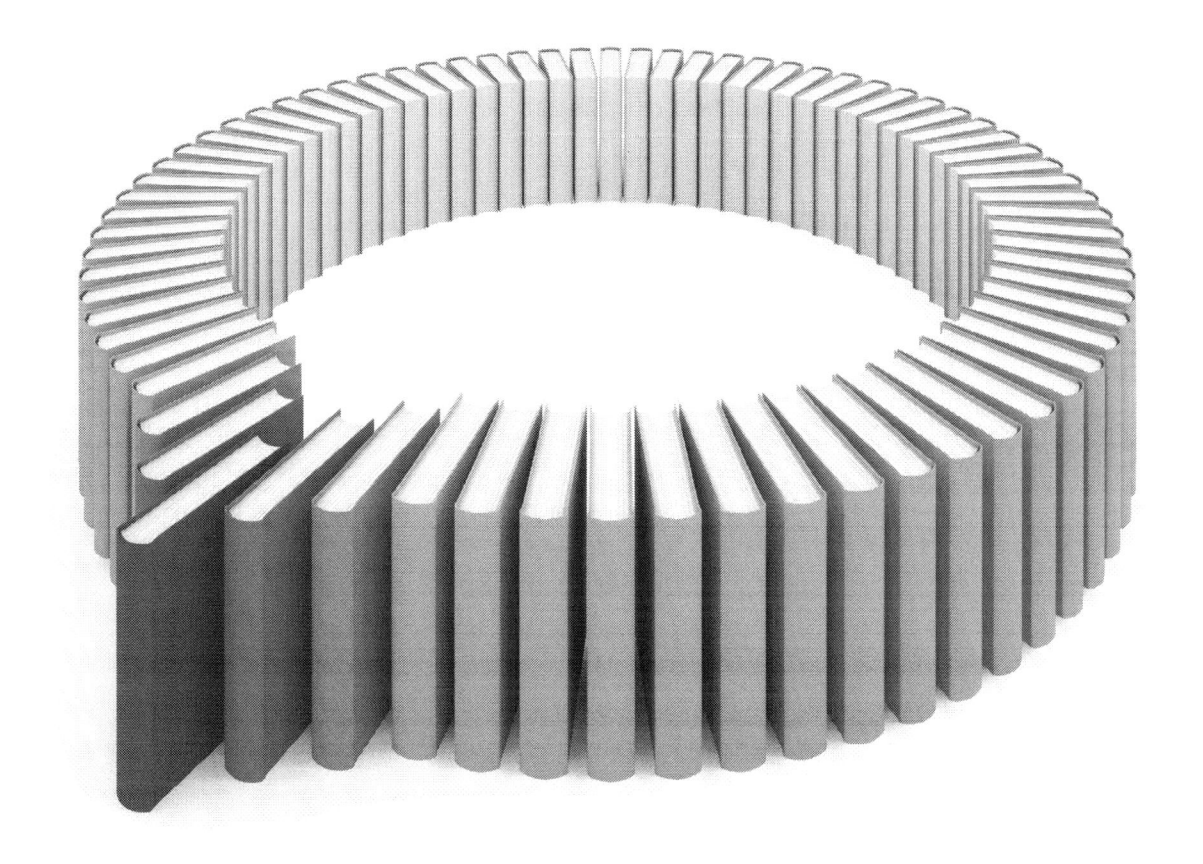

Gain Greater Insight into Your SAS® Software with SAS Books.

Discover all that you need on your journey to knowledge and empowerment.

support.sas.com/bookstore
for additional books and resources.

§.sas
THE POWER TO KNOW®

SAS and all other SAS Institute Inc. product or service names are registered trademarks or trademarks of SAS Institute Inc. in the USA and other countries. ® indicates USA registration. Other brand and product names are trademarks of their respective companies. © 2013 SAS Institute Inc. All rights reserved. S107969US.0613